# Lecture Notes in Computational Science and Engineering

67

Ismail H. Tuncer • Ülgen Gülcat
David R. Emerson • Kenichi Matsuno
*Editors*

# Parallel Computational Fluid Dynamics 2007

## Implementations and Experiences on Large Scale and Grid Computing

With 333 Figures and 51 Tables

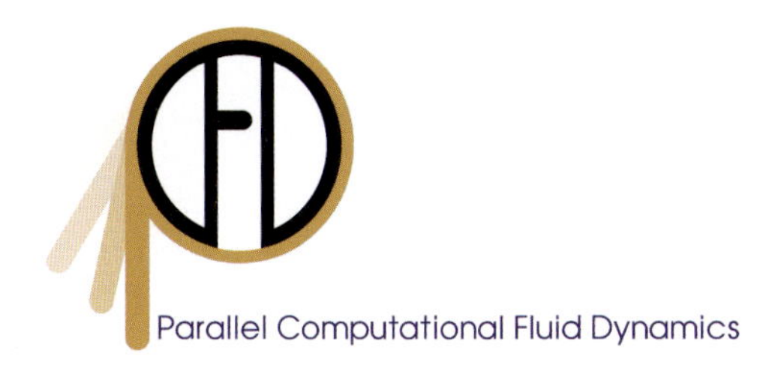

Ismail H. Tuncer
Middle East Technical University
Aerospace Engineering Department
06531 Ankara
Turkey
tuncer@ae.metu.edu.tr

Ülgen Gülcat
Faculty of Aeronautics
& Astronautics
Istanbul Technical University
ITU-Maslak
34469 Istanbul
Turkey
gulcat@itu.edu.tr

David R. Emerson
Science & Technology Facilities Council
Daresbury Laboratory
Daresbury Science & Innovation Campus
Keckwick Lane
Daresbury, Cheshire WA4 4AD
United Kingdom

Kenichi Matsuno
Kyoto Institute of Technology
Dept. Mechanical
& Systems Engineering
Matsugasaki
Kyoto
Sakyo-ku
606-8585 Japan

ISBN 978-3-540-92743-3 e-ISBN 978-3-540-92744-0

Lecture Notes in Computational Science and Engineering ISSN 1439-7358

Library of Congress Control Number: 2008942375

Mathematics Subject Classification Numbers (2000): 65XX and 68XX



*Cover design:* deblik, Heidelberg

Printed on acid-free paper

9 8 7 6 5 4 3 2 1

springer.com

# Preface

Parallel CFD 2007 was held in Antalya, Turkey, from May 21 to 24, 2007. It was the nineteenth in an annual series of international conference meetings focusing on Computational Fluid Dynamics related research performed on parallel computers and attracted over 130 participants from 13 countries.

During the conference, 7 invited papers and 79 contributed papers were presented in three parallel sessions. The host country, Turkey, contributed 19 papers and demonstrated the growing local interest in high performance computing. The topics covered in these sessions ranged from hardware and software aspects of high performance computing to multidisciplinary CFD applications using dedicated PC clusters or GRID computing. The invited papers presented the state-of-the-art CFD research interests of high performance computer centers of Europe, Japan, and the United States. In addition, two short courses on lattice Boltzmann methods and Parallel and Distributed Scientific Computing were organized. Tutorials and hands-on sessions on these short courses attracted large audiences. The related documentation on these sessions is available on the conference web site http://parcfd.ae.metu.edu.tr

These proceedings include about 75% of the oral presentations at the conference. All published papers have been peer reviewed. This volume provides the full versions of the invited as well as the contributed papers.

The editors would like to take this opportunity to express their appreciation and thanks to all of the authors for their contribution, the members of the scientific committee for reviewing the papers, and our sponsors; the Scientific and Technological Research Council of Turkey, Istanbul Technical University and Middle East Technical University for their valuable support.

The Editors

# Table of Contents

# Numerical Simulation of a Spinning Projectile Using Parallel and Vectorized Unstructured Flow Solver

**Marvin Watts,[a] Shuangzhang Tu,[a] Shahrouz Aliabadi[b]**

[a] *Assistant Professor, Jackson State University, School of Engineering, Jackson, MS 39217, USA*

[b] *Northrop Grumman Professor of Engineering, Jackson State University, School of Engineering, Jackson, MS 39204, USA*

Keywords: Finite Volume Method; LU-SGS Preconditioner; GMRES; Vectorization;

## 1. INTRODUCTION

The finite volume method (FVM) is the most widely used numerical method by computational fluid dynamics (CFD) researchers to solve the compressible Navier-Stokes equations. A successful FVM solver should be accurate, efficient and robust. High-order spatial discretization must be used for accuracy. Implicit time integration is usually adopted to obtain better efficiency, especially for high Reynolds number flows. For large-scale applications, the solver should be parallelized and even vectorized to be able to run on parallel and vector computer platforms.

The implicit time integration scheme results in a huge nonlinear system of equations for large-scale applications. Usually, the Newton-Raphson iterative method is used to solve these nonlinear systems. Inside each Newton-Raphson nonlinear iteration, a large, sparse and usually ill-conditioned linear system must be solved. Since the last decade, the Generalized Minimal RESidual (GMRES) solver introduced by Saad [1] has been widely used in solving large sparse linear systems. The beauty of the GMRES solver lies in its excellent convergence characteristics. In addition, the GMRES solver involves only matrix-vector multiplication, thus a Jacobian-free implementation is possible. With a Jacobian-free approach, we do not have to store the large sparse Jacobian matrix and the memory requirement for large applications can be significantly reduced. Like other iterative methods, the performance of the GMRES solver is highly related to the preconditioning technique. Knoll and Keyes [2] have made an extensive

I.H. Tuncer et al., *Parallel Computational Fluid Dynamics 2007*,
DOI: 10.1007/978-3-540-92744-0_1, 

survey of the Jacobian-free Newton-Krylov (JFNK) method where various preconditioning techniques have been reviewed. Though the GMRES solver itself can be Jacobian-free (matrix free), the preconditioning technique is usually not matrix-free. Luo et al. [3] and Sharov et al. [4] proposed a matrix-free preconditioning approach for the GMRES solver. In their approach, the Jacobian obtained from the low-order dissipative flux function is used to precondition the Jacobian matrix obtained from higher-order flux functions. Using the approximate Lower Upper-Symmetric Gauss-Seidel (LU-SGS) factorization of the preconditioning matrix, their preconditioning is truly matrix-free. With the Jacobian-free GMRES solver and the matrix-free LU-SGS preconditioning, the majority of the memory consumed by the FVM code depends solely on the size of the Krylov space in the GMRES solver.

For details concerning the vectorization or implementation of the implicit time integration, Jacobian-free GMRES, and matrix-free LU-SGS preconditioning methods in *CaMEL_Aero* see [5-7]. In the following section we describe an application of *CaMEL_Aero* to the simulation of a spinning projectile. As a final point, we summarize with conclusions.

## 2. NUMERICAL SIMULATION

Simulations were performed on the Cray X1E. To obtain transient solutions more quickly, all simulations are treated as steady state computations initially, and later as time accurate computations to allow the unsteady features of the flow to develop. Performance statistics are given for each example to demonstrate the efficiency, scalability, and speed of the present solver. The total CPU time excluding the problem setup and preprocessing time for the parallel simulation is recorded in the simulation. The speed, $T_c$, is evaluated according to the following formula;

$$T_c = \left(n_{proc} / \left(n_{elem} n_{ts} n_{it} n_k\right)\right) T_{run}, \qquad (1)$$

where $n_{proc}$, $n_{elem}$, $n_{ts}$, $n_{it}$, and $n_k$ are the numbers of processors, elements, time steps, nonlinear iterations within each time step, and size of Krylov space, respectively, and $T_{run}$ is the CPU time. Additionally, numerical and experimental aerodynamic coefficients are given to demonstrate the accuracy of the present solver.

### 2.1. Computational Speed and Accuracy In Predicting Aerodynamic Forces

A 0.50-cal. (1 cal. = 12.95 mm) spinning projectile is subjected to high speed, viscous flow at various Mach numbers and 0 degree angle of attack. The projectile is 4.46 cal. in length and 1.0 cal. in diameter (maximum). This projectile has a 0.16-cal.-long and 0.02-cal.-deep groove and a 9° filleted boat tail. Figure 1 shows the projectile surface grid.

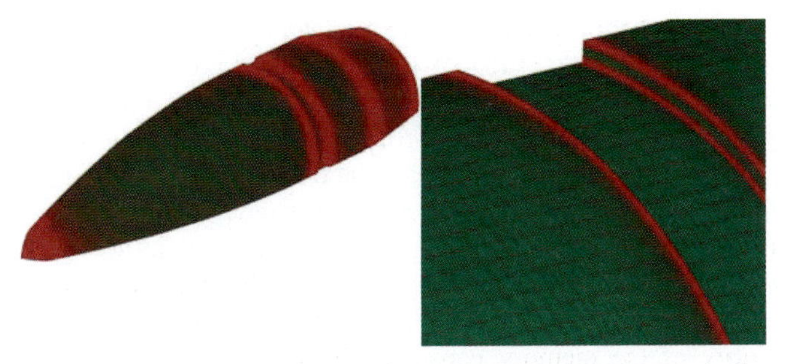

Figure 1: Surface mesh used for the CFD calculation of supersonic flows past a spinning projectile. A close-up view of the groove section is also shown.

Grid lines are clustered at regions of high curvature. A structured grid, shown in Figure 2, consisting of 40,151,112 hexahedral elements was created.

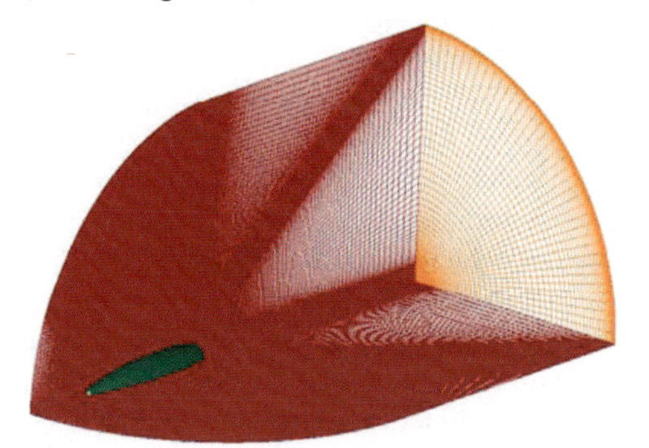

Figure 2: Volume mesh used for the CFD calculation of supersonic flows past a spinning projectile.

Though we use a structured mesh here, we do use a data structure that is suitable for arbitrarily unstructured meshes in our code.

The outflow boundary is three projectile body lengths downstream of the projectile base, the inflow boundary is only half the body length upstream of the projectile nose, and the circumferential boundary is three body lengths away from the model. The mesh contained an initial boundary layer spacing of approximately 0.65 micrometers in height.

All experimental and numerical data used in this comparison were obtained from the U.S. Army Research Laboratory (ARL) [8] and covered a wide range of Mach numbers including subsonic, transonic, and supersonic flow regimes. Effects of 0° and 2° angles were investigated, but results are only presented at 0° angle of attack. The simulations carried out at ARL used a 2.7 million element hexahedral mesh. This mesh contained a radial surface layer around the projectile body 1.07 micrometers in height.

For all cases investigated, the free-stream pressure and temperature are set to 101 kPa and 288 K, respectively. Using the perfect gas assumption, the resulting free-stream density is 1.225 kg/m$^3$. The Reynolds number for all cases was determined using the maximum projectile diameter (1.0 cal. or 0.01295 m) as the characteristic length. For the projectile body, the boundary condition is set to be no-slip, and rotating about the x-axis. The Mach numbers and corresponding roll rates are given in Table 1.

| Mach No. | Roll Rate (rad/s) |
|---|---|
| 1.10 | 6172.5 |
| 1.25 | 7014.2 |
| 1.50 | 8417.0 |
| 2.00 | 11222.8 |
| 2.70 | 15150.6 |

Table 1: Supersonic flow past a spinning projectile. Mach numbers and resulting roll rates used in CFD calculations.

The steady-state computations were done using the Cray X1E. To accelerate the convergence toward steady-state solution, local-time-stepping is used. One nonlinear iteration is performed for the Navier-Stokes equations and one nonlinear iteration for the Spalart-Allmaras turbulence equation per time step. For Mach number cases 1.10 – 2.0, the Krylov space is set to 10 in the GMRES solver. 64 MSPs were used for these computations. Each computation achieved a minimum convergence of three orders of magnitude, which required approximately 8000 – 16,000 iterations. However, the Krylov space is set to 15 in the GMRES solver for Mach number case 2.7. Furthermore, this steady-state case was restarted from a previous unsteady calculation. The performance timings given below for this case reflect only a partial computation of 2000 time steps. The final computed drag coefficient, convergence order, total CPU time, and time-cost for each problem, according to Eq. 1, are given in Table 2.

| Mach No. | $C_D$ | Time Steps | Convergence order | Time-Cost (Speed) (seconds) |
|---|---|---|---|---|
| 1.10 | 0.376 | 16000 | 3.602836 | 1.915e-6 |
| 1.25 | 0.363 | 9000 | 3.249719 | 1.921e-6 |
| 1.50 | 0.335 | 8000 | 3.520705 | 1.921e-6 |
| 2.00 | 0.308 | 8000 | 3.571653 | 1.979e-6 |
| 2.70 | 0.254 | 2000 | 4.010311 | 1.880e-6 |

Table 2: Supersonic flow past a spinning projectile. CFD calculation results and timings for various Mach numbers.

Figure 3 shows the steady-state Mach number distribution for each case. Figure 3 (f) compares the current computed drag coefficients with experimental data and other numerical solutions.

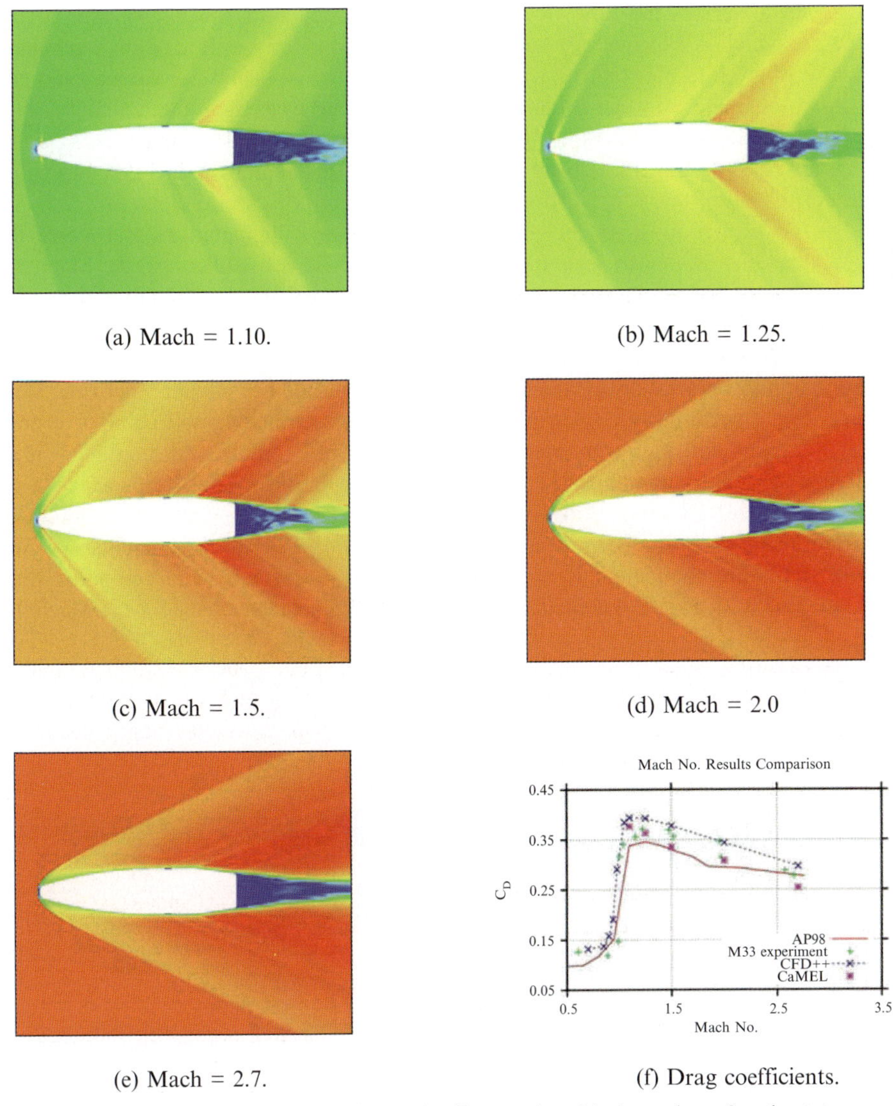


(a) Mach = 1.10. (b) Mach = 1.25.

(c) Mach = 1.5. (d) Mach = 2.0

(e) Mach = 2.7. (f) Drag coefficients.

Figure 3: Supersonic flows past a spinning projectile at various Mach numbers. Steady-state Mach number distribution and drag coefficient comparison.

Refer to [8] for information about AP98 and M33 experimental results. The "CaMEL" data points refer to the results from our current finite volume solver. Keeping in mind that the geometry used to obtain the experimental data is not exactly identical to the geometry used for numerical computations, the results obtained using our FV solver are in good agreement with the available comparison data.

In all cases, the flow field exhibits a detached bow shock upstream of the projectile, which is expected for supersonic blunt body flows, followed by a subsonic region leading up to the stagnation point. A number waves (shocks or expansions) can be seen to emanate from the spinning projectile. At the base-edge, the boundary layer separates, resulting in a free shear layer, recirculation zone, and trailing wake. Due to the complexity of the interactions taking place in this region, it is often classified separately as base flow, which is briefly discussed later in this section.

To further illustrate the need for time accurate computations to capture the flow characteristics vital to accurate aerodynamic coefficient predictions, the Mach 2.7 case was continued as an unsteady calculation for approximately 80 nondimensional units in time. This time period is approximately the time for the projectile to travel 18 body lengths. Two nonlinear iterations are performed for the Navier-Stokes equations and one nonlinear iteration for the Spalart-Allmaras turbulence equation per time step. Thanks to the performance of the preconditioner, the seemingly small number of non-linear iterations still yields an average convergence of 2 orders of magnitude within each time step (sufficient to ensure time-accurate calculations). The Krylov space is set to 10 in the GMRES solver and we perform one outer GMRES iteration per nonlinear iteration inside each time-step. In addition, we used 64 MSPs for this calculation. The time-cost of this application is approximately 1.85e-6 seconds. The computed drag coefficient is 0.25. Figure 4 depicts the steady-state and unsteady wake for the Mach 2.7 case.

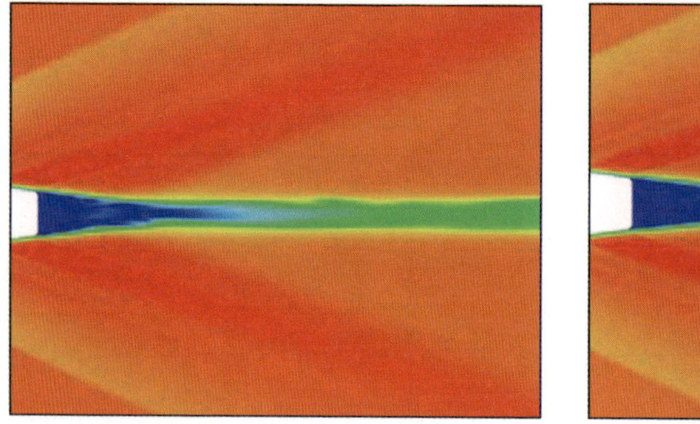
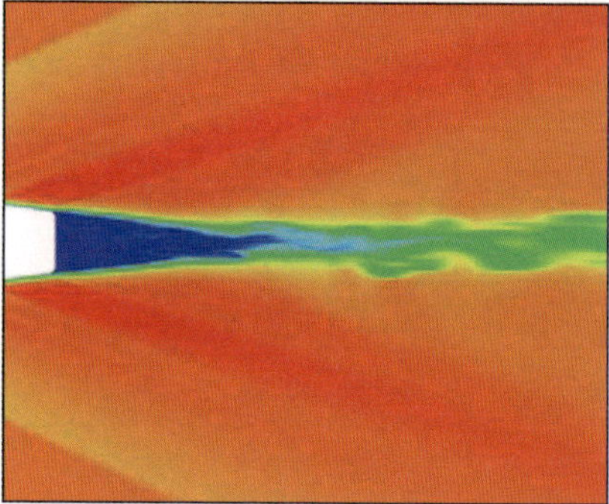

Figure 4: Supersonic flow (Mach = 2.7) past a spinning projectile. Left: steady state wake, and right: unsteady wake.

The convergence history of the drag coefficient due to pressure (pressure drag) and the drag coefficient due to friction (skin friction drag) for the unsteady Mach 2.7 case is shown in Figure 5.

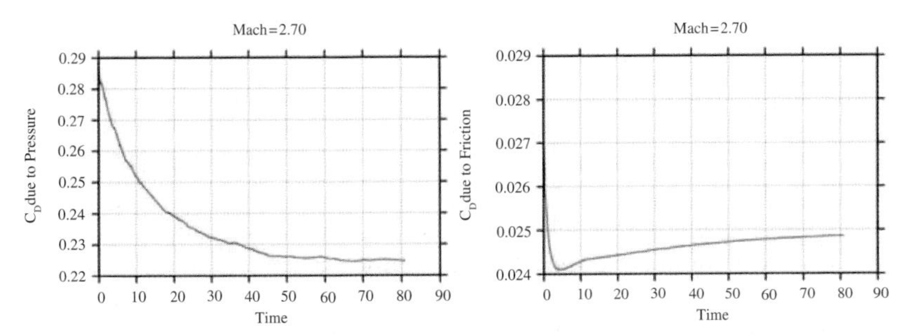


Figure 5: Supersonic flow (Mach = 2.7) past a spinning projectile. Left: unsteady drag coefficient due to pressure, and right: unsteady drag coefficient due to friction.

The difficulty involved in accurately predicting the aerodynamic coefficients for spinning projectiles, which are subject to Magnus forces, is well documented [9-12]. An added challenge for simulation of projectiles comes in the form of base flow. Base flow refers to a recirculation region located down stream of the projectile base-edge, and is the result of boundary layer separation. Within this region the pressure drops significantly lower than that of the free stream pressure. According to [13], for certain projectile configurations the base drag can contribute as much as 50% of the total drag. As such, it is vitally important that both grid resolution and the numerical scheme in use be finely tuned to capture this phenomenon. However, a clear understanding of base flow physics has yet to be established, due to the complicated interactions between the separated shear layers, recirculation zone, trailing wake, turbulence, etc. Many researchers are currently investigated the characteristics of base flow [13-15].

## 3. CONCLUSIONS

In this paper, we analyze the performance of *CaMEL_Aero*. We presented simulations of 3D flows around a spinning projectile at various Mach numbers to demonstrate the accuracy, applicability, and performance of the methods described in [5-7]. These cases also show the robustness of *CaMEL_Aero* at high Reynolds number flows.

In summary, we have developed a parallel and vectorized numerical solver which, implemented on the Cray X1E or Linux cluster family of supercomputers, is capable of solving problems with the following features:

- Unstructured (hexahedral, tetrahedral, prismatic or pyramid elements) or structured meshes.
- Transonic and supersonic flows.
- Viscous flows.
- Implicit time integration.
- Jacobian-free GMRES solver
- Matrix-free LU-SGS preconditioning

Additionally, the typical speed, $T_c$, for a tetrahedral mesh simulation is approximately 1.0e-6 seconds on the Cray X1E for solving the Navier-Stokes equations and 3.4e-7 seconds for the turbulence equation. The speed for a hexahedral mesh simulation is approximately 1.3 - 2.0 times that of a tetrahedral mesh.

**Acknowledgments**

This work is funded in part by the Army High Performance Computing Research Center (AHPCRC) under the auspices of the Department of the Army, Army Research Laboratory contract numbers DAAD19-01-2-0014 and DAAD19-03-D-0001. Partial support for this publication made possible through support provided by Northrop Grumman Ship Systems and National Science Fundation.

**References**

1. Y. Saad. Iterative methods for sparse linear systems. PWS Publishing Company, 1996.
2. D. Knoll, D. Keyes. Jacobian-free Newton-Krylov methods: a survey of approaches and applications. Journal of Computational Physics 2004; 193: 357-397.
3. H. Luo, J. Baum, R. LÄohner. A fast matrix-free implicit method for compressible flows on unstructured grids. Journal of Computational Physics 1998; 146: 664-690.
4. D. Sharov, H. Luo, J. Baum, R. Lohner. Implementation of unstructured grid GMRES+LU-SGS method on shared-memory cache-based parallel computers. AIAA paper 2000-0927, 2000.
5. S. Tu, S. Aliabadi, A. Johnson, M. Watts. A robust parallel implicit finite volume solver for high-speed compressible flows. AIAA paper 2005-1396, 2005.
6. S. Tu, M. Watts, A. Fuller, R. Patel and S. Aliabadi, "Development and Performance of CaMEL_Aero, a Truly Matrix-free, Parallel and Vectorized Unstructured Finite Volume Solver for Compressible Flows," in Proceedings of 25th Army Science Conference, Nov. 2006, Orlando, FL.
7. S. Tu, S. Aliabadi, A. Johnson and M. Watts, "High Performance Computation of Compressible Flows on the Cray X1," in Proceedings of the Second International Conference on Computational Ballistics, pp. 347-356, on May 18 - 20, 2005, Cordoba, Spain.
8. Silton, S. I. Navier-Stokes Computations for A Spinning Projectile From Subsonic To Supersonic Speeds. AIAA 2003-3936, 2003.
9. DeSpirito, J., Edge, L., Vaughn, M. E. Jr., and Washington, W. D. CFD Investigation of Canard-Controlled Missiles with Planar and Grid Fins In Supersonic Flow, AIAA 2002-4509, 2002.
10. Sahu, J., Edge, H., Dinavahi, S., and Soni, B. Progress on Unsteady Aerodynamics of Maneuvering Munitions. Users Group Meeting Proceedings, Albuquerque, NM, June 2000.
11. Sahu, J. High Performance Computing Simulations for a Subsonic Projectile with Jet Interactions. Proceedings of the Advanced Simulations and Technologies Conference, San Diego, CA, April 2002.
12. Sahu, J. Time-Accurate Numerical Prediction of Free Flight Aerodynamics of a Finned Projectile. AIAA 2005-5817, 2005.
13. Kawai, S. Computational Analysis of the Characteristics of High Speed Base Flows. Ph.D. Thesis, University of Tokyo, 2005.
14. Subbareddy, P., and Candler, G. Numerical Investigations of Supersonic Base Flows Using DES. 43rd AIAA Aerospace Sciences Meeting and Exhibit, Reno, NV, 10-13 Jan, 2005.
15. Sahu, J., and Nietubicz, C. Navier-Stokes Computations of Projectile Base Flow with and without Mass Injection. AIAA Journal 1985; 23(9):1348-1355.

# Development of a framework for parallel simulators with various physics and its performance

**Kenji Ono,[a, b] Tsuyoshi Tamaki,[c] Hiroyuki Yoshikawa[c]**

[a]*Functionality Simulation and Information team, VCAD System Research Program, RIKEN, 2-1 Hirosawa, Wako, 351-0198, Japan*

[b]*Division of Human Mechanical Systems and Design, Faculty and Graduate School of Engineering, Hokkaido University, N13, W8, Kita-ku, Sapporo, 060-8628, Japan*

[c]*Fujitsu Nagano Systems Engineering, 1415 Midori-cho, Tsuruga, Nagano, 380-0813, Japan*

Keywords: Object-Oriented Framework; Parallel Computation; MPI; Data Class

## 1. Application Framework

An object-oriented framework with class libraries is designed to enhance the software development and to manage various physical simulators. The proposed framework provides an efficient way to construct applications using the inheritance mechanism of object-oriented technology (OOT). In addition, the inheritance plays an important role to build applications with a unified behavior. It is expected that the framework brings efficiency for software development and makes easy to operate the developed applications. The framework also delivers high-level conceptual parallel programming environment based on the parallelism of domain decomposition.

### 1.1. Outline of Framework Library

A proposed system is an object-oriented framework named SPHERE (Skeleton for PHysical and Engineering REsearch), which is designed to apply building unsteady

I.H. Tuncer et al., *Parallel Computational Fluid Dynamics 2007*,
DOI: 10.1007/978-3-540-92744-0_2, 

physical simulators. SPHERE would provide various benefits both developers and end-users as described by following sub-sections.

*1.1.1. Basic features of SPHERE*

This framework supplies both a control structure and basic functions that are essential for time evolutional physical simulations. Since a simulation code can be conceptually divided into three parts like, pre-process, main-process, and post-process, programmers can describe their applications using a common structure. Thus, the mechanism that has the skeleton of the control structure should be established and be utilized as a solver base class depicted in Fig. 1.

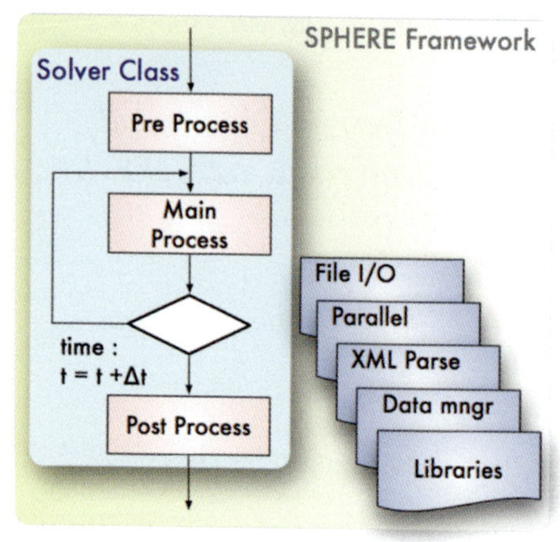


Fig. 1. Control structure of a solver class and provided libraries on a proposed framework. SPHERE framework supplies a skeleton for unsteady physical simulators that require time evolution as well as steady simulation.

A programmer makes a target application class inheriting from the base class and constructs the application by implementing user's code on the derived class. Although the framework is written by the manner of OOT, a procedural programming style is taken over because many scientific applications are described by C/Fortran. The class abstraction that corresponds to the conventional procedural process enables us to port existing software resources to the program on the framework easily. The maximum unit of a class can be assumed by the program unit working as a solver, and defines as a solver class. In addition, SPHERE provides convenient functions as libraries to construct solvers. These functions are consisting of common functions, such as data management, parsing of various parameters described by XML language, file I/O, and so on.

*1.1.2. Realization of a unified user interface using a user-defined base class*

Making and utilizing the base class that has more concrete and lump functions, the programmer can develop solver classes that have a unified user interface and a behavior. Both the abstraction of common functions and the versatility of the implemented methods are required for the user-defined base class to realize this idea. To do so, the mechanism of class inheritance in OOT plays an important role.

*1.1.3. Application management and execution environment*

Programmers can describe their specific solver class using the proposed framework. The written solver class is equivalent to a program unit as an application in the usual sense. The framework has a structure that enables us to register user's solver classes on it. SPHERE behaves as a monolithic application that can execute the registered solver classes by invoking a solver keyword described in an XML parameter file. This mechanism brings the extension of solver coupling into view.

*1.1.4. Source code portability*

The current framework supports major platforms like UNIX (Linux), Windows, and Mac OSX. The interoperability of compilers turns the mixed language programming into reality. Although the main function of the framework is required to be written by C++, C and Fortran language are available as functions or subroutines. This feature helps us to port existing Fortran program to the solver class. While system call is written by C/C++, the process that requires high performance can be written by Fortran language.

*1.1.5. Achievement of high performance*

On scientific applications with OOT, it should be taken care to achieve high performance. One may describe the code with the operator overload technique that permits to carry out operations between the classes in order to increase flexibility of the code description. Giving high readability of source code to us, this operator overload technique brings a serious defect on performance due to the internal generation of temporal class objects [1]. To remedy this problem, expression templates technique [2] is proposed and employed in POOMA [3]. Instead, the current framework keeps the performance high by simple implementation, which suppresses the operator overload and passes the address of array variable to subroutines and functions directly.

*1.1.6. Easy parallelization from serial code based on domain decomposition*

SPHERE provides a high-level parallel programming environment based on the domain decomposition method. An extension to a parallel code can be realized by inserting methods into the serial source code. Several useful patterns for parallelization are prepared and are delivered by the framework. This simple procedure is so powerful to build up the parallel code.

**1.2. Organization of framework**

The framework takes on the functions between an operating system and applications as shown in Fig. 2. The MPI library is employed to help the parallelization and is used to construct data/parallel manager classes. Several libraries are incorporated in the framework such as file I/O library, libxml2 library that parses the parameter described in XML language and so on. Other libraries, of course, can be added to the framework.

SolverBase class, which is a base class provided by the framework, has all functions of libraries incorporated and the procedure of the solver control. The classes Ω and ß just above on SolverBase class in Fig. 2 indicate the user-defined base classes; and four different kind of solver classes are build on them.
As mentioned above, although this framework is designed with OOT, the framework SPHERE has advantages of high ability of porting from and high affinity with the existing C/Fortran codes, introducing the abstraction based on conventional procedure and mixed language programming. These merits are unique feature that is not found in other OOT frameworks like SAMRAI [4], Overture [5], and Cactus [6].

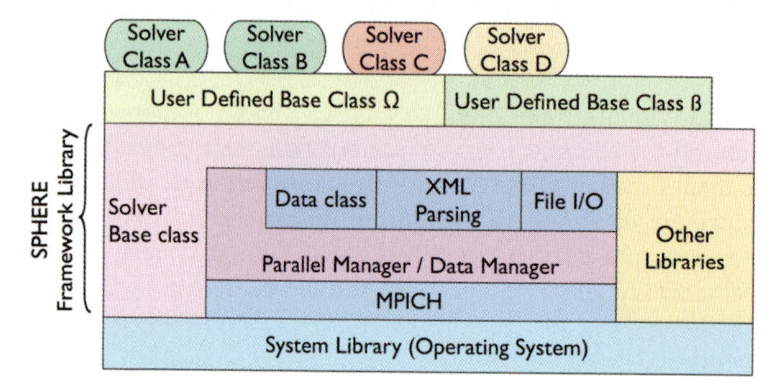


Fig. 2. Diagram of SPHERE architecture. User-defined base classes are useful class that includes common functions for specific applications such as flow simulators, structural simulators, and electro-magnetic solvers. Each solver class A~D indicates a real application in the usual sense.

## 2. Application packaging by inheritance of a user-defined base class

As shown in Fig. 2, SPHERE framework consists of libraries and provides its functions as a base class to the programmers who will write an application program using the inheritance mechanism. SPHERE is made up of highly abstracted generic program components that support the development of time evolutional simulator. More concrete and functionally lumped processes should be described with those basic functions to make a specific application. Consequently, the authors employed the approach that the programmer could design the user-defined base class that has versatility in a way for limited applicable scope; programmer can write the specific application inheriting the user-defined base class.
In Fig. 3, SklSolverBase class is a super class, which is able to use all the function of the incorporated libraries. On the other hand, FlowBass class is the user-defined base class that is derived from SklSolverBase class and is designed for flow simulators. This FlowBase class has functions such as the solver control, boundary control and treatment, preprocess of voxel data, the retention and reading parameters in XML files. The user-defined base class includes basic user interface of applications.
This approach offers both a programmer and end-users several merits. First, the development of application can be efficient. The base classes enable us to write source

code with high-level description. The programmers can concentrate their effort on algorithm and maintain their codes without great effort. Second, the unified user interface will deliver for the end-users. In this context, since the behavior of the application on the framework is almost same regardless of the different types of physics or methods utilized, the barrier and the learning curve at the introducing the applications will be greatly reduced. As a demerit, meanwhile, the current way has no small effect on the developers, who are enforced in coding manner and programming style as well as the concept of OOT.

In Fig. 3, five solver classes are derived from FlowBase and registered on SPHERE. Each keyword of CBS, CVC, RSC, and PBC means the shape approximation by binary voxel for an object on the Cartesian mesh with staggered variable arrangement, by volume fraction on the Cartesian with collocated, by Signed Distance Function (SDF) on non-uniformly Cartesian with collocated, and by binary voxel on Octree with collocated, respectively. In addition, _H, _CP, _IC keywords indicate heat solver, compressible flow solver, and incompressible flow solver, respectively. Thus, the proposed framework has two aspects; a development environment and a run-time environment.

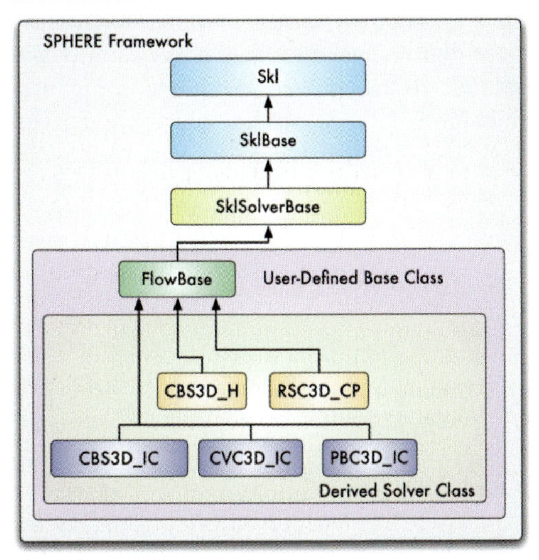


Fig. 3. A user-defined base class and derived solver classes. While framework provides Skl, SklBase, and SklSolverBase classes, other classes are written by application developers.

## 3. Data Class

Parallel process, which the framework takes on, is one of the key functions in SPHERE. In the case of distributed parallel computation, generally, programming and the implementation of boundary conditions tend to be complex. SPHERE provides an abstracted programming environment where the description of the low-level process is hidden by introducing a concept of data class. While parallel process will be discussed

for a domain decomposition method on a structured grid system in this paper, the proposed framework has no limit to incorporate any data structure, calculation algorithm, and parallel algorithm.
Data class is a collection of class to provide useful functions, which are designed for accessing to one-dimensional array data inside the data class as if the data class is multi-dimensional array. Data class library consists of two layers, i.e., data class itself and data manager class which keeps it up, to provide flexible programming environment. An abstraction of data class is designed so that the provided operations and the functions are confined to the management, the operation, and the parallelization for the array. The selection of implemented functions are paid attention to following issues; the portability of legacy Fortran programs, the affinity on the program development with mixed language, the suppression of inherent overhead for the object-oriented language, and the flexibility of programming.

### 3.1. Role and function of data class

Data class, which is one of a class libraries inside SPHERE, is in charge of management of arrays such as creating, deleting, allocating arrays on memory space, and so on. A base class of data class is implemented as C++ template class. N-dimensional class inherits from the base class that has only one-dimensional array and provides interface to access array data with n-dimensional indices. In the concrete, as shown in Fig. 4, a set of data class has SklArayidxN (N stands for 1, 2, 3, 4) classes, which are derived from SklAray (Base) class. These classes are also distinguished by the variable type of scalar and vector.

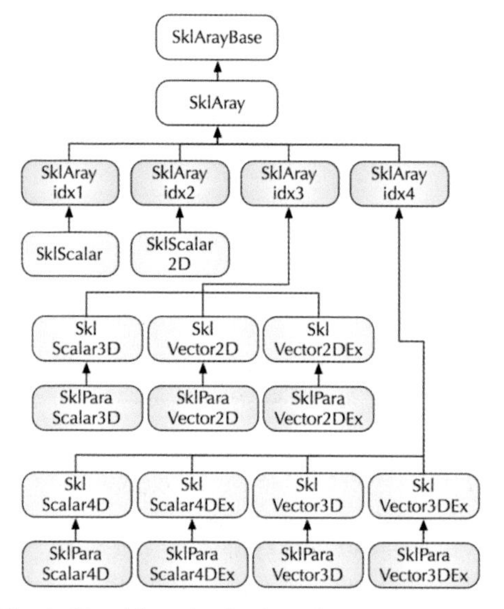


Fig. 4. Class hierarchy for data class. SklArayBase is implemented as an abstract class with pure virtual functions.

In the case of parallel solver, this data class has functions of data communication between each sub-domain. The basic idea is borrowed from Ohta [7], but the concept is extended by the combination of data class and manager class so that SPHERE can give more flexibility to manage various types of data array. Derived parallel version has additional information of guide cell, which acts as buffer region to synchronize data between adjacent domains.
Data class provides the functions such as the acquisition of top address of array data inside data class, the operator to access n-dimensional array, synchronization according to the domain decomposition manner.

### 3.2. Role and function of manager class

Manager class has two types; one is for serial code and the other is for parallel. Serial manager class performs the creation, the deletion, and the registration of a data class. Parallel manager class has a function of management of parallel environment in addition to one of serial manager class.

### 3.3. Functions for parallel process

In the case of parallel code, the parallel manager and related classes are derived from SklBase class as shown in both Fig. 3 and Fig. 5. SklParaManager class has SklParaNodeInfo class, SklVoxelInfo class, and SklParaIF class in its inside (see Fig. 6). SklParaIF is an abstract class and real communication procedures are implemented on SklparaMPI class. Of course, other communication libraries like PVM and LAM are another option. SklParaNodeInfo and SklVoxelInfo classes have the information such as the number of cells for a whole computational domain, the number of nodes, the number of domains, and rank number of each domain.

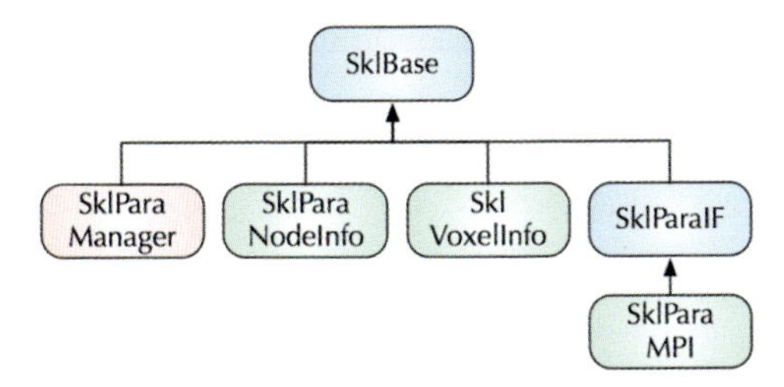


Fig. 5. Class hierarchy for manager class. Both SklBase and SklParaIF are abstract class. SklParaManager class includes SklParaNodeInfo, SklVoxelInfo, and SklParaIF class.

Fig. 6 shows the behaviour of both a solver class and SklParaManager classes. Node numbers indicate each sub-domain. A solver class in SPHERE has the class objects of SklParaManager class as one of the member, and requests parallel instructions to SklParaManager objects. SklParaManager class takes in charge of all communications as far as parallel process in the solver class. When synchronizing, solver class object issues the synchronization instruction to SklParaManager object, and then SklParaManager objects communicate each other.

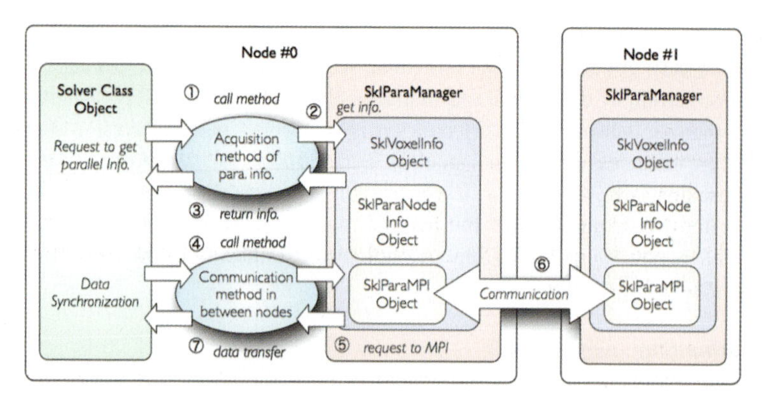


Fig. 6. Role of parallel manager class. SklParaManager class is in charge of all communications between sub-domains and includes low-level implementation for communication.

## 4. Performance Evaluation

To confirm the performance of the developed application on the framework, a benchmark program was used to evaluate on several parallel machines.

### 4.1. Test Environment and Preliminarily Benchmark Code

Benchmark is performed in several environments as shown in Table 1. RSCC [8], which is the acronym for "Riken Super Combined Cluster," is Linux cluster system and consists of 1,024 nodes (2048 CPUs) in total. SCore is used as a clustering middleware.

Table 1. Specification for evaluated environments.

| | RSCC | SGI Altix 4700 | Xeon Cluster |
|---|---|---|---|
| CPU | Xeon (Prestonia) | Itanium2 9000 (Montecito) | Xeon 5160 (Woodcrest) |
| Clock [GHz] | 3.06 | 1.6 | 3.0 |
| CPU (Core) / Node | 2 (2) | 2 (4) | 2 (4) |
| Node | 1,024 | 32 | 32 |
| Memory/node | 1GB | 4GB | 8GB |
| Middleware | Score 5.6.1 | SGI ProPack 5 | SGI ProPack 5 |
| Interconnect | InfiniBand (8Gbps) | NUMA link (8.5GB/s) | InfiniBand (8Gbps) |
| Compiler | Fujitsu Linux Parallel Package | Intel C++/Fortran Compiler 9.1 | Intel C++/Fortran Compiler 9.1 |
| MPI Library | Included in SCore | SGI MPT 1.14 | Voltaire MPI |

A program used för the benchmark is a code to solve an elliptic partial differential equation by Jacobi relaxation method, which is commonly found in an incompressible flow solver or an electro-magnetic solver. This code has the characteristics of memory-bound, that is, the memory bandwidth has much impact on the performance in addition to CPU ability because the number of arithmetic operation is comparable with the number of load/store in the source code. This original code is parallelized by MPI library based on the domain decomposition method.
The benchmark code on the framework [8] is ported from the original code which is parallelized by the similar MPI coding.

### 4.2. Parallel Performance

Table 2 shows the comparison results of the measured timing on RSCC for both the original code without the framework and SPHERE code. It was found that the speed of operation on SPHERE code goes down around five percent in comparison with the original code. This point is thought as reasonable because the framework has the inherent overhead for its parallel process in exchange for convenience in writing code.

Table 2. Measured performance in GFLOPS for both the original and SPHERE code with different size (Original / SPHERE code).

| Number of CPU core | 256x128x128 (M) | 512x256x256 (L) | 1024x512x512(XL) |
|---|---|---|---|
| 32 | 11.13 / 10.21 | 11.34 / 11.50 | 11.22 / 12.57 |
| 64 | 22.42 / 20.99 | 21.19 / 22.85 | 21.48 / 23.98 |
| 128 | 38.53 / 35.91 | 44.44 / 44.68 | 40.59 / 48.94 |
| 256 | 65.36 / 62.56 | 83.12 / 83.40 | 80.18 / 90.93 |
| 512 | 109.00 / 107.14 | 158.09 / 160.28 | 155.10 / 180.95 |
| 1024 | 150.57 / 147.25 | 264.93 / 251.17 | 316.11 / 297.90 |

Next, Table 3 shows the performance on different kind of machines for SPHERE code. SGI Altix 4700 system shows the best performance among them due to the fastest interconnect speed and also, achieves high speed up ratio; this performance is over 30 percent of its theoretical performance. The Xeon (Woodcrest) cluster demonstrated second best performance. In this result, no gain is observed from one core to two cores because this system shares a system bus with two cores and the benchmark code is characterized by memory-bound.

### 5. Concluding Remarks

An object-oriented framework for physical simulators is developed to enhance efficiency of the software development and to manage various applications, and then, a preliminarily benchmark is performed on several parallel environments. Finally, it was found that the newly designed framework worked well and exhibited reasonably good performance in comparison to the original benchmark code without the framework. This framework will be deliver from our website [9].

Acknowledgement
This study was partially supported by Industrial Technology Research Grant Program in '04 from New Energy and Industrial Technology Development Organization (NEDO) of Japan and, by computational resources of the RIKEN Super Combined Cluster (RSCC). We would like to thanks to Mr. Takeda, SGI Japan, Ltd. for the cooperation of the benchmark.

Table 3. Measured performance in GFLOPS and speed up ratio of SPHERE code for 512x256x256 (L size) grid points. Values with * mark are for reference. Those are results for 256x128x128 (M size) due to the restriction of amount of main memory. The figures shown in parenthesis indicate the speed up ratio of parallelization.

| Number of CPU core | RSCC | SGI Altix 4700 | Xeon Cluster |
|---|---|---|---|
| 1 | 0.55* (1.00) | 2.57 (1.00) | 1.33 (1.00) |
| 2 | 0.97* (1.76) | 3.45 (1.34) | 1.35 (1.02) |
| 4 | 1.58 (2.87) | 6.84 (2.67) | 2.35 (1.71) |
| 8 | 3.08 (5.60) | 13.67 (5.33) | 5.08 (3.81) |
| 16 | 5.65 (10.27) | 26.67 (10.39) | 10.32 (7.74) |
| 32 | 11.73 (21.33) | 54.29 (21.16) | 20.82 (15.61) |
| 64 | 23.30 (42.36) | 111.73 (43.55) | 44.54 (33.39) |
| 128 | 45.55 (82.82) | 268.51 (104.65) | 88.80 (66.57) |

**Reference**

1. Bulka, D., and Mayhew, D., Efficient C++ Performance Programming Techniques, Addison-Wesley, (1999).
2. Veldhuizen, T., "Expression Templates," C++ Report, Vol. 7 No. 5 June, pp. 26-31 (1995).
3. http://www.nongnu.org/freepooma
4. Hornung, R.D., and Kohn, S.R., "Managing Application Complexity in the SAMRAI Object-Oriented Framework," in Concurrency and Computation: Practice and Experience (Special Issue), 14, pp. 347-368 (2002).
5. Henshaw, W.D., "Overture: An Object-Oriented Framework for Overlapping Grid Applications," AIAA conference on Applied Aerodynamics (2002), also UCRL-JC-147889.
6. http://www.cactuscode.org/
7. Ohta, T., and Shirayama, S., "Building an Integrated Environment for CFD with an Object-Orientd Framework," Transactions of JSCES, No. 19990001, in Japanese (1999).
8. http://accc.riken.go.jp/
9. http://vcad-hpsv.riken.jp/

# Experience in Parallel Computational Mechanics on MareNostrum

G. Houzeaux[a*], M. Vázquez[a], R. Grima[a], H. Calmet[a], and J.M. Cela[a]

[a]Department of Computer Applications in Science and Engineering
Barcelona Supercomputing Center
Edificio C1-E201, Campus Nord UPC
08034 Barcelona, Spain

We present in this paper the experience of the authors in solving very large problems of computational mechanics on a supercomputer. The authors are researchers of the Barcelona Supercomputing Center-Centro Nacional de Supercomputación (BSC-CNS), a brand new research center in Spain, which hosts the fastest supercomputer in Europe and the fifth in the world: MareNostrum (TOP500, November 2006). A brief presentation of MareNostrum is given in the first section. In the next section, we describe the physical problems we are faced with, followed by the section on the associated numerical strategies. Next section presents the parallelization strategy employed; emphasis will be put on the important aspects to obtain good speed-up results on thousands of processors. We follow the discussion by describing some applications together with speed-up results. We finalize the paper by describing the performance tool used at BSC-CNS and their application in computational mechanics.

## 1. MareNostrum

Early in 2004 the Ministry of Education and Science (Spanish Government), Generalitat de Catalunya (local Catalan Government) and Technical University of Catalonia (UPC) took the initiative of creating a National Supercomputing Center in Barcelona. The consortium signed with IBM an agreement to build one of the fastest computer in Europe MareNostrum. MareNostrum is a supercomputer made off PowerPC 970 processors mounted on JS21 blades, working under Linux system and connected through a Myrinet interconnection. Summary of the system: Peak Performance of 94,21 Teraflops; 10240 IBM Power PC 970MP processors at 2.3 GHz (2560 JS21 blades); 20 TB of main memory; 280 + 90 TB of disk storage; Interconnection networks: Myrinet and Gigabit Ethernet; Linux: SuSe Distribution.

According to the TOP500 list of November 2006 [8], MareNostrum is actually the fastest supercomputer in Europe and the fifth one in the world. See Figure 1.

*The research of Dr. Houzeaux has been partly done under a *Ramon y Cajal* contract with the Spanish *Ministerio de Educación y Ciencia*

I.H. Tuncer et al., *Parallel Computational Fluid Dynamics 2007*,
DOI: 10.1007/978-3-540-92744-0_3, 

Figure 1. MareNostrum supercomputer

## 2. Physical problems

The physical problems described here are the incompressible Navier-Stokes equations, the compressible Navier-Stokes equations and the wave propagation equation.

Let $\boldsymbol{u}$ be the velocity, $p$ the pressure, $\rho$ the density, $\mu$ the viscosity. The velocity strain rate $\boldsymbol{\varepsilon}(\boldsymbol{u})$ and the stress tensor $\boldsymbol{\sigma}$ are

$$\begin{aligned}\boldsymbol{\varepsilon}(\boldsymbol{u}) &= \frac{1}{2}(\nabla\boldsymbol{u} + \nabla\boldsymbol{u}^t) - \frac{1}{3}\nabla\cdot\boldsymbol{u}, \\ \boldsymbol{\sigma} &= -p\boldsymbol{I} + 2\mu\boldsymbol{\varepsilon}(\boldsymbol{u}),\end{aligned}$$

where $\boldsymbol{I}$ is the $d$-dimensional identity matrix, that is $\boldsymbol{I}_{ij} = \delta_{ij},\ i,j = 1,\ldots,d$. The traction $\boldsymbol{\sigma}\cdot\boldsymbol{n}$ is the force acting on a unit fluid surface element with unit outwards normal $\boldsymbol{n}$.

**Incompressible Navier-Stokes equations.**

The solution of the incompressible Navier-Stokes equations consist in finding $\boldsymbol{u}$ and $p$ such that they satisfy the following equations [7]:

$$\begin{aligned}\rho\partial_t\boldsymbol{u} + \rho(\boldsymbol{u}\cdot\nabla)\boldsymbol{u} - \nabla\cdot[2\mu\boldsymbol{\varepsilon}(\boldsymbol{u})] + \nabla p &= \rho\boldsymbol{g}, \\ \nabla\cdot\boldsymbol{u} &= 0,\end{aligned} \tag{1}$$

where $\boldsymbol{g}$ is the gravity vector. This system must be provided with appropriate initial and boundary conditions. The latter one can be:

- Dirichlet (inflow, no-slip wall): prescription of $\boldsymbol{u}$;
- Neumann (outflow): prescription of the traction $\boldsymbol{\sigma}\cdot\boldsymbol{n}$, which for uniform flows is exactly minus the pressure;
- Mixed (symmetry, wall law): prescription of the tangential traction $\boldsymbol{\sigma}\cdot\boldsymbol{n}-(\boldsymbol{n}\cdot\boldsymbol{\sigma}\cdot\boldsymbol{n})\boldsymbol{n}$ and normal velocity $\boldsymbol{u}\cdot\boldsymbol{n}$.

**Compressible Navier-Stokes equations.**

Let $T$ be the temperature, $E$ the total energy and $k$ the thermal conductivity. The compressible flow equations are [7]:

$$\begin{aligned} \partial_t \rho \boldsymbol{u} + \nabla \cdot (\rho \boldsymbol{u} \otimes \boldsymbol{u}) - \nabla \cdot [2\mu \varepsilon(\boldsymbol{u})] + \nabla p &= \rho \boldsymbol{g}, \\ \frac{\partial \rho}{\partial t} + \nabla \cdot (\rho \boldsymbol{u}) &= 0, \\ \frac{\partial E}{\partial t} + \nabla \cdot (\boldsymbol{u} E - k \nabla T - \boldsymbol{u} \cdot \boldsymbol{\sigma}) &= \rho \boldsymbol{u} \cdot \boldsymbol{g}, \end{aligned} \tag{2}$$

This system is closed with a law for the viscosity and thermal conductivity and with a state equation. For example for air,

$$\frac{\mu}{\mu_\infty} = \frac{T_\infty + 110}{T + 110} \left(\frac{T}{T_\infty}\right)^{3/2}, \quad \frac{k}{k_\infty} = \frac{T_\infty + 133.7}{T + 133.7} \left(\frac{T}{T_\infty}\right)^{3/2}, \tag{3}$$

$$p = \rho R T, \quad E = \rho \left( C_v T + \frac{1}{2} \boldsymbol{u}^2 \right), \tag{4}$$

where $C_v$ is the specific heat at constant volume and $R$ is the gas constant. We have for air at 20 $^o$ C and 1 atm $C_v = 718\,\mathrm{J/kgK}$, $R = 287\,\mathrm{J/kgK}$. The boundary conditions are more complex as they can depend on the local characteristics of the flow [5].

**Wave propagation.**

The wave propagation equation can describe, among other propagation problems, the propagation of sound in a media [4]:

$$\partial^2_{t^2} u - c_0^2 \Delta u = f, \tag{5}$$

where $c_0^2$ is the speed of sound in the media, and $f$ is a forcing term. The boundary conditions are:

- Dirichlet: prescription of $u$;
- Neumann: prescription of $\nabla u \cdot \boldsymbol{n}$;
- First order absorbing: $\partial_t u + \nabla u \cdot \boldsymbol{n} = 0$.

In compressible flows, it can be shown that the density satisfies Equation 5 (Lighthill's equation) with

$$f = \nabla \cdot [\nabla \cdot (\rho \boldsymbol{u} \otimes \boldsymbol{u} - \boldsymbol{\sigma} + (p - c_0^2 \rho) \boldsymbol{I})].$$

In incompressible flows, for which acoustic waves are filtered, a perturbation of the equilibrium pressure and density would lead to a similar equation for the perturbed pressure [6]. In this case, this equation is used as a postprocess of the incompressible Navier-Stokes solutions to compute the approximate associated sound propagation.

### 2.1. Numerical strategy

The numerical simulation software studied here is an in-house code named Alya. The space discretization is based on a Finite Element method. The stabilization method depends on the equation considered. The incompressible Navier-Stokes equations as well as the wave equation are stabilized using the variational subgrid scale approach [2,3]. For the former case, the tracking of the subgrid scale is performed in the time and non-linear terms. That is, the subgrid scale ($\tilde{\boldsymbol{u}}$) equation is integrated in time and it is taken into account in the convection velocity ($\boldsymbol{u} \leftarrow \boldsymbol{u}_h + \tilde{\boldsymbol{u}}$). For the linear wave equation, only time tracking is done. The compressible Navier-Stokes equations are stabilized using the Characteristic Based Split Galerkin method (CBS) [5].

As for the time discretization: implicit or explicit? Both strategies are available. For the compressible flow equations, fully explicit method and fractional step techniques (which has better stability properties in low Mach regions) are used. The wave equation is also integrated explicitly in time. For the incompressible Navier-Stokes equations, both a monolithic scheme and a predictor corrector method is used, solving a Poisson-like equation for the pressure. The choice for one strategy or another is mainly motivated by the physics of the problem. For example, in acoustic problems, the critical time step may be close to the physical one. Therefore, an explicit scheme is a good candidate. As will be shown at the end of the paper, scalability results are similar for explicit and implicit schemes.

## 3. Parallelization strategy

The parallelization strategy is based on a Master-Slave technique. The Master is in charge of reading the mesh, of performing the partitioning and the output (for example of the convergence residuals). The slaves are in charge of the construction of the local residuals and possibly the local matrices and of the solution of the resulting system in parallel.

### 3.1. Master-Slave technique

Two strategies are possible, and are illustrated in Figure 2:

- For small problems (say $<$ 5 millions of elements): The Master partitions the mesh, sends the partition to the Slaves, and the simulation follows.
- For large problems: The Master partitions and writes the partitioning in restart files. This strategy is necessary for the partitioning may require a lot of RAM, only available on specific servers. The restart files are then copied to MareNostrum. The simulation can then start. Instead of receiving the problem data through MPI, both the master and the slaves read the partition data from the restart files. Note that a special strategy is needed when the number of subdomains is large. In fact, the gpfs file system available on MareNostrum cannot handle the reading and writing of too many files located in the same file system.

### 3.2. Types of communications

In a finite element implementation, only two kinds of communications are necessary between subdomains. The first type of communication consists in exchanging arrays

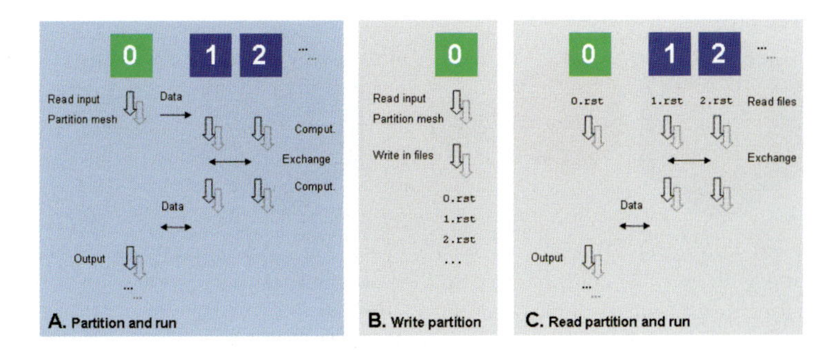


Figure 2. Two parallelization strategies. 0 square represents the master, and $1, 2\ldots$ squares represent the slaves. (Left) For small problems. (Right) For large problems.

between neighbors with `MPI_Sendrecv`. The strategy is the following:

- Explicit solvers:
  1. For each slave, compute elemental RHS for each element.
  2. For each slave, assemble (scatter) elemental RHS into global RHS.
  3. Exchange RHS of boundary nodes, the nodes belonging to more than one subdomain, and sum the contribution.
- Implicit solvers:
  1. For each slave, compute elemental LHS and RHS for each element.
  2. For each slave, assemble (scatter) elemental RHS and LHS into global LHS and RHS.
  3. Exchange RHS of boundary nodes, the nodes belonging to more than one subdomain, and sum the contribution.
  4. The operations of an iterative solver consists in matrix-vector multiplication. For each slave, compute matrix-vector multiplication.
  5. Exchange the results on the boundary nodes, as was done for the RHS.

The second type of communication is global and of reduce type with `MPI_reduce`. It is used to compute:

- The critical time step for explicit solvers: it is the minimum over the slaves.
- The convergence residual: the sum over all the nodes of the domain. It is needed by iterative solvers as well as to control the convergence of the non-linearity iteration loop or time loop.

More information can be found on the explicit code in [1].

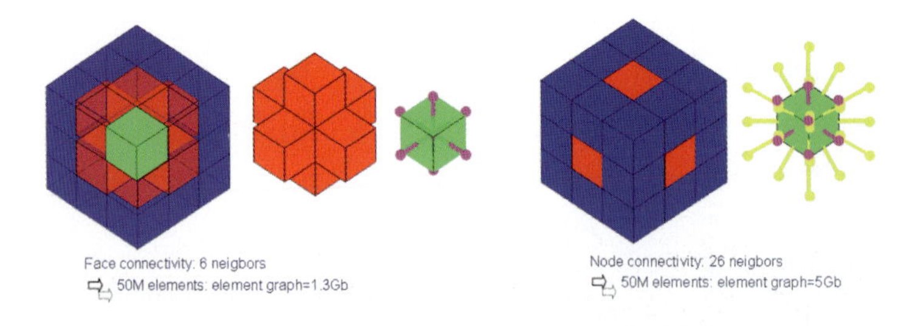


Figure 3. Two possible element graph strategies. (Left) Adjacency based on node sharing. (Right) Adjacency based on face sharing.

### 3.3. Critical points of the parallelization strategy

In order to have an efficient algorithm to run on thousands of processors, some important aspects of the parallelization must be carefully treated.

The **mesh partitioning** is performed using METIS [9]. The main input data of METIS are the element graph and the weight of the vertices of the graph. For this last aspect, the number of Gauss integration points is used to control the load balance of hybrid meshes. For the element graph, two options are possible, as illustrated in Figure 3. The correct option consists in considering as adjacent elements to an element $e$ all the elements sharing a node with $e$. However, this graph can be quite memory consuming as shown in the figure for a mesh of hexahedra. The other strategy, which has been giving good load balance results consists in taking as adjacent elements to $e$ only the elements sharing a face with $e$. This strategy requires much less memory. Figure 4 shows some partition examples.

The second important aspects is the **node numbering** of the local (slave) meshes. In the previous subsection we mentioned that for explicit as well as for implicit solvers, boundary nodal arrays must be exchanged between slaves. In order to perform an efficient data exchange, the local nodes are divided into three categories, as illustrated in Figure 5.

- Interior nodes. They are the nodes which are not shared by another subdomain.
- Boundary nodes. They are the nodes for which subdomains contribution must be summed when computing a RHS or a matrix-vector product.
- Own boundary nodes. Boundary nodes are divided into own boundary nodes and others boundary nodes. This is useful for computing the convergence residual so that the residuals of the boundary nodes are accounted for only once before performing the sum over the subdomains.

Next aspect concerns the **communication scheduling**. Figure 6 (Left) illustrates the importance of this scheduling on a simple example. For this example, each subdomain

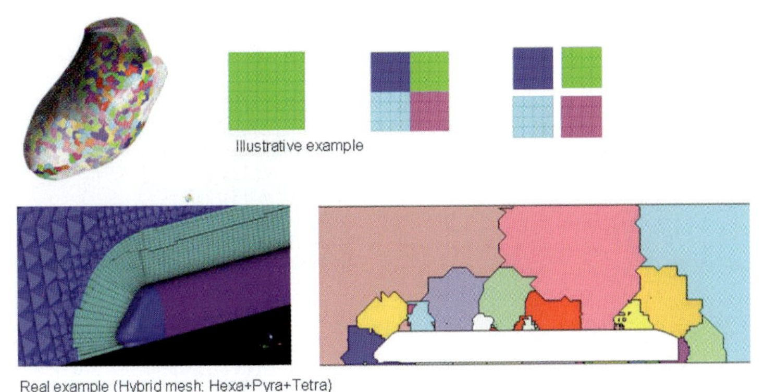


Figure 4. (Top) (Left) Partition of a ventricle into 1000 subdomains. (Top) (Right) Illustrative example of a simple partition. (Bot.) Hybrid mesh and corresponding partition.

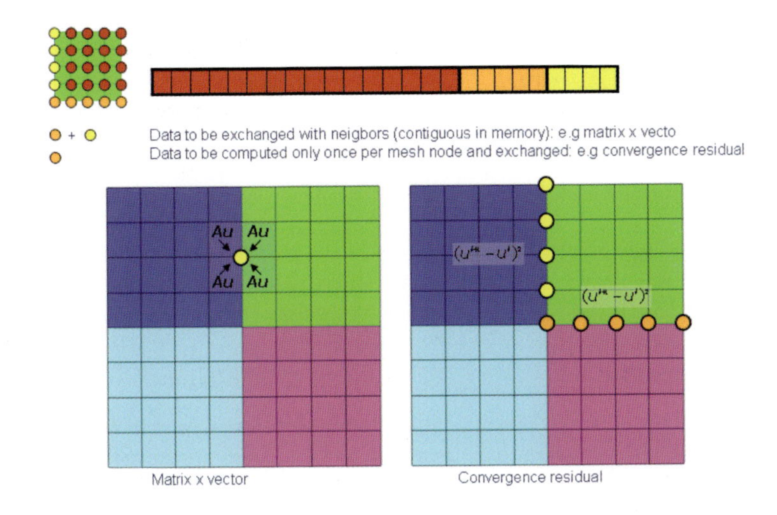


Figure 5. Local node numbering. In red: interior nodes. In orange and yellow: boundary nodes. In orange: own boundary nodes.

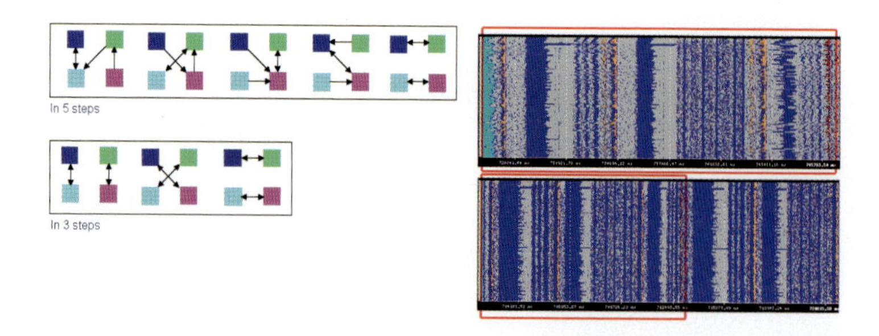


Figure 6. Scheduling strategy. (Top) (Left) Simple example: Bad communication scheduling. (Bot.) (Left) Simple example: Optimum communication scheduling. (Right) One iteration of a CFD simulation on 128 processors. $x$-axis: time. $y$-axis: process number. (Top) No special scheduling strategy. (Bot.) Good scheduling strategy. Blue/Dark grey: CPU active. Gray,Orange/Light grey: communications phase. Box: one iteration.

has to communicate with all the others. On the left part of the figure, we observe a very bad scheduling. For example, the top right subdomain cannot communicate with the bottom left subdomain until this last one has communicated with the top left one. The idea to optimize the scheduling is that at each communication step, every subdomain should communicate. The right part of the figure illustrates the optimum scheduling for this example. Figure 6 (Right) shows a typical example of bad and good orderings of the communications and the associated load asynchronism of work between the slaves. The figure shows how a good communication scheduling can drastically reduce the time needed to carry out one iteration (red rectangle).

Finally, the last aspect is the **transparency** of the parallelization. The parallelization is done in such a way that the solution of a set of partial differential equations (basically assembly of matrix and right-hand side) is almost not affected by the parallelization.

## 4. Scalability results

Figure 7 shows some results of scalability for the compressible and incompressible flow equations and for the wave equation. The first and last ones are solved explicitly. The incompressible flow equations are solved implicitly with a GMRES solver. Note that in this case, the average number of elements per processor is as low as 2000 for 100 processors. We observe that very good scalability is obtained for both strategies, up to 2500 processors for the compressible Navier-Stokes equations.

## 5. Monitoring

The study of the performance of the code is necessary to reduce cache misses, to increase the IPC (instructions per cycle), and to identify load imbalance problems due to a

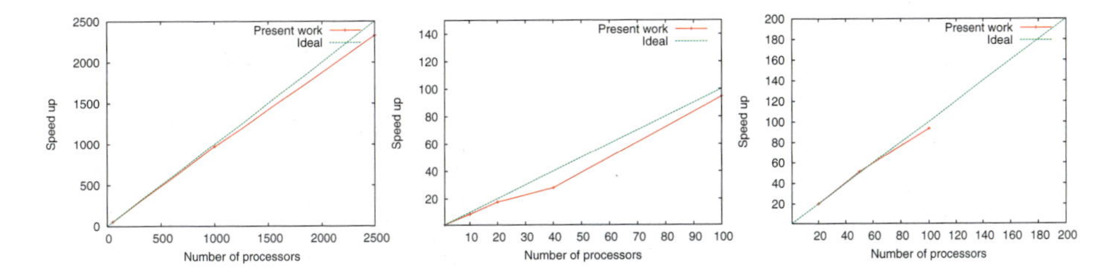


Figure 7. Speed-up. (Left) Compressible flow equations (explicit) 22M element hybrid mesh. (Mid.) Incompressible flow equations (implicit GMRES), 200K element Q1 mesh. (Right) Wave equation (explicit), 1M Q1 element mesh.

bad communication. The computational mechanics code Alya has been analyzed with a graphical performance tool developed at BSC-CNS, PARAVER [10]. Figure 8 shows an example. It consists of the explicit solution of the compressible Navier-Stokes equations during 10 time steps on 200 processors. The $x$-axis is time. The $y$-axis is the processor, the top line representing the master. Light blue color indicates a working CPU. Dark blue color indicates that CPU is idle. The yellow lines represents local communications (`MPISend`, `MPIRecv` and `MPI_SendRecv`), while the pink lines represents global communications performed with `MPI_Allreduce`. On the top figure we can observe the total computation. We can easily identify the 10 time steps on the left part as well as the postprocess of the solution on the right part. In this example, all the slaves send their solution with `MPISend` to the master (received with `MPIRecv`) which output the results in files. We see that the total time of communication costs the same time as 10 time steps. The bottom figure shows a zoom of one iteration. We can identify an `MPI_Allreduce` at the beginning of the iteration to compute the critical time step. Then comes the element loop in each slaves. At the end of the element loop, slaves exchange their RHS with `MPI_SendRecv` (yellow lines), then the solution is updated and the convergence residual is computed (last pink line). From the graphical interface, one can select some specific zones and compute statistics. For this particular case, communications represent from 6% to 10% of the CPU time over all the slaves.

## 6. Conclusion

We have presented in this paper the supercomputing strategy followed at BSC-CNS to solve computational mechanics problems. The numerical simulations show very good speed-up for both explicit and implicit solvers, and using up to 2500 processors. Since then, the authors have carried out encouraging tests on 10000 CPU's.

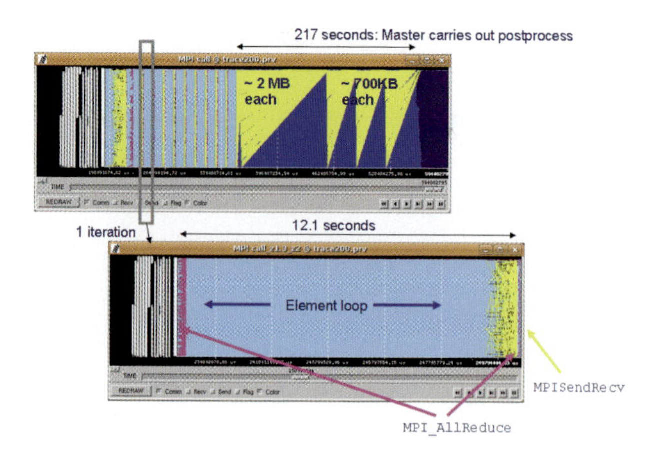


Figure 8. Visualization of a trace with PARAVER. (Top) All computation. (Bot.) Zoom on one iteration.

## REFERENCES

1. M. Vázquez, G. Houzeaux, R. Grima and J.M. Cela, Applications of Parallel Computational Fluid Mechanics in MareNostrum Supercomputer: Low-Mach Compressible Flows, PARCFD conference, Antalya, Turkey, May 21-24 (2007).
2. R. Codina and J. Principe and O. Guasch and S. Badia, Time dependent subscales in the stabilized finite element approximation of incompressible flow problems, submitted to Comp. Meth. Appl. Mech. Eng. (2006).
3. G. Houzeaux and J. Principe, A Variational Subgrid Scale Model for Transient Incompressible Flows, sumbitted to Int. J. Comp. Fluid Dyn. (2006).
4. G.C. Cohen, Higher-Order Numerical Methods for Transient Wave Equations, Scientific Computation, Springer (2002).
5. O.C. Zienkiewicz, P.Nithiarasu, R. Codina , M. Vazquez and P. Ortiz. The characteristic - based - split procedure: an efficient and accurate algorithm for fluid problems. Int. J. Numer. Meth. Fluids, 31, 359-392; (1999).
6. L. D. Landau and E. M. Lifshitz, Fluid Mechanics 2ed., Course of Theoretical Physics vol. 6, Butterworth-Heinemann (1987).
7. G.K. Batchelor, An Introduction to Fluid Dynamics, Cambridge University Press; 1970.
8. TOP500, Supercomputer sites: http://www.top500.org/list/2006/11/100
9. METIS, Family of Multilevel Partitioning Algorithms: http://glaros.dtc.umn.edu/gkhome/views/metis
10. PARAVER, the flexible analysis tool: http://www.cepba.upc.es/paraver

# New approaches to modeling rarefied gas flow in the slip and transition regime

**X. J. Gu and D. R. Emerson**

*Computational Science and Engineering Department, STFC Daresbury Laboratory, Warrington WA4 4AD, United Kingdom*

Keywords: rarefied gas, moment method, velocity slip, temperature jump, MEMS

## 1. Introduction

Rarefied gas dynamics, which has been explored for more than a century, studies gas flow where the mean free path, $\lambda$, is comparable with a typical length, $L$, of the device being considered. Traditionally, it has been used to study high-speed high-altitude flow applications, such as space re-entry vehicles, and flows under ultra-low pressure (vacuum) conditions, where $\lambda$ has a large value. However, recent technological developments have enabled major advances in fabricating miniaturized devices such as micro-electro-mechanical systems (MEMS). Gas flow in micro-scale devices can suffer from rarefaction effects because the characteristic length of the device is so small that it is comparable to the mean free path of the gas, even under atmospheric conditions.

The extent of the rarefaction can be measured by the Knudsen number, $Kn$, which is the ratio of $\lambda$ to $L$. The different rarefaction regimes can be summarized according to the value of the Knudsen number [1]: (i) no slip ($Kn \leq 10^{-3}$); (ii) slip ($10^{-3} < Kn \leq 10^{-1}$); (iii) transition ($10^{-1} < Kn \leq 10$); and (iv) free molecular flow $Kn > 10$, respectively. Different approaches have been employed by various researchers to capture and describe the non-equilibrium phenomena that arise due to an insufficient number of molecular collisions occurring under rarefied conditions. These approaches can be broadly divided into two specific categories: the microscopic and the macroscopic. In this paper, different approaches are reviewed and the emphasis will be on the development of the macroscopic method and the application of high performance parallel computers for engineering applications.

## 2. Microscopic and macroscopic approaches to modeling rarefied gas flow

Microscopically, the behavior of a rarefied gas can readily be described by kinetic theory and the Boltzmann equation, which governs the molecular distribution

I.H. Tuncer et al., *Parallel Computational Fluid Dynamics 2007*,
DOI: 10.1007/978-3-540-92744-0_4, © Springer-Verlag Berlin Heidelberg 2009

function, $f(\xi, \boldsymbol{x}, t)$, where $\boldsymbol{x}$ and $\xi$ are the position and velocity vectors, respectively, of a molecule at time $t$ [2]. The Boltzmann equation, expressed by

$$\frac{\partial f}{\partial t} + \xi_i \frac{\partial f}{\partial x_i} = Q(f, f), \tag{1}$$

is the central equation in kinetic theory, the properties of which can be used to guide the development of kinetic and macroscopic models for rarefied gas flow. Here $Q(f, f)$ is the collision integral. Equation (1) tells us that there is only one equilibrium state and its distribution function is expressed by the Maxwellian, $f_M$. The five collision invariants, $\phi(\xi)$, which satisfy

$$\int \phi(\xi) Q(f, f) d\xi = 0, \tag{2}$$

lead to the conservation laws for mass, momentum and energy. The most interesting feature of the Boltzmann equation is that the macroscopic irreversible process of entropy is imbedded in it, known as the *H*-theorem. Although there has been some advancement in the numerical simulation of several rarefied gas dynamics problems using the linearized Boltzmann equation, the theoretical and numerical treatment of the nonlinear Boltzmann equation, for practical applications, remains a formidable challenge due to the complicated structure of the collision term and its high dimensionality.

A rarefied gas can also be described using macroscopic quantities. A macroscopic model can be built up from its microscopic expression - the Boltzmann equation. The first widely used approach to construct a macroscopic model is the Chapman-Enskog method of expansion using powers of *Kn* [3]. The zeroth-order expansion yields the Euler equations and the first-order results in the Navier-Stokes-Fourier (NSF) equations. Higher order expansions yield the Burnett equations (second-order), Super-Burnett equations (third-order), and so on. However, the Chapman-Enskog expansion is only valid when the Knudsen number is small and the higher order expansion violates the underlying physics behind the Boltzmann equation [2,4,5]. In association with velocity-slip and temperature-jump wall-boundary conditions, the NSF equations, which govern the first five lowest moments of the molecular distribution function, can be used to predict flows in the slip regime fairly accurately [1]. As the value of *Kn* increases beyond 0.1, entering the transition regime, non-equilibrium effects begin to dominate the flow and *Knudsen layers* (a kinetic boundary layer) start to play a significant role [2]. More moments of $f$ are then required to accurately describe the flow.

Grad [6] proposed the moment method as an approximation solution procedure for the Boltzmann equation. The molecular distribution function is expanded in *Hermite* polynomials, the coefficients of which are linear combinations of the moments of the molecular distribution function. The infinite set of Hermite coefficients is equivalent to the molecular distribution function itself and no kinetic information is lost in such a system. In practice, the molecular distribution function has to be truncated. By choosing a sufficient number of Hermite coefficients, a general solution of the Boltzmann equation can be approximated. The governing equations of the moments involved in the Hermite coefficients can be derived from the Boltzmann equation. However, the set of moment equations is not closed, since the fluxes and the collision terms are unknown.

The truncated distribution function is used to calculate the collision terms and higher moments in the flux terms as functions of the chosen moments to close the equation set. However, for Maxwell molecules, the collision terms can be evaluated without any knowledge of the exact form of the distribution function [7].

The recently developed lattice Boltzmann method (LBM) is constructed from simplified kinetic models that incorporate the essential physics of microscopic processes so that the macroscopic averaged property obeys the desired macroscopic equations [8]. Unlike the solution of the Boltzmann equation, the LBM utilizes a minimal set of velocities in phase space so that the averaging process is greatly simplified. The LBM lies between the macroscopic and microscopic approaches and is often referred to as a *mesoscopic* method. The ability of the LBM to capture non-equilibrium phenomena in rarefied gas flows largely depends on the lattice model [9].

In contrast, the direct simulation Monte Carlo (DSMC) method proposed by Bird [10] describes a gas flow by the motion of a number of "computational molecules" following the laws of classical mechanics, each of which represents a large number of real molecules. The method is powerful to simulate mixtures and reactions but, due to its stochastic nature, it is computationally expensive to apply the DSMC method to low speed flows.

## 3. Development of the macroscopic approach: the regularized higher order moment method

In Grad's moment method, once the distribution function, *f*, is known, its moments with respect to $\xi$ can be determined. For example, the density, $\rho$, and the momentum, $\rho u_i$, can be obtained from

$$\rho = \int f d\xi \qquad \text{and} \qquad \rho u_i = \int \xi_i f\, d\xi \tag{3}$$

where $\xi_i$ and $u_i$ represent the particle and fluid velocity, respectively. An intrinsic or peculiar velocity is introduced as

$$c_i = \xi_i - u_i \tag{4}$$

so that the moments with respect to $u_i$ can be conveniently calculated. A set of $N$ moments are then used to describe the state of the gas through

$$\rho_{i_1 i_2 \ldots\ldots i_N} = \int c_{i_1} c_{i_2} \ldots\ldots c_{i_N} f\, d\xi\,. \tag{5}$$

Any moment can be expressed by its trace and traceless part [11]. For example, the pressure tensor can be separated as follows:

$$p_{ij} = p\delta_{ij} + p_{<ij>} = p\delta_{ij} + \sigma_{ij} = \int c_i c_j f\, d\xi\,, \tag{6}$$

where $\delta_{ij}$ is the Kronecker delta function, $p = p_{kk}/3$ is the pressure, and $\sigma_{ij} = p_{<ij>}$ is the deviatoric stress tensor. The angular brackets denote the traceless part of a symmetric tensor. Furthermore, the thermal energy density is given by

$$\rho\varepsilon = \frac{3}{2}\rho\frac{k}{m}T = \frac{1}{2}\int c^2 f\, d\xi\,. \tag{7}$$

The temperature, $T$, is related to the pressure and density by the ideal gas law, $p = \rho(k/m)T = \rho RT$, where $k$ is Boltzmann's constant, $m$ the mass of a molecule and $R$ the gas constant, respectively. The heat flux vector is

$$q_i = \frac{1}{2}\int c^2 c_i f \, d\boldsymbol{\xi} . \tag{8}$$

The molecular distribution function, $f$, can be reconstructed from the values of its moments. Grad expressed $f$ in Hermite polynomials as [6]:

$$f = f^M \sum_{n=0}^{\infty} \frac{1}{n!} a_A^{(n)} H_A^{(n)} = f^M \left( a^{(0)} H^{(0)} + a_i^{(1)} H_i^{(1)} + \frac{1}{2!} a_{ij}^{(2)} H_{ij}^{(2)} + \frac{1}{3!} a_{ijk}^{(3)} H_{ijk}^{(3)} + \ldots.. \right), \tag{9}$$

where $H_A^{(n)}$ is the Hermite function and $a_A^{(n)}$ are the coefficients. To accurately describe the state of a gas an infinite number of moments is required to reconstruct the distribution function. However, for gases not too far from equilibrium, a finite number of moments should provide an adequate approximation. All the moments expressed in the truncated distribution function are regarded as the slow moment manifold [12]. The values of the rest of the moments are determined by the values of the moments in the slow manifold.

In his seminal paper in 1949, Grad [6] chose the first and second moments of $f$ to construct the well known G13 moment equations. Recently, Karlin *et al.* [13] and Struchtrup and Torrilhon [14] regularized Grad's 13 moment equations (R13) by applying a Chapman-Enskog-like expansion to the governing equations of moments higher than second order. The R13 equations have been successfully used by Torrihon and Struchtrup [15] to study shock structures up to a Mach number of 3.0. To apply the moment equations to confined flows, such as those found in microfluidic channels, wall boundary conditions are required. The authors [16] obtained a set of wall-boundary conditions for the R13 equations using Maxwell's kinetic wall boundary conditions [17] and DSMC data. In addition, they also proposed a numerical strategy to solve the moment equations for parabolic and elliptic flows within a conventional finite volume procedure. However, the R13 equations were not able to predict the nonlinear velocity profile in planar Couette flow in the transition regime [16,18]. To capture the non-equilibrium phenomena of flows in the transition regime, the authors expanded the molecular phase density function to fourth-order accuracy in Hermite polynomials [16] and regularized the Grad-type 26 moment equations (R26) [19].

The first 26 moments or their deviation from Grad's moment manifold are used in the present study. The governing equations for the moments can derived from the Boltzmann equation. For Maxwell molecules, they are [11]:

$$\frac{\partial \rho}{\partial t} + \frac{\partial \rho u_k}{\partial x_k} = 0 , \tag{10}$$

$$\frac{\partial \rho u_i}{\partial t} + \frac{\partial \rho u_i u_j}{\partial x_j} + \frac{\partial \sigma_{ij}}{\partial x_j} = -\frac{\partial p}{\partial x_i} , \tag{11}$$

$$\frac{\partial \rho T}{\partial t} + \frac{\partial \rho u_i T}{\partial x_i} + \frac{2}{3R}\frac{\partial q_i}{\partial x_i} = -\frac{2}{3R}\left( p \frac{\partial u_i}{\partial x_i} + \sigma_{ij} \frac{\partial u_j}{\partial x_i} \right) , \tag{12}$$

$$\frac{\partial \sigma_{ij}}{\partial t}+\frac{\partial u_k \sigma_{ij}}{\partial x_k}+\frac{\partial m_{ijk}}{\partial x_k}=-\frac{p}{\mu}\sigma_{ij}-2p\frac{\partial u_{<i}}{\partial x_{j>}}-\frac{4}{5}\frac{\partial q_{<i}}{\partial x_{j>}}-2\sigma_{k<i}\frac{\partial u_{j>}}{\partial x_k}, \tag{13}$$

$$\begin{aligned}\frac{\partial q_i}{\partial t}+\frac{\partial u_j q_i}{\partial x_j}+\frac{1}{2}\frac{\partial R_{ij}}{\partial x_j}=&-\frac{2}{3}\frac{p}{\mu}q_i-\frac{5}{2}p\frac{\partial RT}{\partial x_i}-\frac{7}{2}\sigma_{ik}\frac{\partial RT}{\partial x_k}-RT\frac{\partial \sigma_{ik}}{\partial x_k}+\frac{\sigma_{ik}}{\rho}\frac{\partial p}{\partial x_k}\\&+\frac{\sigma_{ij}}{\rho}\frac{\partial \sigma_{jk}}{\partial x_k}-\frac{q_k}{5}\left(7\frac{\partial u_i}{\partial x_k}+2\frac{\partial u_k}{\partial x_i}\right)-\frac{2q_i}{5}\frac{\partial u_k}{\partial x_k}-\frac{\partial \Delta}{6\partial x_i}-m_{ijk}\frac{\partial u_j}{\partial x_k},\end{aligned} \tag{14}$$

$$\begin{aligned}\frac{\partial m_{ijk}}{\partial t}+\frac{\partial u_l m_{ijk}}{\partial x_l}+\frac{\partial \phi_{ijkl}}{\partial x_l}=&-\frac{3}{2}\frac{p}{\mu}m_{ijk}-3RT\frac{\partial \sigma_{<ij}}{\partial x_{k>}}-3R\sigma_{<ij}\frac{\partial T}{\partial x_{k>}}-\frac{12}{5}q_{<i}\frac{\partial u_j}{\partial x_{k>}}\\&+3\frac{\sigma_{<ij}}{\rho}\left(\frac{\partial \sigma_{k>l}}{\partial x_l}+\frac{\partial p}{\partial x_{k>}}\right)-3m_{l<ij}\frac{\partial u_{k>}}{\partial x_l}-\frac{3}{7}\frac{\partial R_{<ij}}{\partial x_{k>}},\end{aligned} \tag{15}$$

$$\begin{aligned}\frac{\partial R_{ij}}{\partial t}+\frac{\partial u_k R_{ij}}{\partial x_k}+\frac{\partial \psi_{ijk}}{\partial x_k}=&-\frac{7}{6}\frac{p}{\mu}R_{ij}-\frac{2}{3}\frac{p}{\rho\mu}\sigma_{k<i}\sigma_{j>k}-\frac{28}{5}RT\frac{\partial q_{<i}}{\partial x_{j>}}-\frac{56}{5}q_{<i}\frac{\partial RT}{\partial x_{j>}}\\&+\frac{28}{5}\frac{q_{<i}}{\rho}\frac{\partial p}{\partial x_{j>}}-4RT\left(\sigma_{k<i}\frac{\partial u_k}{\partial x_{j>}}+\sigma_{k<i}\frac{\partial u_{j>}}{\partial x_k}-\frac{2}{3}\sigma_{ij}\frac{\partial u_k}{\partial x_k}\right)+\frac{14}{3}\frac{\sigma_{ij}}{\rho}\left(\frac{\partial q_k}{\partial x_k}+\sigma_{kl}\frac{\partial u_k}{\partial x_l}\right)\\&-2RT\frac{\partial m_{ijk}}{\partial x_k}-9m_{ijk}\frac{\partial RT}{\partial x_k}-\left(\frac{6}{7}R_{<ij}\frac{\partial u_{k>}}{\partial x_k}+\frac{4}{5}R_{k<i}\frac{\partial u_k}{\partial x_{j>}}+2R_{k<i}\frac{\partial u_{j>}}{\partial x_k}\right)+2\frac{m_{ijk}}{\rho}\frac{\partial p}{\partial x_k}\\&-2\phi_{ijkl}\frac{\partial u_k}{\partial x_l}-\frac{14}{15}\Delta\frac{\partial u_{<i}}{\partial x_{j>}}-\frac{2}{5}\frac{\partial \Omega_{<i}}{\partial x_{j>}}\end{aligned} \tag{16}$$

and

$$\begin{aligned}\frac{\partial \Delta}{\partial t}+\frac{\partial \Delta u_i}{\partial x_i}+\frac{\partial \Omega_i}{\partial x_i}=&-\frac{2}{3}\frac{p}{\mu}\Delta-\frac{2}{3}\frac{p}{\mu}\frac{\sigma_{ij}\sigma_{ij}}{\rho}-8RT\frac{\partial q_i}{\partial x_i}-4\left(2RT\sigma_{ij}+R_{ij}\right)\frac{\partial u_i}{\partial x_j}\\&+8\frac{q_i}{\rho}\frac{\partial p}{\partial x_i}-28q_i\frac{\partial RT}{\partial x_i}-\frac{4}{3}\Delta\frac{\partial u_i}{\partial x_i},\end{aligned} \tag{17}$$

in which $m_{ijk}$, $R_{ij}$, $\Delta$, $\phi_{ijkl}$, $\psi_{ijk}$ and $\Omega_i$ are higher moments or their deviation from Grad's moment manifold, i.e.

$$\begin{aligned}&m_{ijk}=\rho_{<ijk>}, \quad R_{ij}=\rho_{<ij>rr}-7RT\sigma_{ij}, \quad \Delta=\rho_{rrss}-15pRT,\\&\phi_{ijkl}=\rho_{<ijkl>}, \quad \psi_{ijk}=\rho_{rr<ijk>}-9RTm_{ijk}, \quad \Omega_i=\rho_{rrssi}-28RTq_i.\end{aligned} \tag{18}$$

The variables, $\phi_{ijkl}$, $\psi_{ijk}$ and $\Omega_i$ in equations (15)-(17) are unknown. Using the Chapman-Enskog expansion of the molecular distribution function around the fourth order approximation as a pseudo-equilibrium state, the 26 moment field equations are closed by

$$\phi_{ijkl} = -\frac{4\mu}{C_1\rho}\frac{\partial m_{<ijk}}{\partial x_{l>}} - \frac{12}{C_1}\frac{\mu}{\rho}\sigma_{<ij}\frac{\partial u_k}{\partial x_{l>}} + \frac{4\mu}{C_1 p\rho} m_{<ijk}\left(\frac{\partial \sigma_{l>m}}{\partial x_m} + \frac{\partial p}{\partial x_{l>}}\right) - 4\frac{\mu R}{C_1 p} m_{<ijk}\frac{\partial T}{\partial x_{l>}}$$
$$-\frac{12}{7}\frac{\mu R_{<ij}}{C_1 p}\frac{\partial u_k}{\partial x_{l>}} - \frac{C_2}{C_1}\frac{\sigma_{<ij}\sigma_{kl>}}{\rho}, \tag{19}$$

$$\psi_{ijk} = -\frac{27}{7Y_1}\frac{\mu}{\rho}\frac{\partial R_{<ij}}{\partial x_{k>}} - \frac{27\mu}{Y_1\rho}\sigma_{<ij}\frac{\partial RT}{\partial x_{k>}} - \frac{108}{5Y_1}\frac{\mu}{\rho} q_{<i}\frac{\partial u_j}{\partial x_{k>}} + \frac{27}{7Y_1}\frac{\mu}{p}\frac{R_{<ij}}{\rho}\left(\frac{\partial \sigma_{k>m}}{\partial x_m} + \frac{\partial p}{\partial x_{k>}}\right)$$
$$-\frac{54}{7}\frac{\mu R_{<ij}}{Y_1 p}\frac{\partial RT}{\partial x_{k>}} - \frac{\mu}{Y_1\rho}\left(\frac{54}{7} m_{m<ij}\frac{\partial u_m}{\partial x_{k>}} + 8 m_{<ijk}\frac{\partial u_{m>}}{\partial x_m} - 6 m_{ijk}\frac{\partial u_m}{\partial x_m}\right) \tag{20}$$
$$+\frac{6\mu}{Y_1 p}\frac{m_{ijk}}{\rho}\left(\frac{\partial q_m}{\partial x_m} + \sigma_{ml}\frac{\partial u_m}{\partial x_l}\right) - \left(\frac{Y_2}{Y_1}\frac{\sigma_{<li}m_{jkl>}}{\rho} + \frac{Y_3}{Y_1}\frac{q_{<i}\sigma_{jk>}}{\rho}\right)$$

and

$$\Omega_i = -\frac{7}{3}\frac{\mu}{\rho}\frac{\partial \Delta}{\partial x_i} - \frac{56}{5}\frac{\mu}{\rho}\left(q_j\frac{\partial u_i}{\partial x_j} + q_j\frac{\partial u_j}{\partial x_i} - \frac{2}{3} q_i\frac{\partial u_j}{\partial x_j}\right) - 28\frac{\mu}{\rho}\sigma_{ij}\frac{\partial RT}{\partial x_j} - 7\frac{\mu}{p}\Delta\frac{\partial RT}{\partial x_i}$$
$$-18\frac{\mu}{p} R_{ij}\frac{\partial RT}{\partial x_j} - 4\frac{\mu}{\rho}\frac{\partial R_{ij}}{\partial x_j} + \frac{56}{3}\frac{\mu}{p}\frac{q_i}{\rho}\left(\frac{\partial q_j}{\partial x_j} + \sigma_{jk}\frac{\partial u_j}{\partial x_k}\right) + 4\frac{\mu}{p}\frac{R_{ij}}{\rho}\left(\frac{\partial p}{\partial x_j} + \frac{\partial \sigma_{jk}}{\partial x_k}\right) \tag{21}$$
$$+\frac{7}{3}\frac{\Delta}{\rho}\frac{\mu}{p}\left(\frac{\partial p}{\partial x_i} + \frac{\partial \sigma_{ij}}{\partial x_j}\right) - 8\frac{\mu}{\rho} m_{ijk}\frac{\partial u_j}{\partial x_k} - \frac{2}{15}\left(\frac{5 m_{ijk}\sigma_{jk} + 14 q_j\sigma_{ij}}{\rho}\right),$$

in which $C_1(=1.8731)$, $C_2(=-0.3806)$, $Y_1(=1.5704)$ and $Y_2(=-1.004)$ are collision term constants [7].

## 4. Numerical procedure and parallelization of the regularized higher order moment equations

The foregoing set of 26 moment equations (10)-(17), denoted as the R26 equations, is a mixed system of first and second order partial differential equations (PDEs). When they are employed to study parabolic or elliptic flows in confined geometries, it is not clear how many and which wall boundary conditions are required. In the moment system, the higher moment provides the transport mechanism for the moment one order lower. For a gas flow close to the equilibrium state, a sufficient number of molecular isotropic collisions cause the flow to behave as a continuum and a gradient transport mechanism (GTM) prevails. As the value of $Kn$ increases, nongradient transport mechanisms (NGTM) occur in addition to the gradient transport mechanism. In fact, both GTM and NGTM coexist in the transition regime. Gu and Emerson [16] decomposed higher moments explicitly into their GTM and NGTM components in their study of the R13 equations. The approach is readily extended to the R26 equations. As the value of the GTM component of a moment can be calculated from the value of the relevant moment one order lower, the governing equations for the NGTM components of the moments are readily obtained from the moment equations. As a result, a system of the second order is constructed in the conventional convection-diffusion format with appropriate source terms. The characteristics of the system are then determined by the flow conditions and the requirement for the boundary conditions is well defined.

The numerical methods for solving the convective and diffusion equations are well documented for both high and low speed flows [20,21]. In the present study, the finite volume approach is employed to discretize the PDEs. The diffusive and source terms are discretized by a central difference scheme. For the convective terms, a range of upwind schemes including QUICK, SMART [22], and CUBISTA [23] are available in our in-house code THOR [24] and the CUBISTA scheme was selected for the present study. The coupling of the velocity and pressure fields is through the SIMPLE algorithm [25]. A collocated grid arrangement is used and the interpolation scheme of Rhie and Chow [26] is used to eliminate any resultant non-physical pressure oscillations. The general purpose CFD code THOR can deal with highly complex geometries through the use of multi-block grids and body-fitted coordinates. It has been parallelized using standard grid partitioning techniques and communication is through the use of the MPI standard. The parallel performance of the code was shown in ref [24].

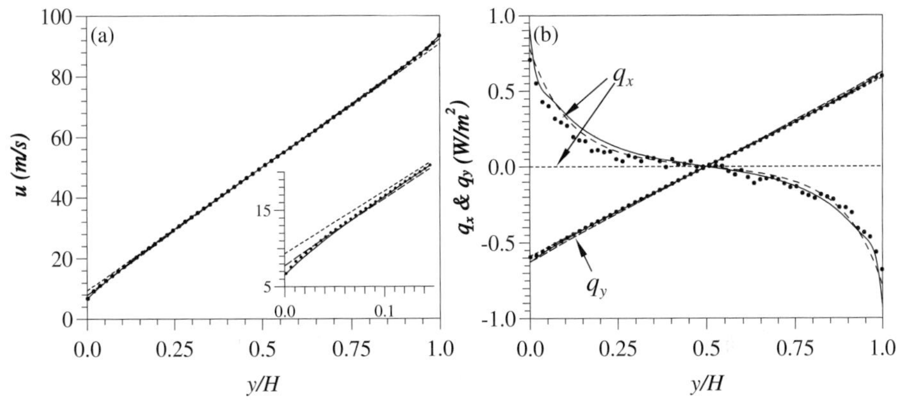


**Figure** 1. Predicted (a) velocity and (b) heat fluxes profiles at $Kn = 0.1$ for Couette flow with initial conditions: wall temperature $T_w = 273\ K$ and wall velocity $u_w = 100\ m/s$. Symbol, DSMC; ——, R26; ––––, R13; -------, NSF.

## 5. Applications of the regularized higher order moment equations

The R26 equations have been solved numerically with the method proposed by the authors along with wall boundary conditions for 1D steady state Couette flows, 2D Poiseuille flows, and flow past a cylinder for a range of values of *Kn*. Clearly the computational requirements will be substantially higher than the NSF equations alone and parallel computing is therefore essential for anything but the simplest of problems. However, the 26 moment approach can capture the various non-equilibrium phenomena, such as nonlinear velocity profiles and non-gradient heat fluxes in the Knudsen layer, for *Kn* up to a value of unity, as shown in figure 1 for planar Couette flow. The distance between the two parallel plates is $H$, the walls are parallel to the $x$-direction and $y$ is the direction perpendicular to the plates. Figure 1(a) shows the computed velocity profiles at the upper limit of the slip regime, $Kn$=0.1. The velocity profile predicted by the NSF and R13 equations is linear. In the core part of the flow (i.e. outside of the Knudsen layer), the NSF, R13 and R26 are all in agreement with DSMC data. However, in the region close to the wall, the DSMC results exhibit a nonlinear velocity profile. Neither the NSF nor the R13 equations follow the nonlinearity whilst the solid curve line for the R26 equations is in excellent agreement with the DSMC data.

An interesting non-equilibrium phenomenon that occurs in planar Couette flow is the appearance of a heat-flux *without* the presence of a temperature gradient and this is illustrated in figure 1(b), which shows both the tangential and normal heat fluxes, $q_x$ and $q_y$, respectively As indicated by the DSMC data, a significant amount of tangential heat flux, $q_x$, is generated at the upper limit of the slip regime. This is shown in figure 1(b) where both the R13 and R26 models are in good agreement with the DSMC data. However, the NSF equations are not able to predict this aspect of the flow.

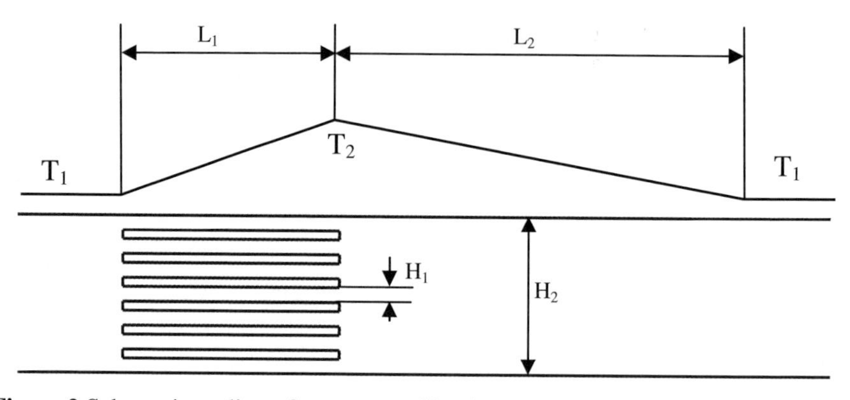


**Figure** 2 Schematic outline of a one stage Knudsen pump which uses the principle of thermal transpiration with no moving parts is provided.

Under rarefied conditions, the temperature gradient in channels with different values of Knudsen number will generate a differential pressure gradient. The engineering application of this rarefaction effect is the Knudsen pump, which transports the flow without any moving parts, as shown figure 2. The 26 moment equations have been used to model this simplified 1-stage Knudsen pump. The computational domain consists of 34 blocks with 0.3 million grid points and four processors were used to perform the computations. The computed velocity vectors are presented in figure 3. The rarefied gas starts to creep from the cold section towards the hot due to thermal transpiration. As the width of the seven narrow channels is a tenth of the wide channel, the Knudsen number ratio of the narrow over wide channel is about 10. A net gas flow is generated in the direction shown in figure 3. However, there are two recirculation zones close to the walls of the wide channel, which are the combined results of thermal transpiration of the wide and narrow channels in the opposite directions. The gas velocity in the Knudsen pump is usually low. The maximum gas velocity in the present is less than 0.1 m/s, a value that is so low that it will be extremely expensive for DSMC to perform the simulation of a Knudsen pump.

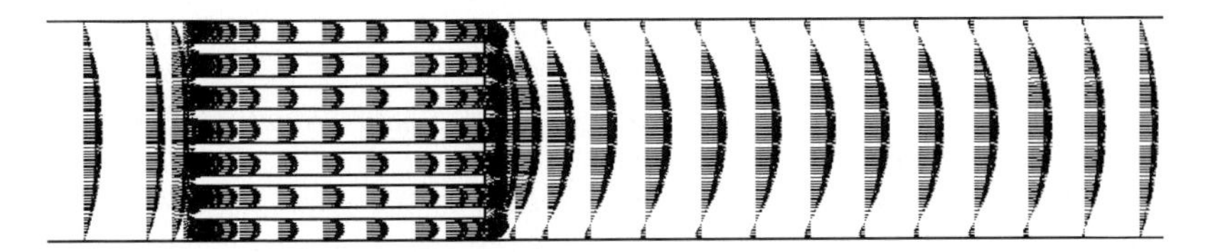

**Figure** 3 Computed velocity vectors for a Knudsen pump. $L_1$ = 70 μm, $L_2$ = 140 μm, $H_1$= 1μm, $H_1$= 10 μm, $T_1$= 300 K and $T_2$= 340 K.

## 6. Summary

The different methods for studying rarefied gas dynamics in low speed flows were reviewed in the present paper. They are broadly classified as two approaches: microscopic and macroscopic. The capability of the moment method is demonstrated through applications of geometrically simple but complex flows. The paper has shown that the regularized 26 moment system can capture non-equilibrium effects for Knudsen numbers up to a unity.

**Acknowledges**

The paper has benefited from discussions with Drs. R. W. Barber, Y. H. Zhang and Mr S. Mizzi. The authors would like to thank the Engineering and Physical Sciences Research Council (EPSRC) for their support of Collaborative Computational Project 12 (CCP12).

**References:**

1. M. Gad-el-hak, *The Fluid Mechanics of Microdevices - The Freeman Scholar Lecture*, J. Fluids Eng. 121:5-33. 1999.
2. C. Cercignani, *The Boltzmann Equation and Its Applications* (Springer, New York, 1988).
3. S. Chapman and T. G. Cowling, *The Mathematical Theory of Non-uniform Gases* (Cambridge University Press, 1970).
4. H. Grad, *Asymptotic Theory of the Boltzmann Equation*, Phys. Fluids 6:147-181, 1963.
5. A. V. Bobylev, *The Chapman-Enskog and Grad Method for Solving the Boltzmann Equation*, Sov. Phys.-Dokl. 27:29-31, 1982.
6. H. Grad, *On the Kinetic Theory of Rarefied Gases*, Commun. Pure Appl. Math. 2:331-407, 1949.
7. C. Truesdell and R. G. Muncaster, *Fundamentals of Maxwell's Kinetic Theory of a Simple Monotomic Gas* (Academic Press, New York, 1980).

8. S. Chen and G. D. Doolen, *Lattice Boltzmann Method For Fluid Flows*, Ann. Rev. Fluid Mech. 30:329-64, 1998.
9. X. Shan, X.-F. Yuan and H. Chen, *Kinetic Theory Representation of Hydrodynamics: A Way Beyond the Navier-Stokes Equation*, J. Fluid Mech. 550:413-441, 2006.
10. G. Bird, *Molecular Gas Dynamics and the Direct Simulation of Gas Flows* (Claredon Press, Oxford, 1994).
11. H. Struchtrup, *Macroscopic Transport Equations for Rarefied Gas Flows* (Springer-Verlag, Berlin-Heidelberg, 2005).
12. A. N. Gorban, I.V. Karlin and A. Y. Zinovyev, *Constructive Methods of Invariant Manifolds for Kinetic Problems, Phys.* Rep. 396:197-403, 2004.
13. I. V. Karlin, A. N. Gorban, G. Dukek and T. F. Nonenmacher, *Dynamic Correction to Moment Approximations*, Phys. Rev. E. 57:1668-1672, 1998.
14. H. Struchtrup and M. Torrihon, *Regularization of Grad's 13 Moment Equations: Derivation and Linear Analysis*, Phys. Fluids 15:2668-2680, 2003.
15. M. Torrihon and H. Struchtrup, *Regularized 13 Moment Equation: Shock Structure Calculations and Comparison to Burnett Models*, J. Fluid Mech. 513: 171-198, 2004.
16. X. J. Gu and D. R. Emerson, *A Computational Strategy for the Regularized 13 Moment Equations With Enhanced Wall-boundary Conditions*. J. Comput. Phys. 225:263-283, 2007.
17. J. C. Maxwell, *On Stresses in Rarified Gases Arising From Inequalities of Temperature*, Phil. Trans. Roy. Soc. (Lond.) 170:231-256, 1879.
18. X. J. Gu, R. W. Barber and D. R. Emerson, *How Far Can 13 Moments Go in Modeling Micro-Scale Gas Phenomena?* Nanoscale and Microscale Thermophysical Eng. 11:85-97, 2007.
19. X. J. Gu and D. R. Emerson, *A high-order moments approach for capturing nonequilibrium phenomena in the transition regime,* in preparation, 2007.
20. E. F. Toro, 1999 *Riemann Solvers and Numerical Methods for Fluids Dynamics: A Practical Introduction*, 2nd ed., (Springer, Berlin, 1999).
21. J. H. Ferziger and M. Perić, *Computational Methods for Fluid Dynamics*, 2nd Ed, (Springer-Verlag, Berlin,-Heidelberg, 1999).
22. P. H. Gaskell and A. K. C. Lau, *Curvature-compensated convective-transport - SMART, A new boundedness-preserving transport algorithm*. Int. J. Numer. Meth. Fluids 8:617-641,1988.
23. M. A. Alves, P. J. Oliveira and F. T. Pinho, *A convergent and universally bounded interpolation scheme for the treatment of advection*. Int. J. Numer. Meth. Fluids 41:47-75, 2003.
24. X. J. Gu and R. W. Barber and D. R. Emerson, *Parallel Computational Fluid Dynamics – Advanced Numerical Methods, Software and Applications* (B. Chetverushkin, A. Ecer, J. Periaux, N. Satofuka, P. Fox, Eds., Elsevier Science, pp.497-504, 2004).
25. S. V. Patankar, *Numerical Heat Transfer and Fluid Flow*. (McGraw-Hill, New York, 1980).
26. C. M. Rhie and W. L. Chow, *Numerical study of turbulent flow past an airfoil with trailing edge separation*, AIAA J., 21:1525-1532, 1983.

A parallel scientific software for heterogeneous hydrogeoloy

J. Erhel [a] *, J.-R. de Dreuzy [b], A. Beaudoin [c], E. Bresciani [a], and D. Tromeur-Dervout [d]

[a]INRIA,35042 Rennes Cedex, France

[b]Geosciences Rennes, UMR CNRS 6118, France

[c]LMPG, University of Le Havre, France

[d]CDCSP, University of Lyon 1, France

## 1. Introduction

Numerical modelling is an important key for the management and remediation of groundwater resources [28]. As opposed to surface water and to highly karstic geologies, groundwater does not flow in well-identified open streams but is like water flowing in the voids of a sponge. Groundwater is highly dependent on the percentage of pores (porosity), size and connectivity of the pores that controls the permeability of the medium. Its quality depends on the biochemical reactivity of the crossed geological media and on the kinetics of the biochemical reactions. These parameters (porosity, permeability, reactivity) are highly variable. Several field experiments show that the natural geological formations are highly heterogeneous, leading to preferential flow paths and stagnant regions. The contaminant migration is strongly affected by these irregular water velocity distributions. Transport of contaminant by advection and dispersion induce a large spreading of the particles generally called plume. The characterization of the plume remains a much debated topic [7,16,25]. Fractures and heterogeneous sedimentary units cannot be identified whatever the remote data, whether geophysical, geological or hydraulic. Because data give a rather scarce description of the medium hydraulic properties, predictions rely heavily on numerical modelling. Numerical modelling should integrate the multi-scale geological heterogeneity, simulate the hydraulic flow and transport phenomena and quantify uncertainty coming from the lack of data. Analytical techniques like homogenization or perturbations are in general not relevant. Modelling must thus be performed in a probabilistic framework that transfers the lack of data on prediction variability [1]. Practically, random studies require running a large number of simulations, for two reasons. First, non intrusive Uncertainty Quantification methods rely on sampling of data. Second, the questions addressed must consider a large panel of parameters (Peclet number, variance of probabilistic models, etc). The hydraulic simulations must be performed on domains of a large size, at the scale of management of the groundwater resource or at the scale of the homogeneous medium type in terms of geology. This domain must be discretized at a fine resolution to take into account the scale of geological heterogeneities. Also, large time ranges must be considered in order to determine an asymptotic behaviour. High performance computing is thus necessary to carry out these large scale simulations.

Therefore we have developed the parallel scientific platform HYDROLAB, which integrates in a modular structure physical models and numerical methods for simulating flow and transport

*this work was partly funded by the Grid'5000 grant from the French government

I.H. Tuncer et al., *Parallel Computational Fluid Dynamics 2007*,
DOI: 10.1007/978-3-540-92744-0_5, © Springer-Verlag Berlin Heidelberg 2009

in heterogeneous porous and fractured media. It includes stochastic models and Uncertainty Quantification methods for dealing with random input data. It relies as much as possible on existing free numerical libraries such as parallel sparse linear solvers. This policy increases the global reliability and efficiency of HYDROLAB. Our software is also fully parallel and follows a powerful object-oriented approach. Most existing codes for groundwater flow and transport contaminant do not gather all these original features. To our knowledge, there is no public parallel code performing groundwater simulations in 3D fracture networks. Popular free software for reservoir modeling, environmental studies or nuclear waste management, such as MODFLOW [19], MT3DMS [29] and TOUGH2 [24] are not easy to use for highly heterogeneous porous media; moreover, MODFLOW and MT3DMS are not parallel. Other software, such as IPARS [26] or ALLIANCES [23], are private or commercial.

Our paper is organized as follows: we first describe the physical models processed by HYDROLAB; then we describe the numerical methods and algorithms implemented in the software with details about parallel features. After a short synopsis on HYDROLAB, we discuss performance results on parallel clusters of various numerical experiments. The paper ends up with concluding remarks about future work.

## 2. Physical models

### 2.1. Geometry and data

We consider two types of geological domains, either porous media or fractured media. Permeability values span more than ten orders of magnitude from $10^{-12}$ m/s in non fractured crystalline rocks or clays to $10^{-1}$ m/s for gravels [14]. Variability can be huge not only at the transition between geological units, like between sand and clay, but also within units. Since data is lacking, the physical model of porous media is governed by a random permeability field, following a given distribution law. In order to cover a broad-range of structures, we consider finitely correlated, fractal or multifractal distribution laws. The geometry is kept very simple and is a deterministic 2D or 3D regular polyhedron. In rocks, the fractures can represent only 0.01 % of the volume although they are the sole water carrying structures. Fractured media are by nature very heterogeneous and multi-scale, so that homogenisation approaches are not relevant. In an alternative approach, called the discrete approach, flow equations are solved on the fracture network, explicitly taking into account the geometry of the fractured medium, and thus the structure of the connected network. According to recent observations of fractured rocks, fractures are characterized by a variety of shapes and a broad range of lengths and apertures. The physical model is thus based on a complex and random geometry, where the geological medium (the surrounding matrix) is a cube; fractures are ellipses with random distributions of eccentricity, length, position and orientation. The main feature of our model is the fracture length distribution, which is given by a power law. We first consider an impervious matrix, because almost no water flows in it [9,6]. For some applications, it is also important to consider the coupling between the matrix and the fractures.

### 2.2. Governing equations

Well tests are the first source of information on the hydraulic properties of underground media. Numerical modelling of these tests involves computation of transient flow in large domains. We assume here that the medium is saturated and that the water density is constant. Flow is governed by the classical Darcy law and the mass conservation law:

$$\begin{cases} \epsilon v = -K\nabla h, \\ s\partial h/\partial t + \nabla.(\epsilon v) = q \end{cases} \tag{1}$$

where the Darcy velocity $v$ and the hydraulic head $h$ are unknown, $K$ is a given permeability field, $\epsilon$ is a given porosity, $s$ is a given storativity parameter and $q$ is a given source term. An initial condition and boundary conditions close the system. For steady-state flows, the time derivative disappears. In order to characterize plumes of contaminant, numerical simulations consist in computing the velocity field over large spatial domains and simulating solute transport over large temporal scales [16]. We consider here one inerte solute and a steady-state flow. Governing equations are:

$$\frac{\partial(\epsilon c)}{\partial t} + \nabla.(\epsilon c v) - \nabla.(\epsilon D \nabla c) = 0 \tag{2}$$

where $c$ is the unknown solute concentration, $v$ is the previous Darcy velocity and $D$ is the given dynamic dispersion tensor. Boundary conditions and an initial condition complete the equation.

## 3. Numerical methods

### 3.1. Uncertainty Quantification methods

The physical model is not complex in itself since equations are linear but both the broad geological heterogeneity and the lack of measures require dealing with uncertain physical and geometrical properties. The permeability field is a random variable and the geometry of fracture networks is random. Hydrogeological studies aim at describing the random velocity field and the random dispersion of solute. They must rely on Uncertainty Quantification methods. Currently, we use a classical Monte-Carlo approach [8]. It is non intrusive in the sense that each sample is a deterministic simulation, using classical spatial and temporal discretizations. It is easy to implement and can be applied to any random field, either data or geometry. However, only the mean of random output can be attained with few samples; other statistical results require much more samples. Therefore, more sophisticated UQ methods can be considered, such as non intrusive spectral methods or intrusive Galerkin methods [27,1].

### 3.2. Spatial and temporal discretizations

For each sample, we have to discretize the set of Partial Differential Equations. In flow computations, we use a method of lines in the case of transient equations and get a system of ODEs after spatial discretization. For porous media, the geometry is simple and the mesh is a regular grid. In both transient and steady-state cases, we choose a finite volume method, which computes accurately the velocity and ensures mass conservation. For fracture networks, the geometry is complex, so that it is quite difficult to generate a mesh. We have designed a two-fold mesh generator, with a preprocessing step before meshing each fracture separately [3]. We first discretize the intersections, so that they are uniquely defined in the mesh. Then, by considering these discretized intersections as boundaries, we mesh each fracture with a 2D mesh generator. We consider steady-state flows and use a hybrid mixed finite element method, based on the unstructured mesh. Again, the method computes accurately the velocity and ensures mass conservation. The unknowns in the hybrid method are the hydraulic head in each cell and on each face, so that the continuity equations are easy to implement at the intersections.

At each time step, we get a large sparse symmetric positive definite linear system. Direct solvers are efficient for medium size systems but suffer from too big memory requirements for large size systems. Therefore, we prefer to use iterative solvers. Algebraic multigrid methods appear to be very efficient, either as such or as preconditioners of Conjugate Gradient method [4].

In transport computations, we choose a random walker method, because it does not introduce numerical dispersion in advection-dominated problems. We generate a bunch of particles for the initial condition and solve a stochastic differential equation to compute their trajectories. In the case of pure advection, this is simply a deterministic particle tracker method, whereas dispersion is modeled by a random motion. We use an explicit first-order Euler scheme to solve the equation.

### 3.3. Parallel algorithms

In each simulation, the spatial and temporal dimensions are very large, in order to study multi-scale effects and asymptotic behaviour. Numerical modelling must overcome two main difficulties, memory size and runtime, in order to solve very large linear systems and to simulate over a large number of time steps. High performance computing is thus necessary to carry out these large scale simulations.

Because memory requirements are very high, it is mandatory to distribute data from the beginning to the end of the simulation. Therefore, everything is done in parallel, including data generation, mesh generation, flow computation, transport computation. Parallel algorithms are based on a subdomain decomposition. In porous media, we subdivide the domain into regular blocks, which can be slices or rectangles in 2D. Each processor owns a subdomain and ghost cells surrounding the domain. This classical tool reduces communication costs between processors. In fracture networks, we subdivide the network into groups of fractures, assigned to different processors. Computations are done in parallel and communications occur for synchronizing data at the intersections.

Once the data and the mesh are generated, it is possible to launch flow computations and transport computations. In flow computations, we use a parallel sparse linear solver. We have developed interfaces to several libraries, such as SUPERLU [22], PSPASES [18], HYPRE [12]. The distributed matrix and right-hand side are input data to the solver. Parallel computations inside the solver may redistribute the matrix but the output solution is distributed as the right-hand side. The velocity field is computed in each processor from the hydraulic head returned by the solver. Thanks to this structure, we can compare different sparse solvers. We require a linear complexity with the matrix size $N$, a robust solver for heterogeneous permeability fields, parallel scalability. We observe that direct solvers are robust and scalable but have a complexity in $O(N^{1.5})$ [4]. Preconditioned Conjugate Gradient methods have also a poor complexity [10]. Domain decomposition methods seem very promising [17], thus we are developing such methods [11], extending the Aitken-Schwarz methodology [15]. Algebraic multigrid methods have a linear complexity and are robust [10]. Therefore, in this paper, we compare geometric and algebraic multigrid methods, using HYPRE.

In transport computations, we have designed a simple but efficient parallel algorithm which allows to consider large temporal domains [2]. Our parallel algorithm takes advantage of a subdomain decomposition and of the non interaction between the particles. This hybrid parallel strategy is quite original, since most of previous work uses either one or the other [20],[5],[21]. This strategy allows to run simulations on very large computational domains. Actually, we use the domain decomposition and data distribution arising from permeability generation and flow computation [3]. Communications with processors owning neighboring subdomains must occur when particles exit or enter the subdomain. We add a global synchronisation point after the exchanges of particles, in order to define a global state where each processor knows the total number of particles still in the domain. By the way, it guarantees that all RECV operations have been successful and that the algorithm may execute new SEND operations. Thus neither deadlock nor starvation may occur and the algorithm terminates safely. We also take advantage of the independence of the particles and launch them by bunches. The idea is to initiate a pipeline of particles. At each global synchronisation point, the master processors inject a new bunch of particles, until all particles have been injected into the domain. All processors track the particles in their subdomain, then exchange particles on their boundaries, finally synchronize globally to define the total number of particles in activity.

## 4. Parallel software HYDROLAB

We have developed a fully parallel object-oriented software platform, called HYDROLAB, which provides a generic platform to run Monte-Carlo numerical simulations of flow and transport in highly heterogeneous porous media and in fracture networks. The software is organized in three main parts, respectively dedicated to physical models, numerical methods and various utilitaries. Physical models include heterogeneous porous media and discrete fracture networks. We plan to develop a coupled model for porous fractured media. The physical model is composed of several tasks: the generation of the random geometry and the random permeability field, the mesh generation, the spatial discretization of the flow operator. The numerical methods include PDE solvers, ODE solvers, linear solvers, multilevel methods, a random walker, UQ methods. Utilitaries are generic modules including Input/Output, results management, parallel tools, grid tools, geomotry shapes, visualization, etc. This modularity and genericity allows a great flexibility and portability, enhancing the development of new applications. The software is written in C++ and is implemented on machines with Unix or Windows systems. Graphical functions are written with OpenGL and parallel programming relies on the MPI library. The software integrates open-source libraries such as sparse linear solvers for flow computation.

Our commitment is to design free packages available for downloading on the website of the platform (http://hydrolab.univ-rennes1.fr/).

## 5. Numerical results for 2D porous media

We have done multiparametric Monte-Carlo simulations for solving steady-state flow and transport problems with 2D heterogeneous porous media [8]. We have done many experiments on a parallel cluster of the Grid'5000 computing resource installed at INRIA in Rennes and we got good performances with parallel sparse linear solvers and our parallel particle tracker. Some results can be found in [3], [4] and [2].

### 5.1. Numerical experiments

In this paper, we study the case of a 2D porous medium where the permeability field $K$ follows a log-normal correlated law, where $Y = ln(K)$ is defined by a mean $m$ and a covariance function $C$ given by $C(r) = \sigma^2 \exp(-\frac{|r|}{\lambda})$, where $\sigma^2$ is the variance of the log hydraulic conductivity, $|r|$ represents the separation distance between two points and $\lambda$ denotes the correlation length scale. The length $\lambda$ is typically in the range $[0.1m, 100m]$ and the variance $\sigma^2$ is in the interval $[0, 7]$. These two ranges encompass most of the generally studied values.

We consider a steady-state flow and solve first the flow equations, then the transport equations of an inerte solute with the computed velocity. In flow equations (1) and transport equations (2), we assume a constant porosity $\epsilon = 1$. In flow equations, boundary conditions are no-flow on upper and lower sides and specified-head $h = 0$ on left side, $h = 1$ on right side. In transport equations, we consider only molecular diffusion, assumed homogeneous and isotropic. The ratio of diffusion to advection is measured by the Peclet number defined by $Pe = \frac{\lambda U}{D_m}$, where $D_m$ is the molecular diffusion coefficient, with a typical range of $[100, \infty]$, where $\infty$ means pure advection. This range covers advection-dominated transport models, whereas diffusion-dominated models require smaller values. In transport equations, the initial condition at $t = 0$ is the injection of the solute. In order to overcome the border effects, the inert solute is injected at a given distance

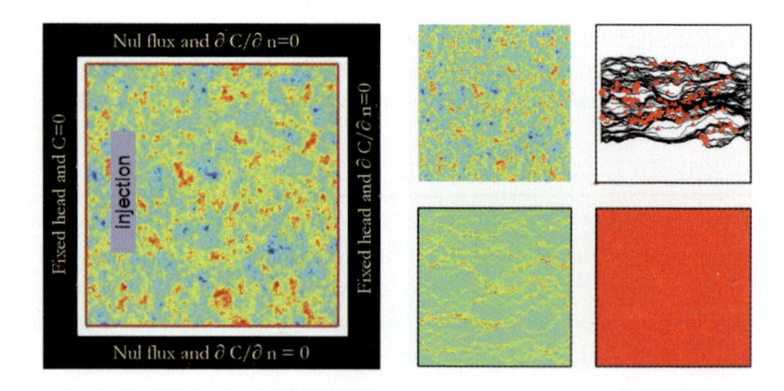


Figure 1. Numerical conditions (left) and simulation result for heterogeneous porous media (right): permeability field (top left), solute transport (top right), horizontal velocity field (bottom left), vertical velocity field (bottom right).

| Parameter | Values |
|---|---|
| $\sigma$ | $1, 2, 2.5, 3$ |
| $\lambda$ | 10 |
| $Pe$ | $100, 1000, \infty$ |
| $\Delta x$ | 1 |
| $n$ | $128, 256, 512, 1024, 2048, 4096$ |
| $tol$ | $10^{-11}$ |
| $N_p$ | 2000 |
| solver | SMG,AMG, TRACKER |

Table 1
Physical and numerical parameters used for heterogeneous 2D porous medium with log-normal correlated permeability

of the left side. Boundary conditions are no-mass-flux on upper and lower sides, specified zero concentration on left side, free mass outflow on right side. An example of simulation output is depicted in Figure 1.

### 5.2. Complexity analysis

Many physical parameters have a strong impact on the computations, such as the variance, the correlation length or the Peclet number. Also the computational requirements are governed by numerical parameters such as linear solver, problem size and convergence thresholds. Here, we consider a square domain with as many cells in each direction. The problem size is defined by the mesh step $\Delta x$, the number of cells $n$ in each direction of the mesh and by the number of particles $N_p$ in the random walker. In iterative sparse linear solvers, another numerical parameter is the threshold *tol* for convergence. Physical and numerical parameters are summarized in Table 1. The number of particles is large enough to ensure convergence of the random walker and the mesh step is small enough to get an accurate flow field. The correlation length is also fixed. Here, we consider multigrid sparse linear solvers: SMG and Boomer-AMG from the HYPRE library [13], [12].

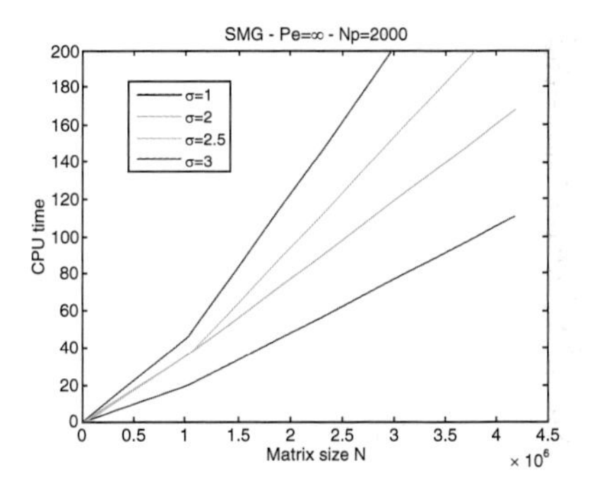


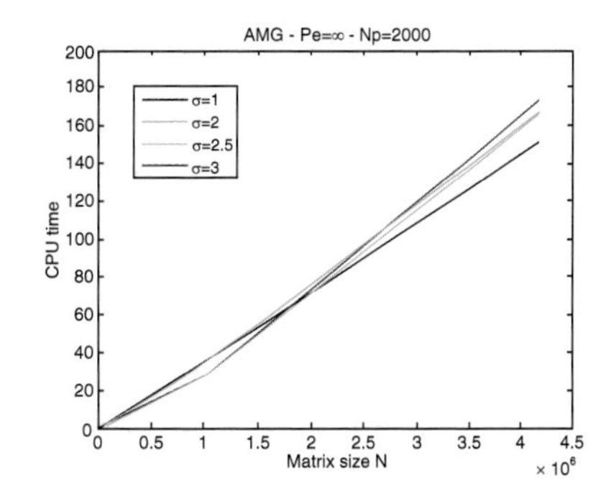


Figure 2. CPU time of flow computations using SMG geometric multigrid (left) and AMG algebraic multigrid (right). The matrix size is $N = n^2$.

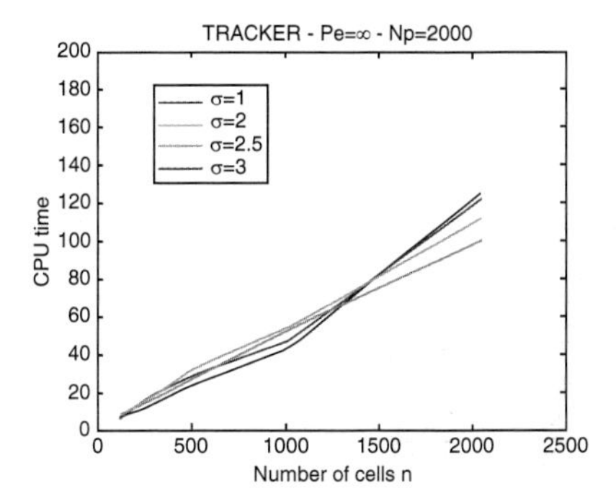


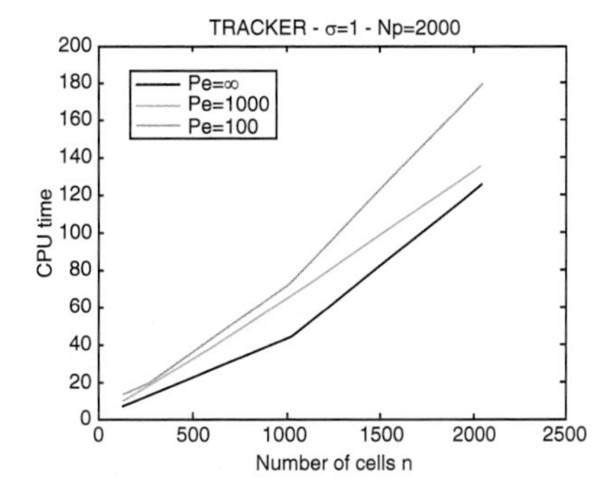


Figure 3. CPU time of transport computations using a particle tracker. Impact of the variance (left) and of the Peclet number (right). The number of cells is each direction is $n$.

One of the objectives of the simulations is to analyze the effect of heterogeneity and Peclet number on solute dispersion. Therefore, it is very important to develop a software with computational times as much independent of the variance $\sigma$ and the Peclet number $Pe$ as possible. Another objective is to estimate the asymptotic dispersion; therefore, we have to run the transport simulation with very large times; because all particles of the random walker must stay in the domain, this implies to run flow and transport simulations in very large computational domains. We analyze the effect of the size of the computational domain.

In Figure 2, we plot the CPU time of AMG and SMG multigrid solvers versus the matrix size $N = n^2$, for different values of $\sigma$. We observe in both cases a linear complexity, as expected from the theory. However, the behaviors are different when $\sigma$ varies. The geometric multigrid solver SMG is more efficient for $\sigma = 1$, the geometric solver SMG and the algebraic solver AMG are equivalent for $\sigma = 2$ and AMG becomes more efficient for $\sigma = 2.5$ and $\sigma = 3$. Therefore, the best choice depends on the value of $\sigma$.

In Figure 3, we plot the CPU time of the particle tracker versus the number of cells $n$ in each

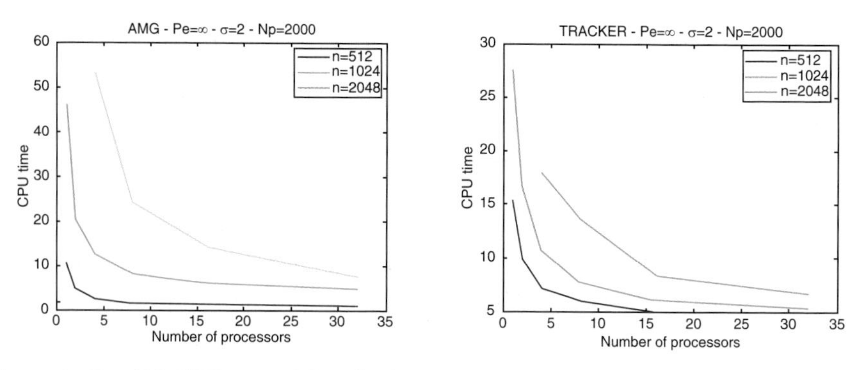


Figure 4. Parallel CPU time of flow (left) and transport (right) computations using respectively algebraic multigrid (AMG) a particle tracker. Impact of the matrix size $N = n^2$. The heterogeneity is driven by $\sigma = 2$ and transport is pure advection with $Pe = \infty$.

direction. The tracker is nearly independent of the variance $\sigma$. We observe a complexity which is roughly linear with $n$ (thus with $\sqrt{N}$). For small values of $n$, the transport computation is slower than the flow computation, but the CPU times become equivalent for $n = 2048$ and the transport will become much faster than the flow for larger $n$ because of the different complexities.

Whereas flow computations are obviously independent of the Peclet number, we observe that the computational time of transport computations increases with the Peclet number. However, for these large values of $Pe$, the transport computations remains cheaper than the flow computations with a large number of cells.

### 5.3. Scalability analysis

From the sequential results, it is clear that the algebraic multigrid solver is very efficient, has a linear complexity with the number of cells in the mesh and is not sensitive to the heterogeneity. Therefore, we choose this solver for parallel flux computations. We run experiments on a cluster composed of two nodes of 32 computers each. Each computer is a 2.2 Ghz AMD Opteron bi-processor with 2 Go of RAM. Inside each node, computers are interconnected by a Gigabit Ethernet Network Interface, and the two nodes are interconnected by a Gigabit Ethernet switch. In Figure 4, we plot the CPU time of flux (using AMG) and transport (using our particle tracker) computations with a varying number of processors $P$. Here, the heterogeneity is $\sigma = 2$ and we consider a pure advection case ($Pe = \infty$). We observe good speed-ups for the different sizes of the computational domain and a quite good scalability for both parallel modules. We did more experiments with various $\sigma$ and $Pe$ and got very similar results.

## 6. Current and future work

We have designed a parallel modular software, called HYDROLAB, for solving flow and transport problems in heterogeneous and fractured geological media. Our performance results show that our software is very efficient on parallel clusters for dealing with random 2D heterogeneous porous media. We are extending this work to 3D domains and have designed multilevel methods based on subdomain decompositions. As far as fracture networks are concerned, we are now able to generate a mesh for any network and to solve steady-state flow problems [3]. We are doing more experiments for improving parallel performances.

Future work concern transient flow problems in porous and fractured media and transport problems in fracture networks. Also, we plan to investigate non intrusive Uncertainty Quantification methods, in order to improve efficiency and to get more statistical information than with Monte-Carlo simulations.

## REFERENCES


1. I. Babuska, R. Tempone, and G. Zouraris. Solving elliptic boundary value problems with uncertain coeffcients by the finite element method: the stochastic formulation. *Computer methods in applied mechanics and engineering*, 194:1251–1294, 2005.
2. A. Beaudoin, J-R. de Dreuzy, and J. Erhel. An efficient parallel particle tracker for advection-diffusion simulations in heterogeneous porous media. In A.-M. Kermarrec, L. Boug, and T. Priol, editors, *Euro-Par 2007, LNCS 4641*, pages 705–714. Springer-Verlag, Berlin, Heidelberg, 2007.
3. A. Beaudoin, J-R. de Dreuzy, J. Erhel, and H. Mustapha. Parallel simulations of underground flow in porous and fractured media. In G.R. Joubert, W.E. Nagel, F.J. Peters, O. Plata, P. Tirado, and E. Zapata, editors, *Parallel Computing: Current and Future Issues of High-End Computing*, volume 33 of *NIC Series*, pages 391–398. NIC, 2006.
4. A. Beaudoin, J. Erhel, and J.-R. de Dreuzy. A comparison between a direct and a multigrid sparse linear solvers for highly heterogeneous flux computations. In *Eccomas CFD 2006*, volume CD, 2006.
5. J.-R. Cheng and P. Plassmann. The accuracy and performance of parallel in-element particle tracking methods. In *Proceedings of the Tenth SIAM Conference on Parallel Processing for Scientific Computing*, pages 252 – 261. Portsmouth, VA, 2001.
6. C. Clauser. Permeability of crystalline rock. *Eos Trans. AGU*, 73:237–238, 1992.
7. G. Dagan, A. Fiori, and I. Jankovic. Flow and transport in highly heterogeneous formations: 1. conceptual framework and validity of first-order approximations. *Water Resources Research*, 9, 2003.
8. J.-R. de Dreuzy, A. Beaudoin, and J. Erhel. Asymptotic dispersion in 2D heterogeneous porous media determined by parallel numerical simulations. *Water Resource Research*, to appear.
9. J-R. de Dreuzy, P. Davy, and O. Bour. Hydraulic properties of two-dimensional random fracture networks following power law distributions of length and aperture. *Water Resources Research*, 38(12), 2002.
10. R. Detwiler, S. Mehl, H. Rajaram, and W. Cheung. Comparison of an algebraic multigrid algorithm to two iterative solvers used for modeling ground water flow and transport. *groundwater*, 40(3):267–272, 2002.
11. J. Erhel, A. Frullone, D. Tromeur-Dervout, and J.-R. de Dreuzy. Aitken-Schwarz DDM to solve Darcy flow in heterogeneous underground media. In *Parallel Matrix Algorithms and Applications (PMAA06)*, 2006.
12. R. D. Falgout, J. E. Jones, and U. Meier Yang. Pursuing scalability for Hypre's conceptual interfaces. *ACM Trans. Math. Softw.*, 31(3):326–350, 2005.
13. R.D. Falgout, J.E. Jones, and U.M. Yang. *Numerical Solution of Partial Differential Equations on Parallel Computers*, chapter The Design and Implementation of Hypre, a Library of Parallel High Performance Preconditioners, pages 267–294. Springer-Verlag, 2006.
14. Alan R. Freeze and John A. Cherry. *Groundwater.* Prentice Hall, 1979.
15. M. Garbey and D. Tromeur-Dervout. On some Aitken-like acceleration of the Schwarz method. *Internat. J. Numer. Methods Fluids*, 40(12):1493–1513, 2002.

16. L. Gelhar. *Stochastic Subsurface Hydrology.* Engelwood Cliffs, New Jersey, 1993.
17. I. G. Graham, P. O. Lechner, and R. Scheichl. Domain decomposition for multiscale PDEs. *Numerische Mathematik*, 106(4):589–626, 2007.
18. A. Gupta, F. Gustavson, M. Joshi, G. Karypis, and V. Kumar. Pspases: An efficient and scalable parallel sparse direct solver. In Tianruo Yang, editor, *Kluwer International Series in Engineering and Computer Science*, volume 515, 1999.
19. A.W. Harbaugh, E.R. Banta, M.C. Hill, and M.G. McDonald. MODFLOW-2000, the U.S. geological survey modular ground-water model – user guide to modularization concepts and the ground-water flow process. Open-File Report 00-92, U.S. Geological Survey, 2000.
20. E. Huber, D. Spivakovskaya, H. Lin, and A. Heemink. The parallel implementation of forward-reverse estimator. In P. Wesseling, E. Onate, and J. Periaux, editors, *ECCOMAS CFD.* TU Delft, The Netherlands, 2006.
21. B. Kaludercic. Parallelisation of the lagrangian model in a mixed eulerian-lagrangian cfd algorithm. *Journal of Parallel and Distributed Computing*, 64:277–284, 2004.
22. X. S. Li and J. W. Demmel. SuperLU-DIST: A scalable distributed-memory sparse direct solver for unsymmetric linear systems. *ACM Transactions on Mathematical Software (TOMS)*, 29(2):110–140, 2003.
23. P. Montarnal, A. Dimier, E. Deville, E. Adam, J. Gaombalet, A. Bengaouer, L. Loth, and C. Chavant. Coupling methodology within the software platform Alliances. In E. Oate M. Papadrakakis and B. Schrefler, editors, *Int. Conf. on Computational Methods for Coupled Problems in Science and Engineering COUPLED PROBLEMS 2005*, Barcelona, 2005. CIMNE.
24. K. Pruess, C. Oldenburg, and G. Moridis. Tough2 user's guide, version 2.0. report LBNL-43134, Lawrence Berkeley National Laboratory, 1999.
25. P. Salandin and V. Fiorotto. Solute transport in highly heterogeneous aquifers. *Water Resources Research*, 34:949–961, 1998.
26. J. Wheeler. Ipars user's manual. Technical report, University of Texas at Austin, 2000.
27. D. Zhang and Z. Lu. An efficient high-order perturbation approach for flow in random porous media via Karhunen-Love and polynomial expansions. *Journal of Computational Physics*, 194:773–794, 2004.
28. C. Zheng and G. D. Bennett. *Applied Contaminant Transport Modeling; second edition.* John Wiley & Sons, New-York, 2002.
29. C. Zheng and P. Wang. MT3DMS: a modular three-dimensional multi-species model for simulation of advection, dispersion and chemical reactions of contaminants in groundwater systems: documentation and user's guide. Serdp-99-1, U.S. Army Engineer Research and Development Center, 1999.

# Aerodynamic Shape Optimization Methods on Multiprocessor Platforms

K.C. Giannakoglou[a] *, I.C. Kampolis[a], P.I.K. Liakopoulos[a], M.K. Karakasis[a], D.I. Papadimitriou[a], T. Zervogiannis[a], V.G. Asouti[a]

[a]National Technical University of Athens (NTUA),
School of Mechanical Engineering, Lab. of Thermal Turbomachines,
Parallel CFD & Optimization Unit, P.O. Box 64069, Athens 157 10, GREECE

**Abstract:** An overview of modern optimization methods, including Evolutionary Algorithms (EAs) and gradient–based optimization methods adapted for Cluster and Grid Computing is presented. The basic tool is a Hierarchical Distributed Metamodel–Assisted EA supporting Multilevel Evaluation, Multilevel Search and Multilevel Parameterization. In this framework, the adjoint method computes the first and second derivatives of the objective function with respect to the design variables, for use in aerodynamic shape optimization. Such a multi–component, hierarchical and distributed scheme requires particular attention when Cluster or Grid Computing is used and a much more delicate parallelization compared to that of conventional EAs.

## 1. FROM PARALLEL CFD TO PARALLEL OPTIMIZATION

During the last decade, the key role of CFD-based analysis and optimization methods in aerodynamics has been recognized. Concerning the flow analysis tools, adequate effort has been put to optimally port them on multiprocessor platforms, including cluster and grid computing. Nowadays, there has been a shift of emphasis from parallel analysis to parallel optimization tools, [1]. So, this paper focuses on modern stochastic and deterministic optimization methods as well as hybrid variants of them, implemented on PC clusters and enabled for grid deployment.

The literature on parallel solvers for the Navier–Stokes is extensive. Numerous relevant papers can be found in the proceedings of the past ParCFD and other CFD–oriented conferences. We do not intend to contribute further in pure parallel CFD, so below we will briefly report the features of our parallel flow solver, which is an indispensable part of our aerodynamic shape optimization methods and the basis for the development of adjoint methods. Our solver is a Favre–averaged Navier–Stokes solver for adaptive unstructured grids, based on finite–volumes, with a second–order upwind scheme for the convection terms and parallel isotropic or directional agglomeration multigrid, [2]. For a parallel flow analysis, the grid is partitioned in non-overlapping subdomains, as many as the available processors. The grid (or graph) partitioning tool is based on a single-pass

*e-mail: kgianna@central.ntua.gr

I.H. Tuncer et al., *Parallel Computational Fluid Dynamics 2007*,
DOI: 10.1007/978-3-540-92744-0_6, © Springer-Verlag Berlin Heidelberg 2009

multilevel scheme including EAs, graph coarsening algorithms and heuristics, [3]. This software is as fast as other widely used partitioning tools, [4], and produces optimal partitions with evenly populated subdomains and minimum interfaces. An updated version of the software overcomes the restrictions of recursive bisection and is suitable for use in heterogeneous clusters.

The aforementioned Navier–Stokes solver is used to evaluate candidate solutions in the framework of various shape optimization methods, classified to those processing a single individual at a time and population–based ones. It is evident that the parallelization mode mainly depends upon the number of candidate solutions that can simultaneously and independently be processed. Optimization methods which are of interest here are the *Evolutionary Algorithms* (being by far the most known population–based search method) and *gradient–based methods.* EAs have recently found widespread use in engineering applications. They are intrinsically parallel, giving rise to the so–called Parallel EAs (PEAs). *Metamodel–Assisted EAs* (MAEAs,[5], [6]) and *Hierarchical Distributed MAEAs* (HDMAEAs, [6]) are also amenable to parallelization. We recall that MAEAs are based on the interleaving use of CFD tools and surrogates for the objective and constraint functions, thus, reducing the number of calls to the costly CFD software and generate optimal solutions on an affordable computational budget. On the other hand, this paper is also concerned with *adjoint–based* optimization methods. The adjoint approaches serve to compute the gradient or the Hessian matrix of the objective function for use along with gradient–based search methods, [7], [8]. These approaches can be considered to be either stand–alone optimization methods or ingredients of HDMAEA. Since substantial progress has been made in using hybridized search methods, it is important to conduct a careful examination of their expected performances on multiprocessor platforms.

## 2. PARALLEL EVOLUTIONARY ALGORITHMS (PEAs)–AN OVERVIEW

The recent increase in the use of EAs for engineering optimization problems is attributed to the fact that EAs are general purpose search tools that may handle constrained multi–objective problems by computing the Pareto front of optimal solutions at the cost of a single run. When optimizing computationally demanding problems using EAs, the CPU cost is determined by the required number of evaluations. In the next section, smart ways to reduce the number of costly evaluations within an EA will be presented. Reduction of the wall clock time is also important and one might consider it as well. PEAs achieve this goal on parallel computers, PC clusters or geographically distributed resources, by partitioning the search in simultaneously processed subproblems. PEAs can be classified into *single–population* or *panmictic EAs* and *multi–population EAs.* In a *panmictic EA*, each population member can potentially mate with any other; standard way of parallelization is the concurrent evaluation scheme, with centralized selection and evolution (master–slave model). A *multi–population EA* handles partitioned individual subsets, according to a topology which (often, though not necessarily) maps onto the parallel platform and employs decentralized evolution schemes; these can be further classified to *distributed* and *cellular* EAs, depending on the subpopulations' structure and granularity.

*Distributed* (DEAs) rely upon a small number of medium–sized population subsets (demes) and allow the regular inter–deme exchange of promising individuals and intra–

deme mating. By changing the inter–deme communication topology (ring, grid, etc.), the migration frequency and rate as well as the selection scheme for emigrants, interesting variants of DEAs can be devised. To increase diversity, different evolution parameters and schemes at each deme can be used (*nonuniform* DEAs). *Cellular* EAs (CEAs) usually associate a single individual with each subpopulation. The inter–deme communication is carried out using overlapping neighborhoods and decentralized selection schemes. The reader is referred to [9], [10] for a detailed overview of PEAs. A list of popular communication tools for PEAs and a discussion on how PEAs' parallel speedup should be defined can be found in [11]. These excellent reviews, however, do not account for what seems to be an indispensable constituent of modern EAs, namely the use of metamodels.

## 3. METAMODEL–ASSISTED EAs (MAEAs)

Although PEAs are capable of reducing the wall clock time of EAs, there is also a need to reduce the number of calls to the computationally expensive evaluation software (sequential or parallel CFD codes), irrespective of the use of parallel hardware. To achieve this goal, the use of low–cost *surrogate* evaluation models has been proposed. Surrogates, also referred to as *metamodels*, are mathematical approximations (response surfaces, polynomial interpolation, artificial neural networks, etc.) to the discipline–specific analysis models; they can be constructed using existing datasets or knowledge gathered and updated during the evolution. The reader should refer to [12] for general approximation concepts and surrogates and to [5] for the coupling of EAs and metamodels. There is, in fact, a wide range of techniques for using metamodels within EAs, classified to *off-line* and *on–line* trained metamodels, [5].

In EAs assisted by *off-line* trained metamodels, the latter are trained separately from the evolution using a first sampling of the search space, based on the theory of design of experiments. The deviation of the fitness values of the "optimal" solutions resulting by an EA running exclusively on the metamodel from those obtained using the "exact" CFD tool determines the need for updating the metamodel and iterating.

In EAs assisted by *on-line* trained local metamodels, for each newly generated individual a locally valid metamodel needs to be trained on the fly, on a small number of neighboring, previously evaluated individuals. Evaluating the generation members on properly trained metamodels acts as a filter restricting the use of the CFD tool to a small population subset, i.e. the top individuals as pointed by the metamodel. This will be referred to as the Inexact Pre–Evaluation (IPE) technique; IPE has been introduced in [5], [13] and then extended to DEAs, [14], and hierarchical EAs (HEAs), [6]. Note that, the first few generations (usually no more than two or three) refrain from using metamodels and once an adequate piece of knowledge on how to pair design variables and system performances is gathered, the IPE filter is activated.

Let us briefly describe the IPE technique for multi–objective optimization problems. Consider a $(\mu, \lambda)$ EA with $\mu$ parents, $\lambda$ offspring, $n$ design variables and $m$ objectives. The EA handles three population sets, namely the parent set ($\mathcal{P}_{\mu,g}$, $\mu = |\mathcal{P}_{\mu,g}|$), the offspring ($\mathcal{P}_{\lambda,g}$, $\lambda = |\mathcal{P}_{\lambda,g}|$) and the archival one ($\mathcal{P}_{\alpha,g}$) which stores the $\alpha = |\mathcal{P}_{\alpha,g}|$ best individuals found thus far. At generation $g$, $\lambda$ local metamodels are trained and, through them, approximate objective function values $\tilde{\mathbf{f}}(\mathbf{x}) \in \mathbb{R}^m$ for all $\mathbf{x} \in \mathcal{P}_{\lambda,g}$ are computed.

Any approximately evaluated offspring $\mathbf{x}$ is assigned a provisional cost value $\phi(\mathbf{x}) = \phi(\tilde{\mathbf{f}}(\mathbf{x}), \{\tilde{\mathbf{f}}(\mathbf{y}) \mid \mathbf{y} \in \mathcal{P}_{\lambda,g} \setminus \{\mathbf{x}\}\}) \in \mathbb{R}$ where $\phi$ can be any cost assignment function, based on either sharing techniques, [15], or strength criteria, [16]. The subset $\mathcal{P}_{e,g} = \{\mathbf{x}_i,\ i = 1, \lambda_e : \phi(\mathbf{x}_i) < \phi(\mathbf{y}),\ \mathbf{y} \in \mathcal{P}_{\lambda,g} \setminus \mathcal{P}_{e,g}\}$ of the $\lambda_e < \lambda$ most "promising" individuals is singled out. For the $\mathcal{P}_{e,g}$ members, the exact objective function values $\mathbf{f}(\mathbf{x}) \in \mathbb{R}^m$ are computed; this, practically, determines the CPU cost per generation and, consequently, this all we need to parallelize. A provisional Pareto front, $\hat{\mathcal{P}}_{\alpha,g}$, is formed by the non-dominated, exactly evaluated individuals of $\mathcal{P}_{e,g} \cup \mathcal{P}_{\alpha,g}$. A cost value $\phi(\mathbf{x}) = \phi(\hat{\mathbf{f}}(\mathbf{x}), \{\hat{\mathbf{f}}(\mathbf{y}) \mid \mathbf{y} \in \mathcal{P}_{\lambda,g} \cup \mathcal{P}_{\mu,g} \cup \hat{\mathcal{P}}_{\alpha,g} \setminus \{\mathbf{x}\}\})$ is assigned to each individual $\mathbf{x} \in \mathcal{P}_{\lambda,g} \cup \mathcal{P}_{\mu,g} \cup \hat{\mathcal{P}}_{\alpha,g}$, where $\hat{\mathbf{f}}(\mathbf{x}) = \tilde{\mathbf{f}}(\mathbf{x})$ or $\mathbf{f}(\mathbf{x})$. The new set of non-dominated solutions, $\mathcal{P}_{\alpha,g+1}$, is formed by thinning $\hat{\mathcal{P}}_{\alpha,g}$ through distance-based criteria, if its size exceeds a predefined value $a_{max}$, [16]. The new parental set, $\mathcal{P}_{\mu,g+1}$, is created from $\mathcal{P}_{\lambda,g} \cup \mathcal{P}_{\mu,g}$ by selecting the fittest individuals with certain probabilities. Recombination and mutation are applied to generate $\mathcal{P}_{\lambda,g+1}$. This procedure is repeated until a termination criterion is satisfied. It is important to note that the number $\lambda_e$ of exact and costly evaluations per generation is neither fixed nor known in advance (only its lower and upper bounds are user–defined). So, care needs to be exercised when parallelizing MAEAs based on the IPE technique (see below).

## 4. HIERARCHICAL, DISTRIBUTED MAEAs (HDMAEAs)

We next consider the so–called *Hierarchical* and *Distributed* variant of MAEAs (HDMAEAs), proposed in [6]. An HDMAEA is, basically, a multilevel optimization technique, whereby the term multi–level implies that different search methods, different evaluation software and/or different sets of design variables can be employed at each level. Any EA–based level usually employs DEAs (or, in general PEAs) as explained in the previous sections; also, within each deme, the IPE technique with locally trained metamodels is used (MAEAs). Multilevel optimization algorithms can be classified as follows:

(a) *Multilevel Evaluation* where each level is associated with a different evaluation software. The lower levels are responsible for detecting promising solutions at low CPU cost, through less expensive (low–fidelity) problem–specific evaluation models and delivering them to the higher level(s). There, evaluation models of higher fidelity and CPU cost are employed and the immigrated solutions are further refined. HDMAEAs are multilevel evaluation techniques using PEAs at each level.

(b) *Multilevel Search* where each level is associated with a different search technique. Stochastic methods, such as EAs, are preferably used at the lower levels for the exploration of the design space, leaving the refinement of promising solutions to the gradient–based method at the higher levels. Other combinations of search tools are also possible.

(c) *Multilevel Parameterization* where each level is associated with a different set of design variables. At the lowest level, a subproblem with just a few design variables is solved. At the higher levels, the problem dimension increases. The most detailed problem parameterization is used at the uppermost level.

Given the variety of components involved in a multilevel algorithm (the term HDMAEA will refer to the general class of the multilevel algorithms, since EAs are key search tools) and the availability of parallel CFD evaluation software, parallelization issues of the optimization method as a whole needs to be revisited.

## 5. ADJOINT–BASED OPTIMIZATION METHODS & PARALLELIZATION

Although gradient–based methods assisted by the adjoint approach are stand-alone optimization methods, they are herein considered one of the components of the aforementioned general optimization framework. They require the knowledge of the first and (depending on the method used) second derivatives of the objective function $F$ with respect to the design variables $b_i$, $i=1,N$. The adjoint method is an efficient strategy to compute $\nabla F$ and/or $\nabla^2 F$, [7], [17].

In mathematical optimization, it is known that schemes using only $\nabla F$ are not that fast. So, schemes that also use $\nabla^2 F$ (approximate or exact Newton methods) are preferred. However, in aerodynamic optimization, the computation of the exact Hessian matrix is a cumbersome task, so methods approximating the Hessian matrix (the BFGS method, for instance) are alternatively used. We recently put effort on the computation of the exact Hessian matrix with the minimum CPU cost and, thanks to parallelization, minimum wall clock time. Our method, [8], [18], consists of the use of direct differentiation to compute $\nabla F$, followed by the adjoint approach to compute $\nabla^2 F$. The so–called discrete *direct–adjoint* approach is presented below in brief. An augmented functional $\hat{F}$ is first defined using the adjoint variables $\hat{\Psi}_n$, as follows

$$\frac{d^2\hat{F}}{db_i db_j} = \frac{d^2 F}{db_i db_j} + \hat{\Psi}_m \frac{d^2 R_m}{db_i db_j} \tag{1}$$

where $\frac{d^2F}{db_idb_j} = \frac{d^2F}{db_jdb_i}$ and $\frac{d^2R_m}{db_idb_j} = \frac{d^2R_m}{db_jdb_i} = 0$. We may also write

$$\frac{d^2\Phi}{db_i db_j} = \frac{\partial^2\Phi}{\partial b_i \partial b_j} + \frac{\partial^2\Phi}{\partial b_i \partial U_k}\frac{dU_k}{db_j} + \frac{\partial^2\Phi}{\partial U_k \partial b_j}\frac{dU_k}{db_i} + \frac{\partial^2\Phi}{\partial U_k \partial U_m}\frac{dU_k}{db_i}\frac{dU_m}{db_j} + \frac{\partial\Phi}{\partial U_k}\frac{d^2U_k}{db_i db_j} \tag{2}$$

where $\Phi$ stands for either $F$ or $R_m=0, m=1,M$; $R_m$ are the discrete residual equations, arising from some discretization of the flow pde's; $M$ is the product of the number of grid nodes and the number of equations per node. From eq. 1, we obtain

$$\begin{aligned}\frac{d^2\hat{F}}{db_i db_j} &= \frac{\partial^2 F}{\partial b_i \partial b_j} + \hat{\Psi}_n \frac{\partial^2 R_n}{\partial b_i \partial b_j} + \frac{\partial^2 F}{\partial U_k \partial U_m}\frac{dU_k}{db_i}\frac{dU_m}{db_j} + \hat{\Psi}_n \frac{\partial^2 R_n}{\partial U_k \partial U_m}\frac{dU_k}{db_i}\frac{dU_m}{db_j} + \frac{\partial^2 F}{\partial b_i \partial U_k}\frac{dU_k}{db_j} \\ &+ \hat{\Psi}_n \frac{\partial^2 R_n}{\partial b_i \partial U_k}\frac{dU_k}{db_j} + \frac{\partial^2 F}{\partial U_k \partial b_j}\frac{dU_k}{db_i} + \hat{\Psi}_n \frac{\partial^2 R_n}{\partial U_k \partial b_j}\frac{dU_k}{db_i} + \left(\frac{\partial F}{\partial U_k} + \hat{\Psi}_n \frac{\partial R_n}{\partial U_k}\right)\frac{d^2 U_k}{db_i db_j}\end{aligned} \tag{3}$$

The last term is eliminated by satisfying the adjoint equation

$$\frac{\partial F}{\partial U_k} + \hat{\Psi}_n \frac{\partial R_n}{\partial U_k} = 0 \tag{4}$$

and the remaining terms in eq. 3 give the expression for the Hessian matrix elements. However, to implement the derived formula, we need to have expressions for $\frac{dU_n}{db_j}$. These can be computed by solving the following $N$ equations (*direct differentiation*)

$$\frac{dR_m}{db_i} = \frac{\partial R_m}{\partial b_i} + \frac{\partial R_m}{\partial U_k}\frac{dU_k}{db_i} = 0 \tag{5}$$

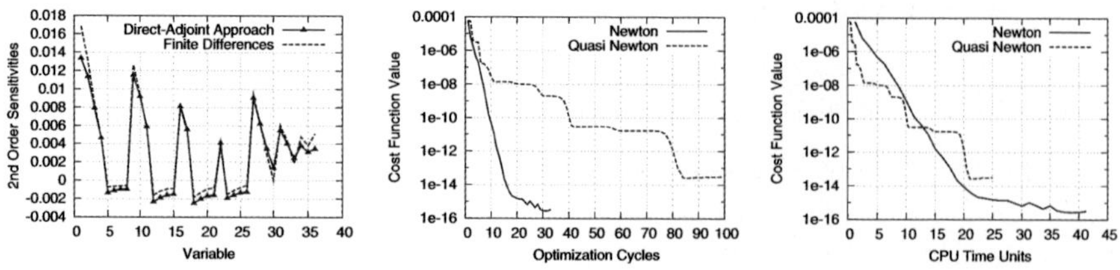


Figure 1. Optimization of a compressor cascade using exact– and quasi–Newton method, based on parallel adjoint and direct solvers (eight design variables). Left: Hessian matrix values computed using the *direct–adjoint* approach and finite differences; only the lower triangular part of the symmetric Hessian matrix is shown. Mid: Convergence of the exact– and quasi–Newton (BFGS) methods, in terms of optimization cycles; the computational load per cycle of the former is 10 times higher. Right: Same comparison in terms of (theoretical) wall clock time on an eight–processor cluster; the flow and the adjoint solvers were parallelized using subdomaining (100% parallel efficiency is assumed) whereas the eight direct differentiations were run in parallel, on one processor each.

for $\frac{dU_k}{db_i}$. To summarize, the total CPU cost of the Newton method is equal to that of $N$ system solutions to compute $\frac{dU_n}{db_j}$, eqs. 5, plus one to solve the adjoint equation, eq. 4 and the solution of the flow equations ($N+2$ system solutions, SSs, per step, in total). In contrast, BFGS costs 2 SSs per descent step. The application of the method (as stand alone optimization tool) on the inverse design of a compressor cascade is shown and discussed in fig. 1.

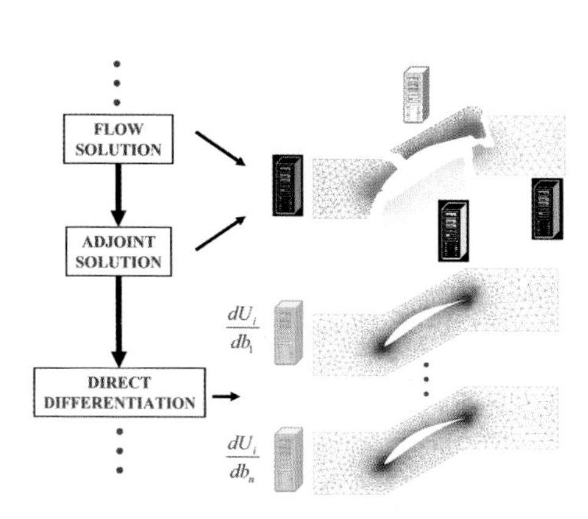


Figure 2: Parallelization of a Newton cycle based on the direct-adjoint method.

Working with either discrete or continuous adjoint approaches and any parallel flow solver (such as the aforementioned one, [2]), the parallelization of the adjoint solver for $\nabla F$ is based on subdomaining in a similar manner to that used for the associated flow solver. At this point and in view of what follows, this will be considered as fine–grained parallelization. What is new and interesting is the parallelization of the proposed adjoint method for the computation of $\nabla^2 F$. Among the $N+2$ SSs needed, $N$ of them corresponding to the computation of $\frac{dU_k}{db_i}$ (direct differentiation, eqs. 5) may run concurrently, due to the independence of the $N$ equations (coarse–grained parallelization; needless to say that subdomaining can additionally be used if more processors are available). The remaining two system solutions (for the adjoint and flow equations) are fine–grained parallelized using subdomaining. The process is illustrated in fig 2.

## 6. PARALLELIZATION OF HDMAEA

### 6.1. HDMAEAs and Cluster Computing

In small commodity clusters, one may use a customized configuration for each application; this case is of no interest in this paper. In larger clusters or HPC systems, the presence of a local resource manager that keeps an updated list of the available processors along with their characteristics, becomes necessary. Using these pieces of information, the *Local Resource Management Software* (LRMS) allows the optimal use of the cluster, by assigning each evaluation request to the appropriate computational resource(s) according to user–defined specifications (i.e. RAM, CPU etc.).

The deployment of the optimization software within a cluster requires linking it to the LRMS. In this respect, DEAs or HDMAEAs use a central evaluation manager allowing concurrent job submissions from any deme. The evaluation manager executes in a separate thread inside the optimization algorithm process and receives evaluation requests from all demes. These requests are submitted to the centralized LRMS *first–in–first–out* queue and each new job is assigned to the first node that becomes available and meets the requirements set by the user. *Condor* middleware was chosen as LRMS, [19]. The *DRMAA* library was used to interface the optimization software with *Condor*, by communicating arguments related to file transfer, I/O redirection, requirements for the execution host etc.

In a distributed optimization scheme (DEA , HDMAEA), upon the end of an evaluation, the evaluation manager contacts the corresponding deme and updates the objectives' and constraints' values of the evaluated individual. Therefore, all evaluation specific details are hidden from the EA core. A unique database (per evaluation software) for all demes is maintained to prevent the re-evaluation of previously evaluated candidate solutions.

Let us lend some more insight into the role of the evaluation manager by comparing evaluation requests launched by a conventional EA and the HDMAEA. In a conventional $(\mu, \lambda)$ EA, $\lambda$ evaluations must be carried out per generation; they all have the same hardware requirements and, theoretically, the "same" computational cost; this is in favor of parallel efficiency. In contrast, maintaining a high parallel efficiency in hierarchical optimization (*Multilevel Evaluation* mode) is not straightforward, due to the use of evaluation tools with different computational requirements. For instance, in aerodynamic optimization, a Navier–Stokes solver with wall functions on relatively coarse grids can be used at the low level (for the cheap exploration of the search space) and a much more accurate but costly flow analysis tool with a better turbulence model on very fine grids at the high level. The latter has increased memory requirements. In such a case, given the maximum amount of memory available per computational node, a larger number of subdomains may be needed. In heterogeneous clusters (with different CPU and RAM per node), a different number of processors might be required for the same job, depending on the availability of processors by the time each job is launched. In the case of synchronous inter–level communication, a level may suspend the execution temporarily, until the adjacent level reaches the generation for which migration is scheduled and this usually affects the assignment of pending evaluations. Also, the IPE technique in a MAEA changes dynamically the number of "exact" evaluations to a small percentage of the population (as previously explained, the percentage $\lambda_e/\lambda$ is not constant but changes according to the population dynamics), limiting thus the number of simultaneous evaluation jobs and causing some processors to become idle. For all these reasons, the use of an evaluation manager in HDMAEAs is crucial. Note that any call to the adjoint code within a HDMAEA (*Multilevel Search* mode) is considered as an "equivalent" evaluation request.

### 6.2. Grid Enabled HDMAEAs

The modular structure of the HDMAEA and the utilization of an evaluation manager allows its deployment in the Grid, leading to a Grid Enabled HDMAEA (GEHDMAEA). Grid deployment requires a link between the cluster and the grid level. A middleware which is suitable for this purpose must: (a) match grid resources to job–specific requirements, (b) connect the LRMS to the grid framework to allow job submission using security certificates for user authentication and (c) allow grid resource discovery by collecting information regarding resource types and availabilities within the clusters. In the developed method, this is accomplished via *Globus Toolkit 4* (*GT*4, [20]).

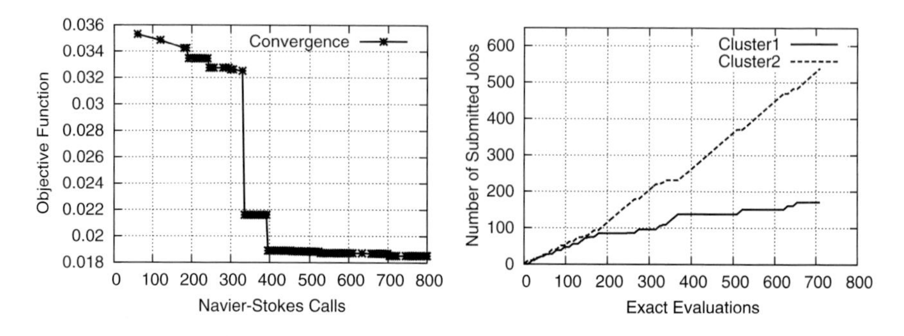


Figure 3. Optimization of an axial compressor stator using the Grid Enabled HDMAEA. Left: Convergence history. Right: History of the number of submitted jobs to each cluster.

The execution of a job through the *Globus Toolkit* starts with the verification of user's credentials before submitting a job to a grid resource. A user may submit a job to a cluster in two different ways. The submission can be made through either the front-end of the cluster and, then, passed on to the LRMS, or through a gateway service by submitting a job from a host that has *GT*4 installed to a remote grid resource. For the second option, the user's credentials are also transmitted for authentication to the remote resource (via *GT*4) before the job is accepted for execution.

All grid resources that have *GT*4 installed publish information about their characteristics which can be accessed using the *http* protocol. The discovery and monitoring of resources via a tool that probes all resources for their characteristics by accessing the corresponding information service of the *GT*4, i.e. a *Meta–Scheduler*, is needed. The *Meta–Scheduler* unifies the grid resources under a common queue and associates each job submitted to the queue with the first available grid resource satisfying the requirements set by the user. Herein, *GridWay*, [21], was used as *Meta–Scheduler*. *GridWay* has an API that enables GEHDMAEA to forward all job requests from the evaluation server to the common grid queue. *GridWay* collects information regarding the available resources from the information services of the $GT4$ installed on each resource and keeps the complete list available to all authorized users. When submitting jobs to *GridWay*, each job is forwarded from the global grid queue to the $GT4$ of the matched grid resource and, then, to the LRMS, if this resource is a cluster. The LRMS (*Condor*) performs the final matchmaking between the job and the available resources within the cluster, followed by the job execution.

The optimization process in a grid environment is carried out the same way as within a cluster. For illustrative purposes we present a design problem of an axial compressor stator with minimum total pressure loss coefficient $\omega$, at predefined flow conditions (more details on this case are beyond the scope of this paper). A number of constraints were imposed regarding the minimum allowed blade thickness at certain chordwise locations as well as the flow exit angle (by, consequently, controlling the flow turning). The convergence history of the optimization process is presented in fig. 3. The grid environment in which the optimization software was deployed consists of two clusters. The description of the clusters along with the number of jobs submitted, during the airfoil design, per cluster is given in the following table. The number of submitted jobs to each cluster are shown in fig. 3 (right).

| | Host types | Number of Hosts | Jobs Submitted |
|---|---|---|---|
| Cluster1 | PIII and P4 up to 2.4GHz | 14 | 154 |
| Cluster2 | P4 and Pentium D up to 3.4GHz | 30 | 550 |

## 7. ACKNOWLEDGEMENTS

Parts of this synthetic work and development during the last years in the lab were supported by the $PENED01$ Program (Measure 8.3 of the Operational Program Competitiveness) under project number $01ED131$ and by the Operational Program for Educational and Vocational Training II (EPEAEK II) (Program PYTHAGORAS II); in both, funding was 75% by the European Commission and 25% by National Resources. Some of the PhD students were supported by grants from the National Scholarship Foundation.

## REFERENCES

1. D. Lim, Y-S. Ong, Y. Jin, B. Sendhoff, B-S. Lee, Efficient Hierarchical Parallel Genetic Algorithms using Grid Computing, J. Fut. Gener. Comput. Syst., 23(4):658–670, 2007.
2. N. Lambropoulos, D. Koubogiannis, K. Giannakoglou, Acceleration of a Navier-Stokes Equation Solver for Unstructured Grids using Agglomeration Multigrid and Parallel Processing, Comp. Meth. Appl. Mech. Eng., 193:781-803, 2004.
3. A. Giotis, K. Giannakoglou, An Unstructured Grid Partitioning Method Based on Genetic Algorithms, Advances in Engineering Software, 29(2):129-138, 1998.
4. G. Karypis, V. Kumar, Fast and High Quality Multilevel Scheme for Partitioning Irregular Graphs, SIAM J. Scientific Computing, 20(1):359-392, 1998.
5. K. Giannakoglou, Design of Optimal Aerodynamic Shapes using Stochastic Optimiza-

tion Methods and Computational Intelligence, Int. Review J. Progress in Aerospace Sciences, 38:43-76, 2002.
6. M. Karakasis, D. Koubogiannis, K. Giannakoglou, Hierarchical Distributed Evolutionary Algorithms in Shape Optimization, Int. J. Num. Meth. Fluids, 53:455-469, 2007.
7. D. Papadimitriou, K. Giannakoglou, A Continuous Adjoint Method with Objective Function Derivatives Based on Boundary Integrals for Inviscid and Viscous Flows, Computers & Fluids, 36:325-341, 2007.
8. D. Papadimitriou, K. Giannakoglou, Direct, Adjoint and Mixed Approaches for the Computation of Hessian in Airfoil Design Problems, Int. J. Num. Meth. Fluids, to appear.
9. E. Cantu-Paz, A Survey of Parallel Genetic Algorithms", Calculateurs Paralleles, Reseaux et Systemes Repartis, 10(2):141-171, 1998.
10. M. Nowostawski, R. Poli, Parallel Genetic Algorithm Taxonomy, Proc. 3rd Int. Conf. on Knowledge-based Intelligent Information Engineering Systems KES'99:88-92, IEEE, 1999.
11. E. Alba, M. Tomassini, Parallelism and Evolutionary Algorithms, IEEE Trans. Evol. Comp., 6(5), Oct. 2002.
12. A. Keane, P. Nair, *Computational Approaches for Aerospace Design. The Pursuit of Excellence*, John Wiley & Sons, Ltd, 2005.
13. K. Giannakoglou, A. Giotis, M. Karakasis, Low–Cost Genetic Optimization based on Inexact Pre-evaluations and the Sensitivity Analysis of Design Parameters, Inverse Problems in Engineering, 9:389-412, 2001.
14. M. Karakasis, K. Giannakoglou, Inexact Information aided, Low-Cost, Distributed Genetic Algorithms for Aerodynamic Shape Optimization, Int. J. Num. Meth. Fluids, 43(10-11):1149-1166, 2003.
15. N. Srinivas, K. Deb, Multiobjective Optimization Using Nondominated Sorting in Genetic Algorithms, Evolutionary Computation, 2(3):221–248, 1995.
16. E. Zitzler, M. Laumans, L. Thiele, SPEA2: Improving the Strength Pareto Evolutionary Algorithm, Report, Swiss Federal Institute of Technology (ETH), Computer Engineering and Communication Networks Lab., May 2001.
17. D. Papadimitriou, K. Giannakoglou, Total Pressure Losses Minimization in Turbomachinery Cascades, Using a New Continuous Adjoint Formulation, Journal of Power and Energy (Special Issue on Turbomachinery), to appear, 2007.
18. K. Giannakoglou, D. Papadimitriou, Adjoint Methods for gradient- and Hessian-based Aerodynamic Shape Optimization, EUROGEN 2007, Jyvaskyla, June 11-13, 2007.
19. D. Thain and T. Tannenbaum, M. Livny, Distributed computing in practice: the Condor experience, Concurrency - Practice and Experience, 17(2-4):323-356, 2005.
20. I. Foster, Globus Toolkit Version 4: Software for Service-Oriented Systems, International Conference on Network and Parallel Computing, LNCS: 2-1317(2-4), Springer-Verlag, 2006.
21. E. Huedo, R. Montero, I. Llorente, A Framework for Adaptive Execution on Grids, Journal of Software - Practice and Experience, 34:631-651, 2004.

# Non-Sinusoidal Path Optimization of Dual Airfoils Flapping in a Biplane Configuration

M. Kaya[a*] and I. H. Tuncer[a†]

[a]Aerospace Engineering Department
Middle East Technical University, Ankara, Turkey

The path of dual airfoils in a biplane configuration undergoing a combined, non-sinusoidal pitching and plunging motion is optimized for maximum thrust and/or propulsive efficiency. The non-sinusoidal, periodic flapping motion is described using Non-Uniform Rational B-Splines (NURBS). A gradient based algorithm is then employed for the optimization of the NURBS parameters. Unsteady, low speed laminar flows are computed using a Navier-Stokes solver in a parallel computing environment based on domain decomposition. The numerical evaluation of the gradient vector components, which requires unsteady flow solutions, is also performed in parallel. It is shown that the thrust generation may significantly be increased in comparison to the sinusoidal flapping motion.

## 1. INTRODUCTION

Based on observations of flying birds and insects, and swimming fish, flapping wings have been recognized to be more efficient than conventional propellers for flights of very small scale vehicles, so-called micro-air vehicles (MAVs) with wing spans of 15 $cm$ or less. The current interest in the research and development community is to find the most energy efficient airfoil adaptation and flapping wing motion technologies capable of providing the required aerodynamic performance for a MAV flight.

Recent experimental and computational studies investigated the kinematics, dynamics and flow characteristics of flapping wings, and shed some light on the lift, drag and propulsive power considerations[1,2]. Water tunnel flow visualization experiments on flapping airfoils conducted by Lai and Platzer[3] and Jones et al.[4] provide a considerable amount of information on the wake characteristics of thrust producing flapping airfoils. Anderson et al.[5] observe that the phase shift between pitch and plunge oscillations plays a significant role in maximizing the propulsive efficiency. A recent work by Lewin and Haj-Hariri[6] indicates that the aerodynamic forces generated by flapping insects are very sensitive to the wing kinematics. NavierStokes computations performed by Tuncer et al.[7–9] and by Isogai et al.[10] explore the effect of flow separation on the thrust generation and the propulsive efficiency of a single flapping airfoil in combined pitch and plunge oscillations.

Jones and Platzer[11] recently demonstrated a radiocontrolled micro air vehicle propelled by flapping wings in a biplane configuration The experimental and numerical studies by Jones et al.[11] and Platzer and Jones[12] on flapping-wing propellers points at the gap between numerical flow solutions and the actual flight conditions over flapping wings.

*Graduate Research Assistant, e-mail: mkaya@ae.metu.edu.tr
†Professor, e-mail: tuncer@ae.metu.edu.tr

I.H. Tuncer et al., *Parallel Computational Fluid Dynamics 2007*,
DOI: 10.1007/978-3-540-92744-0_7, 

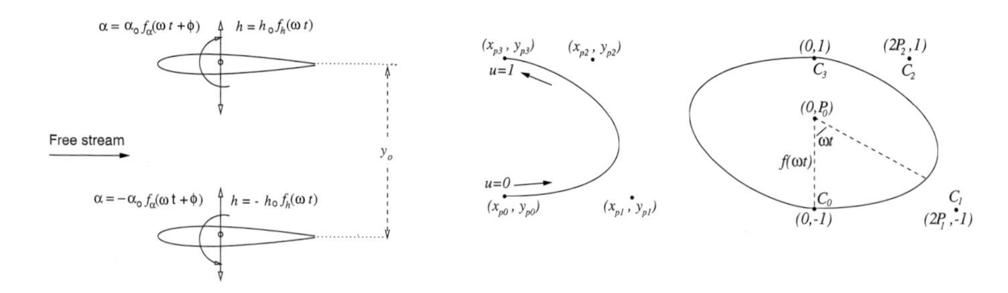


Figure 1. Out-of-phase flapping motion of two airfoils in a biplane configuration

Figure 2. Flapping path defined by a $3^{rd}$ degree NURBS

Tuncer and Kaya[13–15] and Kaya et al.[16] investigated the optimization of sinusoidal motion parameters of a single flapping airfoil and dual airfoils flapping in a biplane configuration for maximum thrust and/or propulsive efficiency. Kaya and Tuncer[17,18] showed that the thrust generation of a non-sinusoidally flapping airfoil may significantly be increased in comparison to an airfoil undergoing sinusoidal flapping motion.

It should be noted that in the sinusoidal motion, the pitch and plunge positions are based on the projection of a vector rotating on a unit circle, and the maximum plunge and pitch velocities occur at the mean plunge and pitch positions. In the present study, the periodic flapping motion (Figure 1) is relaxed using a closed curve based on a $3^{rd}$ degree Non-Uniform Rational B-Splines (NURBS) instead of the unit circle[18]. The parameters defining the NURBS is then optimized for maximum thrust and propulsive efficiency.

## 2. NURBS Based Periodic Path

A smooth curve $S$ based on a $n^{th}$ degree rational Bezier segment is defined as follows[19]:

$$S(u) = (x(u), y(u)) = \frac{\sum_{i=0}^{n} W_i B_{i,n}(u) C_i}{\sum_{i=0}^{n} W_i B_{i,n}(u)} \qquad 0 \leq u \leq 1 \tag{1}$$

where $B_{i,n}(u) \equiv \frac{n!}{i!(n-i)!} u^i (1-u)^{n-i}$ are the classical $n^{th}$ degree Bernstein polynomials, and $C_i = (x_{pi}, y_{pi})$, are called control points with weights, $W_i$. Note that $S(u=0) = C_o$ and $S(u=1) = C_n$. A closed curve which describes the upstroke and the downstroke of a flapping path is then defined by employing a NURBS composed of two $3^{rd}$ degree rational Bezier segments. The control points and their corresponding weights are chosen such that the non-sinusoidal periodic function ($y$ coordinates of the closed curve) is between $-1$ and 1. The periodic flapping motion is finally defined by 3 parameters. The first parameter $P_o$ defines the center of the rotation vector on the closed curve. The remaining two parameters, $P_1$ and $P_2$ are used to define the $x$ coordinates of the control points, which are $C_1 = (2P_1, -1)$ and $C_2 = (2P_2, 1)$ (Figure 2). The parameters $P_1$ and $P_2$ define the flatness of the closed NURBS curve.

The $x$ and $y$ coordinates on the periodic NURBS curve may be obtained as a function of $u$:

$$x(u) = \frac{2P_1 u(1-u)^2 + 2P_2 u^2(1-u)}{(1-u)^3 + u(1-u)^2 + u^2(1-u) + u^3} \quad y(u) = \frac{-(1-u)^3 - u(1-u)^2 + u^2(1-u) + u^3}{(1-u)^3 + u(1-u)^2 + u^2(1-u) + u^3} \tag{2}$$

The non-sinusoidal periodic function, $f$, is then defined as, $f(u(\omega t)) = y(u) \equiv f(\omega t)$, where

$$\tan(\omega t) = -\frac{x(u)}{y(u) - P_o} \tag{3}$$

For a given $\omega t$ position, Equation 3 is solved for $u$. Once $u$ is determined, $y(u) \equiv f(\omega t)$ is evaluated using Equation 2.

## 3. Numerical Method

Unsteady, viscous flowfields around flapping airfoils in a biplane configuration are computed by solving the Navier-Stokes equations on moving and deforming overset grids[16]. A domain decomposition based parallel computing algorithm is employed. PVM message passing library routines are used in the parallel solution algorithm. The computed unsteady flowfields are analyzed in terms of time histories of aerodynamic loads, and unsteady particle traces.

### 3.1. Flow Solver

The strong conservation-law form of the 2-D, thin-layer, Reynolds averaged Navier-Stokes equations is solved on each subgrid. The discretized equations are solved in parallel by an approximately factored, implicit algorithm. The background grid (Figure 3) is partitioned into two overlapping subgrids at the symmetry plane. The computational domain is then decomposed into a total of four subgrids for parallel computing[20,21]. The holes in the background grid formed by the overset airfoil grids are excluded from the computations by an *i-blanking* algorithm. The conservative flow variables are interpolated at the intergrid boundaries formed by the overset grids[20].

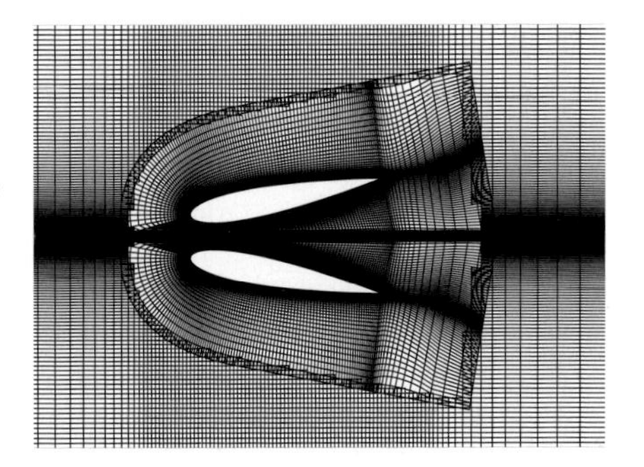

Figure 3: Moving and deforming overset grid system

### 3.2. Flapping Motion

The out of phase flapping motion of the airfoils are imposed by moving the airfoils and the C-type grids around them over the background grid (Figure 3). The airfoil grids are deformed as they come close to the symmetry line between the airfoils. The maximum deformation is such that the distance between airfoils is constrained at 5% chord length. The sinusoidal flapping motion of the upper airfoil in plunge, $h$, and in pitch, $\alpha$, is defined by:

$$\begin{aligned} h(t) &= h_0\, f_h(\omega t) \\ \alpha(t) &= \alpha_0\, f_\alpha(\omega t + \phi) \end{aligned} \tag{4}$$

where $h_o$ and $\alpha_o$ are the plunge and pitch amplitudes, $f$ is a periodic function based on NURBS, $\omega$ is the angular frequency which is given in terms of the reduced frequency, $k = \frac{\omega c}{U_\infty}$. $\phi$ is the phase shift between plunge and pitching motions. The pitch axis is located at the mid-chord. The flapping motion of the lower airfoil is in counter-phase. The average distance between the dual airfoils is set to $y_0 = 1.4$.

An important parameter for analyzing the performance of the flapping airfoils is the instantaneous effective angle of attack. The effective angle of attack due to the pitching and plunging velocities at the leading edge, is defined by:

$$\alpha_{eff}(t) = \alpha(t) - \arctan\left(\frac{\dot{h}(t) + \frac{1}{2}c\dot{\alpha}(t)\cos(\alpha(t))}{U_\infty - \frac{1}{2}c\dot{\alpha}(t)\sin(\alpha(t))}\right) \tag{5}$$

where $\frac{1}{2}c$ is the distance between the leading edge and the pitch axis location.

### 3.3. Optimization

The objective function is taken as a linear combination of the average thrust coefficient, $C_t$, and the propulsive efficiency, $\eta$, over a flapping period:

$$O^l[C_t,\eta] = (1-\beta)\frac{C_t^l}{C_t^{l-1}} + \beta\frac{\eta^l}{\eta^{l-1}} \tag{6}$$

where

$$C_t = -\frac{1}{T}\int_t^{t+T} C_d\,dt\;, \qquad \eta = \frac{C_t\,U_\infty}{C_P}\;, \qquad C_P = \frac{1}{T}\int_t^{t+T}\int_S C_p(\vec{V}\cdot d\vec{A})\,dt \tag{7}$$

where $T$ is the period of the flapping motion. $C_P$ is the power coefficient which accounts for the average work required to maintain the flapping motion. $l$ denotes the iteration step during optimization process. Note that $\beta = 0$ sets the objective function to the normalized thrust coefficient.

Optimization process is based on following the direction of the steepest ascent of the objective function, $O$. The direction of the steepest ascent is given by the gradient vector of the objective function, $\vec{\nabla}O(\vec{v}) = \frac{\partial O}{\partial V_1}\vec{v_1} + \frac{\partial O}{\partial V_2}\vec{v_2} + \cdots$, where $V_i$'s are the optimization variables, and the $\vec{v_i}$'s are the corresponding unit vectors in the variable space.

The components of the gradient vector are then evaluated numerically by computing the objective function for a perturbation of all the optimization variables one at a time. It should be noted that the evaluation of these vector components requires an unsteady flow solution over a few periods of the flapping motion until a periodic flow behavior is reached. Once the unit gradient vector is evaluated, an optimization step, $\Delta\vec{S} = \varepsilon\frac{\vec{\nabla}O}{|\vec{\nabla}O|}$, is taken along the vector. This process continues until a local maximum in the optimization space is reached. The stepsize $\varepsilon$ is evaluated by a line search along the gradient vector, at every optimization step.

In the optimization studies, the location of the points $P_1$ and $P_2$ are constrained within the range 0.2 to 5.0, and $P_o$ in the range $-0.9$ to 0.9, in order to define a flapping motion which does not produce large acceleration that causes numerical difficulties in reaching periodic solutions.

### 3.4. Parallel Processing

In the solution of unsteady flows, a parallel algorithm based on domain decomposition is implemented in a master-worker paradigm. The moving and deforming overset grid system is decomposed into its subgrids first, and the solution on each subgrid is computed as a separate process in the computer cluster. The background grid may also partitioned to improve the static load balancing. Intergrid boundary conditions are exchanged among subgrid solutions at each time step of the unsteady solution. `PVM` (version 3.4.5) library routines are used for inter–process communication. In the optimization process, the components of the gradient vector which require unsteady flow solutions with perturbed optimization variables, are also computed in parallel. Computations are performed in a cluster of Linux based computers with dual Xeon and Pentium-D processors.

## 4. Results

In this optimization study, the optimization variables are chosen as the NURBS parameters defining the plunging and pitching paths, $P_{oh}$, $P_{1h}$, $P_{2h}$, $P_{o\alpha}$, $P_{1\alpha}$ and $P_{2\alpha}$ (Figure 2). As the airfoil is set to a flapping motion in plunge and pitch on a closed path defined by NURBS, the NURBS parameters are optimized for maximum thrust and propulsive efficiency.

The optimization variables are optimized for a fixed value of the reduced flapping frequency, $k \equiv \frac{wc}{U_\infty} = 1.5$. The unsteady laminar flowfields are computed at a low Mach number of 0.1 and a Reynolds number of 10000. In a typical optimization process, parallel computations take about

Table 1
Optimization cases and initial conditions based on sinusoidal path optimization

| Case | $\beta$ | $h_o$ | $\alpha_o$ | $\phi$ | $P_{oh}$ | $P_{1h}$ | $P_{2h}$ | $P_{o\alpha}$ | $P_{1\alpha}$ | $P_{2\alpha}$ | $\eta$ | $C_t$ |
|---|---|---|---|---|---|---|---|---|---|---|---|---|
| **1** | 0.0 | 0.53 | $11.6^o$ | $93.7^o$ | 0. | 1. | 1. | 0. | 1. | 1. | 0.41 | 0.45 |
| **2** | 0.5 | 0.51 | $12.9^o$ | $92.6^o$ | 0. | 1. | 1. | 0. | 1. | 1. | 0.45 | 0.39 |

Table 2
Non-sinusoidal optimization results

| Case | $\beta$ | $h_o$ | $\alpha_o$ | $\phi$ | $P_{oh}$ | $P_{1h}$ | $P_{2h}$ | $P_{o\alpha}$ | $P_{1\alpha}$ | $P_{2\alpha}$ | $\eta$ | $C_t$ |
|---|---|---|---|---|---|---|---|---|---|---|---|---|
| **1** | 0.0 | 0.53 | $11.6^o$ | $93.7^o$ | $-0.34$ | 0.99 | 1.50 | 0.16 | 1.18 | 1.10 | 0.34 | 0.59 |
| **2** | 0.5 | 0.51 | $12.9^o$ | $92.6^o$ | 0.21 | 1.17 | 1.15 | 0.14 | 1.16 | 1.16 | 0.43 | 0.46 |

30 – 100 hours of wall clock time using 10 – 16 Pentium 4, 2.4GHz processors. Flowfields are analyzed in terms of unsteady particle traces, variations of the unsteady thrust/drag coefficient and effective angle of attack in a flapping period. Particle traces are obtained by a simple and efficient integration of the particle pathlines within the flow solver as the unsteady flowfield is computed.

In this study, the $P_1$ and $P_2$ values are constrained within the range 0.2 to 5.0, and $P_o$ in the range $-0.9$ to 0.9, in order to define a proper flapping motion which does not impose excessively large accelerations. A flapping motion with very large instantaneous accelerations causes numerical difficulties in reaching periodic flow solutions, and is, in fact, not practical due to the inertial effects.

The optimization cases studied and the initial conditions are given in Table 1. The initial values of the optimization variables correspond to the optimum sinusoidal flapping motion for the given $\beta$ value, which are obtained from a previous optimization study[16]. It should be noted that the NURBS parameters in the initial conditions, $P_o = 0.0$, $P_1 = 1.0$, $P_2 = 1.0$, define a sinusoidal flapping motion. The non-sinusoidal optimization results obtained in this study are given in Table 2.

In a steepest ascent process, as the optimization variables are incremented along the gradient vector, the objective function increases gradually from an initial value, and reaches a maximum. As seen from Tables 1 and 2, the initial value of the objective function in Case 1 is the average thrust coefficient, $C_t = 0.45$, which is the maximum value for the sinusoidal flapping motion. The maximum value for this case is $C_t = 0.59$. The corresponding propulsive efficiency is about 34% in contrast to the starting value of $\eta = 41\%$.

The optimum non-sinusoidal flapping motion for Case 1 is given in Figure 4. It is noted that the non-sinusoidal flapping motion differs from the sinusoidal motion. The pitching motion now mostly occurs at the minimum and maximum plunge positions, and the airfoil stays at almost a constant incidence angle before reaching top and bottom plunge positions. Figure 5 shows the drag coefficient and effective angle of attack variations in time. As seen from the figure, the non-sinusoidal flapping motion produces higher effective angle of attack values for short durations about mid-plunge positions, and causes a much higher thrust production. The thrust peak is about 5 times higher than that of the sinusoidal flapping motion. In agreement with the previous study[14], the maximum effective angle of attack occurs around mid-plunge locations which is also the instant when the thrust is maximum.

The non-sinusoidal and sinusoidal unsteady flowfields for Case 1 are depicted in Figures 6 and 7 in terms of particle traces. Particles are shed along a vertical line at the leading edge of the airfoil. As seen in the figures, the optimum non-sinusoidal flapping motion produces a higher vortical flowfield where the strong leading edge vortices are formed during the upstroke and

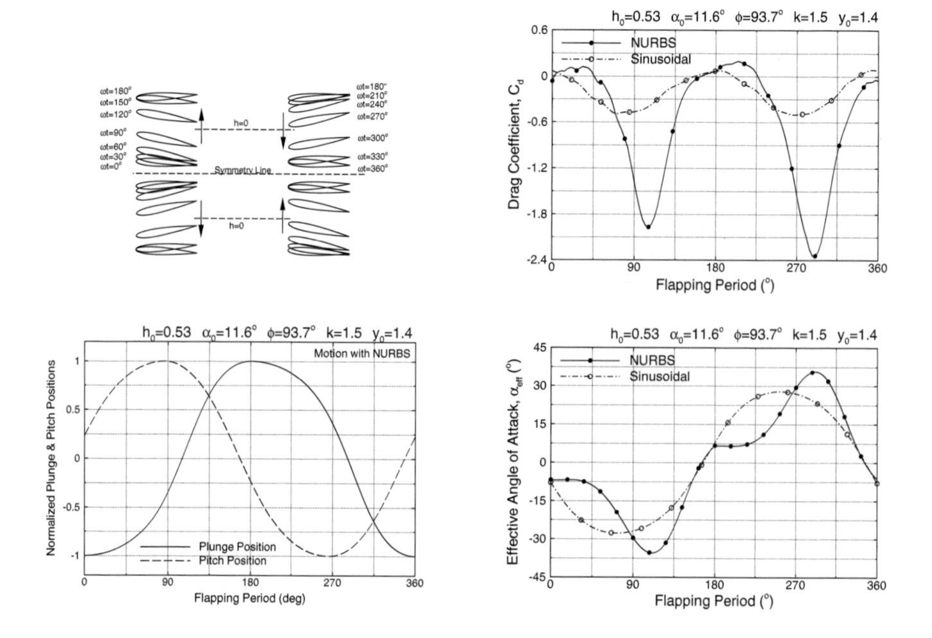


Figure 4. Non-sinusoidal flapping motion for Case 1

Figure 5. Unsteady drag coefficient and effective angle of attack for Case 1

the downstroke, and are shed into the wake. The vortices generated in the optimum sinusoidal flapping motion are observed to be weaker.

In Case 2, the thrust and the propulsive efficiency are optimized simultaneously by setting $\beta = 0.5$, which sets the objective function as a linear combination of the average thrust coefficient and the propulsive efficiency with equal weights. As seen in Table 2, the non-sinusoidal flapping motion based on NURBS produces about 20% higher thrust than the sinusoidal motion without much of a loss in the propulsive efficiency.

## 5. Conclusions

In this study, the flapping path of dual airfoils flapping in a biplane configuration is defined by a $3^{rd}$ degree NURBS. The NURBS parameters are successfully optimized for maximum thrust and/or propulsive efficiency using a gradient based optimization method and an unsteady Navier-Stokes solver. The study shows that the thrust generation of the airfoils flapping in a biplane configuration on a non-sinusoidal path may be enhanced in comparison to the sinusoidal flapping. However, it is achieved at the expense of the propulsive efficiency. The optimum non-sinusoidal flapping motion for maximum thrust imposes a plunging motion with an almost constant incidence angle during the upstroke and the downstroke while the pitching occurs rapidly at the minimum and maximum plunge amplitudes. The thrust producing flows are observed to be highly vortical with the formation of strong leading edge vortices during the upstroke and the downstroke.

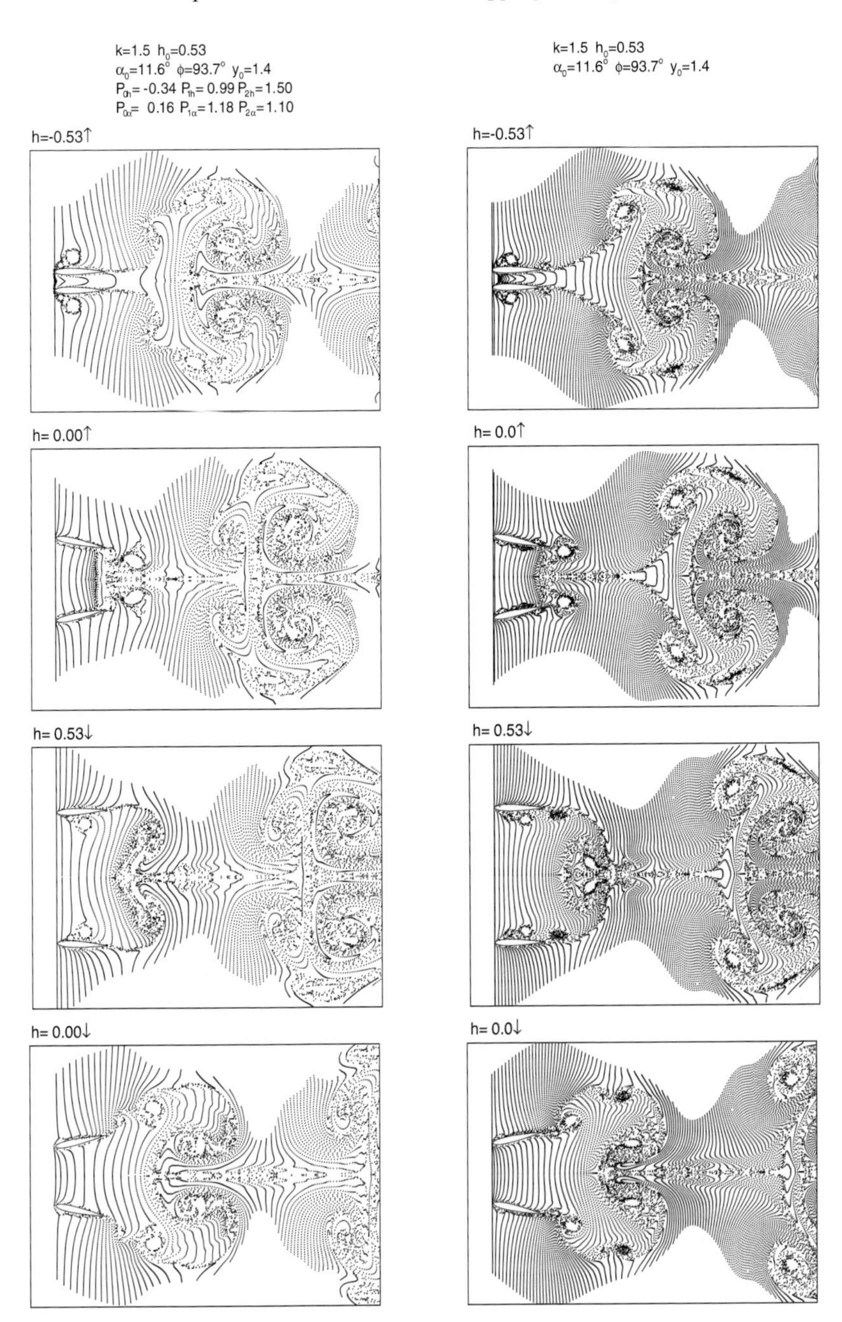


Figure 6. The unsteady flowfield for non-sinusoidal flapping motion in Case 1

Figure 7. The unsteady flowfield for sinusoidal flapping motion in Case 1

## REFERENCES


1. T.J. Mueller (editor), *Fixed and Flapping Wing Aerodynamics for Micro Air Vehicles*, AIAA Progress in Aeronautics and Astronautics, Vol 195, Reston, VA, 2001.
2. W. Shyy, M. Berg and D. Lyungvist, "Flapping and Flexible Wings for Biological and Micro Air Vehicles", *Pergamon Progress in Aerospace Sciences*, Vol 35, p: 455-505, 1999.
3. J.C.S. Lai and M.F. Platzer, "Jet Characteristics of a Plunging Airfoil", *AIAA Journal*, Vol 37, p: 1529-1537, 1999.
4. K.D. Jones, C.M. Dohring and M.F. Platzer, "An Experimental and Computational Investigation of the Knoller-Beltz Effect", *AIAA Journal*, Vol 36, p: 1240-1246, 1998.
5. J.M. Anderson, K. Streitlen, D.S. Barrett and M.S. Triantafyllou, "Oscillating Foils of High Propulsive Efficiency", *Journal of Fluid Mechanics*, Vol 360, p: 41-72, 1998.
6. G.C. Lewin and H. Haj-Hariri, "Reduced-Order Modeling of a Heaving Airfoil", *AIAA Journal*, Vol 43, p: 270-277, 2005.
7. I.H. Tuncer and M.F. Platzer, "Thrust Generation due to Airfoil Flapping", *AIAA Journal*, Vol 34, p: 324-331, 1996.
8. I.H. Tuncer, J. Lai, M.A. Ortiz and M.F. Platzer, "Unsteady Aerodynamics of Stationary/Flapping Airfoil Combination in Tandem", *AIAA Paper*, No 97-0659, 1997.
9. I.H. Tuncer and M.F. Platzer, "Computational Study of Flapping Airfoil Aerodynamics", *AIAA Journal of Aircraft*, Vol 35, p: 554-560, 2000.
10. K. Isogai, Y. Shinmoto Y. and Y. Watanabe, "Effects of Dynamic Stall on Propulsive Efficiency and Thrust of a Flapping Airfoil", *AIAA Journal*, Vol 37, p: 1145-1151, 2000.
11. K.D. Jones and M.F. Platzer, "Experimental Investigation of the Aerodynamic Characteristics of Flapping-Wing Micro Air Vehicles", *AIAA Paper*, No 2003-0418, Jan 2003.
12. M.F. Platzer and K.D. Jones, "The Unsteady Aerodynamics of Flapping-Foil Propellers", $9^{th}$ *International Symposium on Unsteady Aerodynamics, Aeroacoustics and Aeroelasticity of Turbomachines, Ecole Centrale de Lyon*, Lyon, France, Sept 2000.
13. I.H. Tuncer and M. Kaya, "Optimization of Flapping Airfoils for Maximum Thrust", *AIAA Paper*, No 2003-0420, Jan 2003.
14. I.H. Tuncer and M. Kaya, "Optimization of Flapping Airfoils For Maximum Thrust and Propulsive Efficiency", *AIAA Journal*, Vol 43, p: 2329-2341, Nov 2005.
15. M. Kaya and I.H. Tuncer, "Parallel Optimization of Flapping Airfoils in a Biplane Configuration for Maximum Thrust", *Proceedings of Parallel CFD 2004 Conference*, Gran Canaria, Canary Island, Spain May 24-27, 2004.
16. M. Kaya, I.H. Tuncer, K.D. Jones and M.F. Platzer, "Optimization of Flapping Motion of Airfoils in Biplane Configuration for Maximum Thrust and/or Efficiency", *AIAA Paper*, No 2007-0484, Jan 2007.
17. M. Kaya and I.H. Tuncer, "Path Optimization of Flapping Airfoils for Maximum Thrust Based on Unsteady Viscous Flow Solutions", $3^{rd}$ *Ankara International Aerospace Conference*, Ankara, Aug 2005.
18. M. Kaya and I.H. Tuncer, "Non-Sinusoidal Path Optimization of Flapping Airfoils", *AIAA Journal*, Vol 45, p: 2075-2082, 2007.
19. L. Piegl and W. Tiller, *The NURBS Book*, $2^{nd}$ ed., Springer-Verlag, Berlin, 1997.
20. I.H. Tuncer, "A 2-D Unsteady Navier-Stokes Solution Method with Moving Overset Grids", *AIAA Journal*, Vol. 35, No. 3, March 1997, pp. 471-476.
21. I.H. Tuncer and M. Kaya, "Parallel Computation of Flows Around Flapping Airfoils in Biplane Configuration", *Proceedings of Parallel CFD 2002 Conference*, Kansai Science City, Japan, May 20-22, 2002

# Parallel Computation of 3-D Viscous Flows on Hybrid Grids

Murat Ilgaz[a] and Ismail H. Tuncer[b]

[a]Aerodynamics Division, Defense Industries Research and Development Institute, 06261, Ankara, Turkey

[b]Department of Aerospace Engineering, Middle East Technical University, 06531, Ankara, Turkey

In this study, a newly developed parallel finite-volume solver for 3-D viscous flows on hybrid grids is presented. The boundary layers in wall bounded viscous flows are discretized with hexahedral cells for improved accuracy and efficiency, while the rest of the domain is discretized by tetrahedral and pyramidal cells. The computations are performed in parallel in a computer cluster. The parallel solution algorithm with hybrid grids is based on domain decomposition which is obtained using the graph partitioning software METIS. Several validation cases are presented to show the accuracy and the efficiency of the newly developed solver.

## 1. INTRODUCTION

The developments in computer technology in the last two decades prompted the use of parallel processing in Computational Fluid Dynamics (CFD) problems. With parallel processing, the computation time to solve a CFD problem is significantly reduced by having many operations, each being a part of the solution algorithm, performed at the same time. Today, parallel processing is necessary for an efficient CFD algorithm/solver, which aims at the solution of complex 3-D flows over realistic configurations.

Although numerical methods developed for the solution of Euler and Navier-Stokes equations initiated in the structured grid context, the use of unstructured grids has become very popular after 1990s and has contributed to the solution of flow fields around complex geometries significantly. The unstructured grids offer the most flexibility in the treatment of complex geometries. On the other hand, the generation of tetrahedral cells in viscous boundary layers is proven to be difficult. In those regions, strong solution gradients usually occur in the direction normal to the surface, which requires cells of very large aspect ratio. It is well known that the structured grids are superior in capturing the boundary layer flow. However, the generation of structured grids for complex geometries is rather difficult and time consuming. Although some of the difficulties can be overcome by multiblock grid methods, the complexity of the flow solver is then increased.

The hybrid or mixed grid concept has become promising since it carries favorable features of both structured and unstructured grids. A 3-D hybrid grid usually consists of a

I.H. Tuncer et al., *Parallel Computational Fluid Dynamics 2007*,
DOI: 10.1007/978-3-540-92744-0_8, © Springer-Verlag Berlin Heidelberg 2009

mix of hexahedral, tetrahedral, prismatic and pyramidal cells. The hexahedral elements are mostly employed near the solid walls to better resolve the viscous boundary layer flows, while the rest of the domain consists of tetrahedral cells. In the last decade, the use of hybrid grids along with parallel processing has attracted much attention and were employed for the solution of viscous flows [1–6].

In the present work, a finite-volume method for the solution of 3-D Navier-Stokes equations on hybrid unstructured grids is developed. The parallel solution algorithm is based on domain decomposition. The flow solver employs an explicit Runge-Kutta time-stepping scheme and the Roe upwind flux evaluations. METIS, a well known graph partitioning software, is used for the decomposition of hybrid grids. Several flow cases are presented for validation of the newly developed flow solver.

## 2. GOVERNING EQUATIONS

The non-dimensional integral conservation form of the Navier-Stokes equations for a 3-D flow through a volume $\Omega$ enclosed by the surface $S$ in a compact form suitable for finite volume numerical discretization is

$$\frac{d}{dt}\int_{\Omega} \mathbf{Q}\, d\Omega + \int_{S} \mathbf{F}\cdot d\mathbf{S} = \int_{S} \mathbf{G}\cdot d\mathbf{S} \tag{1}$$

where $\mathbf{Q}$ shows the conservative variables, $\mathbf{F}$ is the convective flux and $\mathbf{G}$ is the diffusive flux all given by

$$\mathbf{Q} = \begin{pmatrix} \rho \\ \rho\mathbf{V} \\ \rho E \end{pmatrix}, \qquad \mathbf{F} = \begin{pmatrix} \rho\mathbf{V} \\ \rho\mathbf{V}\otimes\mathbf{V} + p\overline{\overline{\mathbf{I}}} \\ \rho E\mathbf{V} + p\mathbf{V} \end{pmatrix}, \qquad \mathbf{G} = \frac{M_\infty}{Re_\infty}\begin{pmatrix} 0 \\ \overline{\overline{\boldsymbol{\tau}}} \\ \overline{\overline{\boldsymbol{\tau}}}\mathbf{V} + \mathbf{q} \end{pmatrix}. \tag{2}$$

Here $\mathbf{V}$ is the velocity vector, $p$ is the pressure, $\overline{\overline{\boldsymbol{\tau}}}$ is the stress tensor and $\mathbf{q}$ is the heat flux vector.

### 2.1. Spalart-Allmaras Turbulence Model

One-equation turbulence model of Spalart-Allmaras [7] has been used in this study. The non-dimensional integral form of the model equation is

$$\frac{d}{dt}\int_{\Omega} \tilde{\nu}\, d\Omega + \int_{S} \mathbf{C}\cdot d\mathbf{S} = \int_{S} \mathbf{D}\cdot d\mathbf{S} + \int_{\Omega} S_1\, d\Omega + \int_{\Omega} S_2\, d\Omega \tag{3}$$

where $\tilde{\nu}$ is the turbulent working variable. Here $\mathbf{C}$ is the convective term, $\mathbf{D}$ is the conservative diffusive term, $S_1$ is the production term and $S_2$ is the non-conservative diffusive term with destruction term all given by

$$\mathbf{C} = \tilde{\nu}\,\mathbf{V}, \qquad \mathbf{D} = \frac{M_\infty}{Re_\infty}\frac{1}{\sigma}(\nu + \tilde{\nu})\nabla\tilde{\nu}, \tag{4}$$

$$S_1 = C_{b1}\,(1 - f_{t2})\,\{S\, f_{v3}(\chi) + \frac{M_\infty}{Re_\infty}\frac{\tilde{\nu}}{\kappa^2 d^2}\, f_{v2}(\chi)\}\,\tilde{\nu}, \tag{5}$$

$$S_2 = \frac{M_\infty}{Re_\infty}\{\frac{C_{b2}}{\sigma}\,\nabla\tilde{\nu}\cdot\nabla\tilde{\nu} - (C_{w1}\, f_w - \frac{C_{b1}}{\kappa^2}\, f_{t2})(\frac{\tilde{\nu}}{d})^2\}. \tag{6}$$

## 3. NUMERICAL METHOD

The proposed numerical method is based on the cell-centered finite volume formulation. The finite volume method is constructed on separate discretization in space and time, the methodology of which is explained in terms of spatial and temporal discretization.

### 3.1. Spatial Discretization

The physical domain is first subdivided into a number of elements or control volumes in order to construct a finite volume spatial discretization. In the present study, hybrid grids are employed. Hybrid grids are comprised of a combination of tetrahedral prismatic, pyramidal and hexahedral elements (Fig. 1).

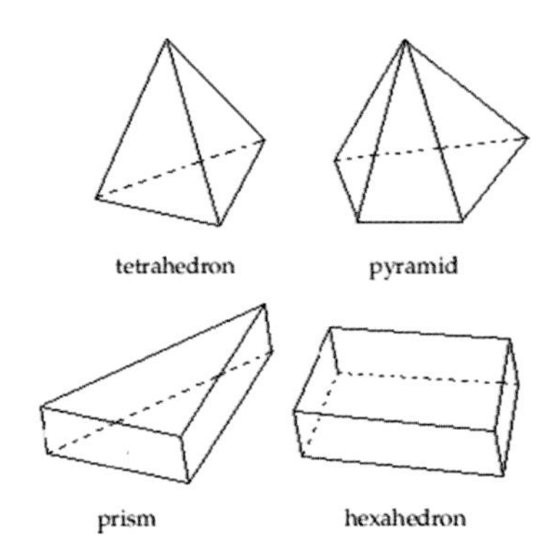


Figure 1. Elements of hybrid grids.

The finite volume spatial discretization can be written as

$$\frac{\partial \mathbf{Q}}{\partial t} = -\frac{1}{\Omega}\int_S \mathbf{F}\cdot d\mathbf{S} + \frac{1}{\Omega}\int_S \mathbf{G}\cdot d\mathbf{S} \tag{7}$$

where $\Omega$ is any control volume and the surface integral denotes the fluxes through control surface, $S$. Specifically, for the grid element volumes mentioned above, Eq. (7) can be approximated as

$$\frac{d\mathbf{Q}}{dt} = -\frac{1}{\Omega_e}\sum_{j=1}^{N} \mathbf{F}_{ci}^{j}\,\Delta S_j + \frac{1}{\Omega_e}\sum_{j=1}^{N} \mathbf{G}_{ci}^{j}\,\Delta S_j. \tag{8}$$

Here $\Omega_e$ is the volume of the grid element, $\Delta S_j$ is the surface area of the $j$th face, $\mathbf{F}_{ci}^{j}$ is the numerical flux through the centroid of $j$th face and $N$ is the number of faces. This approach is called as *cell-centered* approach, where the fluxes are evaluated through the boundaries of the grid elements and the conservative variables are updated at the cell centroids.

The first-order accurate inviscid fluxes are calculated at the faces of the cells using the approximate Riemann solver of Roe [8]. The viscous fluxes are approximated at the faces of the cells by first computing the velocity gradients at cell centroids, then averaging the values with neighbor cell centroids at the shared cell face. The velocity gradients at the cell centroids are calculated using the divergence theorem. For the turbulent flow solutions, non-dimensionalized integral form of Spalart-Allmaras model equation has been solved separately from the flow equations. For the spatial discretization of the conservative part of the equation, simple upwinding has been used by implementing the turbulent working variable as a passive scalar in the formulation. The face-averaged values of the conservative flow variables are calculated using the same formulation used for the inviscid fluxes. The diffusive part of the equation has been modeled by the same methodology used for the computation of viscous fluxes.

### 3.2. Temporal Discretization

The finite volume temporal discretization is performed based on Eq. (8) after the numerical fluxes are evaluated:

$$\frac{d\mathbf{Q}}{dt} = -\frac{1}{\Omega_e}\mathbf{R}_e. \tag{9}$$

Here the total numerical flux at the control volume is called as the residual. From Eq. (9), a basic explicit scheme can be derived

$$\Delta\mathbf{Q}^n = -\frac{\Delta t}{\Omega_e}\mathbf{R}_e^n \tag{10}$$

where

$$\Delta\mathbf{Q}^n = \mathbf{Q}^{n+1} - \mathbf{Q}^n.$$

In the present study, an explicit three-stage Runge-Kutta time-stepping scheme is employed for the flow equations while an explicit one-stage Euler scheme is utilized for the turbulent model equation.

## 4. PARALLEL PROCESSING

The parallel processing methodology employed in the present study is based on domain decomposition. The hybrid grid is partitioned using METIS software package. A graph file is necessary for the hybrid grids, which is actually the neighbor connectivity of the control volume cells. The partitioning of the graph is performed using *kmetis* program. During the partitioning, each cell is weighted by its number of faces so that each partition has about the same number of total faces to improve the load balancing in parallel computations.

*Parallel Virtual Machine* (PVM) message-passing library routines are employed in a *master-worker* algorithm. The *master* process performs all the input-output, starts up pvm, spawns worker processes and sends the initial data to the workers. The *worker* processes first receive the initial data, apply the inter-partition and the flow boundary conditions, and solve the flow field within the partition. The flow variables at the partition boundaries are exchanged among the neighboring partitions at each time step for the implementation of inter-partition boundary conditions.

## 5. RESULTS AND DISCUSSION

The method developed is first implemented for the parallel solution of laminar flow over a flat plate for which the analytical Blasius solution is available, then a turbulent flow over a flat plate is solved using the one-equation Spalart-Allmaras turbulence model. The results are compared to the analytical and theoretical data.

### 5.1. Case 1: Laminar Flow over a Flat Plate

As a first test case, laminar flow over a flat plate at zero angle of attack is selected. The freestream conditions and the computational mesh with partitions are given in Table 1 and Figure 2, respectively. The computational mesh consists of 11403 nodes and 12005 cells. The flat plate is placed between 0 and 1 and the boundary layer region is meshed with hexahedral cells while pyramidal and tetrahedral cells are used at the farfield.

Table 1
Freestream conditions for Case 1.

| Mach Number | Angle of Attack, *deg* | Reynolds Number |
|---|---|---|
| 0.3 | 0.0 | 35000 |

Figure 3 shows the Mach contours and velocity vectors inside the boundary layer on the flat plate at $y = 0$ plane. The comparisons of u-velocity and v-velocity profiles at three different y-planes for two different x-locations are given in Figures 4, 5, 6 and 7, respectively, along with the analytical Blasius solutions. The results compare quite well with the Blasius solutions at different y-planes and x-locations. Moreover, the use of hexahedral cells near the wall region leads to the better resolution of the boundary layers while the use of tetrahedral cells outside greatly reduces the overall number of cells.

### 5.2. Case 2: Turbulent Flow over a Flat Plate

A turbulent flow over a flat plate at zero angle of attack is selected as a second test case. The freestream conditions are given in Table 2. The computational mesh consists of 41407 nodes and 33807 cells. The flat plate is placed between 0 and 1 and the boundary layer region is meshed with hexahedral cells while pyramidal and tetrahedral cells are used at the farfield.

Table 2
Freestream conditions for Case 2.

| Mach Number | Angle of Attack, *deg* | Reynolds Number |
|---|---|---|
| 0.5 | 0.0 | 2000000 |

Figure 8 shows the Mach contours and velocity vectors inside the boundary layer on the flat plate at $y = 0$ plane. The turbulent boundary layer, which is well-resolved with hexahedral cells, is captured. The comparison of the calculated universal velocity distribution with the theoretical one is given in Figure 9. As seen, present calculations are in good agreement with those of theoretical values, which verify the accuracy of the present approach.

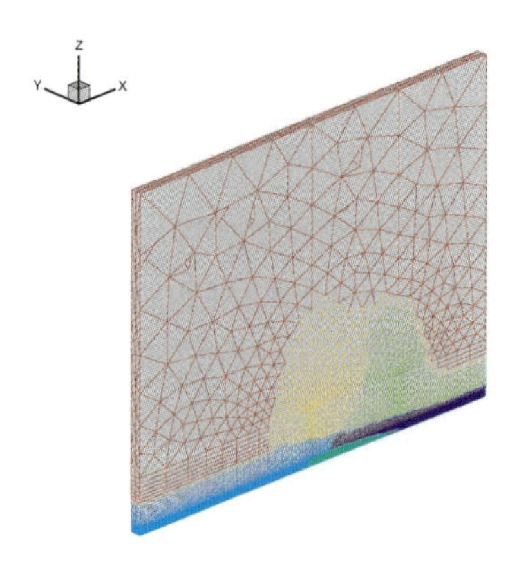


Figure 2. Computational mesh with partitions for Case 1.

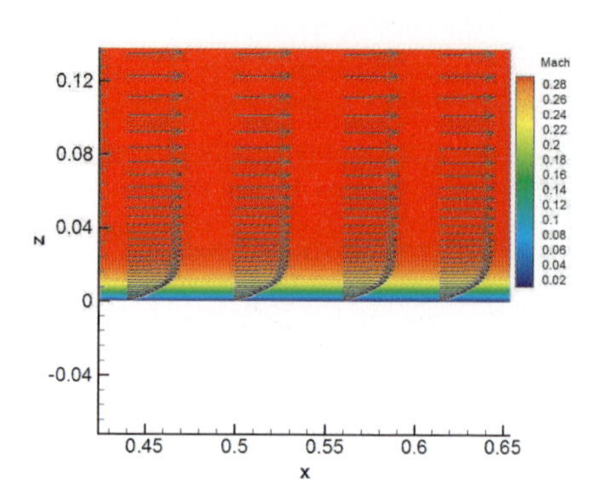


Figure 3. Mach number contours and velocity vectors at y=0 plane for Case 1 (Zoomed view).

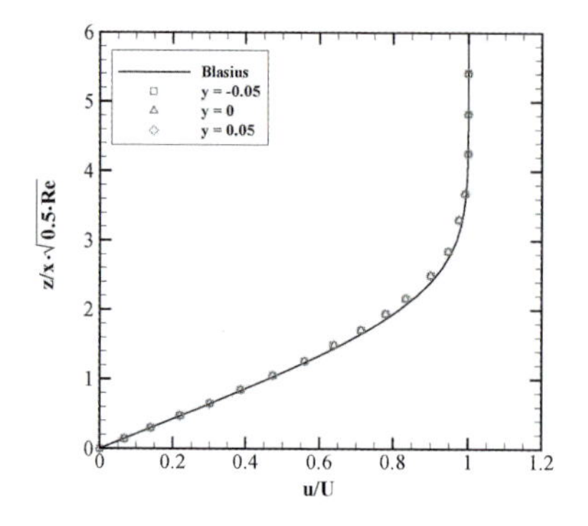


Figure 4. u-velocity profile in the boundary layer at x=0.44 for Case 1.

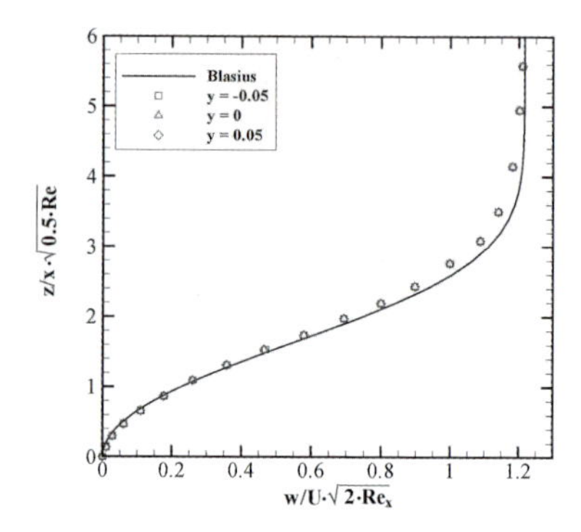


Figure 5. v-velocity profile in the boundary layer at x=0.44 for Case 1.

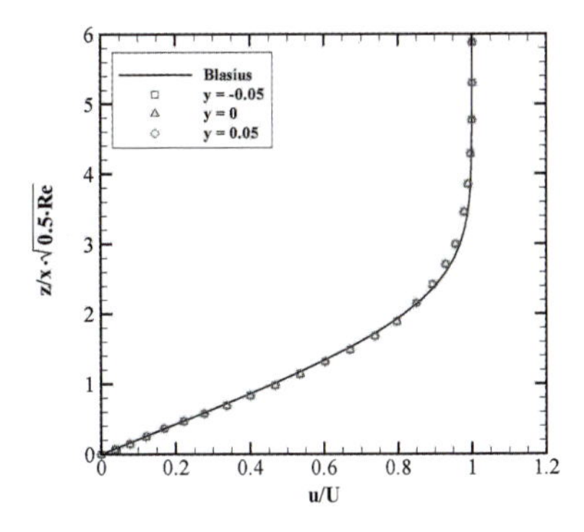


Figure 6. u-velocity profile in the boundary layer at x=0.66 for Case 1.

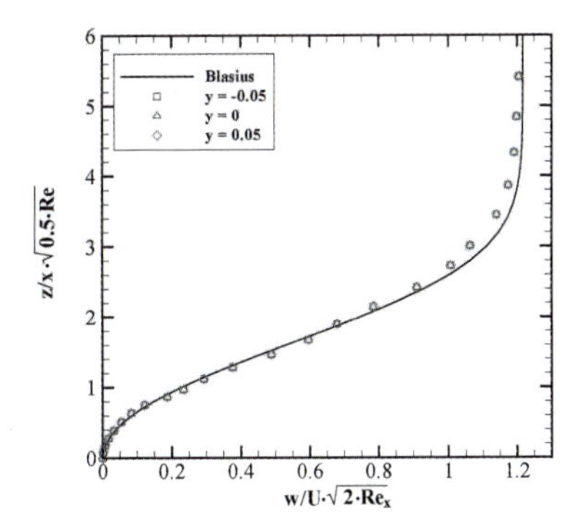


Figure 7. v-velocity profile in the boundary layer at x=0.66 for Case 1.

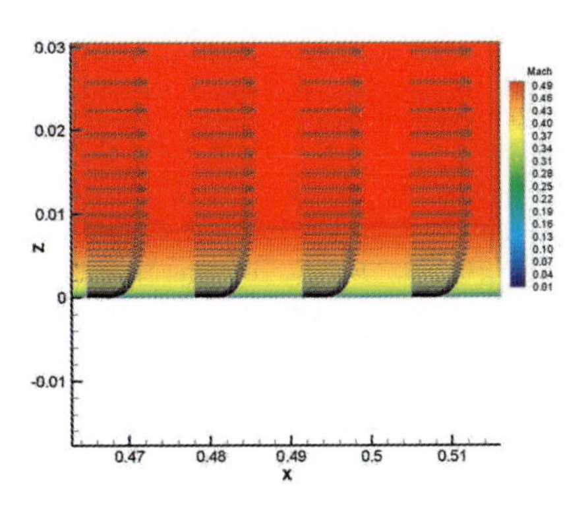


Figure 8. Mach number contours and velocity vectors at y=0 plane for Case 2 (Zoomed view).

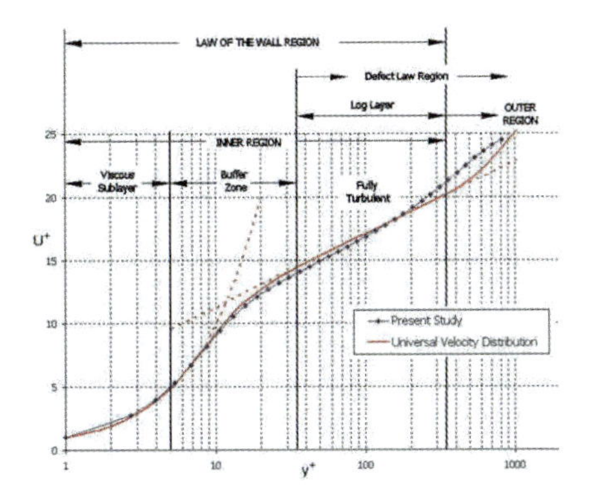


Figure 9. Universal velocity distribution for Case 2.

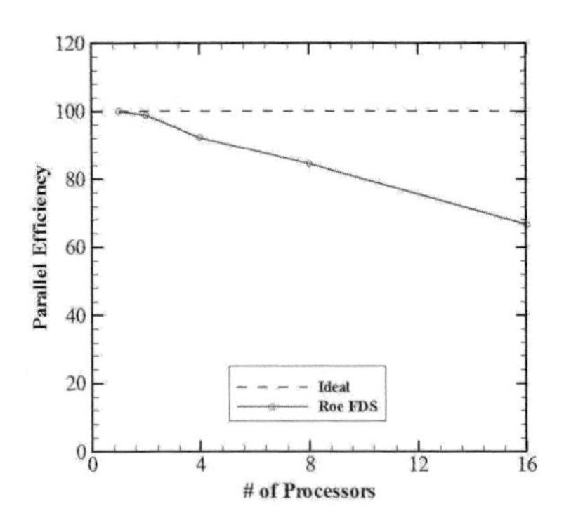


Figure 10. Parallel efficiency.

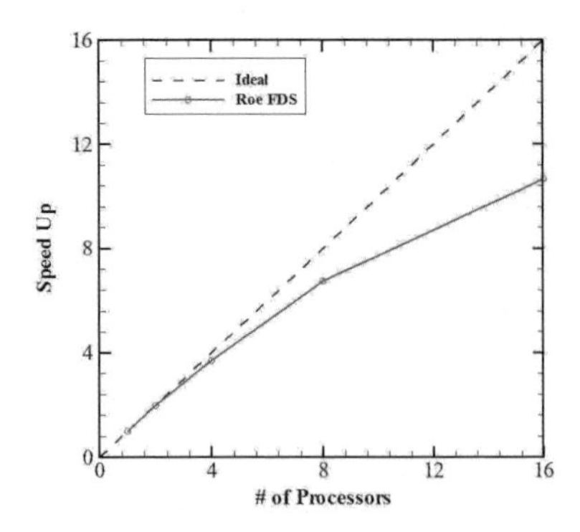


Figure 11. Computational speed-up.

The parallel computations are performed on *AMD Opteron* clusters running on Linux. Dual AMD Opteron processors, each of which has *dual core*, operate at 2.4Ghz with 1MB L2 cache and 2GB of memory for each. The parallel efficiency and the computational speed-up are given in Figures 10 and 11, respectively. It is seen that the parallel efficiency of the calculations decreases almost linearly with the number of processors. For 16 processors, a speed-up of more than 10 is achieved.

## 6. CONCLUDING REMARKS

A parallel finite-volume method for 3-D viscous flows on hybrid grids is presented. In order to accurately resolve the boundary layers in wall bounded viscous flow solutions, hexahedral and prismatic grid cells are employed in the boundary layer regions normal to the solid surfaces while the rest of the domain is discretized by tetrahedral and pyramidal cells. The laminar and turbulent flow cases considered validates the accuracy and robustness of the method presented. It is also shown that the computation time is significantly improved by performing computations in parallel.

## REFERENCES


1. T. Minyard and Y. Kallinderis, AIAA-95-0222 (1995).
2. A. Khawaja, Y. Kallinderis and V. Parthasarathy, AIAA-96-0026 (1996).
3. M. Soetrisno, S. T. Imlay, D. W. Roberts and D. E. Taflin, AIAA-97-0623 (1997).
4. A. Haselbacher and J. Blazek, AIAA-99-3363 (1999).
5. R. Chamberlain, D. Lianos, N. Wyman and F. Merritt, AIAA-2002-1023 (2002).
6. M. Mani, A. Cary and S.V. Ramakrishnan, AIAA-2004-0524 (2004).
7. S. R. Spalart, S. A. Allmaras, AIAA-1992-0439 (1992).
8. P. L. Roe, J. Comp. Phys., 43, (1981), 357.

# Implementation of parallel DSMC method to adiabatic piston problem

N. Sengil[a]* and F.O. Edis[a]

[a]Faculty of Aeronautics and Astronautics, Istanbul Technical University, Maslak 34469, Istanbul, Turkiye

## 1. INTRODUCTION

In the last 25 years a number of Micro Electro Mechanical System (MEMS) have been developed. These MEMS devices not only include the mechanical systems but also the fluids. Knowledge about fluid flows in this scale is not as mature as the mechanical properties of the MEMS [1]. As their dimensions are between 1 mm and 1 micron [2], gas flows related with the MEMS devices have higher Knudsen numbers (Kn) similar to high atmosphere flights. If Kn is higher than 0.1, instead of the classical continuum based Euler or Navier-Stokes (N-S) equations, deterministic or stochastic atomistic models should be used. This is due to the departure from local thermodynamic equilibrium with increasing Kn number [3]. Consequently, both the linear relation between shear stress and velocity gradient, and linear relation between heat conduction and temperature gradient are lost.

DSMC method pioneered by Bird [4] is a stochastic atomistic simulation method that can be used for high Kn number gas simulations. Our application of DSMC starts with the division of the computational domain into smaller cells. Linear dimensions of these cells are of the same order as the mean-free-path ($\lambda$) which is average distance between consecutive molecule collisions of the gas. A group of physical gas molecules are represented by one representative molecule. Every representative molecule carries position, velocity, cell number and if applicable, internal energy information on it. Representative molecule movements and collisions are separated from each other. As a first step, molecules move according to their velocities and initial conditions. Their velocities, positions and cell numbers are updated after each movement step. In the collision step, stochastic approach is used and molecule velocities are updated according to the collision model chosen. Next step is the calculation of the macroscopic gas flow properties for each cell from the microscopic molecule information. As an unsteady flow adiabatic piston problem requires the

*Part of this work has been performed under the Project HPC-EUROPA (RII3-CT-2003-506079), with the support of the European Community - Research Infrastructure Action under the FP6 "Structuring the European Research Area" Programme.

I.H. Tuncer et al., *Parallel Computational Fluid Dynamics 2007*,
DOI: 10.1007/978-3-540-92744-0_9, © Springer-Verlag Berlin Heidelberg 2009

use of ensemble averages of those macroscopic values.

Callen's adiabatic piston [5] can be described as a closed cylindrical system that permits no mass, momentum and energy transfer with it's surrounds. This closed system is divided two parts by a frictionless piston as in Figure 1.

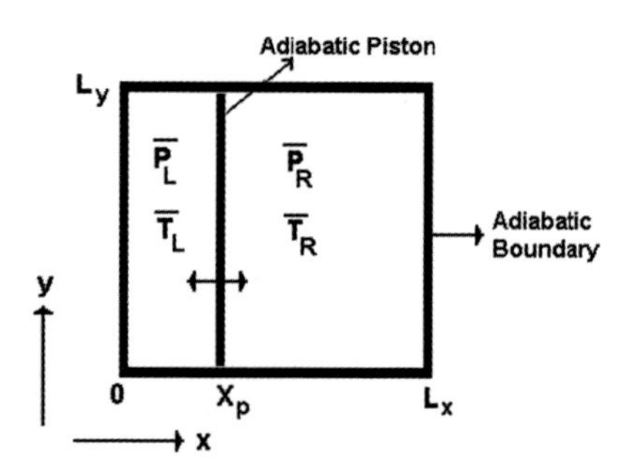


Figure 1. Adiabatic Piston

This piston can move from high pressure side to low pressure side. But it does not conduct heat between two compartments located both side of the piston. Because of the gas viscosity, the piston performs a dumped oscillatory motion. First, system reaches a "Mechanical Equilibrium" state. Although the pressures are equal in both side of pistons, the temperatures are not in this state. "Mechanical Equilibrium" temperatures which depends on the initial conditions and the kinetic properties of the gas cannot be predicted by elementary thermodynamic relations [5]. Finally, system reaches a "Thermodynamic Equilibrium" in a much longer time span. In this state not only pressures but also the temperatures and densities are equal. This interesting problem of the adiabatic piston has recently draw a lot of attention. Consequently many hydrodynamical models are developed by different researchers [6–8]. Hydrodynamic models are validated with the Molecular Dynamic simulations and similar results are reported [9].

In this paper a micro scale 2-D adiabatic piston with a length and width of 1 $\mu$m is studied using DSMC method. Piston movement is based on the pressure difference between both sides of the piston. The important aspects of DSMC simulation of adiabatic piston problem are the unsteady nature of the problem, existence of deforming domain, and high computational requirement. As an unsteady flow, adiabatic piston problem requires the use of ensemble averages. Deforming meshes and moving boundary condition treatment are also to be employed to solve the problem. To keep computation time low, up to 32 processors running parallel are utilized.

The final part includes piston position in "Mechanical Equilibrium" obtained with the DSMC solver and comparisons with reference study [9] along with the solution times of the parallel implementation.

## 2. HYDRODYNAMIC MODEL

If dissipative processes are not taken into account a simple hydrodynamic model can be build starting from the Newtonian equation of motion.

$$M_p \frac{d^2 X_p}{dt^2} = L_y \left( P_L - P_R \right) \tag{1}$$

Here piston mass and piston width are shown by $M_p$ and $L_y$ while $P_L$ and $P_R$ refer to pressures at the left and right sides of the piston. Finally $X_p$ and $t$ are the piston position and time. Being an adiabatic system and slow piston velocity the process is said to be isentropic. Consequently, isentropic process equation $PV^{\gamma} = P_0 V_0^{\gamma} = const = C_1$ can also be embedded into the equation of motion. $\gamma$ (specific heat ratio) is ratio of $c_p$ (constant-pressure) and $c_v$ (constant-volume) specific heats. "0" refers to initial values at $t = 0$. When gas fluid velocity in the piston direction is taken into account a "renormalized" piston mass ($\hat{M}$) is given by [9] as $\hat{M} = M_p + (m\,N/3)$ assuming a linear velocity profile in both side of the piston. $N$ is total number of molecules at both sides of the piston and $m$ is molecule mass. Using these equations and energy conservation principle, a general form of the piston equation of motion is presented by [9] as below.

$$\hat{M} \frac{d^2 X_p}{dt^2} = L_y \left( \bar{P}_L - \bar{P}_R \right) - L_y \left( \frac{\bar{\Gamma}_L}{X_p} + \frac{\bar{\Gamma}_R}{(L_x - X_p)} \right) \frac{dX_p}{dt} \tag{2}$$

$\bar{P}$ and $\bar{\Gamma}$ are the spatially averaged hydrostatic pressure P and viscosity coefficient $\Gamma = \zeta + \eta$ respectively. In a dilute gas, bulk viscosity coefficient is assumed that $\zeta \approx 0$. But shear viscosity coefficient $\eta$ is dependent on the gas temperature $\eta = \eta_0 \sqrt{T}$ while $\eta_0 = (5/16)(R/\pi)^{1/2}(m/d^2)$ [10]. Here $R$ is specific gas constant meanwhile $d$ is molecule diameter. Gas mass in both sides of the piston is taken equal $M_g = (N/2)m$. Taking into account these results and using the isentropic process relations, a new closed dimensionless hydrodynamic model can be derived from Equation ( 2) as below.

$$\frac{d^2 x_p}{d\tau^2} = \frac{x_0^{\gamma-1}}{x_p^{\gamma}} - \frac{(1-x_0)^{\gamma-1}}{(1-x_p)^{\gamma}} - \mu \left( \frac{x_0^{\gamma/2-0.5}}{x_p^{\gamma/2+0.5}} + \frac{(1-x_0)^{\gamma/2-0.5}}{(1-x_p)^{\gamma/2+0.5}} \right) \frac{dx_p}{d\tau} \tag{3}$$

In the Equation ( 3) some dimensionless variables and coefficients are defined. Those variables are $x_p = \frac{X_p}{L_x}$, $x_0 = \frac{X_p(0)}{L_x}$, and $\tau = \frac{1}{L_x}\sqrt{\frac{R\,T_0\,M_g}{\hat{M}}}\,t$. $\mu$ is a dimensionless viscosity coefficient given by $\mu = \frac{L_y \eta_0}{\sqrt{R\,\hat{M}\,M_g}}$. $L_x$ and $T_0$ are piston length and initial gas temperature.

## 3. DIRECT SIMULATION MONTE CARLO METHOD

### 3.1. Basics of the DSMC method

For gas molecule collisions and sampling purposes gas flow area is divided to computational cells like in the CFD meshes. The linear dimensions of the cell's are at the same order with the local mean-free-path of the gas flow [11]. To keep computation time low, representative molecules are employed instead of physical molecules. One representative molecule represents many real physical molecules. To keep the statistical errors in reasonable limits, each cell is provided with an average of 20 representative molecules. Additionally, body fitted quadrilateral structural meshes are used to keep computational algorithms simple and faster.

Two main data structures are used in this DSMC solver. First one is constructed to keep the cell's macroscopic data. Those data consist of macroscopic values like pressure, temperature, density, flow velocity and cell volume. Initially, representative molecules are placed into the cells. Their numbers and velocities are calculated from the cells' initially imposed densities and temperatures. For initial velocity components Maxwell velocity distribution is used. Molecules' initial positions are chosen as random in the cell. Second data structure is used to store microscopic values like individual molecule velocity components, molecule positions and molecule internal energies if applicable. Once representative molecules are initially placed into the gas flow zone, DSMC solver moves the representative molecules to their new positions using a time step ($\Delta t$) and a given molecule velocity. Time step is determined such that a molecule cannot cross many cells within each time step [12]. In this DSMC solver VHS (Variable Hard Sphere) collision model and NTC (No Time Counter) scheme [4] are used. Borgnakke-Larsen model is implemented for polyatomic molecule collisions. Finally, macro values are obtained from the molecule velocity and number density information. These data are also used for completing and updating the boundary conditions at every predetermined time period.

### 3.2. Moving boundaries, deforming cells

In the adiabatic piston problem, because of the piston movements gas flow boundary should be moved accordingly. Additionally cell shapes deform continuously. In order to handle those two additional requirements a new remeshing algorithm should be used. So in each DSMC time step piston moves to a new position which calculated from the piston equations of motion. Piston acceleration ($a_p$), piston velocity ($U_p$) and piston position ($X_p$) are given by,

$$a_p = \frac{(P_L - P_R)L_y}{M_p} \tag{4}$$

$$U_p^{'} = U_p + a_p\,\Delta t \tag{5}$$

$$X_p^{'} = X_p + U_p\,\Delta t + 1/2\,a_p\,(\Delta t)^2 \tag{6}$$

At the end of each time step ($\Delta t$), cell data structure is updated with new macro values and cell physical coordinates. In this step, representative molecules moved to their new positions. If molecules are found outside of the system boundaries , they are reflected from the walls. Adiabatic piston system is surrounded by walls that imperviable to energy. So molecules are reflected specularly without an energy exchange with the walls. That means molecule's parallel velocity components ($u_{p1}$,$u_{p2}$) to the wall don't change. But the normal component of velocity ($u_n$) changes its sign.

$$u_n^{'} = -u_n \qquad u_{p1}^{'} = u_{p1} \qquad u_{p2}^{'} = u_{p2} \tag{7}$$

Assuming molecule mass is much smaller ($m << M_p$) than the piston mass, reflection from the moving piston requires a special treatment of the molecule velocities.

$$u_n^{'} = 2U_p - u_n \qquad u_{p1}^{'} = u_{p1} \qquad u_{p2}^{'} = u_{p2} \tag{8}$$

In these equations $U_p$ refer to piston velocity. $u_{p1}$ is normal velocity component in x-direction, while $u_{p1}$ and $u_{p2}$ are parallel velocity components in y-direction and z-direction in cartesian coordinate system. After completing the reflections, all the molecules are placed into their new cells created for the new boundaries. Then collision and sampling steps are completed.

### 3.3. Unsteady gas flow, ensemble averages

Gas flow in the adiabatic piston problem is an unsteady gas flow until it reaches "Thermodynamic Equilibrium". So gas properties like pressure, temperature, density, and flow velocities change with time. In DSMC method those properties are calculated from molecule velocities, masses and number densities using ensemble averages obtained from the separate runs. Although increased number of the separate runs reduces standard deviation ($\sigma \propto 1/\sqrt{S}$), it costs more computation time. Here $S$ represents the number of separate runs. So the number of separate runs should be kept at an optimum number in terms of statistical error and computational time. In this study 32 separate runs were employed and obtained satisfactory results in agreement with the result taken from the literature.

### 3.4. Parallelization of the DSMC method

Because of the high computational cost of the DSMC method [14] certain measures are need to be taken to shorten the computation time. In this study two compartments of the adiabatic piston are appointed to different processors.

When the pressures in the left and right compartments are calculated, this data is shared between processors. Still each processor should run 32 separate runs to obtain the ensemble averages. Using more processors, it is possible to lessen the work load per processor. So each processor runs $32/N_P$ separate runs. $N_P$ is the number of processors allocated to each subsystem of the adiabatic piston. As a result of this parallelization, computation is realized in a shorter period of time.

## 4. TEST PROBLEM

To verify the unsteady 2-D parallel DSMC method, a test problem is run on both DSMC solver and a hydrodynamic model taken from the literature [9]. An adiabatic piston with a length ($L_x$) and width ($L_y$) of 1 $\mu m$ is chosen. Same amount of Nitrogen gas ($M_g$) is supplied to both compartment. In spite of the fact that initial temperatures ($T_L = 300\,K, T_R = 300\,K$) are same in both compartments, but pressures ($P_L = 100\,kPa, P_R = 25\,kPa$) are not. Piston's initial position is determined as ($X_p(0) = 0.2L_x$). The mass of the piston ($M_p$) is 0.44x10$^{-13}$ kg, while the gas mass ($M_g$) in one side of the piston is 2.2x10$^{-13}$ kg. Both compartment contain same amount of gas. Nitrogen is used in this problem. So mass of the molecule ($m$), specific gas constant ($R$) and molecule diameter ($d$) are taken 46.5x10$^{-27}$ kg, 296.95 $J\,kg^{-1}\,K^{-1}$, 4.17x10$^{-10}$ $m$ in order.

### 4.1. Solution with 2-D parallel DSMC method

DSMC solutions are calculated using 2, 4, 8, 16 and 32 processors working parallel. Although the runs are performed with different number of processors, results are derived from the same number of ensemble averages that is 32 for each case. Additionally, macroscopic gas properties are not only a function of time but also a function of space in DSMC method. To calculate the pressures on both side of the piston only the closest cells in y-direction are taken into account. Their averages in y-direction are used for the piston acceleration calculations. Results from DSMC solver are exhibited in Figure 2 for different number of processors.

### 4.2. Solution with hydrodynamic model

Taking the specific heat ratio ($\gamma = 1.4$) for the Nitrogen gas, and placed into the Equation ( 3), it is found that,

$$\ddot{x} = \left( \frac{x_0^{0.4}}{x_p^{1.4}} - \frac{(1-x_0)^{0.4}}{(1-x_p)^{1.4}} \right) - \mu \left( \frac{x_0^{0.2}}{x_p^{1.2}} - \frac{(1-x_0)^{0.2}}{(1-x_p)^{1.2}} \right) \dot{x} \tag{9}$$

Here $\dot{x} = \frac{dx_p}{d\tau}$ and $\ddot{x} = \frac{d^2x_p}{d\tau^2}$. If test problem values are applied, dimensionless time parameter ($\tau = \frac{1}{L_x}\sqrt{\frac{R\,T_0\,M_g}{\hat{M}}}\,t = (3.2\text{x}10^8)\,t$) and friction coefficient ($\mu = \frac{L_y \eta_0}{\sqrt{R\,\hat{M}\,M_g}} = 0.22$) are found. When this ODE (Ordinary Differential Equation) is solved with MATLAB, piston movement history in the x-direction is derived. Solution is presented in the Figure 3 together with DSMC solution.

Parallel efficiency figures and total solution times of the DSMC solver are summoned in the Table 1.

The formula used to calculate efficiency figure is given [15] as,

$$E = \frac{t_s}{t_p \times n} \tag{10}$$

Table 1
Parallel efficiency figures.

| Number of Processors | 2 | 4 | 8 | 16 | 32 |
|---|---|---|---|---|---|
| Solution Time (min) | 200 | 99 | 51 | 25 | 12.3 |
| Efficiency (percent) | 100 | 101 | 98 | 100 | 102 |

In this equation $t_s$ represents execution time using one processor while $t_p$ and $n$ represent execution time using multiprocessors and the number of processors.

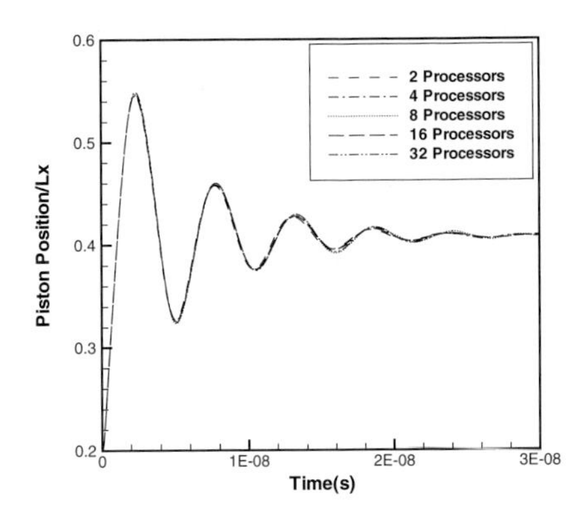


Figure 2. Piston movement history derived with DSMC method using different processors.

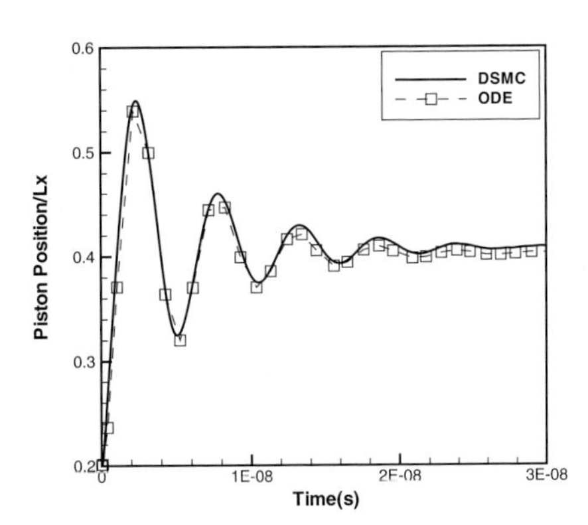


Figure 3. Piston movement history derived with DSMC method and ODE (Hydrodynamic model).

## 5. CONCLUSIONS

In this work DSMC method is implemented to adiabatic piston problem. As an unsteady gas flow, adiabatic piston problem requires moving boundaries, deforming meshes and ensemble averages to be employed in the DSMC solver. In order to shorten the calculation time, DSMC solver is run on the multi-processors working parallel. Results taken from the different number of processors working parallel are compared with each other.

Similar piston movement history in x-axis are monitored starting from initial conditions to the "Mechanical Equilibrium" state. To compare the results obtained from the DSMC method, same test problem is solved using a Hydrodynamical Model taken from literature. Both methods are very dissimilar in nature and used different approaches and assumptions to solve the problem. But they both predict the piston position in "Mechanical Equilibrium" very closely that is $X_p/L_x \cong 0.4$.

Momentum based piston movement will also be studied in future work in addition to pressure driven piston movement. Not only piston movement, but also piston temperatures will be investigated and compared with hydrodynamical models.

## REFERENCES

1. E. Piekos and K.S.Breuer, Transactions of the ASME, 118 (1996) 464.
2. M. Gad-el-Hak, Mech.Ind., 2001.2, (2001) 313.
3. G.E. Karniadakis and A. Beskok, Micro Flows Fundamentals and Simulation, Springer-Verlag, New York, 2002.
4. G.A. Bird, Molecular Gas Dynamics and the Direct Simulation of Gas Flows, Oxford University Press, Oxford, 1994.
5. H.B. Callen, Thermodynamics and an Introduction to Thermostatistics,Wiley, New York, 1985.
6. B. Crosignani, B.Di Porto and M. Segev, Am.J.Physics, 64 1996 610.
7. C. Gruber and L. Frachebourg, Physica 272 1999 392.
8. M.M. Mansour, C. Broeck and E. Kestemont,Europhysics Letters 69(4) 2005 510.
9. M.M. Mansour, A.L. Garcia and E. Baras, Physical Review E 73 2006 016121 .
10. C. Shen, Rarefied Gas Dynamics Fundamentals, Simulations and Micro Flows, Springer-Verlag, Berlin, 2005.
11. W.L. Liou and Y. Fang, Microfluid Mechanics Principles and Modeling, McGraw Hill, New York, 2006. .
12. G.J. Beau, Comput. Methods Appl.Mech.Engnr. 174 1999 319.
13. M.G. Kim, H.S. Kim H.S. and O.J. Kwon, Int.Journal for Num.Methods in Fluids 44 2004 1317.
14. S. Dietrich and I.D. Boyd, Journal of Comp.Physics 126 1996 328.
15. B. Wilkinson and M. Allen, Parallel Programming, Prentice-Hall, New Jersey, 1999.

Efficient parallel Algorithm for multiconstrained optimization of wing-body configurations

S. Peigin[a], B.Epstein[b]

[a] Israel Aircraft Industries, Israel

[b]The Academic College of Tel-Aviv Yaffo, Israel

## 1. INTRODUCTION

It is generally expected in aircraft industry that automatic or even semi-automatic optimizers may essentially shorten the process of aerodynamic design, especially at the preliminary design stage, which may cost above 100 million USD [1]. Alongside the reduction in the design costs, accurate optimizers may improve the quality of design making the project more competitive.

This is why, in the last years, CFD driven aerodynamic shape design has drawn increased interest [2]-[4]. Nevertheless, the impact of the existing CFD based shape optimization methods on practical aircraft design is rather limited, to the best of our knowledge.

With the purpose to overcome these difficulties, in this paper an accurate, robust and computationally efficient approach to the multipoint constrained optimization of wing-body aircraft configurations is presented. It is based on the complete aerodynamic model (Navier-Stokes computations), expands the globality of optimum search (by employing mixed deterministic/probabilistic Genetic Algorithms), and handles simultaneously a large number of constraints. The present algorithm essentially extends the capabilities of the method for aerodynamic design of 3D isolated transport-type wings, previously developed by the authors [5].

In industry, the success of the method should be assessed through practical optimization of realistic aerodynamic configurations. In this paper, we illustrate the capabilities of the method by applying it to optimization of transport-type aircraft configurations by the example of ARA M-100 NASA test case wing-body shape. It was demonstrated that the proposed method allows to ensure a low drag level in cruise regime, to accurately handle a required large number of constraints and to achieve a good off-design performance in take-off conditions and high Mach zone.

## 2. Statement of the Problem

The design goal is to optimize a given aircraft configuration for minimum drag at specified design conditions subject to constraints places upon the solution. This objective is accomplished through the solution of the multipoint constrained optimization problem which is driven by CFD estimations and genetic search. It is assumed that the drag coefficient $C_D$ can be employed as a reliable indicator of aerodynamic performance, and the objective function of the multipoint problem is chosen to be a weighted combination of drag coefficients at $K$ prescribed design points:

$$f^{objective} = \sum_{k=1}^{K} w_k C_D(k)$$

I.H. Tuncer et al., *Parallel Computational Fluid Dynamics 2007*,
DOI: 10.1007/978-3-540-92744-0_10, © Springer-Verlag Berlin Heidelberg 2009

The optimal solution is also required to satisfy the following uniform (in respect to $k$) constraints imposed in terms of sectional airfoils properties: relative thickness $(t/c)_i$, relative local thickness $(\Delta y/c)_{ij}$ at the given chord locations $(x/c)_{ij}$ (beam constraints), relative radius of leading edge $(R/c)_i$, trailing edge angle $\theta_i$.

The aerodynamic constraints placed upon the solution separately at each design point $k$ are prescribed lift coefficient $C_L^*(k)$ and maximum allowed pitching moment $C_M^*(k)$

$$C_L(k) = C_L^*(k), \quad C_M(k) \geq C_M^*(k), \quad 1 \geq k \geq K \tag{1}$$

As a gas-dynamic model for calculating $C_D$, $C_L$ and $C_M$ values, the full Navier-Stokes equations are used. Numerical solution of the full Navier-Stokes equations was provided by the multiblock code NES [6] which employs structured point-to-point matched grids. The code ensures high accuracy of the Navier-Stokes computations and high robustness for a wide range of flows and geometrical configurations already on relatively coarse meshes and thus allows to reduce dramatically the volume of CFD computations.

## 3. Optimization Algorithm

The optimal search is performed by Genetic Algorithms (GAs) in combination with a Reduced-Order Models (ROM) method. The main features of the method include a new strategy for efficient handling of nonlinear constraints in the framework of GAs, scanning of the optimization search space by a combination of full Navier-Stokes computations with the ROM method and multilevel parallelization of the whole computational framework.

### 3.1. Choice of the Search Space

In the case of wing-body optimization for minimum drag, the main object of optimization are lifting surfaces of the aircraft. Thus the whole surface of the aircraft may be divided into two parts. The first part of the surface (which contains all the points located on the aircraft fuselage) does not change in the course of optimization process and thus does not necessitate any global parameterization.

The object of optimization is the second part of the surface which contains all the points located on the wing and thus requires the parameterization. In this work it is assumed that:

1) The geometry is described by the absolute Cartesian coordinate system $(x, y, z)$, where the axes $x$, $y$ and $z$ are directed along the streamwise, normal to wing surface and span directions, respectively;

2) Wing planform is fixed;

3) Wing surface is generated by a linear interpolation (in the span direction) between sectional 2D airfoils;

4) The number of sectional airfoils $N_{ws}$ is fixed;

5) The wing-body boundary (the wing root airfoil) is not subject to change in the optimization process;

6) Shape of sectional airfoils is determined by Bezier Splines. In the absolute coordinate system, the location of the above profiles is defined by the corresponding span positions of the trailing edge on the wing planform, twist angles $\{\alpha_i^{tw}\}$ and dihedral values $\{\gamma_i^{dh}\}$ (relatively to the root section).

### 3.2. Strategy of Optimum Search

There are two main classes of search algorithms. The first class constructs a search path based on the analysis of local data taken from the vicinity of the current point. As an example of such an approach, the gradient based search techniques may be mentioned. The second class determines the optimum by analyzing the global information about the objective function coming from the whole search space.

As to the considered problem of aerodynamic shape optimization, the aerodynamic practice shows that a significant number of local extrema exist for the same design conditions. This means that the globality is a crucial property of the corresponding strategy of optimum search.

This why we turn to the second class of the optimization methods. Specifically, as a basic search algorithm, a variant of the floating-point GA [7] is used. The mating pool is formed through the use of tournament selection. This allows for an essential increase in the diversity of the parents. We employ the arithmetical crossover and the non-uniform real-coded mutation. To improve the convergence of the algorithm we also use the elitism principle.

The constraints handling in the framework of GAs search was based on the approach which was previosly proposed by the authors [8]. Contrary to the traditional approach, we employ search paths which pass through both feasible and infeasible points.

### 3.3. Improvement of Computational Efficiency

One of the main weaknesses of GAs lies in their low computational efficiency. That means that special efforts should be made in order to significantly improve the computational efficiency of the algorithm. This was accomplished through the application of the following two approaches.

In the framework of the first approach, we use Reduced-Order Models theory in the form of Local Approximation Method (LAM). The idea is to approximate the cost function using information from the specially constructed local data base.

The data base is obtained by solving the full Navier-Stokes equations in a discrete neighbourhood of a basic point positioned in the search space. Specifically a mixed linear-quadratic approximation is employed. One-dimensionally, the one-sided linear approximation is used in the case of monotonic behaviour of the approximated function, and the quadratic approximation is used otherwise.

Besides, in order to enhance the global character of the search, we perform iterations in such a way that in each iteration, the result of optimization serves as the initial point for the next iteration step (further referred to as optimization step).

In the framework of the second approach we use an embedded multilevel parallelization strategy [9] which includes: Level 1 - Parallelization of full Navier-Stokes solver; Level 2 - Parallel CFD scanning of the search space; Level 3 - Parallelization of the GAs optimization process; Level 4 - Parallel optimal search on multiple search domains.

The first parallelization level is based on the geometrical decomposition principle. All processors are divided into two groups: one master-processor and $N_s$ slave-processors. A large body of computational data demonstrated that the above approach for parallel implementation of the multiblock full Navier-Stokes solver, enables one to achieve high level of parallel efficiency while retaining high accuracy of calculations, and thus to significantly reduce the execution time for large-scale CFD computations.

The first level of parallelization is embedded with the second level, which performs parallel scanning of the search space and thus provides parallel CFD estimation of fitness function on multiple geometries.

The third level parallelizes the GAs optimization work unit. At this level of parallelization, all the processors are divided into one master-processor and $P_s$ slave-processors. The goal of the master-processor is to distribute the initial random populations among the slaves and to get back the results of optimal search ($P_s$ is the number of initial random populations).

The third level of parallelization is embedded with the fourth level, which performs parallel optimal search on multiple search domains. At this level of parallelization all the processors are divided into three groups: one main-processor, $P_m$ master-processors and $P_m \cdot P_s$ of slave-processors (where $P_m$ is equal to the number of domains).

Finally we can conclude that the multilevel parallelization approach allowed us to sustain a high level of parallel efficiency on massively parallel machines, and thus to dramatically improve the computational efficiency of the optimization algorithm.

## 4. Analysis of Results

The method was applied to the problem of multiconstrained optimization of transport-type aircraft configurations, by the example of ARA M-100 NASA wing-body shape [10]. The test case was chosen because the configuration is typical of transonic transport aircraft design, with extensive and reliable experimental and computational data available. On the computer cluster with about 200 IBM-Blade processors, it tooks about 16 wall clock hours.

The geometrical constraints (per section) were placed upon relative thickness, relative leading edge radius and trailing edge angle as well as relative local thickness at two fixed $x/c$ locations (beam constraints). An additional (aerodynamic) constraint was imposed on the value of pitching moment. The values of all the constraints were kept to the level of the original geometry. The corresponding optimal shapes are designated by $CaseARA\_1$ to $CaseARA\_9$. The number of wing sections subject to design was equal to 2 (crank and tip).

In the original ARA M-100 wing-body configuration, a strong shock is present already at $M = 0.77$ and $C_L = 0.5$, which leads to a rather high total drag value - $C_D = 276.0$ aerodynamic counts. One-point optimization allowed to significantly improve the aerodynamic characteristics of the configuration by essentially reducing the total drag. Specifically, with no constraint placed upon the pitching moment ($CaseARA\_1$), the total drag amounts to 227.7 counts. Though the maximum thickness of wing sections remained unchanged, the wing loading for the optimal geometry was strongly redistributed compared to the original one and the resulted pressure distribution is virtually shockless.

Together with the unconstrained pitching moment optimization, we performed two additional optimizations with different values of the constraint on pitching moment: $C_M^* = -0.134$ ($CaseARA\_2$) and $C_M^* = -0.096$ ($CaseARA\_3$). In the latter case, the value of $C_M^*$ was kept to the original level. The results were as follows: the drag penalty due to the imposition of this constraint was equal to about 3 counts for $CaseARA\_2$ and 9 counts for $CaseARA\_3$. The comparison of the corresponding pressure distribution shows that the inclusion of the above constraint resulted (as aerodynamically expected) in a higher loading of the leading edge area of the wing.

In Fig.1, the polar curves are compared with the original polar at $M = 0.77$ thus illustrating the influence of $C_M^*$ on the results of optimization. The local drag gains are retained in a wide range of lift coefficient values from $C_L = 0.2$ to above $C_L = 0.7$. Note that the penalty due to the imposition of the pitching moment constraint remains low in the whole above mentioned $C_L$ range.

In terms of shape modification, the influence of the pitching moment constraint on the wing tip optimal section can be assessed from Fig.2. Note, that a significant cusp (especially in the outboard part of the wing) is present already in the original shape. To achieve its goals, the optimization algorithm (though operating in the automatic mode) features well-known aerodynamic means such as trailing edge cusp and leading edge droop. In doing so, different combinations of these trends are used according to a specified value of $C_M^*$.

As the free-stream Mach value increases, the strength of shock wave over the initial configuration continues to escalate which is illustrated in Fig.3 where the pressure distribution is shown at $M = 0.80$, $C_L = 0.50$. Already at these flight conditions this results in an essential drag rise ($C_D = 296.4$ counts which is equivalent to additional 20 counts of wave drag compared to $M = 0.77$).

The unconstrained one-point optimization ($CaseARA\_4$) allowed to reduce the drag to 232.2 counts while the constrained pitching moment optimization ($CaseARA\_5$) yielded the value of 245.9 counts. The corresponding pressure distribution for $CaseARA\_4$ is depicted in Fig.4.

Similar to optimizations at $M = 0.77$, the analysis of drag polars at $M = 0.80$ (Fig.5) allows to conclude that the gain in drag due to the performed one-point optimization is preserved well beyond the vicinity of the design $C_L$ value. Compared to the optimization cases at $M = 0.77$, the penalty due to the imposition of pitching moment constraint, slightly increases from 9 to 13.7 counts.

Another important (both theoretically and practically) optimization aspect is the uniqueness of optimal solutions. Our experience shows that close values of total drag may correspond to significantly different aircraft geometries. The ill-posedness of the considered optimization problem can be illustrated by the example of $CaseARA\_4$, varying the value of constraint on the leading edge radius $R_L^*$. As it is seen from Fig.6, markedly different shapes corresponding to different $R_L^*$ values, yield very close drag values (232.2 counts vs. 233.2 counts, respectively).

At very high design free-stream Mach numbers ($M = 0.82$ and $M = 0.85$), the supersonic zone on the wing upper surface of the original wing-body becomes dominant, covering the major part of the wing. The corresponding total drag values reflect an additional increase in shock strength, amounting to 309.5 and 331.1 counts, respectively.

The unconstrained one-point optimization ($CaseARA\_6$) at $C_L = 0.5$, $M = 0.82$ reduced the drag to 237.7 counts while the constrained (with respect to $C_M$) optimization at the same design conditions ($CaseARA\_7$) yielded the value of 248.8 counts. This was achieved by a significant deterioration of the shock strength in the flow over the optimal shape.

The analysis of the results reveals that as the design Mach number increases, the drag reduction due to optimization also increases. For example, the constrained optimization at $M = 0.80$ ($CaseARA\_5$), yielded a reduction of 50.5 aerodynamic counts, while for the corresponding optimization at $M = 0.82$ ($CaseARA\_7$), the reduction was equal to 60.7 counts.

This trend extends to an even higher Mach value $M = 0.85$. In the unconstrained $CaseARA\_8$, the drag reduction amounted to 82.1 counts, while in the presence of the constraint on $C_M$, the corresponding drag reduction was equal to 76.2 counts. The corresponding pressure distribution are given in Fig.7 ($CaseARA\_8$).

The influence of $C_M^*$ on optimal geometry may be studied from Fig.8, where the wing tip forms are shown. As a whole, the shapes of $CaseARA\_9$ possess a lower curvature on the upper surface of the trailing edge in comparison with the optimal shapes of the unconstrained optimization of $CaseARA\_8$. The constrained shapes also feature a significant leading edge droop.

Drag polars at $M = 0.85$ are presented in Fig.9. It appeared that good off-design properties of the optimized configurations in terms of lift/drag relation, mentioned for a lower transonic Mach range, are preserved for the considered high Mach free-stream values. Mach drag rise curves at $C_L = 0.5$ for unconstrained and constrained pitching moment optimizations are respectively presented in Fig.10. In both cases, the optimization succeeded to shift the Mach drag divergence point at least up to the corresponding Mach design value. Additionally, the subsonic drag level is also decreased, especially in the first group of optimization cases.

The last 2 cases ($CaseARA\_14$-$CaseARA\_15$) are related to optimization in the presence of additional constraints imposed on the local wing thickness at specified points (beam constraints). The drag reduction at the main design point was respectively equal to 44.2 counts ($CaseARA\_14$) and 63.2 counts ($CaseARA\_15$).

The incorporation of beam constraints significantly influences the form of optimized shapes (see Fig.11, where the corresponding tip sections are depicted). Note, that the change in drag reduction caused by the beam constraints is small, even if the corresponding optimizations produced markedly different wing geometries. Moreover, as it is seen from Fig.12, this conclusion is valid not only pointwise, at the design $C_L$, but in a very wide range of off-design lift coefficients.

In terms of optimal shapes, the feasibility of resulting configurations is achieved through the imposition of natural aerodynamic constraints. The optimization automatically makes use of well-known aerodynamic trends (such as supercritical airfoils and leading edge droop) by targeting minimum drag which is especially non-trivial in the three-dimensional wing-body case. Already the initial ARA wing is supercritical and the optimizer also "discoveres" this effect in order to delay and reduce the transonic drag rise, due to both strong shock and shock-induced boundary layer separation.

Since this may lead to augmented aft loading and relatively high pitching moment, the supercritical effect is moderated by a local twist in the leading edge area (so called leading edge droop). In the optimizations with a constrained pitching moment, the effect of droop is automatically employed as a means of shifting the loading from the rear part of the wing to the leading edge area.

In principle, an experienced aerodynamic designer may successfully combine the above aerodynamic tools in a manual way, in order to achieve good aerodynamic performance. However, this requires a lot of imagination and intuition accompanied numerous computations and tests. In this sense, the automatic optimizer may essentially reduce the design costs and improve the quality of design.

## REFERENCES

1. Jameson, A., Martinelli, L., and Vassberg, J., "Using Computational Fluid Dynamics for Aerodynamics - a Critical Assessment", *Proceedings of ICAS 2002*, Paper ICAS 2002-1.10.1, Optimage Ltd., Edinburgh, Scotland, U.K., 2002.
2. Jameson, A., "Optimum Aerodynamic Design Using Control Theory", *CFD Review*, Wiley, New-York, 1995, pp.495-528.
3. Mohammadi, B., and Pironneau, O., *Applied Shape Optimization for Fluids*, Oxford, Oxford University Press, 2001.
4. Obayashi, S., Yamaguchi, Y. and Nakamura, T., Multiobjective Genetic Algorithm for Multidisciplinary Design of Transonic Wing Planform. *Journal of Aircraft*, Vol. 34, No. 5, 1997, pp.690–693.
5. Epstein, B., and Peigin, S., "Constrained Aerodynamic Optimization od Three-Dimensional Wings Driven by Navier-Stokes Computations", AIAA Journal, Vol. 43, No. 9, 2005, pp.1946-1957.
6. Epstein, B., Rubin, T., and Seror, S., "Accurate Multiblock Navier-Stokes Solver for Complex Aerodynamic Configurations. *AIAA Journal*, Vol. 41, No. 4, 2003, pp.582–594.
7. Michalewicz, Z., *Genetic Algorithms + Data Structures = Evolution Programs*, Springer Verlag, New-York, 1996.
8. Peigin, S. and Epstein, B., "Robust Handling of Non-linear Constraints for GA Optimization of Aerodynamic Shapes", *Int. J. Numer. Meth. Fluids*, Vol. 45, No. 8, 2004, pp.1339–1362.
9. Peigin, S., and Epstein, B., "Embedded Parallelization Approach for Optimization in Aerodynamic Design", *The Journal of Supercomputing*, Vol. 29, No. 3, 2004, pp.243–263.
10. Carr, M., Palister, K., "Pressure Distribution Measured on Research Wing M100 Mounted on an Axisymmetric Body", *AGARD, AR-138*, Addendum, 1984.

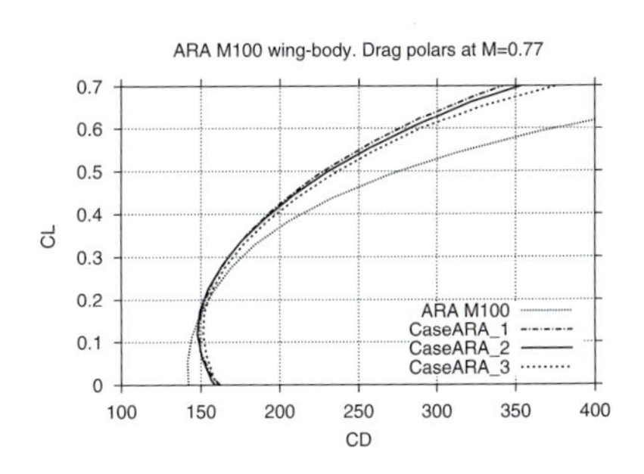


Figure 1. ARA M-100 wing-body. Drag polars at M=0.77. Optimized configurations vs. the original one.

Figure 4. Optimized ARA M-100 wing-body - *CaseARA_4*. Pressure distribution on the upper surface of the wing at M=0.80, $C_L = 0.50$.

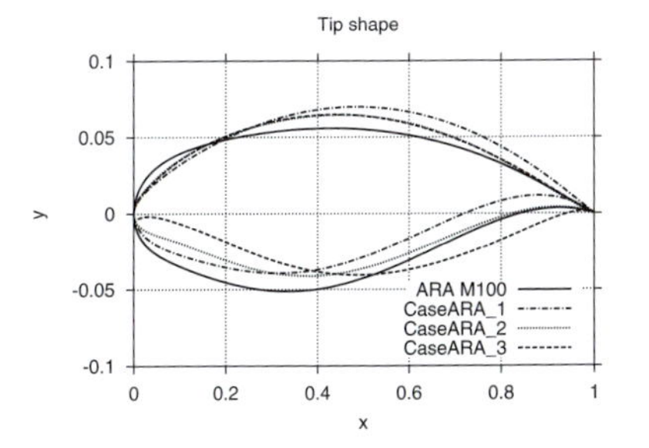


Figure 2. ARA M-100 wing-body. Optimized tip shapes vs. the original one.

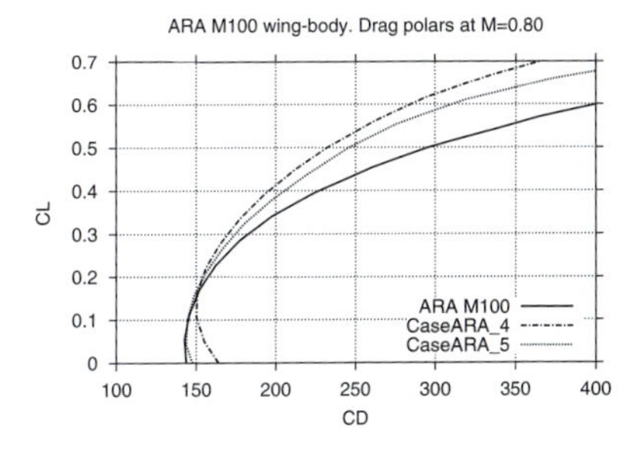


Figure 5. ARA M-100 wing-body. Drag polars at M=0.80. Optimized configurations vs. the original one.

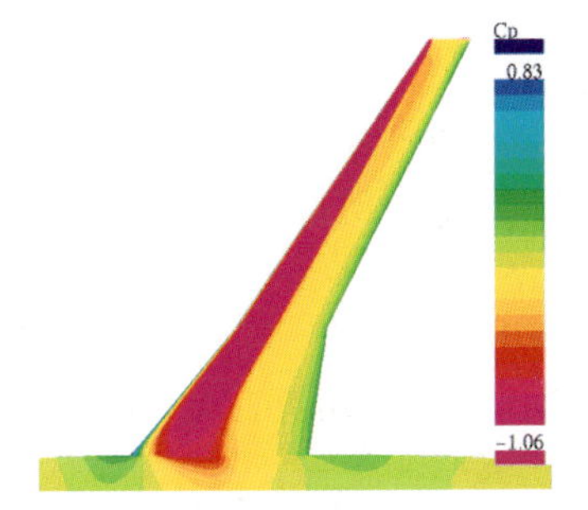


Figure 3. Original ARA M-100 wing-body configuration. Pressure distribution on the upper surface of the wing at M=0.80, $C_L = 0.50$.

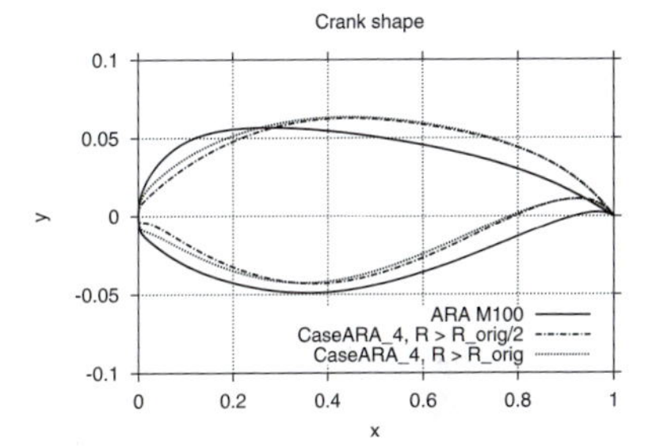


Figure 6. ARA M-100 wing-body. Optimized crank shapes vs. the original one.

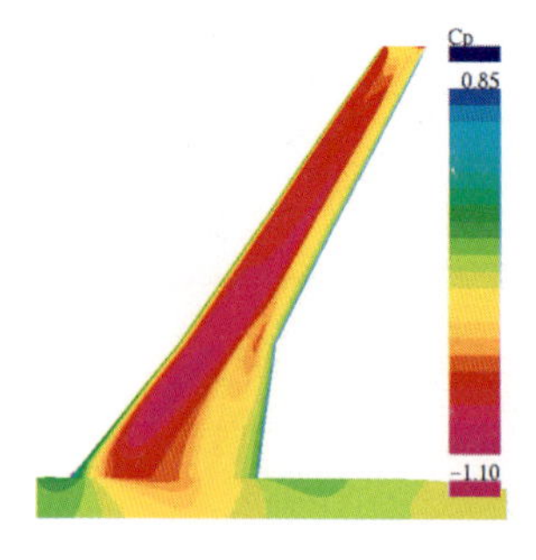


Figure 7. Optimized ARA M-100 wing-body - *CaseARA_8*. Pressure distribution on the upper surface of the wing at M=0.85, $C_L = 0.50$.

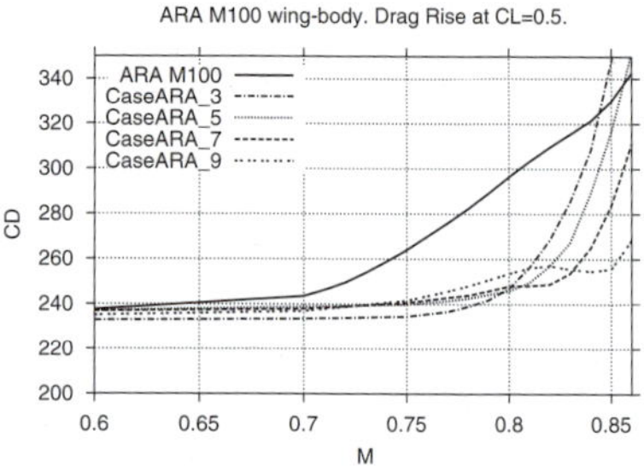

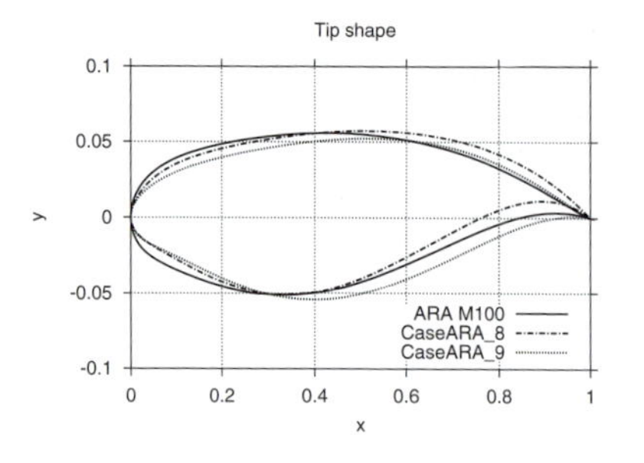


Figure 8. ARA M-100 wing-body. Optimized tip shapes vs. the original one.

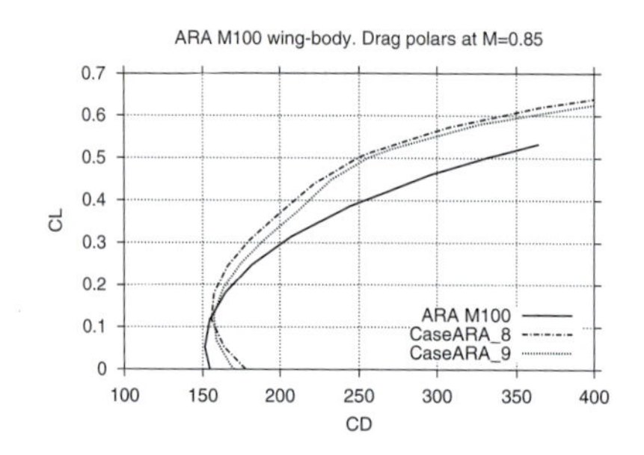


Figure 9. ARA M-100 wing-body. Drag polars at M=0.85. Optimized configurations vs. the original one.

Figure 10. ARA M-100 wing-body. Mach drag divergence at $C_L = 0.5$. Optimized configurations vs. the original one.

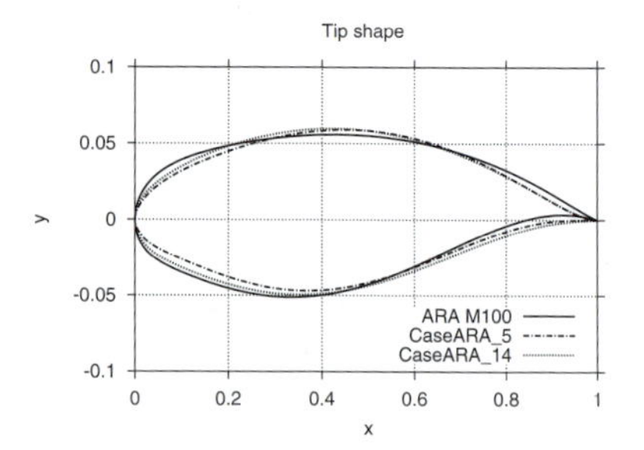


Figure 11. ARA M-100 wing-body. Optimized tip shapes vs. the original one.

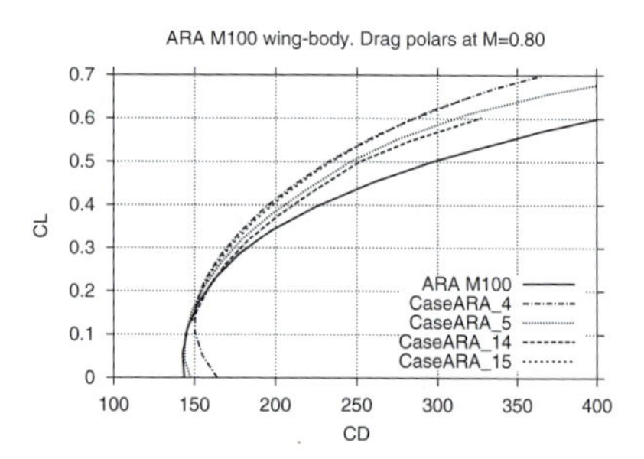


Figure 12. ARA M-100 wing-body. Drag polars at M=0.80. Optimized configurations vs. the original one.

# Parallel Three Dimensional Direct Simulation Monte Carlo for Simulating Micro Flows

**John Benzi and M. Damodaran**

*Division of Thermal and Fluids Engineering,*
*School of Mechanical and Aerospace Engineering,*
*Nanyang Technological University,*
*50 Nanyang Avenue, Singapore 639798*

The Direct Simulation Monte Carlo (DSMC) method is widely considered as the most accurate method for dilute gas flow simulation when they violate continuum hypothesis. However the application of DSMC is a computationally intensive method for realistic problems and hence motivates the usage of parallel computers to reduce the computational time. This work discusses the parallelization aspects of the three dimensional DSMC method for micro flow simulations used in computing the flow field in the narrow gap of a slider air bearing. The performance of the parallel DSMC code on several advanced computing platforms illustrating the portability and scalability of the method are compared and discussed. Superlinear speedup, implying a drastic reduction in the computational time has been obtained for all cases on all the parallel computing platforms. Load balancing aspects have also been considered and a dynamic load balancing scheme has also been implemented which results in further computational savings.

Keywords: Parallel DSMC; MPI; Micro Flows;Load Balancing

## 1. INTRODUCTION

For flows at the micro and nano scale the characteristic system length is comparable to the mean free path. In a modern hard disk drive (HDD) the head-disk interface (HDI) gap is just a few nanometers and is much less than the mean free path of air. Hence in such regions, the Knudsen number tends to be very high and the nature of the airflow deviates significantly from that in a continuum regime. The Knudsen number, $Kn$ is defined as $Kn = \lambda/L$ , where $\lambda$ is the mean free path and $L$ is the characteristic length involved. The global airflow characteristics and particle trajectories in large and small form factor HDD enclosures containing the disk and the slider have been investigated by continuum flow models based on the incompressible Navier-Stokes equations in a separate study by Ali et al [1] However the continuum flow model fails to predict the

I.H. Tuncer et al., *Parallel Computational Fluid Dynamics 2007*,
DOI: 10.1007/978-3-540-92744-0_11, 

flow in the HDI region accurately, as the continuum hypothesis breaks down in this region. With miniaturization of the HDD enclosures, the HDI gap distance is expected to decrease even further as the form factor for the next generation HDDs reduces, motivated primarily by the fact that the smaller the flying height, the greater the recording capacity. Hence there is a need to accurately predict airflow characteristics in the HDI region by an appropriate non-continuum model.

The Direct Simulation Monte Carlo Method (DSMC) has been used extensively for modeling flows in rarified gas dynamics and in micro geometries for modeling high Knudsen number flows. Three dimensional DSMC simulations have been performed for the slider air bearing problem by Huang and Bogy [2]. For reliable simulations, the DSMC method requires that the cell size be less than a mean-free-path and the time step to be less than the mean free time, besides having an average of at least 20–30 simulated particles per cell. This makes DSMC method computationally intensive in terms of both computer memory and time, especially so for the three dimensional case which is best done with a parallel DSMC method. Implementation of the parallel DSMC method and load balancing aspects has been reported in literature especially for high speed rarefied gas dynamics such as Dietrich and Boyd [3], and LeBeau [4]. Parallel DSMC aspects for microflow simulations for the two dimensional case have also been reported such as Fang and Liou [5], and Aktas et al. [6]. In the present work a parallel three dimensional DSMC method is developed for microflow simulations in the narrow gap of a slider air bearing and parallelization aspects are discussed. A detailed performance study is done on various computational platforms illustrating the performance of this parallel DSMC code.

## 2. DIRECT SIMULATION MONTE CARLO METHOD

DSMC is a stochastic method introduced by Bird [7], in which a small number of computational particles is used to accurately capture the physics of a dilute gas governed by the Boltzmann equation. This method models the macroscopic behavior of a fluid based on the discrete molecular character of gases by simulating the motion, interaction and collision of representative particles with in an array of computational cells. The state of the system is given by the positions and velocities of these particles. The particles in the simulation are considered to be hard spheres. The particle-wall interactions are considered to be inelastic and follow the diffuse reflection model.

The flow velocities in typical micro geometries generally being much less than the speed of sound, the *stream* and *vacuum* boundary conditions typically employed in high speed rarefied DSMC calculations [8] are not appropriate. Instead, implicit treatments must be incorporated to specify the inflow and outflow boundary conditions such that the velocities and other flow variables are updated at each iteration as the simulation progresses. At the inflow and outflow of the slider bearing, ambient pressure conditions are imposed by injecting particles from the inflow and outflow boundary, described by the Maxwellian distribution function. The boundary conditions at both the inflow and outflow are fixed by implicit boundary treatment based on characteristic theory based equations [8, 9]. Particles keep accumulating in the computational domain with each time step and once the total number of particles in the domain becomes steady, sampling is done to derive the macroscopic properties of the flow.

## 3. PARALLEL IMPLEMENTATION OF DSMC

For the present study, the DSMC method is implemented efficiently in C++ program using various features like *classes* and dynamic allocation of memory. Parallelization is done with the aid of the Message Passing Interface (MPI) [10] library routines. The computational domain is partitioned into several subdomains and assigned to different processors. All the processors perform the computational steps for the list of particles in its own subdomain as in the case of a serial DSMC program. In a parallel DSMC simulation, data communication between the processors occurs at any time step when a particle crosses its subdomain and enters a new subdomain, upon which all the information pertaining to that particle needs to be communicated. After completing all the time steps, each processor does sampling on its own subdomain and the results are sent to the host processor, which assembles all the separate results into one single output file. The salient features of the parallelization process based on this approach are described below.

### 3.1. DOMAIN DECOMPOSITION APPROACH

The domain of interest which is the narrow gap of the slider air bearing in which the flow is to be modeled is partitioned equally along the $x$ direction as shown in Figure 1 and each subdomain is mapped onto different processors. Uniform rectangular cells are considered for the collision cells. The use of Cartesian cells avoid any overlap of cells between different processors and makes it easier to sort particles, implement collisions. Since the slider wall is inclined at a certain pitch angle, it cuts through some of the cells within the computational domain resulting in partial cells. The partial cells are considered by discarding the volume of the cell outside the slider and accounting only for the part of volume of the cell within in the simulation domain.

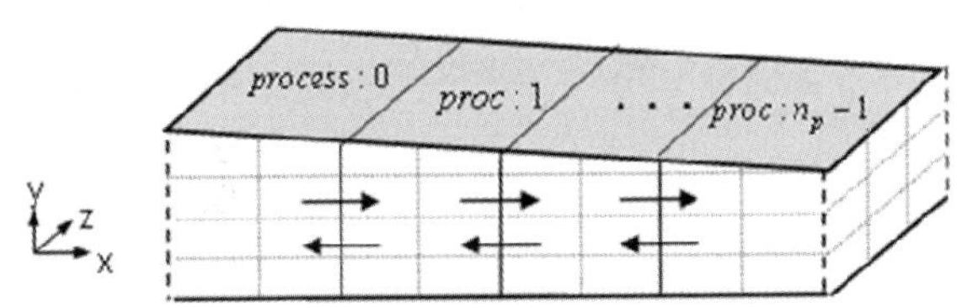


Figure 1. Schematic representation of the partitioning of the computional domain

### 3.2. DATA COMMUNICATION

For an efficient parallel simulation the inter-processor communication time must be reduced. In a DSMC simulation, the communication is associated with the motion of molecules between adjacent subdomains at any time step. If at any time step the particle leaves its subdomain and enters another subdomain, then the particle must be deleted from the current subdomain and added to the existing list of particles in the new subdomain. All the information pertaining to that particle such as its spatial coordinates and velocity components must be communicated to the new host processor. To reduce the communication time associated by using separate *MPI_Send* commands and corresponding *MPI_Recv* commands for passing each of these information, MPI provides a set of packing and unpacking subroutines using *MPI_Pack* and *MPI_Unpack*, respectively which allows the non-contiguous pieces of information related to a particle to be packed into a contiguous section of memory and transfered as a single message rather than sending separate messages for each information.

### 3.3. IMPLEMENTATION ON PARALLEL PLATFORMS

To assess the portability of the parallel DSMC code several platforms have been considered i.e. the *Alpha Server Supercomputer* (SC45) having 44 nodes each with four 1GHz processors having 1 GB memory and 8 MB cache, the *Sun Fire 15000* computing platform which comprises of 32 CPUs with 64 GB onboard memory, each utilizing 900MHz processors and 8 MB cache, the *SGI Onyx 3800* system which comprises of 32 CPUs with 32 GB onboard memory, each having processors of 400MHz and 8 MB cache and the *SGI Origin 2000* system has 32 CPUs with 16 GB onboard memory, each utilizing processors of 250MHz and 4 MB cache.

## 4. RESULTS AND DISCUSSION

### 4.1. PARALLEL PERFORMANCE

To study the performance aspects of the parallel DSMC code, all performance studies are undertaken for a slider bearing geometry of length 3 microns, width 3.3 microns and flying height 25nm consisting of $64 \times 4 \times 50$ cells. To study the scalability effects, three test cases are considered on the same geometry but with different number of particles viz. 200,000, 400,000 and 800,000 particles. Speedup, $S = T_s / T_p$, where $T_s$ and $T_p$ are the serial and parallel simulation times respectively and Efficiency $E = S/p$, where $p$ is the number of processors are considered as performance metrics.

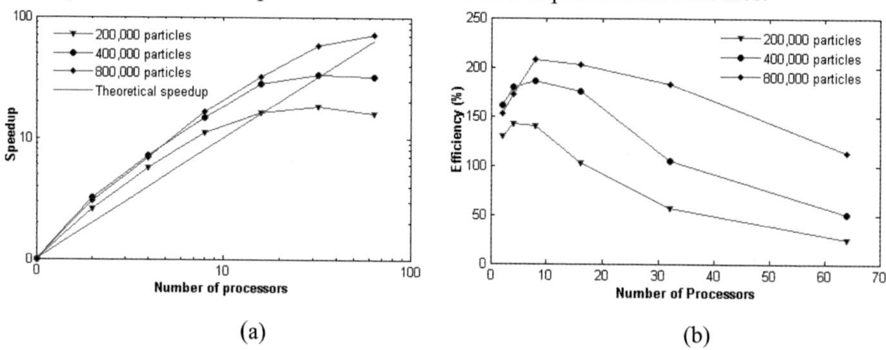


(a) (b)

Figure 2. Performance curves on Alphaserver -SC 45 (a) Speedup (b) Efficiency

Figure 2 shows the parallel performance results for all the three test cases on the *AlphaServer Supercomputer* (SC45).

Super linear speedups are obtained for all the three test cases, wherein the speedup is greater than the ideal speed up. Correspondingly an efficiency greater than 100 % is obtained. This reduces the computational time involved significantly. The superlinear speedup obtained can be attributed to cache effects. With multiple processors the subproblem fits into caches, and results in significant reduction in memory access time. This causes extra speedup apart from that obtained due to reduction in computation time. The best performance is observed for the case with 800,000 particles wherein superlinear speed is observed consistently till 64 processors. For the case with 400,000 particles superlinear speedup is observed up to 32 processors, while for 200,000 particles superlinear speedup is observed only up to 8 processors. With further increase in the number of processsors, the speedup curve drops and subsequently falls below the theoretical speedup as the problem becomes communication bound. This shows that for larger three dimensional DSMC problems, an increased reduction in CPU time can be obtained as the parallel program scales up well up to a higher number of processors.

The performance level of the parallel program on different computers have been studied to check its portability. The parallel DSMC code has been successfully run on other computational platforms like *Sun Fire 15000, SGI Onyx 3800* and *SGI origin 2000*. The test case with 400,000 particles is considered for the purpose. Figure 3 compares the parallel performance curves obtained on these platforms, namely *SunFire* and *SGI systems*, with those obtained for *Alphaserver*. The Superlinear speedups are obtained on all the computers considered. Sustained speedups are obtained for up to 32 processors for the *Alphaserver* and *Sun Fire systems* resulting in significant reduction in computational time. The nodes of both these machines have a significantly higher processor speed and a much higher node memory. The *SGI systems* show superlinear speedup for up to 16 processors after which the speedup curve drops.

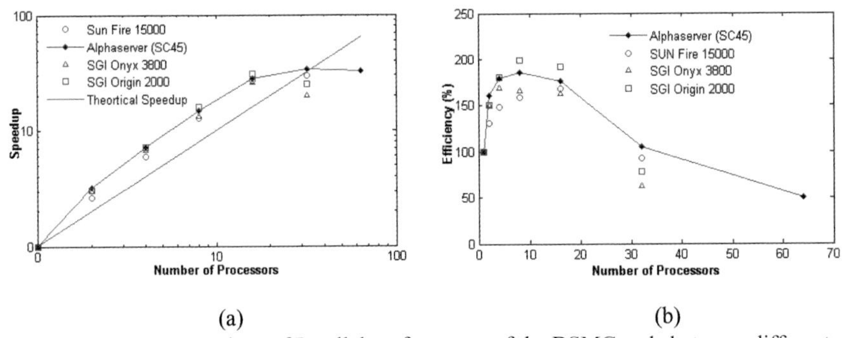


(a) (b)

Figure 3. Comparison of Parallel performance of the DSMC code between different computational platforms (a) Speedup (b) Efficiency

The comparison of the parallel performance obtained on the different computers is summarized in Table 1 for easy reference.

Table 1:Comparison of Speedup (S) and Efficiency (E) on different computing platforms

| | Alphaserver | | Sun Fire 15000 | | SGI Onyx 3800 | | SGI Origin 2000 | |
|---|---|---|---|---|---|---|---|---|
| nproc | S | E | S | E | S | E | S | E |
| 2 | 3.21 | 160.9 | 2.61 | 130.4 | 3.0 | 150.0 | 3.01 | 150.4 |
| 4 | 7.17 | 179.2 | 5.94 | 148.4 | 6.75 | 168.7 | 7.21 | 180.3 |
| 8 | 14.8 | 185.9 | 12.6 | 158.1 | 13.3 | 166.1 | 15.9 | 198.8 |
| 16 | 28.1 | 175.7 | 26.8 | 167.7 | 25.9 | 162.4 | 30.6 | 191.7 |
| 32 | 33.6 | 104.9 | 29.5 | 92.46 | 19.9 | 62.17 | 24.8 | 77.68 |
| 64 | 32.2 | 50.30 | | | | | | |

## 4.2. LOAD BALANCING CONSIDERATIONS

In a DSMC simulation particles keep accumulating in the computational domain with each time step and once steady state is attained, the total number of particles remain about a constant value. Hence there is no single domain partitioning technique that ensures an optimum load balancing through out the simulation. Rather it depends on various factors like the geometry, flow gradients and number of particles in each processor. In the case of a slider bearing the region towards the trailing edge of the slider corresponds to a smaller volume. Therefore the domain decomposition scheme in which the domain is equally divided along the *x* direction (*no load balancing*) as shown in Figure 1 results in an unequal distribution in the number of cells and particles between the processors. Therefore a good load balancing may not be achieved especially at high pitch angles.

A dynamic load balancing is done such that each domain has approximately the same number of cells during the initial phase of simulation and once the system begins to attain a steady state the domain is remapped such that each processor has equal number of particles. Equal number of cells in each domain initially ensures that each processor is loaded with about the same number of particles. To illustrate the computational savings obtained, an example is shown in which the comparison of performance based on elapsed time taken at an interval of every 100 iterations for domain decomposition schemes based on a) no load balancing b) equal number of cells and c) dynamic load balancing using eight processors is shown in Figure 4.

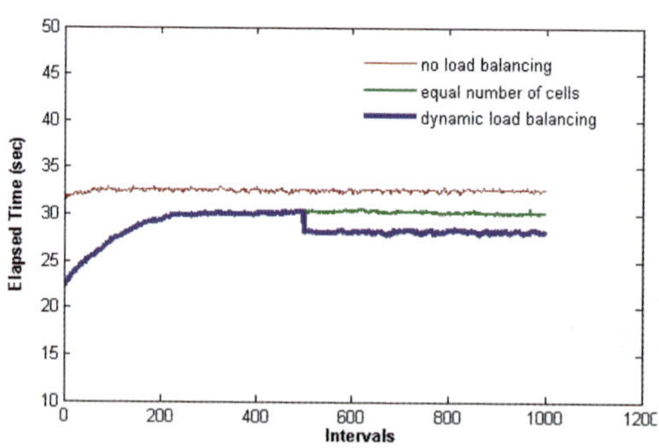


Figure 4. Comparison of elapsed time between different domain decomposition schemes

From Figure 4, it can be seen that the domain decomposition based on approximately equal number of cells performs better initially compared to that with no load balancing, but further down as particle accumulates the performance degrades. The best performance is observed for the dynamic load balancing case as a reasonably good load balance is attained through out the simulation, with computational savings of about 12 percent reduction in time obtained for the entire simulation.

### 4.3. SIMULATED FLOW FIELDS

One of the advantages with a parallel DSMC code is that a larger number of particles can be employed to improve the accuracy of the computed solutions. A study was done to gain insight into the number of particles required per cell for the same number of iterations to obtain accurate solutions for a three dimensional DSMC micro flow simulation. Four cases were considered with different number of particles viz. an average of about 13, 25, 38 and 50 particles per cell. The variation in the solutions of pressure and flow velocity with number of particles on a plane passing through the midsection of the slider is shown in Figure 5. It can be observed from the figure that for the case with an average of 13 particles the solutions are not accurate. For the case with 25 particles per cell, the solution converges and with further increase in the number of particles, the role is just in reducing the statistical fluctuations further.

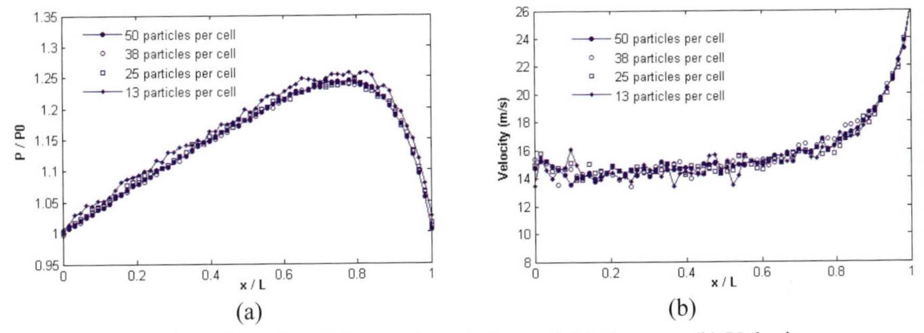


Figure 5. Effect of number of particles on the solutions of (a) Pressure (b) Velocity

The computed pressure field contours in the gap between a slider air bearing of length 3 microns, width 3.3 microns, flying height 25nm and a disk velocity of 25 m/s are shown in Figure 6(a)-(b).

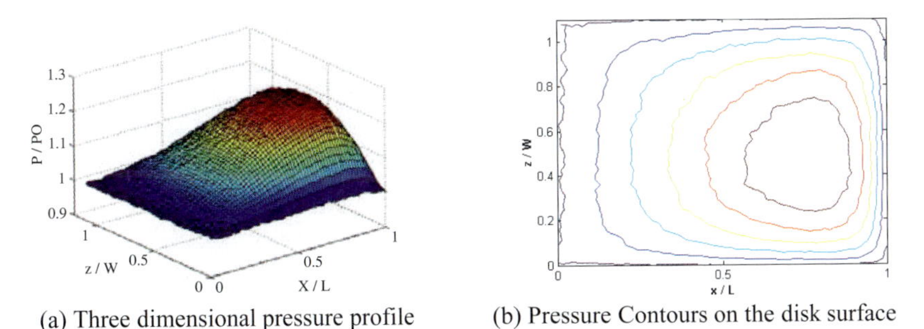


(a) Three dimensional pressure profile (b) Pressure Contours on the disk surface

Figure 6. Computed flowfields in the Head-Disk Interface gap

## 5. CONCLUSION

Typical three dimensional DSMC simulations require massive computing resources especially in terms of the computer memory resources. In this work the parallel performance of a three dimensional DSMC model has been investigated on different computational platforms illustrating the scalability and portability of the parallel program. Superlinear speedup has been observed on all the computers considered. This shows that three dimensional DSMC method can easily be parallelized in an efficient way, especially for cache-based computing machines, resulting in significant reductions in the total computing time. Load balancing aspects have also been studied and a dynamic load balance scheme is implemented to further improve the parallel efficiency. Further improvements to the parallel code to increase the parallel efficiency further are being explored curently.

## Acknowledgment

This work is part of a joint research project funded jointly by Seagate Technology International and Nanyang Technological University.

### REFERENCES

1. Ali S, Damodaran M and Quock Y Ng, Computational Models for Predicting Airflow Induced Particle Contamination in Hard Disk Drive Enclosures, AIAA Paper No. AIAA-2005-5342, 2005
2. Huang W and Bogy D B, Three-dimensional Direct Simulation Monte Carlo method for slider air bearings, Phys. Fluids, Vol 9, pp 1764–1769, 1997
3. Dietrich S and Boyd I D, Scalar and Parallel optimized implementation of the Direct Simulation Monte Carlo method, JCP, Vol 126, pp 328-342, 1996
4. LeBeau G J, A parallel implementation of the direct simulation Monte Carlo method, Comput. Methods Appl. Mech. Engrg., Vol 174, pp 319-337, 1999
5. Fang W and Liou W, Microfluid Flow Computations Using a Parallel DSMC Code, AIAA Paper No. AIAA-2002-1057, 2002
6. Aktas O, Aluru N R and Ravaioli U, Application of a parallel DSMC technique to predict flow characteristics in microfluidic filters, IEEE JMEMS, Vol 10, No 4, 2001
7. Bird G A , Molecular Gas Dynamics and the Direct Simulation of Gas Flows, Oxford : Clarendon, 1994
8. Nance R P, Hash D B and Hassan H A, Role of Boundary Conditions in Monte Carlo Simulation of MEMS Devices, J. Thermophys.Heat Transfer 12, 447–49, 1998
9. Wang M and Li Z, Simulations for gas flows in microgeometries using the direct simulation Monte Carlo method, Int J Heat and Fluid Flow, 25, 975-85, 2004
10. Gropp William , Using MPI : portable parallel programming with the message-passing interface, MIT press, 1994

# A Study on the Prediction of the Aerodynamic Characteristics of an Orbital Block of a Launch Vehicle in the Rarefied Flow Regime Using the DSMC Approach and the Parallel Computation

**Younghoon Kim,[a] YoungIn Choi,[a] Honam Ok,[a] Insun Kim[a]**

*[a]Thermal & Aerodynamics Dept., Korea Aerospace Research Institute, 45 Eoeun-Dong, Youseong-Gu, Daejeon, 305-333, Korea*

Keywords: Aerodynamic Characteristics; Orbital Block; Launch Vehicle; DSMC; Parallel Processing

## 1. Introduction

The major mission object of a launch vehicle is to settle precisely a payload on the designed orbit. For the design process of most launch vehicles, the multi staged rocket system using the propulsion system such as the solid rocket motor or the liquid engine is selected to complete the mission. For the usual launch vehicle, the separation of the lower stage and the payload fairing which is the structure to protect the payload happens firstly before the separation of the payload. The structure after the separation of the lower stage and the payload fairing is called an orbital block consisting of the payload, the propulsion system, the electric bay, etc. The flight conditions of an orbital block such as the velocity, the altitude and the attitude angle are determined according to the mission trajectory. Orbital blocks usually fly at the high altitude belonging to the rarefied flow regime.
In the rarefied flow regime, even though the characteristic length of an orbital block is not small, the Knudsen number defined as the mean free path over the characteristic length is high since the mean free path is very short. For the high Knudsen number regime, the Navier-Stokes equations frequently used in the continuum regime can not be applied. Instead, the Boltzmann equation based on the kinetic theory is used in this high Knudsen number flow. The analytical approach of the Boltzmann equation is limited for some few cases. Therefore DSMC(Direct Simulation Monte Carlo) method analyzing the Boltzmann equation using the probabilistic approach is commonly used. In this

I.H. Tuncer et al., *Parallel Computational Fluid Dynamics 2007*,
DOI: 10.1007/978-3-540-92744-0_12, 

study, the aerodynamic characteristics of an orbital block which flies in the free molecule flow regime are predicted using DSMC method.

## 2. Numerical Approach

For the calculation of the flow field in the rarefied flow regime, DSMC solver SMILE(Statistical Modeling In Low-density Environment)[1] is used. SMILE is built on the majorant collision frequency scheme, And the VHS(Variable Hard Sphere) model is applied for the intermolecular collision. Larsen-Borgnakke phenomenological model is also used for the energy exchange. The molecules which reflect from the surface is defined using Maxwell model. The surface mesh is unstructured triangular type. The simulation molecule number of a cell is less than 10. The parallel computation is performed using 8 processors(Intel 2.66 GHz Dual Core CPU, Linux cluster) excluding the speedup test during which 1~20 processors are used.

## 3. Code Validation

The aerodynamic coefficients of Apollo capsule by NASA[3] are firstly calculated to validate the ability of SMILE to predict the aerodynamic coefficients in the rarefied flow regime. Figure 1 shows the configuration of Apollo capsule using GEOM3D(the preprocessing software of SMILE).[2]
In this calculation, The angle of attack of -25° and the flight velocity of 9.6km/s are fixed. And the variation of the flow conditions as the change of the altitude ranging from 110km to 200km is only considered. Figure 2 shows that the calculated aerodynamic coefficients agree significantly with the reference calculated data by NASA. Figure 2 also shows that as the altitude increases, the axial force coefficient(CA) increases, conversely, the normal force coefficient decreases and the center of pressure moves forward slightly.

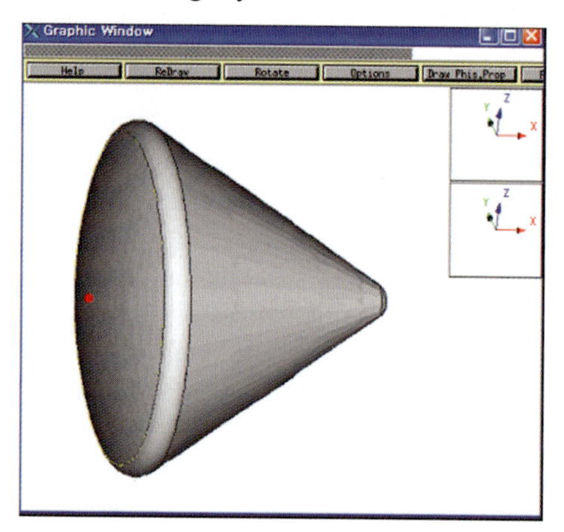


Figure 1. The configuration of Apollo Capsule using GEOM3D

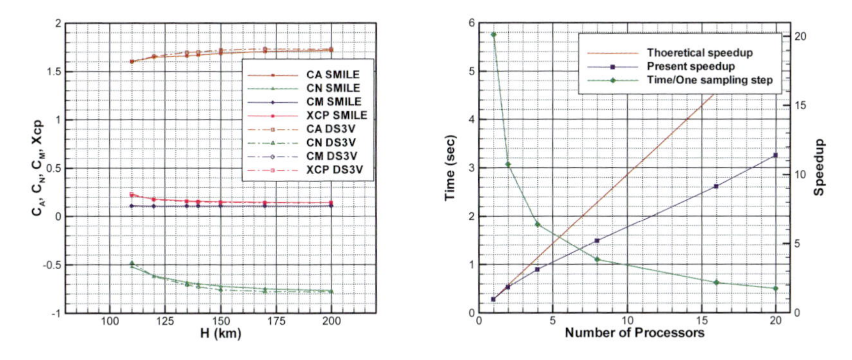


Figure 2. The aerodynamic characteristics of Apollo Capsule (DS3V by NASA [3]) and

the speedup for the altitude of 110km

## 4. Numerical Results

The aerodynamic characteristics of an orbital block are predicted using SMILE DSMC solver. Figure 3 shows the simplified configuration of an orbital block using GEOM3D, the axis system and the moment center coordinate (0.0, 0.0, 0.0). The flight condition, especially Knudsen number in Table 1 represents that the flow characteristics over an orbital block belong to the aspects of the free molecule flow. The atmosphere condition for the given altitude is determined by USSA(U.S. Standard Atmosphere)1976.(Table 1) Table 2 presents the aerodynamic coefficients and the aerodynamic force for the nominal condition. For these high altitudes, the dynamic pressure is very low even though the flight velocity are fast and the reference area are not small since the density is extremely low. Therefore, the aerodynamic forces are exceedingly small. It is, however, possible to perturb the attitude of an orbital block for these small force. For the accurate analysis of the attitude control of an orbital block by the perturbation, the aerodynamic coefficients for the expanded range of both the angle of attack and the side slip angle should be presented. In this study, the additional aerodynamic coefficients are predicted for the ±45° based on the nominal angle of attack and side slip angle. Figure 4 ~ Figure 6 show the variation of the aerodynamic coefficients as the angle of attack and the side slip angle increases for Case 1 ~ Case 3. Cx decreases as both the angle of attack and the side slip angle increases. And Cy and CMz increase as the angle of attack increases. Cy and CMz, however, decrease one with another as the side slip angle increases. Cz and CMy don't seem to be sensitive to the variation of the angle of attack. But Cz and CMy increase remarkably and linearly as the side slip angle increases. The center of pressure (Xcp) moves forward up to 0.2 caliber as both the angle of attack and the side slip angle vary.

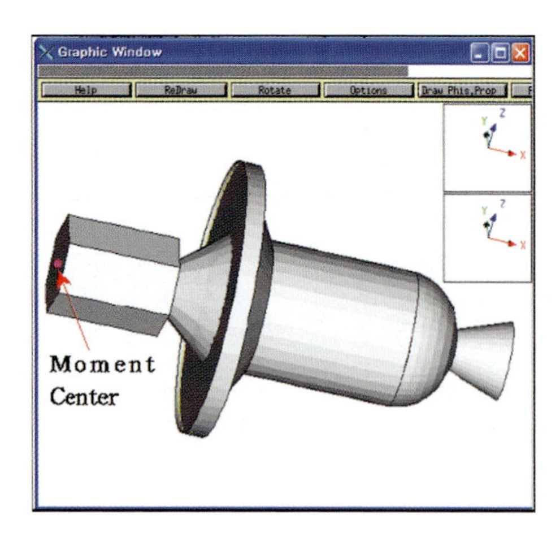


Figure 3. The configuration of an orbital block using GEOM3D

Table 1. The nominal flow conditions for the prediction of the aerodynamic characteristics for an orbital block

| | Angle of Attack | Density ($km/m^3$) | Dynamic Pressure (Pa) | Knudsen number |
|---|---|---|---|---|
| Case 1 | -19.56° | 2.890E-10 | 3.443E-03 | 1.467E+02 |
| Case 2 | -11.09° | 5.776E-11 | 6.585E-04 | 7.016E+02 |
| Case 3 | 0.18° | 2.133E-11 | 2.349E-04 | 1.915E+03 |

Table 2. The aerodynamic coefficients and force of an orbital block for the nominal flow conditions (Ref. length = 2m, Ref. area = $3.145m^2$)

| | Cx | Cy | CMz | Fx (N) Axial Force | Fy (N) Normal Force | Mz (N-m) Pitching Moment |
|---|---|---|---|---|---|---|
| Case 1 | 0.304E+01 | -0.869E+00 | 0.577E+00 | 3.297E-02 | -9.409E-03 | 1.248E-02 |
| Case 2 | 0.320E+01 | -0.491E+00 | 0.303E+00 | 6.627E-03 | -1.017E-03 | 1.256E-03 |
| Case 3 | 0.330E+01 | 0.813E-02 | -0.472E-02 | 2.436E-03 | 6.008E-06 | -6.976E-06 |

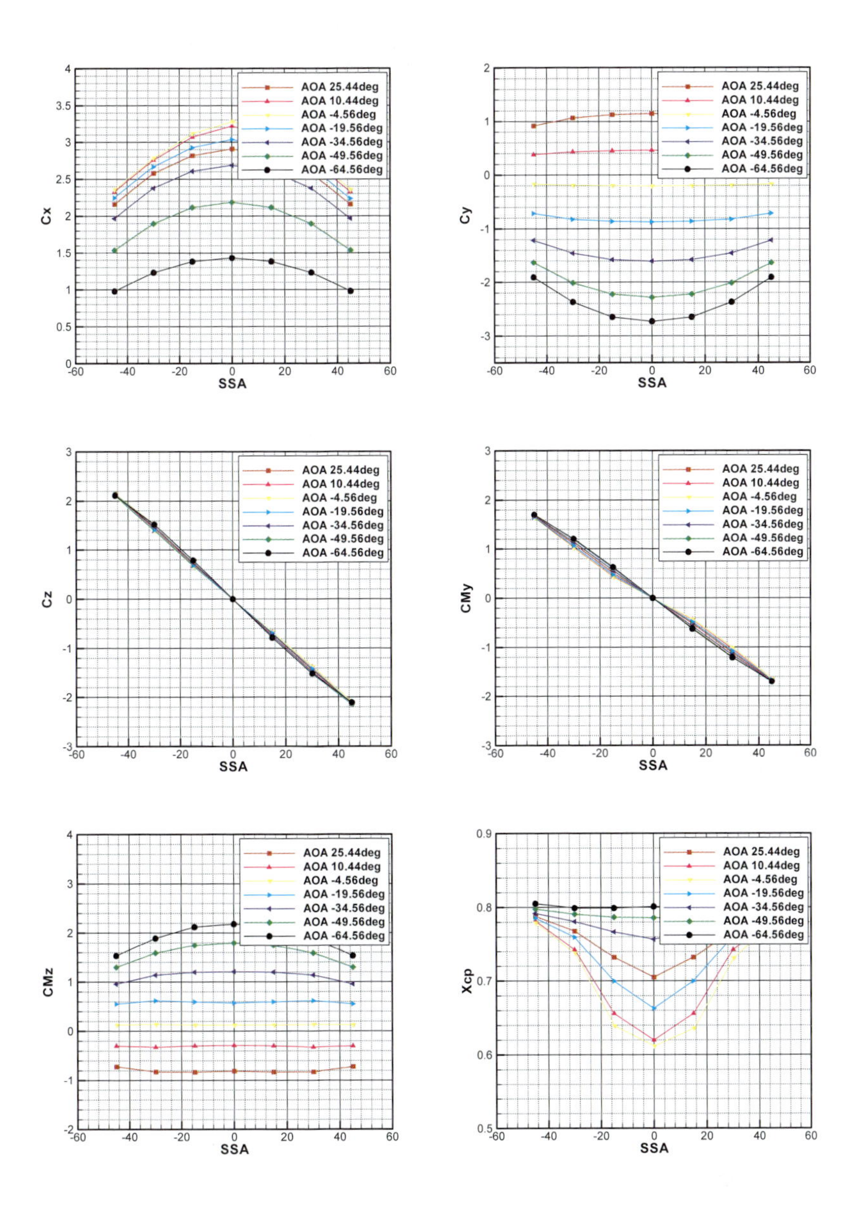


Figure 4. The variation of the aerodynamic coefficients for an orbital block (Case 1)

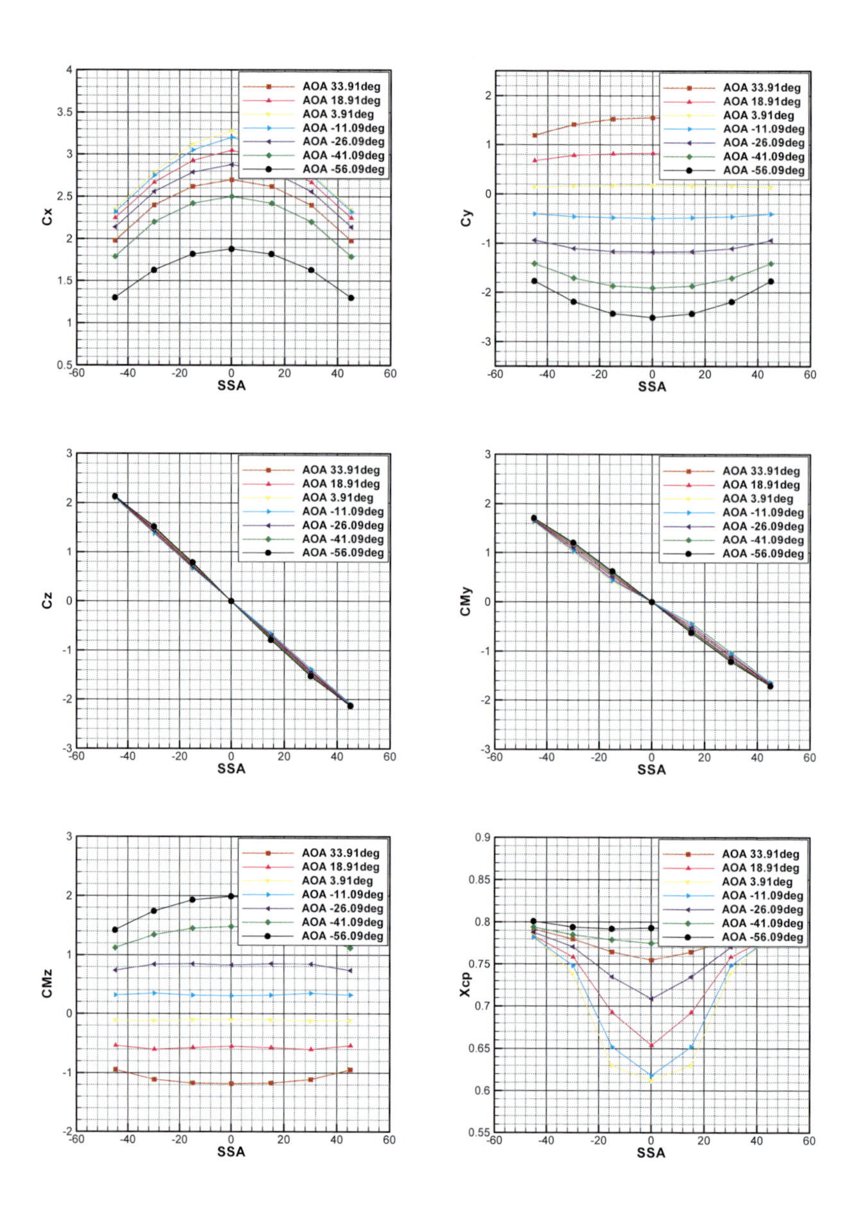


Figure 5. The variation of the aerodynamic coefficients for an orbital block (Case 2)

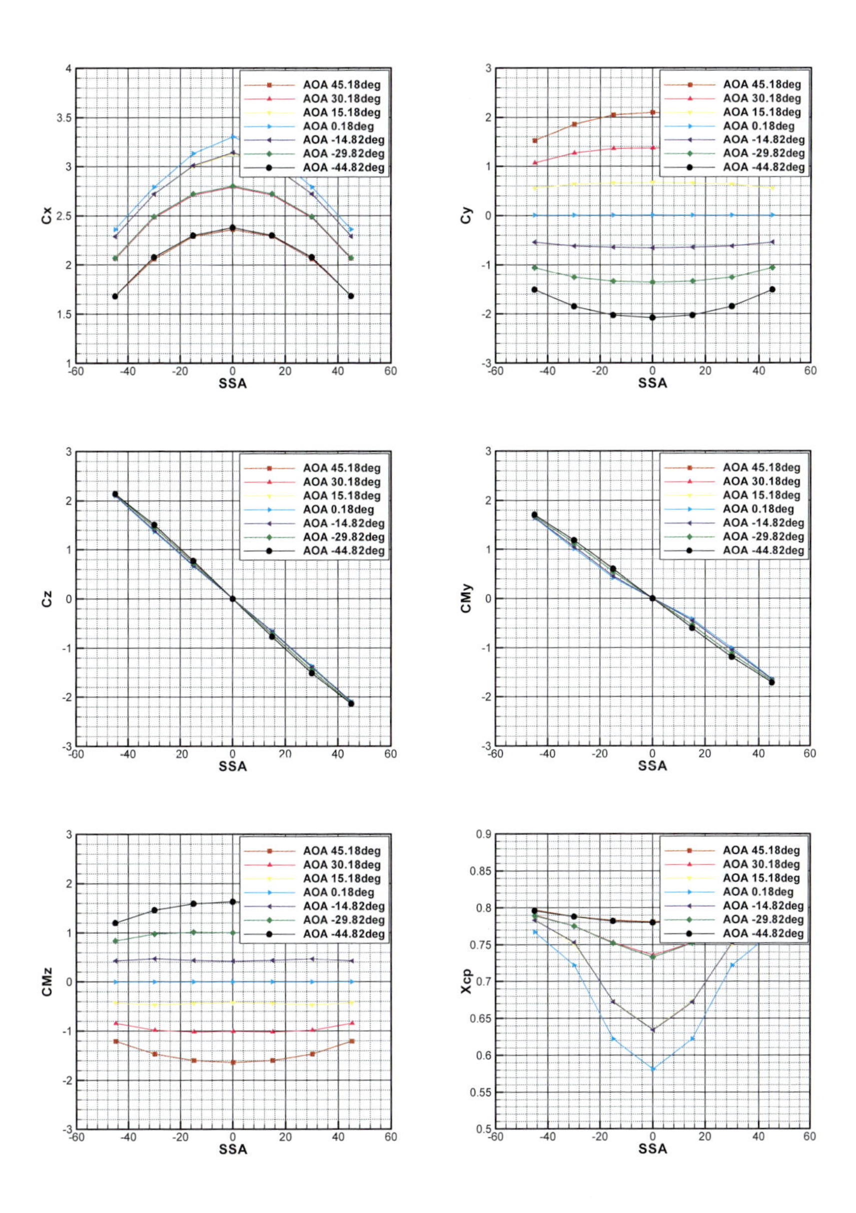


Figure 6. The variation of the aerodynamic coefficients for an orbital block (Case 3)

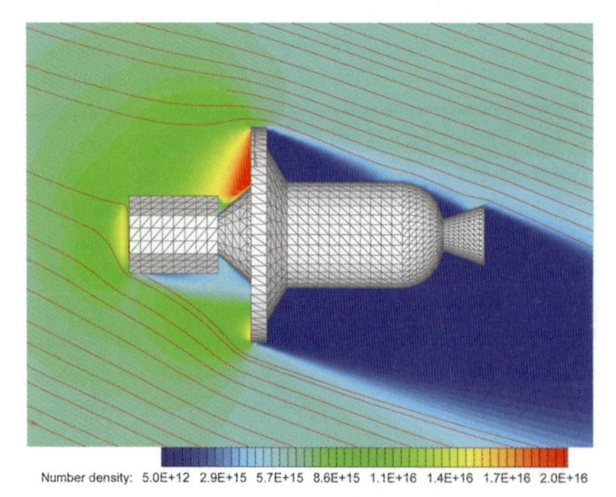


Figure 7. The variation of the number density over an orbital block (Case 1)

## 5. Concluding Remarks

The aerodynamic coefficients of Apollo capsule are calculated using DSMC solver SMILE. In comparison with the data predicted by NASA, the results using SMILE agree well with the reference data. the aerodynamic characteristics of an orbital block which operates on the high altitude in the free molecule regime are also predicted. For the nominal flow conditions, the predicted aerodynamic force is very small since the dynamic pressure is exceedingly low. And the additional aerodynamic coefficients for the analysis of the attitude control are presented as the angle of attack and the side slip angle vary from +45° to -45° of the nominal angle. This study to predict the aerodynamic characteristics of the flying objects in the rarefied flow environment will be useful for the calculation of the satellite lifetime and especially, the prediction of the aerodynamic characteristics for the reentry object.

### REFERENCES

1. M. S. Ivanov, "SMILE System USER MANUAL," Institute of Theoretical and Applied Mechanics, Russia, 2005
2. M. S. Ivanov, "Geometry Modeling System USER MANUAL," Institute of Theoretical and Applied Mechanics, Russia, 2005
3. J. N. Moss, C. E. Flass, and F. A. Greene, "DSMC Simulation of Apollo Capsule Aerodynamics for Hypersonic Rarefied Conditions," AIAA 2006-3577, 2006
4. M.S. Ivanov, G. N. Markelov, S. F. Gimelshein, L. V. Mishina, A. N. Krylov and N. V. Grechko, "High-Altitude Capsule Aerodynamics with Real Gas Effects," Journal of Spacecraft and Rocket, Vol. 35, 1998, No. 1, pp. 16-22.
5. R. G. Wilmoth, R. A. Mitcheltree, and J. N. Moss, "Low-Density Aerodynamics of the Stardust Sample Return Capsule," AIAA-97-2510, 1997

# Parallel Solution of a 3-D Mixed Convection Problem

**V. Ü. Ünal,[a] Ü. Gülçat, [b]**

*[a]Yeditepe University, Physics Department, 81120, Kayışdağı, İstanbul, Turkey*

*[b]Istanbul Technical University, Faculty of Aeronautics and Astronautics, 80626, Maslak, İstanbul,Turkey*

3D Navier-Stokes, implicit parallel solution, domain decomposition, forced-free convection

## 1. INTRODUCTION

As is well known, fully explicit schemes impose severe restrictions on the time step size for analyzing complex viscous flow fields, which are resolved with sufficiently fine grids. To remedy this, implicit flow solvers are used in analyzing such flows. Naturally, the higher order accurate schemes allow one to resolve the flow field with less number of grid points. Resolving the flow field with less number of points gives a great advantage to implicit schemes since the size of the matrix to be inverted becomes small. In this study a second order accurate implicit scheme for solution of full Navier-Stokes equations is developed and implemented. The space is discretized with brick elements while modified version of the two-step fractional method is used in time discretization of the momentum equation. At each time step, the momentum equation is solved only once to get the half time step velocity field. The pressure, on the other hand, is obtained via an auxiliary scalar potential which satisfies the Poisson's equation. For the parallel implicit solution of the matrix equations, modified version of the Domain Decomposition Method, [1,2,3,4] is utilized, and direct inversion in each domain is performed. Super linear speed-ups were achieved, [5].
For high Reynolds number flows in order to stabilize the numerical solution forth order artificial viscosity is added as given in [6]. The parallel solutions results obtained for the forced cooling of a room with chilled ceiling having a parabolic geometry will be given at the end.

I.H. Tuncer et al., *Parallel Computational Fluid Dynamics 2007*,
DOI: 10.1007/978-3-540-92744-0_13, 

The scheme here is made to run on SGI Origin 3000 utilized with 8 processors running Unix operating system. Public version of the Parallel Virtual Machine, PVM 3.3, is used as the communication library. The most recent solutions are obtained on the cluster of work stations and/or PC's.

## 2. FORMULATION

### 2.1. Navier-Stokes Equations

The flow of unsteady incompressible viscous fluid is governed with the continuity equation,

$$\nabla . \mathbf{u} = 0 \tag{1}$$

the momentum (Navier-Stokes) equation

$$\frac{D\mathbf{u}}{Dt} = -\nabla p + \frac{1}{Re}\nabla^2 \mathbf{u} + \frac{Gr}{Re^2} T\mathbf{k} \tag{2}$$

and the energy equation

$$\frac{DT}{Dt} = \frac{1}{RePr}\nabla^2 T . \tag{3}$$

The equations are written in vector form (here on, boldface type symbols denote vector or matrix quantities). The velocity vector, pressure, temperature and time are denoted by $\mathbf{u}=\mathbf{u}(u,v.w)$, $p$, T and t, respectively. The variables are non-dimensionalized using a reference velocity and a characteristic length. Re is the Reynolds number, $Re = UH/\nu$ where U is the reference velocity, $H$ is the characteristic length and $\nu$ is the kinematic viscosity of the fluid. Gr represents the Grashof number, $Gr = g\beta\Delta T H^3 / \nu^2$, where $\beta$ is the coefficient of volumetric expansion. Pr denotes the Prandtl number and **k** denotes the unit vector in z direction.

### 2.2. FEM formulation

The integral form of Eq. (2) over the space-time domain reads as

$$\int_{\Omega}\int_{t} \frac{\partial \mathbf{u}}{\partial t} \mathbf{N}\, d\Omega dt = \int_{\Omega}\int_{t} (-\mathbf{u} . \nabla \mathbf{u} - \nabla p + \frac{1}{Re}\nabla^2 \mathbf{u} + \frac{Gr}{Re^2} T^n \mathbf{k})\, \mathbf{N}\, d\Omega dt \tag{4}$$

where **N** is an arbitrary weighting function. The time integration of both sides of Eq. (4) for half a time step, $\Delta t / 2$, from time step n to n +1/2 gives

$$\int_{\Omega}(\mathbf{u}^{n+1/2}-\mathbf{u}^{n})\,\mathbf{N}\,d\Omega = \frac{\Delta t}{2}\int_{\Omega}(-\mathbf{u}.\nabla\mathbf{u}^{n+1/2}-\nabla p^{n}+\frac{1}{Re}\nabla^{2}\mathbf{u}^{n+1/2}+\frac{Gr}{Re^{2}}T^{n}\,\mathbf{k})\,\mathbf{N}\,d\Omega. \tag{5}$$

At the intermediate time step the time integration of Eq. (4), where the convective and viscous terms are taken at n +1/2 and pressure term at time level n, yields

$$\int_{\Omega}(\mathbf{u}^{*}-\mathbf{u}^{n})\,\mathbf{N}\,d\Omega = \Delta t\int_{\Omega}(-\mathbf{u}.\nabla\mathbf{u}^{n+1/2}-\nabla p^{n}+\frac{1}{Re}\nabla^{2}\mathbf{u}^{n+1/2}+\frac{Gr}{Re^{2}}T^{n}\,\mathbf{k})\,\mathbf{N}\,d\Omega. \tag{6}$$

For the full time step, the averaged value of the pressure and the temperature at time levels n and n+1 are used to give

$$\int_{\Omega}(\mathbf{u}^{n+1}-\mathbf{u}^{n})\,\mathbf{N}\,d\Omega = \Delta t\int_{\Omega}(-\mathbf{u}.\nabla\mathbf{u}^{n+1/2}+\frac{1}{Re}\nabla^{2}\mathbf{u}^{n+1/2} - \nabla\frac{p^{n}+p^{n+1}}{2}+\frac{Gr}{Re^{2}}\frac{T^{n+1}+T^{n}}{2}\mathbf{k})\,\mathbf{N}\,d\Omega. \tag{7}$$

Subtracting (6) from (7) results in

$$\int_{\Omega}(\mathbf{u}^{n+1}-\mathbf{u}^{*})\mathbf{N}\,d\Omega = -\frac{\Delta t}{2}\int_{\Omega}\nabla(p^{n+1}-p^{n})\mathbf{N}\,d\Omega-\frac{\Delta t}{2}\frac{Gr}{Re^{2}}\int_{\Omega}(T^{n+1}-T^{n})\mathbf{kN}\,d\Omega. \tag{8}$$

If one takes the divergence of Eq. (8), the following is obtained;

$$\int_{\Omega}\nabla.\mathbf{u}^{*}\,\mathbf{N}\,d\Omega = -\frac{\Delta t}{2}\int_{\Omega}\nabla^{2}(p^{n+1}-p^{n})\mathbf{N}\,d\Omega-\frac{\Delta t}{2}\frac{Gr}{Re^{2}}\int_{\Omega}\frac{\partial}{\partial z}(T^{n+1}-T^{n})\mathbf{N}\,d\Omega. \tag{9}$$

After subtracting (5) from (6), the integrand of the resulting equation reads,

$$\mathbf{u}^{*} = 2\,\mathbf{u}^{n+1/2}-\mathbf{u}^{n}\,. \tag{10}$$

### 2.3. Numerical Formulation

Defining the auxiliary potential function $\varphi = -\Delta t(p^{n+1}-p^{n})$ and choosing N as trilinear shape functions, discretization of Eq. (5) gives

$$\left(\frac{2\mathbf{M}}{\Delta t}+\mathbf{D}+\frac{\mathbf{A}}{Re}\right)\mathbf{u}_{\alpha}^{n+1/2} = \mathbf{B}_{\alpha}+p_{e}\,\mathbf{C}_{\alpha}+\frac{2\mathbf{M}}{\Delta t}\mathbf{u}_{\alpha}^{n}+\frac{Gr}{Re^{2}}\mathbf{M}\,T^{n}\,\mathbf{k} \tag{11}$$

where $\alpha$ indicates the Cartesian coordinate components x, y and z, **M** is the lumped element mass matrix, **D** is the advection matrix, **A** is the stiffness matrix, **C** is the coefficient matrix for pressure, **B** is the vector due to boundary conditions and **E** is the matrix which arises due to incompressibility.
Equation for the temperature, Eq. (3) is solved explicitly using:

$$\mathbf{M}\,T^{n+1} = \mathbf{M}\,T^{n} - \frac{\Delta t}{2}\left(\frac{\mathbf{A}}{Pr.Re} + \mathbf{D}\right)T^{n} \tag{12}$$

The discretized form of Eq. (9) reads as

$$\frac{1}{2}\mathbf{A}\varphi = -\frac{1}{2}\mathbf{A}\left(p^{n+1} - p^{n}\right)\Delta t = 2\mathbf{E}_{\alpha}\mathbf{u}_{\alpha}^{n+1/2} - \frac{\Delta t}{2}\frac{Gr}{Re^{2}}(T^{n+1} - T^{n})_{e}C_{z} \tag{13}$$

Subtracting Eq. (6) from Eq. (7) and introducing the auxiliary potential function $\varphi$, one obtains the following;

$$\begin{aligned} \mathbf{Mu}_{\alpha}^{n+1} &= \mathbf{Mu}_{\alpha}^{*} + \frac{1}{2}\mathbf{E}_{\alpha}\,\varphi + \frac{\Delta t}{2}\frac{Gr}{Re^{2}}(T^{n+1} - T^{n}) \\ &= 2\,\mathbf{Mu}_{\alpha}^{n+1/2} - \mathbf{Mu}_{\alpha}^{n} + \frac{1}{2}\mathbf{E}_{\alpha}\,\varphi + \frac{\Delta t}{2}\frac{Gr}{Re^{2}}(T^{n+1} - T^{n}) \end{aligned} \tag{14}$$

The element auxiliary potential $\varphi_e$ is defined as

$$\varphi_e = \frac{1}{vol(\Omega_e)}\int_{\Omega_e}\mathbf{N}_i\,\varphi_i\,d\Omega_e, \qquad i = 1,\ldots\ldots\ldots,8, \tag{15}$$

where $\Omega$ is the flow domain and $\mathbf{N}_i$ are the shape functions.
The following steps are performed to advance the solution one time-step.

i. Eq. (11) is solved to find the velocity field at time level n+1/2 with domain de composition,
ii. Eq. (12) is solved explicitly and temperature field $T^{n+1}$ is obtained,
iii. Knowing the half step velocity field and the temperature, Eq. (13) is solved with domain decomposition to obtain the auxiliary potential $\varphi$.
iv. With this $\varphi$, the new time level velocity field $\mathbf{u}^{n+1}$ is calculated via Eq.(14).
v. The associated pressure field $p^{n+1}$ is determined from the old time level pressure field $p^{n}$ and $\varphi$ obtained in the second step.

The above procedure is repeated until the desired time level. In all computations lumped form of the mass matrix is used.

## 3. DOMAIN DECOMPOSITION

The domain decomposition technique, [1,2,3,4,5] is modified and applied for the efficient parallel solution of the momentum, Eq. (11) and the Poisson's Equation for the auxiliary potential function, Eq. (13).

### 3.1. For the Momentum Equation:

Initialization: The momentum, Eq. (11) is solved with a direct solution method, separately in each domain $\Omega_i$ with boundary of $\partial\Omega_i$ and interface $S_j$, with vanishing Neumann boundary condition on the domain interfaces,
At the beginning of each time level: $\mu^k = 0$ and $aw^k = 0$,

$$\overline{\mathbf{A}}\, y_i = f_i \quad \text{in } \Omega_i, y_i = g_i \quad \text{on } \partial\Omega_i, \ \frac{\partial y_i}{\partial n_i} = (-1)^{i-1}\mu^k \quad \text{on } S_j,$$

$$\mu^o = \mu^k, \ w^o = g^o \text{ and } g^o = aw^k - (y_2^o - y_1^o)S_j.$$

where subscript $k$ denotes the interface iteration level at which convergence is obtained, subscript $o$ stands for the initial interface iteration level at the each time level and

$$\overline{\mathbf{A}} = \frac{2\mathbf{M}}{\Delta t} + \mathbf{D} + \frac{\mathbf{A}}{Re} \text{ in Eq. (11) and } y_i = \mathbf{u}_\alpha^{n+1/2}.$$

Handling of unit problem as a steepest decent to obtain converged interface values is described in a detail in [5]. After obtaining the converged flux values, $\mu^k$, at the interface, the finalization of the parallel solution is as follows.
Finalization: Having obtained the correct Neumann boundary condition for each interface, the momentum equation, Eq. (11), is solved via the direct solution method at each sub-domain, together with the correct interface conditions,

$$\overline{\mathbf{A}}\, y_i = f_i \quad \text{in } \Omega_i, \ y_i = g_i \quad \text{on } \partial\Omega_i, \ \frac{\partial y_i}{\partial n_i} = (-1)^{i-1}\mu^k \quad \text{on } S_j$$

### 3.2. Parallel Implementation

During parallel implementation, in order to advance the solution single time step, the momentum equation is solved implicitly with domain decomposition. Solving Eq. (11) gives the velocity field at half time level, which is used at the right hand sides of Poisson's Eq. (13), in obtaining the auxiliary potential. The solution of the auxiliary potential is obtained with again domain decomposition where an iterative solution is also necessary at the interface. Therefore, the computations involving an inner iterative cycle and outer time step advancements have to be performed in a parallel manner on each processor communicating with the neighboring one.

## 4. RESULTS AND DISCUSSION

The scheme was calibrated for Re=1000 for the flow inside a lid driven cavity and with efficient parallelization super linear speed-up was achieved, [5]. After the calibration,

first, the flow in a room as shown in Figure 1 is studied with Reynolds number 1000 based on 1m. length , inflow speed and the kinematic viscosity. The computational domain has the following dimensions: sides x=4m, y=2m (symmetry plane) and maximum height z=2.6m. The ceiling of the room has a parabolic shape with the maximum camber of m=0.1 at p=0.2. The solution is advanced up to the dimensionless time level of 75, where continuous and steady cooling of the room is achieved, as given in, [7]. The initially linear temperature field changes, because of the circulation in the room, so that the cool front moves towards the hot wall. Computations are carried out with the 6-domain partitioning grids with total number of 61x31x35 nodes (Figure 2). The following additional dimensionless numbers are employed: $Pr = 0.7$, $Gr/Re^2 = 0.01$. The elapsed time per time step per grid point is 0.0002 seconds for the dimensionless time step of 0.025.
The equation of the ceiling in x,z coordinates:

$$0<x_c<0.2 : z_c = \frac{m}{p^2}(2px_c - x_c^2), \quad 0.2<x_c<1 \ z_c = \frac{m}{(1-p)^2}((1-2p)+2px_c - x_c^2))$$

*Initial Conditions*: T(t=0)=T(x)=(x-2)/4 inside the cavity,

*Boundary Conditions*: Inflow: u= $\sqrt{2}/2$, v=0 , w= $\sqrt{2}/2$, Outflow: $\partial u / \partial x = 0$
$T_t = -0.7$ at the ceiling, $T_c = -0.5$ cold left wall, $T_h = 0.5$ hot right wall.

The second study is made for the flow Reynolds number of 2000 for the same room with the same initial and boundary conditions while keeping the Pr=0.7 and $Gr/Re^2=0.01$ also the same. The velocity field at the symmetry plane are plotted for the different time steps running between 3500 to 19000 which respectively corresponds to the dimensionless time levels of 87,5 and 475. Almost periodic oscillatory behaviour of the flow field for the Re=2000 case can be predicted from Figure 3a and c. From these figures approximate period of the flow motion is about 160 units in dimensionless time. In conclusion, doubling of the Reynolds number, i.e, raising it from 1000 to 2000, while keeping all the parameters same, changes the flow behaviour completely.

***References***

[1] Dinh, Q.V., Ecer, A., Gülçat, Ü., Glowinski, R., and Periaux, J. (1992) Concurrent Solutions of Elliptic Problems via Domain Decomposition, Applications to Fluid Dynamics, Parallel CFD 92, Rutgers University.
[2] Glowinski, R. and Periaux, J. (1982) Domain Decomposition Methods for Nonlinear Problems in Fluid Dynamics, Research Report 147, Inria, France.
[3] Glowinski, R., Pan, T.W., and Periaux, J. (1995) A One Shot Domain Decomposition/Fictitious Domain Method for the Solution of Elliptic Equations, Parallel CFD 95, Elsevier.
[4] Gülçat, Ü. and Ü. Ünal V. (2000)Accurate Implicit Solution of 3D Navier-Stokes Equations on Cluster of Workstations, Parallel CFD 2000 Conference, Norway.
[5] Gülçat, Ü. and Ü. Ünal V. (2002) Accurate Implicit Parallel Solution of 3D Navier-Stokes Equations, Parallel CFD 2002 Conference, Japan.

[6] Gülçat, Ü. and Aslan A.R. (1997) Accurate 3D Viscous Incompressible Flow Calculations with the FEM, *International Journal for Numerical Methods in Fluids*, Vol.. **25**, 985-1001.
[7] Ünal, Ü.V. and Gülçat, Ü. (2003) Parallel implicit solution of full Navier-Stokes equations, *Lecture Notes In Computer Science*, Vol. **2659**, 622-631, Springer-Verlag.

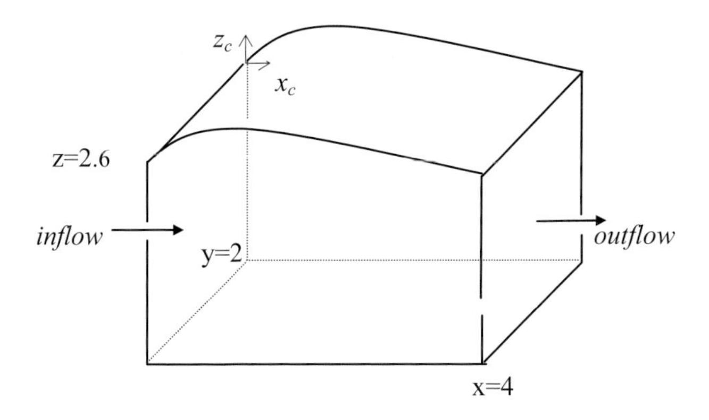


**Fig. 1.** Definition of the real problem

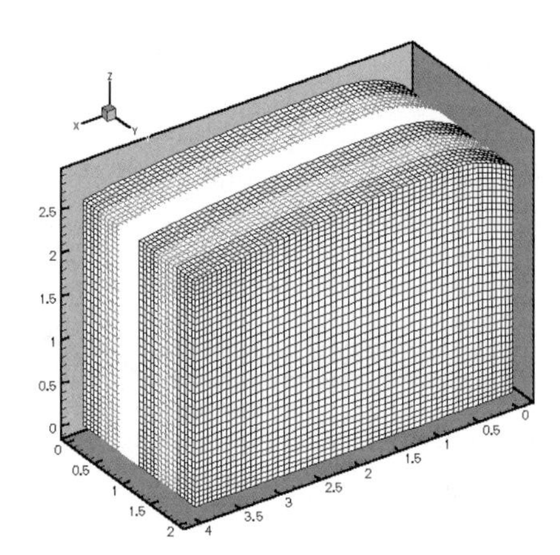

**FIGURE 2** 6-domain partitioning with total number of 61x31x35.

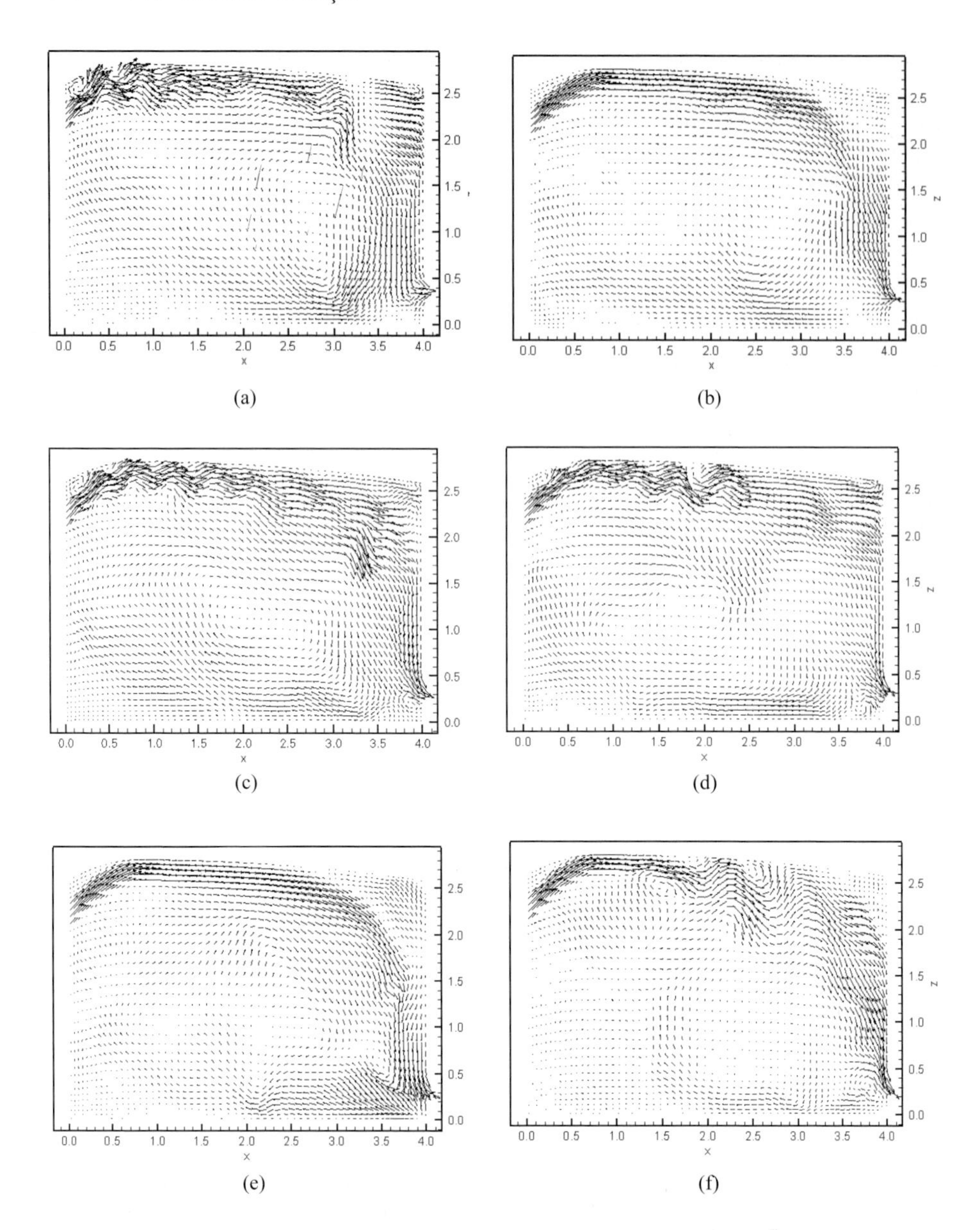


**FIGURE 3** Velocity Distributions at the symmetry plane, for Re=2000, $Gr/Re^2 = 0.01$, dt=0.025 at the time steps: (a) 3500, (b) 8500, (c) 10000, (d) 16000, (e) 17000, (f) 19000

# Computation of Hypersonic Flow of a Diatomic Gas in Rotational Non-Equilibrium past a Blunt Body Using the Generalized Boltzmann Equation

R. K. Agarwal* and R. Chen
Mechanical and Aerospace Engineering Department, Washington University in St. Louis, MO 63130

F. G. Cheremisin
Physical Sciences Division, Dorodnicyn Computing Center, Moscow, Russia

email: rka@me.wustl.edu

The results of 2-D numerical simulations of non-equilibrium hypersonic flow of a diatomic gas, e.g., nitrogen past a 2-D blunt body at low to high Knudsen Numbers are presented. The flow field is computed using the Generalized Boltzmann (or the Wang-Chang Uhlenbeck [1]) Equation (GBE) for Kn varying from 0.1 to 10. In the GBE [2], the internal and translational degrees of freedom are considered in the framework of quantum and classical mechanics respectively. The computational framework available for the classical Boltzmann equation (for a monoatomic gas with translational degrees of freedom) [3] is extended by including the rotational degrees of freedom in the GBE. The general computational methodology for the solution of the GBE for a diatomic gas is similar to that for the classical BE except that the evaluation of the collision integral becomes significantly more complex due to the quantization of rotational energy levels. The solution of GBE requires modeling of transition probabilities, elastic and inelastic cross-sections etc. of a diatomic gas molecule, needed for the solution of the collision integral. An efficient computational methodology has been developed for the solution of GBE for computing the flow field in diatomic gases at high Mach numbers. There are two main difficulties encountered in computation of high Mach number flows of diatomic gases with rotational degrees of freedom using the GBE: (1) a large velocity domain is needed for accurate numerical description of molecular velocity distribution function resulting in enormous computational effort in calculation of the collision integral, and (2) about 50 to 70 energy levels are needed for accurate representation of the rotational spectrum of the gas. These two problems result in very large CPU and memory requirements for shock wave computations at high Mach numbers (> 6). Our computational methodology has addressed these problems, and as a result efficiency of calculations has increased by several orders of magnitude. The code has been parallelized on a SGI Origin 2000, 64 R12000 MIPS processor supercomputer.

**Key Words:** Wang-Chang Uhlenbeck Equation, Nonequilibrium Hypersonic Flows, Rarefied Flow of Diatomic Gas

## INTRODUCTION

In recent years, there has been a resurgence of interest in US Air Force in space access and therefore Air Force is interested in the hypersonic aerodynamics of its future space operations vehicles, long-range-strike vehicles and military-reusable launch vehicles. The size and weight of a hypersonic vehicle and thus its flight trajectory and required propulsion system are largely determined from aerodynamic considerations. Various positions in the flight envelop of such space vehicles may be broadly classified into the continuum, transition and free molecular regimes. Hence, a need exists for a single unified Computational Fluid Dynamics (CFD) code that can treat all the three flow regimes accurately and efficiently. At an altitude greater than 90 km, the Knudsen number $Kn$ is greater than 10, corresponding to the rarefied regime, in which the kinetic model such as DSMC can be used for the simulation. At altitudes under 70 km, $Kn$ is small, corresponding to the continuum regime, in which continuum equations namely the Euler or Navier-Stokes equations can accurately and efficiently model the flow field. For altitudes between 70 and 90 km, the Knudsen numbers are in the neighborhood of unity; the flow fields fall in the transition regime, for which both the Navier-Stokes equations and DSMC method [4-6] have significant limitations. Navier-Stokes equations are no longer accurate because of significant departure from equilibrium and DSMC method becomes computationally prohibitive because of large number of particles required for accurate simulations. A number of alternative approaches have been proposed and investigated in the literature such as the Burnett equations [7, 8], Grad's moment equations [9] or Eu's equations [10]. These equations collectively are known as Extended Hydrodynamics (EHD) equations.

However, the progress has been limited and furthermore the accuracy of the EHD equations remains questionable because of the asymptotic nature of the Chapman-Enskog expansion (e.g. for the Burnett equations), the issues related to regularization and the difficulty in imposing the exact kinetic boundary conditions on the

I.H. Tuncer et al., *Parallel Computational Fluid Dynamics 2007*,
DOI: 10.1007/978-3-540-92744-0_14, © Springer-Verlag Berlin Heidelberg 2009

surface of the body. Physically, gas flows in transition regime, where $Kn$ is O(1), are characterized by the formation of narrow, highly non-equilibrium zones (Knudsen layers) of thickness of the order of molecular mean free path λ; the flow structure is then determined by the fast kinetic processes. Moreover, in case of unsteady flows, an initial Knudsen time interval is of the order $\tau_0 = \lambda/v$, where $v$ is the molecular velocity. Thus, the Knudsen layer can be computed accurately only by directly solving the Boltzmann equation.

Boltzmann equation describes the flow in all the regimes and is applicable for all Knudsen numbers from zero to infinity. In a previous paper [11], the 2-D Classical Boltzmann Solver (CBS) of Professor Cheremisin was applied to simulate the flow field due to supersonic flow of a monoatomic gas past blunt bodies from low to high Knudsen Numbers. A conservative discrete ordinate method, which ensures strict conservation of mass, momentum and energy, was applied in the solver. The effect of Knudsen number varying from 0.01 to 10 was investigated for Mach 3 flows past a blunt body. All the Boltzmann solutions at $Kn$ = 0.01 and 0.1 were compared with the Navier-Stokes and augmented Burnett solutions, and the DSMC solutions obtained using the 'SMILE' code of Professor Ivanov. In the Boltzmann and DSMC computations, the gas was assumed to be composed of hard sphere. In this paper, we extend the work for monoatomic gases reported in [11] to diatomic gases by solving the GBE.

For calculating the non-equilibrium flow of diatomic gases, currently DSMC method is the most widely used technique. In DSMC approach, one usually employs some empirical models [12], which very much resemble a BGK type relaxation model with a constant relaxation time. There are only a few attempts due to Koura [13] and Erofeev [14] for example, which employ the so called "trajectory calculations", where a realistic modeling of inelastic collisions with translational – rotational (T-R) transfer of energy is made. These computations require enormous CPU time. For vibrational excitations, no trajectory calculations with DSMC have been attempted to date. The physical reality is such that the vibrational term becomes important at much higher temperature than the rotational term. For N2 molecule the quanta of vibrational temperature is 3340K, and the quanta of rotational temperature is 2.89K; the corresponding quanta for the O2 molecule are 2230K and 2.1K respectively. Therefore, rotational energy levels are important at all temperatures, but the vibration energy levels are important only at high temperatures. Thus, each vibration energy level splits into a high number of rotational energy levels. The characteristic relaxation times for T-T, T-R, and T-V transfer of energy are in proportion of about 1: 5: 100 (or several thousands). As stated in the paper of Surzhikov et al. [15], a good physical assumption is that the translational and rotational energy are in thermodynamic equilibrium during the vibrational relaxation process.

The solution of GBE at high Mach numbers requires both large memory and CPU time. Following Cheremisin's work for a monoatmic gas [3], we apply a large time step in velocity space (of the order of the mean thermal velocity) ahead of the shock wave in the GBE code. We also use fewer levels (6 to 10) in rotational spectrum to reduce both the memory and CPU requirements. Another source of large CPU time in the currently used time-explicit algorithm for solution of BE is the requirement of a very small time step which results from the condition that the distribution function remains positive after the relaxation stage. For example, for a monatomic gas, at M = 20, the time step is of the order of 0.001 of the mean free path time. In order to overcome this problem, we employ two strategies: (a) employ a filter that excludes the negligibly small contributions to the collision integral, and (b) modify the algorithm to a mixed explicit-implicit scheme to increase the time step by at least a factor of 10. All these improvements are considered in developing an accurate and efficient GBE solver for computing flow field of a diatomic gas at high Mach numbers including the rotational excitations.

## SOLUTION METHODOLOGY

The 1-D Shock Wave (SW) structure and 2-D blunt body flow in polyatomic gases are studied using the GBE [2] that replaces the classical Wang Chang –Uhlenbeck Equation (WC-UE) for the case when energy levels are degenerated. In its mathematical form, the GBE is close to the WC-UE and differs from the later by a factor related to the statistical weights of the levels. It should be noted that the GBE gives the equilibrium spectrum different from WC-UE.

The GBE has the form

$$\frac{\partial f_i}{\partial t} + \boldsymbol{\xi}\frac{\partial f_i}{\partial \mathbf{x}} = R_i \tag{1}$$

The collision operator is given by

$$R_i = \sum_{jkl} \int_{-\infty}^{\infty} \int_0^{2\pi} \int_0^{b_m} (f_k f_l \boldsymbol{\omega}_{ij}^{kl} - f_i f_j) P_{ij}^{kl} g b db d\boldsymbol{\varphi} d\boldsymbol{\xi}_j \tag{2}$$

Here $f_i$ is the distribution function for the energy level $i$, $P_{ij}^{kl}$ is the probability of the transfer from levels $i, j$ to the levels $k, l$, and the factor $\omega_{ij}^{kl} = (q_k q_l)/(q_i q_j)$, $q_i$ being the degeneration of the energy level. For simple levels, the GBE changes to the WC-UE, therefore it can be also considered as a more general form of the later one. In [2] the GBE equation was obtained directly from WC-UE by grouping the $q_i$ single levels that have the same energy and form a degenerated level.

We consider molecular Nitrogen having the Lennard-Jones potential (6, 12) with the depth of the energy hole $\varepsilon = 91K$, degeneration of rotational level $q_i = 2i+1, i = 0,1,..\infty$, and the energy of the level $e_{ri} = \varepsilon_0 i(i+1),\ \varepsilon_0 = 2.9K$. The molecular interaction during the collision consists of two phases. In the first phase, the molecules interact in an elastic manner according to the molecular potential. This stage determines the deviation angle of the relative velocity. In the second stage, the modulus of the relative velocity changes according to the energy conservation equation. For the transition probabilities $P_{ij}^{kl}$ we apply the formulae given in [2] that are obtained by fitting of experimental data of molecular dynamics simulations of interactions of rigid rotors that model $N_2$ molecules.

$$P_{ij}^{kl} = P_0 \omega_{ij}^{kl} [\alpha_0 \exp(-\Delta_1 - \Delta_2 - \Delta_3 - \Delta_4) + \frac{1}{\alpha_0} \exp(-\Delta_3 - \Delta_4)],\text{ where}$$

$$\Delta_1 = |\Delta e_1 + \Delta e_2| / e_{tr0},\ \Delta_2 = 2|\Delta e_2 - \Delta e_1| / e_{tot}$$

$$\Delta_3 = 4|\Delta e_1|/(e_{tr0} + e_{ri}),\ \Delta_4 = 4|\Delta e_2|/(e_{tr0} + e_{rj})$$

$$\Delta e_1 = e_{ri} - e_{rk},\ \Delta e_2 = e_{rj} - e_{rl},\ \alpha_0 = 0.4 e_{tot} / e_{tr0}$$

$$e_{tr0} = mg^2/4,\ e_{tot} = e_{tr0} + e_{ri} + e_{rj}.$$

The energy conservation law in a collision selects virtual collisions with non zero probability. From the equation $mg_{ij}^2/4 + e_{ri} + e_{rj} = mg_{kl}^2/4 + e_{rk} + e_{rl}$, it can be shown that

$P_{ij}^{kl} > 0$, if $g_{kl}^2 \geq 0$, otherwise $P_{ij}^{kl} = 0$.

The elastic collision is a particular case of the collision. The probabilities obey the normalization condition: $\sum_{k,l} P_{ij}^{kl} = 1$. The kinetic equation (1) is solved by the splitting scheme. At a time step $\tau \ll \tau_0$, where $\tau_0$ is a mean inter-collision time, the equation (1) is replaced by the sequence of equations

(a) $$\frac{\partial f_i}{\partial t} + \boldsymbol{\xi} \frac{\partial f_i}{\partial \mathbf{x}} = 0$$

(b) $$\frac{\partial f_i}{\partial t} = R_i$$

The collision operator $R_i$ is evaluated at the uniform grid $S_0$ in the velocity space by the conservative projection method proposed in [3].

## COMPUTED RESULTS

### (a) Shock Wave Structure in Nitrogen at High Mach Numbers

The SW structure is formed as a final stage of the evolution of discontinuity in the initial distribution function. The problem is considered for the interval $-L_1 \le x \le L_2$ with the discontinuity at $x = 0$. The initial distribution function on both sides of the discontinuity is described by the velocities and spectral levels. It has the form

$f_i^{1,2}(\xi,x) = n^{1,2}[m/(2\pi T^{1,2})]^{3/2} \exp[-\frac{m(\xi-u^{1,2})^2}{2T^{1,2}}]\frac{2i+1}{Q_r}\exp(-\frac{e_{ri}}{T^{1,2}})$, where $Q_r$ denotes the statistical sum. Parameters $(n,T,u)^{1,2}$ are defined by the Rankine-Hugoniot relations with $\gamma = 7/5$. At the boundary the initial distribution function is kept constant.

For moderate Mach numbers, the SW structure in Nitrogen can be computed with real value of the spectral energy gap $\varepsilon_0$, but for the hypersonic case the needed number of levels becomes too high (up to 50-70 levels) and therefore the computations become very time consuming. To facilitate this problem it is possible to increase this energy gap and thereby reduce the number of levels. By conducting numerical experiments at M=4, we determined that this increase in spectral energy gap does not influence the results of calculations in any significant way. Figure 1 shows an example of hypersonic SW structure in Nitrogen at Mach 15, computed with 16 levels instead of nearly 70 "natural" levels. These types of calculations compare well with the experiments of Alsmeyer [16] for density "n" for reported Mach numbers up to 10 (not shown here). Figure 2 shows the hypersonic SW structure in a monoatomic gas at M = 25. It should be noted that for this case, the temperature behind the SW increases about 200 times that makes the computation very difficult; it required an important improvement to the algorithm.

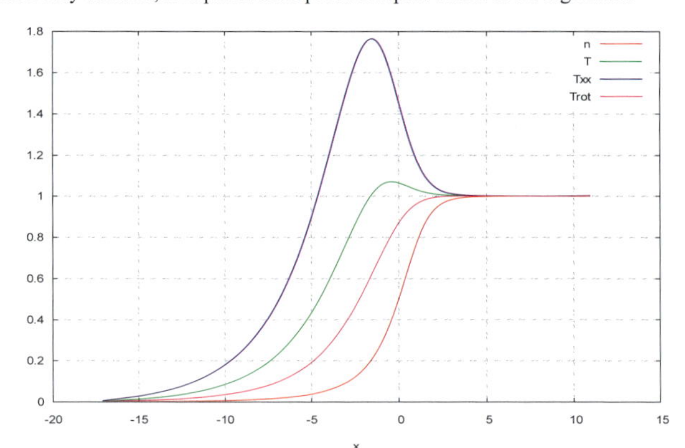


Figure 1: Shock Wave Structure in Nitrogen for M=15

n = density, T = total temp, Txx = translational temp, Trot = rotational temp (normalized)

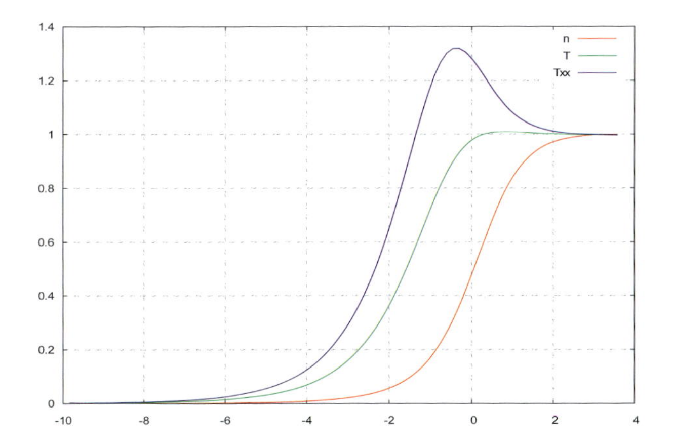


Figure 2: Shock Wave Structure in a monoatomic gas for M= 25
n = density, T = total temp, Txx = translational temp (normalized)

**(b) Flow of Nitrogen past a 2D Blunt Body at High Mach Numbers**

Figures 3-5 show the results of computations for flow of nitrogen past a blunt body in both translational and rotational non-equilibrium at Mach 3 at Kn = 0.1 and 1.0. The computations at Kn = 10 are not presented here but are easy to compute. Both the variations in flow quantities such as density, u- and v- components of velocities, total temperature, rotational temperature, and x-, y-, and z- components of translational temperature as well as some of the contour plots of these quantities in the computational domain are shown.

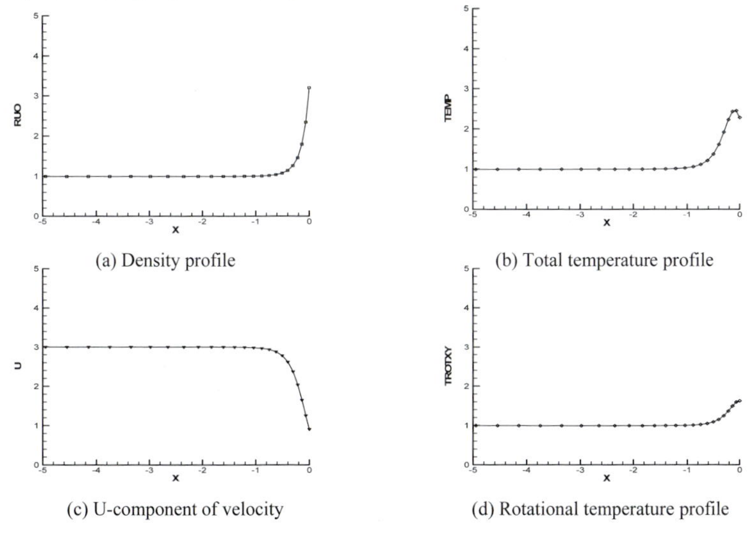


(a) Density profile (b) Total temperature profile

(c) U-component of velocity (d) Rotational temperature profile

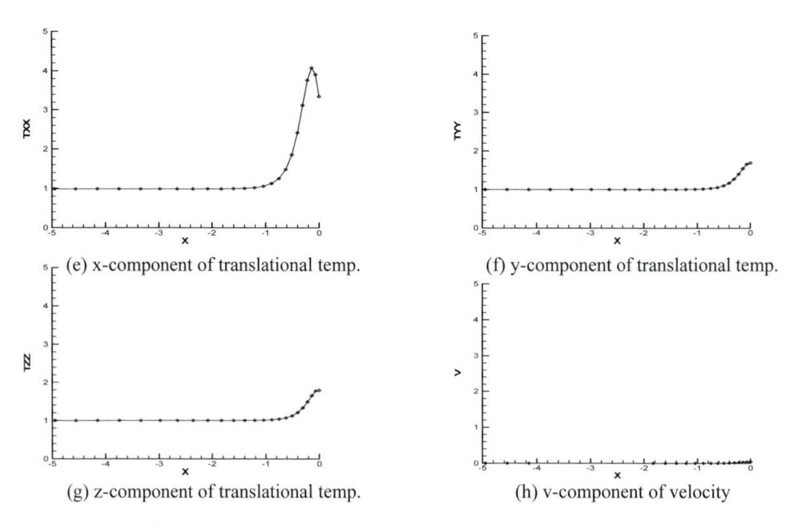


(e) x-component of translational temp.

(f) y-component of translational temp.

(g) z-component of translational temp.

(h) v-component of velocity

Fig. 3: Distributions along the stagnation streamline of the blunt body, M = 3, Kn = 0.1

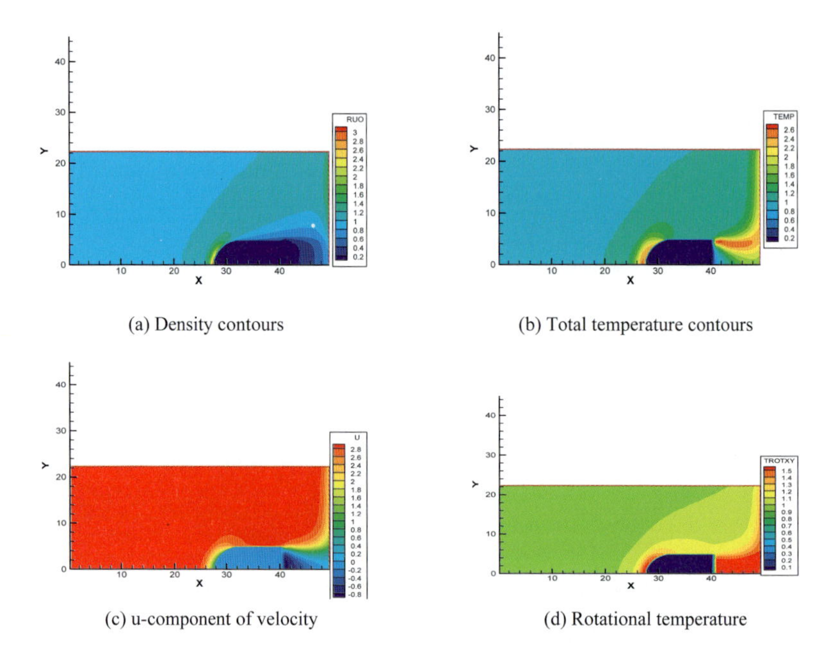


(a) Density contours

(b) Total temperature contours

(c) u-component of velocity

(d) Rotational temperature

Fig. 4: Typical Flowfield contours for nitrogen flow past a blunt body, M = 3, Kn = 0.1

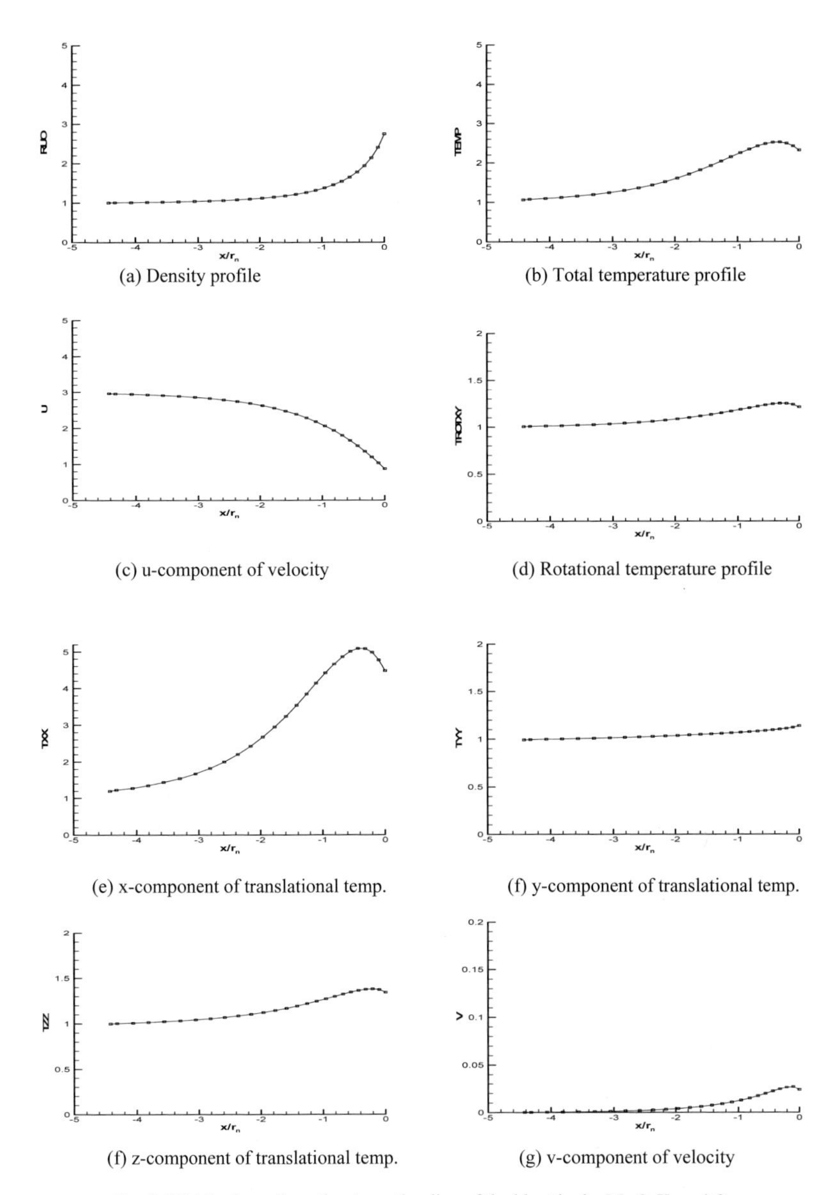


(a) Density profile

(b) Total temperature profile

(c) u-component of velocity

(d) Rotational temperature profile

(e) x-component of translational temp.

(f) y-component of translational temp.

(f) z-component of translational temp.

(g) v-component of velocity

Fig. 5: Distributions along the stagnation line of the blunt body, M =3, Kn = 1.0

## PARALLELIZATION

The computations reported in this paper were performed on four processors of a SGI Origin 2000, 64 R12000 MIPS processor supercomputer (known as Chico) at Washington University. Each processor has 400MHz clock, 8MB Cache and a theoretical peak performance of 800MFLOPS. The grid employed in the calculations has approximately half million points. The CPU time required is 2.481e-08seconds/time step/grid point. The time step for time accurate calculations is 6.667e-04 seconds. There are 10000 time steps per jet cycle. At least ten jet cycles are required to get a periodic solution. It required 45 hours of CPU time on 4-processors to calculate one jet cycle. All the cases were computed using 4-processors. However, one calculation with SST model was performed on 16-processors. Almost linear speed up with parallel efficiency of 93% was obtained.

## ACKNOWLEDGEMENTS

This work was sponsored (in part) by the Air Force Office of Scientific Research. The authors are grateful to AFOSR Program Manager Dr. John Schmisseur. The authors are also grateful to Dr. Eswar Josyula of AFRL for many fruitful discussions and suggestions.

## REFERENCES

[1] Cheremisin, F.G., "Solution of the Wang Chang – Uhlenbeck Master Equation," Doklady Physics, Vol. 47, pp. 872-875, 2002.

[2] Beylich, A.A, "An Interlaced System for Nitrogen Gas," Proc. of CECAM Workshop, ENS de Lyon, France, 2000.

[3] Cheremisin, F. G., "Solution of the Boltzmann Kinetic Equation for High Speed Flows of a Rarefied Gas," Proc. of the 24th Int. Symposium on Rarefied Gas Dynamics, Bari, Italy, 10-16 July, 2004.

[4] Bird, G.A., "Molecular Gas Dynamics and the Direct Simulation of Gas Flows," Oxford Science Publications, New York, NY, 1994.

[5] Oran, E.S., Oh, C.K., and Cybyk, B.Z., "Direct Simulation Monte Carlo: Recent Advances and Application," *Annu. Rev. Fluid Mech.*, Vol. 30, 1998, pp. 403-441.

[6] Ivanov, M.S. and Gimelshein, S.E., "Computational Hypersonic Rarefied Flows," *Annu. Rev. Fluid Mech.*, Vol. 30, 1998, pp. 469-505.

[7] Burnett, D., "The Distribution of Velocities and Mean Motion in a Slight Non-Uniform Gas," *Proc. of the London Mathematical Society*, Vol. 39, 1935, pp. 385-430.

[8] Chapman, S. and Cowling, T.G., "The Mathematical Theory of Non-Uniform Gases," Cambridge University Press, New York, NY, 1970.

[9] Grad, H., "On the Kinetic Theory of Rarefied Gases," *Comm. Pure Appl. Math.*,Vol. 2, 1949, pp.325-331.

[10] Eu, B.C., "Kinetic Theory and Irreversible Thermodynamics", John Wiley & Sons, New York, NY, 1992.

[11] Chen, R., Agarwal, R.K., and Cheremisin, F.G., "A Comparative Study of Navier-Stokes, Burnett, DSMC, and Boltzmann Solutions for Hypersonic Flow Past 2-D Bodies," AIAA Paper 2007-0205, 2007.

[12] Koura, K., "A Set of Model Cross-Sections for the Monte Carlo Simulation of Rarefied Real Gases: Atom-Diatom Collisions," Phys. of Fluids, Vol. 6, pp. 3473-3486, 1994.

[13] Koura, K., "Monte Carlo Direct Simulation of Rotational Relaxation of Diatomic Molecules Using Classical Trajectory Calculations: Nitrogen Shock Wave," Phys. of Fluids, Vol. 9, pp. 3543-3549, 1997.

[14] Erofeev, A.I., "Study of a shock Wave Structure in Nitrogen on the Basis of Trajectory Calculations of Molecular Interactions," Fluid Dynamics, No. 6, pp.134-147, 2002.

[15] Surzhikov, S., Sharikov, I., Capitelli, M., and Colonna, G., "Kinetic Models of Non-Equilibrium Radiation of Strong Air Shock Waves," AIAA Paper 2006-586, AIAA 44th Aerospace Sciences Meeting and Exhibit, Reno, NV., 9-12 January 2006.

[16] Alsmeyer, H., "Density profiles in argon and nitrogen shock waves measured by the absorption of an electron beam," J. Fluid Mech., Vol.74, pp.497-513, 1976.

# Application of Parallel Processing to Numerical Modeling of Two-Phase Deflagration-to-Detonation (DDT) Phenomenon


**Bekir Narin[a], Yusuf Özyörük[b], Abdullah Ulaş[c]**

[a] *Senior Research Engineer*

*The Scientific and Technological Research Council of Turkey, Defence Industries Research and Development Institute, P.K. 16 06261, Mamak, Ankara, Turkey*

[b] *Prof. Dr.*
*Middle East Technical University, Dept.of Aerospace Engineering, Ankara, Turkey*

[c] *Assoc.Prof.*
*Middle East Technical University, Dept.of Mechanical Engineering, Ankara, Turkey*



**ABSTRACT**

This paper describes the development of a one-dimensional, parallel code for analyzing Deflagration-to-Detonation Transition (DDT) phenomenon in energetic, granular solid explosive ingredients. The physical model is based on a highly coupled system of partial differential equations involving basic flow conservation equations and some constitutive equations. The whole system is solved using a high-order accurate explicit finite difference scheme on distributed memory computers with Message Passing Interface (MPI) library routines. In this paper, the physical model for the sources of the equation system is given for such a typical explosive, and several numerical calculations are carried out to assess the developed model, and several parametric studies are carried out to understand the numerical issues involved in solution of such a problem. It is concluded that the results are in good agreement with literature.



Keywords: Deflagration-to-Detonation Transition, 4th Order Runge-Kutta Method, Parallel Processing, Message Passing Interface (MPI)


## 1. Introduction

Simulation of dynamic behavior of the Deflagration-to-Detonation Transition (DDT) phenomenon in reactive gaseous mixtures and in energetic solid explosives is one of the most complex and important problems in solid combustion science. This phenomenon takes place as chemical reactions in such media due to an external thermal or mechanical ignition source. The energetic material starts to deflagrate first, and then this deflagration phenomenon progressively shows transition to detonation. Due to its vital importance for warhead and propulsion researchers, the modelling of the DDT phenomenon in packed, granular, solid energetic materials of rocketry and munition systems has been particularly one of the most active research area in the solid combustion field.

The first recorded studies in this research field were performed by Chapman and Jouget [1], [2], [3], proposing a theory called after their names, Chapman-Jouget (C-J) Theory. This theory assumes that detonation phenomenon occurs in chemical equilibrium and at steady-state conditions. During 1940s, Zeldovich, von Neumann and Döring independently proposed a theory that assumes chemical reaction takes place in a finite region and in finite reaction steps which stacks with the shock wave in the medium during a typical DDT phenomenon [4].


I.H. Tuncer et al., *Parallel Computational Fluid Dynamics 2007*,
DOI: 10.1007/978-3-540-92744-0_15, 

Since 1970s, with the increasing computational capabilities, dynamic simulation studies of DDT phenomenon in both gaseous and condensed phase energetic materials have been performed. In these studies conservation laws were applied with suitable constitutive models, which describe the inter-phase interactions. The DDT phenomenon is described by applying Continuum Mechanics principles for reactive, two-phase flows, and a system of partial differential equations (PDE) is constructed. The resulting equation system is solved by integrating both in spatial and temporal domains using various numerical techniques [5], [6], [7], [8].

These type of problems, due to the inclusion of strong source terms which simulate the interphase interactions, are "stiff", and in calculations too small time steps are required (e.g. on the order of $1x10^{-9}$ to $1x10^{-11}$ s). Because of this, more computation times than those for conventional non-reactive flows are required in DDT calculations even in one-dimensional problems. Therefore in this study, it has been decided to utilize parallel processing with the Message Passing Interface (MPI) libraries.

In the present study, a two-phase reactive model is constructed to simulate the DDT phenomenon in typical granular, energetic solid explosive ingredient called HMX (Cyclotetramethylene-tetranitramine – $C_4H_8N_8O_8$). The governing PDE system is solved employing a 4[th] order Runge-Kutta method. Brief descriptions of the physical and mathematical models are given below which will be followed by a discussion of some numerical simulations.

## 2. The Physical and Mathematical Model

The focus of the present study is the modeling of DDT phenomenon in granular, solid, energetic materials (explosives) confined in a packed bed configuration as depicted in Figure 1.

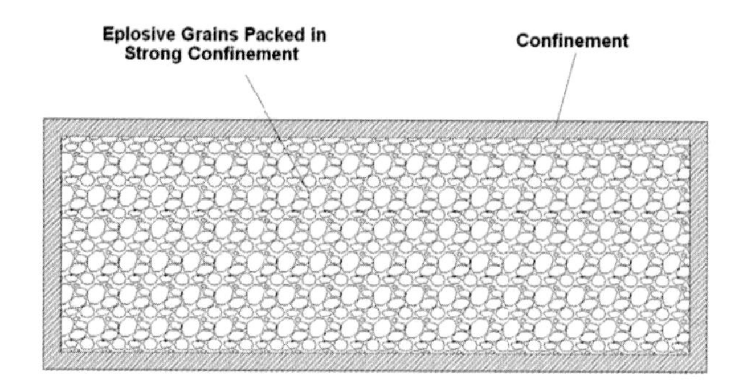


Figure 1. A typical packed bed of granular solid explosive

The DDT process in such an environment starts typically with ignition of first few particles by an external source that may ignite the explosive through either a thermal or mechanical (shock) loading. The calculations presented in the present study are performed by an ignition source of the latter type. The first stage of the DDT process is very slow, and heat conduction between the explosive particles is more effective. After the ignition of first few particles, the neighbouring particles are ignited by heat conduction mechanism as illustrated in Figure 2. In the second stage of the DDT process, the hot gases generated by the combustion of explosive particles penetrate through the porous structure of the unreacted granular explosive bed, as outlined in Figure 3. This process is called as "flame spreading" [5], where convective heat transfer starts dominating.

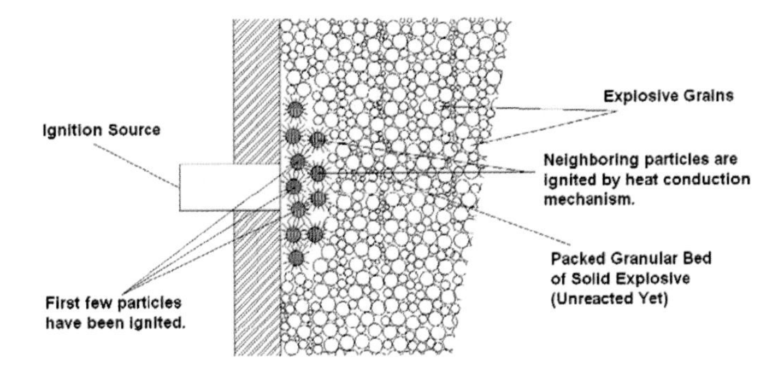


Figure 2. Ignition of explosive bed and first stage of DDT process

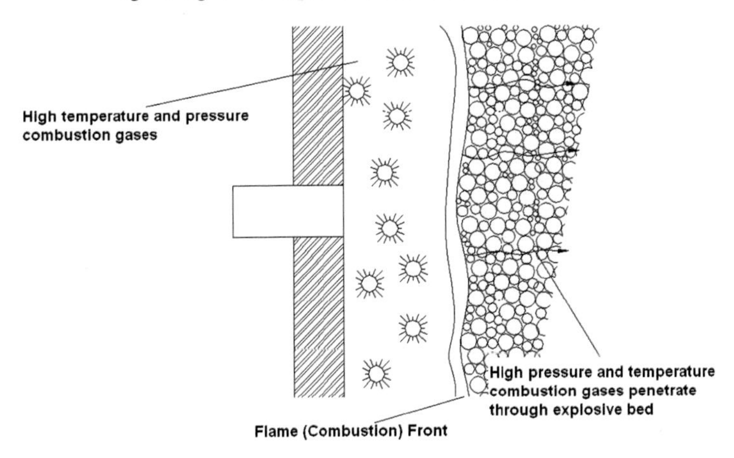


Figure 3. Flame spreading phenomenon

With the effect strong confinement, the convective heat transfer causes to ignite more explosive particles to ignite, leading to increases in gas temperature and pressure by several orders beyond the deflagration limit. Due to the progressive nature of this process, a "steady" detonation wave is reached.

In development of the mathematical model describing the DDT phenomenon it is assumed that each of the solid and gas phases is occupy a given spatial position simultaneously, with corresponding volume fractions those sum to unity. Then the conservation equations for the solid and gas phase masses, momentums and energies are written by accounting for the gains or losses due to mass (solid phase burning and producing gas), and the associated momentum and energy transfers. The rates at which these transfers occur are based on some constitutive models. Hence, to simulate the DDT phenomenon in the investigated granular explosive bed (i.e. HMX), the following two-phase, one-dimensional PDE system is constructed [5], [6], [7], [8].

***Conservation of Mass***

$$\frac{\partial \phi_g \rho_g}{\partial t} + \frac{\partial (\phi_g \rho_g u_g)}{\partial x} = \Gamma_g \tag{1}$$

$$\frac{\partial \phi_p \rho_p}{\partial t} + \frac{\partial (\phi_p \rho_p u_p)}{\partial x} = -\Gamma_g \tag{2}$$

***Conservation of Momentum***

$$\frac{\partial(\phi_g\rho_g u_g)}{\partial t}+\frac{\partial(\phi_g\rho_g u_g^2+\phi_g p_g)}{\partial x}=\Gamma_g u_p - D \tag{3}$$

$$\frac{\partial(\phi_p\rho_p u_p)}{\partial t}+\frac{\partial(\phi_p\rho_p u_p^2+\phi_p p_p)}{\partial x}=-\Gamma_g u_p + D \tag{4}$$

***Conservation of Energy***

$$\frac{\partial(\phi_g\rho_g E_g)}{\partial t}+\frac{\partial(\phi_g\rho_g E_g u_g+\phi_g p_g u_g)}{\partial x}=\Gamma_g E_p - \dot{Q} - D u_p \tag{5}$$

$$\frac{\partial(\phi_p\rho_p E_p)}{\partial t}+\frac{\partial(\phi_p\rho_p E_p u_p+\phi_p p_p u_p)}{\partial x}=-\Gamma_g E_p + \dot{Q} + D u_p \tag{6}$$

***Compaction Dynamics***

$$\frac{\partial(\rho_p)}{\partial t}+\frac{\partial(\rho_p u_p)}{\partial x}=-\frac{\rho_p\phi_g}{\mu_c}(p_p-p_e-p_g) \tag{7}$$

***Evolution of the Solid Propellant Particle Radius***

$$\frac{dr_p}{dt}=-aP_g^n \tag{8}$$

Gas Generation Rate Due to Combustion

$$\Gamma_g=\frac{3}{r_p(t)}\phi_p\rho_p(aP_g^n) \tag{9}$$

Interphase Drag Interaction

$$D=f_{pg}(U_g-U_p) \tag{10.a}$$

$$f_{pg}=\frac{10^4\phi_g\phi_p}{r_p} \tag{10.b}$$

Interphase Heat Transfer Interaction

$$\dot{Q}=h_{pg}(T_g-T_p) \tag{11.a}$$

$$h_{pg}=\frac{10^7\phi_g\phi_p}{r_p^{1/3}} \tag{11.b}$$

Gas Phase EOS (Jones-Wilkins-Lee Form) [5]

$$p_g=Ae^{-R_1\frac{\phi_{p0}\rho_{p0}}{\rho_g}}+Be^{-R_2\frac{\phi_{p0}\rho_{p0}}{\rho_g}}+\omega C_{vg}T_g\rho_g \tag{12.a}$$

$$e_g=C_{vg}(T_g-T_{g0})-E_{ch}+\frac{1}{\phi_{p0}\rho_{p0}}\left[\frac{A}{R_1}e^{-R_1\frac{\phi_{p0}\rho_{p0}}{\rho_g}}+\frac{B}{R_2}e^{-R_2\frac{\phi_{p0}\rho_{p0}}{\rho_g}}\right] \tag{12.b}$$

$C_{vg}=2400\ \mathrm{J/kg\cdot K}$

$A=2.4\times10^{11}$ Pa , $B=5\times10^{8}$ Pa, $R_1=4.2$, $R_2=1.0$, $\omega=0.25$ (Model constants)

*Solid-Phase EOS (Tait) [6]*

$$p_p = (\gamma_p - 1)C_{vp}\rho_p T_p - \frac{\rho_{p0}\varsigma}{\gamma_p} \quad (13.a)$$

$$e_p = C_{vp}T_p + \frac{\rho_{p0}\varsigma}{\gamma_p\rho_p} + E_{ch} \quad (13.b)$$

$C_{vp} = 1500\ \mathrm{J/kg \cdot K}$, $E_{ch} = 5.84\ \mathrm{MJ}$

$\gamma_p = 5$, $\varsigma = 8.98 \mathrm{x} 10^6\ \mathrm{m^2/s^2}$

*Configuration Pressure (Stress occured in the unreacted solid propellant because of configuration change caused by compaction)*

$$p_e = \frac{p_{p0} - p_{g0}}{\phi_{p0}}\phi_p \quad (14)$$

$\phi_{p0} = 0.7$ (Initial solid-phase volume fraction)

In Eqn.'s 1-13, subscripts *g* and *p* are used to define the properties of gas and solid phases, respectively. To simulate the burning and thereby regression of an individual explosive particle, the relation defined in Eqn. 9 is used. In this relation it is assumed that the mass transfer rate from the solid to the gas phase is a function of the particle radius $r_p$, the solid volume fraction $\phi_p$, the solid phase density $\rho_p$, and the gas phase pressure $p_g$. The interphase interactions due to the drag (D) and heat transfer ($\dot{Q}$) are considered as given in Eqn.'s 10 and 11, respectively. The stress exerted on the solid phase because of the configuration change in the explosive bed (i.e. change in volume fractions of each phase) is defined as the "configuration pressure", and Eqn. 14 is used for this. Regression of a single explosive particle is modelled using Eqn. 8. If the particle radius decreases under a specific limit, a burn termination technique is applied. The compaction relation defined in Eqn. 7 drives the solid and gaseous phases toward mechanical equilibrium. Note all the above constitutive relations are determined by obeying the 2nd law of thermodynamics [5], [6], [8].

## 3. Numerical Method

The above PDE system contains equations that are highly coupled due to the source terms that include combustion of explosive particles and interphase interactions. Because of this and also due to the existence of disparate eigenvalues the PDE system is very stiff. In addition, the DDT phenomenon is inherently unsteady. Therefore, very small time scales exist, and numerical integration of the above system is not straight-forward. Accuracy and robustness of the numerical method is important.

Although for numerically stiff equations implicit or upwind schemes may be more feasible, it has been determined that by using the 4th-order Runge-Kutta time integration method with high-order central differencing [9] works sufficiently well. To augment central differencing and prevent excessive dispersion, controlled artificial diffusion terms are added to the equations.

The highly coupled hyperbolic system of equations can be written in the following conservative forms:

$$\frac{\partial \vec{U}}{\partial t} + \frac{\partial \vec{F}}{\partial x} = \vec{S} \quad (15.a)$$

$$\frac{\partial \vec{U}}{\partial t} = -\frac{\partial \vec{F}}{\partial x} + \vec{S} = \vec{H} \quad (15.b)$$

Note that Eqn. (15.b) actually defines a system of ordinary differential equations (ODE). According to 4th order

Runge-Kutta method, this ODE system is integrated as follows:

$$\begin{aligned} U^{(1)} &= U^{n} \\ U^{(2)} &= U^{n} + \Delta t \alpha_2 H^{(1)} \\ U^{(3)} &= U^{n} + \Delta t \alpha_3 H^{(2)} \\ U^{(4)} &= U^{n} + \Delta t \alpha_4 H^{(3)} \end{aligned} \qquad (15.c)$$

Note that in Eqn. 15.c, $\alpha_2 = \alpha_3 = 0.5$ and $\alpha_4 = 1.0$. Then $\vec{U}$ vector is updated according to:

$$U^{n+1} = U^{n} + \frac{\Delta t}{6}\left(H^{(1)} + 2H^{(2)} + 2H^{(3)} + H^{(4)}\right) \qquad (15.d)$$

Numerical solution of DDT problems require quite fines meshes, and the time scales are very low. Therefore, significant computer memory and time is needed, in particular for multi-dimensional problems. Hence parallel computing approach is employed. The domain is divided into subzones of equal grid points and then distributed over processors. Ghost points beyond domain boundaries are introduced for data communication with the neighbouring zones.

## 4. Results for DDT in Granular HMX Bed and Discussions

To test the above defined mathematical model for DDT phenomenon simulation, a well-defined model in open literature for HMX is used. In the calculations a 0.1 m long bed of HMX is considered. It is assumed that the HMX explosive bed is packed to a 70% initial density (i.e. solid phase volume fraction is $\phi_P = 0.7$) with uniform particles of a surface-mean diameter of $200\,\mu$m. In calculations, compaction induced detonation is taken into account, and therefore, at the left boundary moving boundary conditions are applied. At the right boundary, wall boundary conditions are applied [8].

Before comparing the results obtained with the developed algorithm with those of open literature, some numerical investigations are performed to show the grid independency of the solutions. For this purpose, calculations are performed with N = 1000, 2000, 10000, 20000, and 40000 grid point resolutions, and the results are plotted in Figure 4 and Figure 5. In these figures gas and solid phase pressure and temperature profiles for different grid resolutions are shown. It is clear that as the grid resolution is increased, the peak value for the gas phase pressure and temperature profiles show a converging behavior.

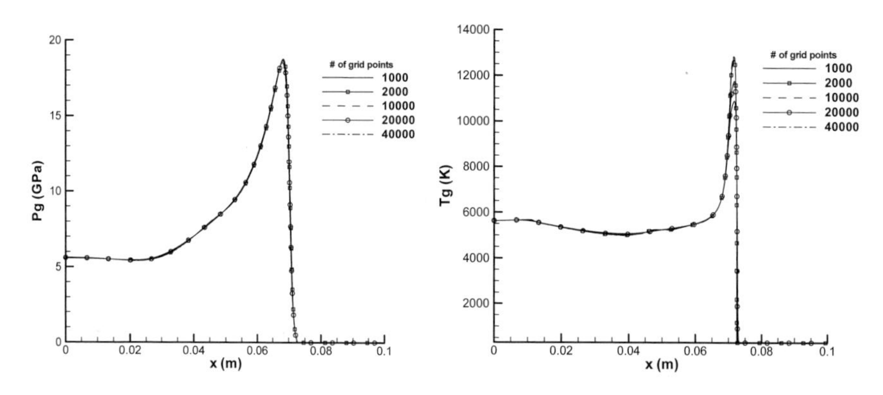


Figure 4. Comparison of gas pressure and temperature profiles for different number of grid points for $t_{final} = 15\,\mu$s, for 10 cm loaded HMX bed

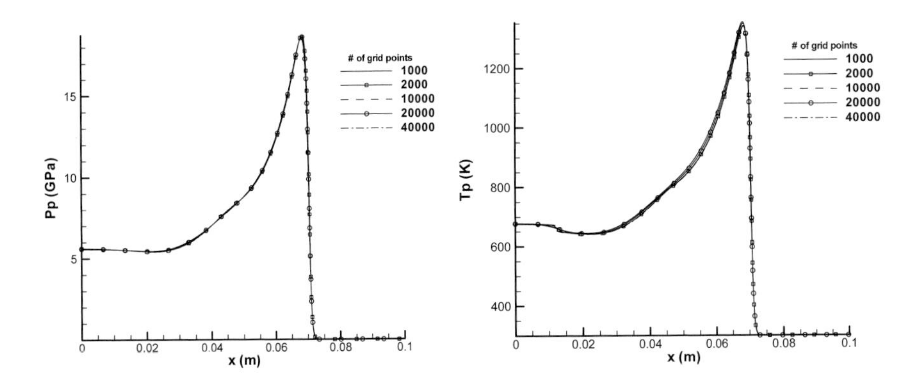


Figure 5. Comparison of solid pressure and temperature profiles for different number of grid points for $t_{final}$ = 15 μs, for 10 cm loaded HMX bed

The grid independent, steady peak values (peaks of detonation waves) are compared in Table 1 with those obtained using a thermochemical code named EXPLO5 [11] and those of the transient calculations of Baer and Nunziato [5]. The transient calculations of the present study and those of Baer and Nunziato are in good agreement for $P_{CJ}$, $T_{CJ}$, $D_{CJ}$ while the $P_{CJ}$ and $T_{CJ}$ values of steady calculations by EXPLO5 differ. This situation is previously stated in literature [5], [12], with a conclusion that the detonation pressure results of transient calculations may differ 10-20% from the steady, chemical equilibrium calculations, while the detonation velocities may agree to within a few percent. It is also indicated that actual gas temperature values may reach above 10000 K [5]. Since the non-steady nature of the physical problem cannot be captured in steady, chemical equilibrium calculations, the detonation temperature and pressure results obtained by such models are drastically lower than those given by transient calculations.

Table 1. Comparison of Results with Those of Open Literature

| | Present Study (Transient Calculation) | EXPLO5 [11] (Steady, Chemical Equilibrium Calculation) | Baer & Nunziato [5] (Transient Calculation) |
|---|---|---|---|
| Detonation Pressure ($P_{CJ}$) | 23.4 GPa | 18,4 GPa | ≈25 GPa |
| Detonation Temperature ($T_{CJ}$) | 10700 K | 4369,2 K | ≈10000 K |
| Detonation Velocity ($D_{CJ}$) | 7200 m/s | 7318 m/s | ≈7200 m/s |

As mentioned above, all calculations are performed with parallel processing by facilitating Message Passing Interface (MPI) libraries. The calculations are performed in parallel processing facilities of Dept. of Aerospace Engineering of Middle East Technical University. The computing platform includes computers with dual Pentium and Xeon processors with 1024 MB memory which are run under Linux operating system. The communication between the processors is provided by a local network with a 1 Gbps switch. The speed-up characteristics of the present code are presented in Figure 6. The figure shows variation of the speed-up by the number of processors used. There is some deviation from the ideal speed up curve.

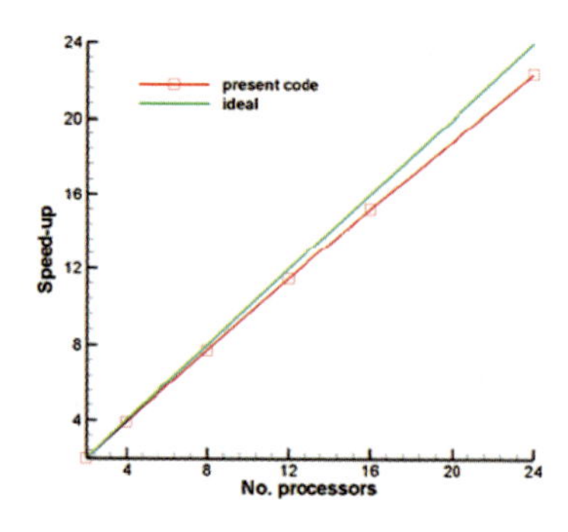


Figure 6. Parallelization speed-up.

## 5. Concluding Remarks

In this study, the development of a one-dimensional, parallel code for analyzing Deflagration-to-Detonation Transition (DDT) phenomenon in energetic, granular solid explosive ingredients is described. The governing equation system and numerical procedure is described and the defined model is tested with typical problem defined in open literature. Speed-up characteristic of the developed parallel algorithm is determined. The results determined with the developed code are compared with those of open literature and reasonable agreement is achieved.

## REFERENCES

[1] Jouget, E., On the Propagation of Chemical Reactions in Gases, J. de Mathematiques Pures et Apliquees 2, pp 5-85, 1906

[2] Jouget, E., On the Propagation of Chemical Reactions in Gases, J. de Mathematiques Pures et Apliquees 1, pp 347-425, 1905

[3] Chapman, D.L., On the Rate of Explosion in Gases, Philos. Mag. 47, pp 90-104, 1899

[4] Yoh, J.J., Thermomechanical and Numerical Modeling of Energetic Materials and Multi-Material Impact, Ph.D. Dissertation, University of Illinois at Urbana-Champaign Graduate College, 2001

[5] Baer, M.R., Nunziato, J.W., A Two-Phase Mixture Theory for the Deflegration-to-Detonation Transition (DDT) in Reactive Granular Materials, International Journal of Multiphase Flow, Vol.12, No.6 , pp 861-889, 1986

[6] Powers, J.M., Stewart, D.S., Krier, H., Theory of Two-Phase Detonation - Part I : Modeling, Combustion and Flame, Vol.80, pp. 264-279, 1990

[7] Powers, J.M., Stewart, D.S., Krier, H., Theory of Two-Phase Detonation - Part II : Structure, Combustion and Flame, Vol.80, pp. 264-279, 1990

[8] Gonthier, K.A., Powers, J.M., A High-Resolution Numerical Method for a Two-Phase Model of Deflagration to Detonation Transition, Journal of Computational Physics, Vol. 163, pp. 376-433, 2000

[9] Hirsch, C., Numerical Computation of Internal and External Flows Vol.1: Fundamentals of Numerical Discretization, John Wiley and Sons Inc., 1999

[10] Toro, E.F., Numerica, A Library of Source Codes for Teaching, Research and Applications, www.numeritek.com, 1999

[11] Suceska, M., Private communication, 2006

[12] Mader, C., Numerical Modeling Explosives and Propellants, CRC Press, 1998

# Highly Scalable Multiphysics Computational Framework for Propulsive Energetic Systems

F.M. Najjar[a*], A. Haselbacher[b], R. Fiedler[a], S. Balachandar[b], and R. Moser[c]

[a]Computational Science & Engineering, University of Illinois at Urbana-Champaign, 1304 West Springfield Avenue, Urbana, IL 68101.

[b]Department of Mechanical & Aerospace Engineering, University of Florida, Gainesville, FL 32611.

[c]Department of Mechanical Engineering, University of Texas, Austin, TX 78712

A computational framework to perform multiphase flow simulation is described and one of its application is focused on modelling of solid-propellant rocket motors. This framework allows to compute the chamber dynamics of the gas mixture, investigate the evolution of $Al$ Lagrangian burning droplets and understand the generation of alumina smoke. Unstructured grids are used to enable meshing of geometrically complex configurations. An efficient parallel droplet-localization algorithm has been developed for unstructured mixed meshes.

## 1. INTRODUCTION

Applications with multiphase flows are found abundantly in nature and engineering devices including particle-laden pipes, cyclones and aerosol transport, to name a few. A particularly interesting example of multiphase (MP) flow is encountered in solid-propellant rocket motors (SRM). Modern SRMs are enriched with aluminum particles to increase specific impulse and to damp combustion instabilities [1]. As the propellant burns, aluminum particles melt and agglomerate collecting as droplets at the propellant-combustion interface. Aluminum ($Al$) droplets are entrained into the shear flow and their initial size typically varies from tens of microns to a few hundred microns. Due the high temperatures typically encountered in the motor chamber, the $Al$ droplets vaporize and react with various oxidizing agents, generating micron-sized aluminum oxide ($Al_2O_3$) particles. Understanding the behavior of $Al$ droplets and $Al_2O_3$ particles is key in the design and analysis of SRM. For example, the impact of oxide particles on the nozzle walls can lead to scouring; and $Al$ droplets can accumulate in the bucket region of submerged nozzles forming slag [2].

Simulating solid-propellant rocket motors (SRM) from first principles with minimal empirical modeling is a challenging task. Modern segmented rocket motors are inherently

*Research supported by the Department of Energy through the University of California under subcontract B523819.

I.H. Tuncer et al., *Parallel Computational Fluid Dynamics 2007*,
DOI: 10.1007/978-3-540-92744-0_16, © Springer-Verlag Berlin Heidelberg 2009

three-dimensional and geometrically complex. A few examples of these complexities include: the time-dependent geometry changes as the propellant burns; the internal flow is nearly incompressible close to the head-end becoming supersonic in the nozzle region; the flow transitions from a laminar state near the head-end to a fully turbulent state farther downstream; the internal flow is multiphase in nature due to metalized propellants; and the flow is chemically reacting. Further, these various physical processes that account for these complexities are often interrelated and therefore require care in their computational modeling.

The current work presents the overarching steps in the development of a MP flow computational framework aimed at investigating the evolution of gas mixture, $Al$ droplets and $Al_2O_3$ smoke in the SRM chamber. This framework is developed for unstructured meshes; and particular thrust has focused on the efficient and fast evolution of the Lagrangian droplet. Thus, a droplet-localization parallel algorithm has developed and implemented for tracking droplets on three-dimensional mixed unstructured grids [7]. One of its distinguishing feature is that information about droplets impacting on boundaries is gathered automatically.

Detailed full-physics multiphase simulations of the flow are performed to investigate the chamber environment for the Space-Shuttle Solid Rocket Booster at its initial burn-time configuration. The paper is organized as follows: Section 2 describes the governing equations and the physical models invoked. Section 3 presents aspects of the numerical parallel modeling. Results obtained from the simulations for the SRM considered are discussed in Section 4. Section 5 provides a summary and conclusions.

## 2. GOVERNING EQUATIONS

We describe the various evolution equations considered to solve the coupled multiphase flow in the SRM chamber including the evolution of the gas mixture, the Lagrangian $Al$ droplet and the formation of $Al_2O_3$ smoke.

The fluid motion is governed by the Navier-Stokes equations, expressed in integral form for a control volume $\Omega$ with boundary $\partial\Omega$ and outward unit normal $\mathbf{n}$,

$$\frac{\partial}{\partial t}\int_{\Omega} \mathbf{u}\, dV + \oint_{\partial\Omega} \mathbf{f}\cdot\mathbf{n}\, dS = \oint_{\partial\Omega} \mathbf{g}\cdot\mathbf{n}\, dS + \int_{\Omega} \mathbf{s}\, dV \tag{1}$$

where $\mathbf{u} = \{\rho, \rho\mathbf{v}, \rho E\}^{\mathrm{T}}$ represents the state vector of conserved variables, $\mathbf{f}$ and $\mathbf{g}$ are the corresponding inviscid and viscous flux tensors, and $\mathbf{s}$ is the source vector. Further details are discussed in [4,5]. The ideal gas law is assumed to apply. The heat flux vector is given by Fourier's law $\mathbf{q} = -\kappa\nabla T = -\frac{\gamma R}{\gamma - 1}\frac{\mu}{\mathrm{Pr}}\nabla T$. The variation of the dynamic viscosity with temperature is given by Sutherland's law.

The Lagrangian evolution equations for the droplet position vector $\mathbf{r}^p$, the constituent species $m_i^p$, the momentum vector $\mathbf{v}^p = \{u^p, v^p, w^p\}^{\mathrm{T}}$, and energy $E^p = C^p T^p + \frac{1}{2}(\mathbf{v}^p)^2$ are

$$\frac{d}{dt}\left\{\begin{array}{c} \mathbf{r}^p \\ m_i^p \\ m^p\mathbf{v}^p \\ m^p E^p \end{array}\right\} = \left\{\begin{array}{c} \mathbf{v}^p \\ \dot{m}_i^p \\ \mathbf{f}^p \\ q^p \end{array}\right\} \tag{2}$$

where the force $\mathbf{f}^p$, and energy source $q^p$ are given by $\mathbf{f}^p = \mathbf{f}^p_{drag} + \mathbf{v}^p \frac{dm^p}{dt}$ and $q^p = q^p_{Ther} + E^p \frac{dm^p}{dt} + \mathbf{f}^p_{drag} \cdot \mathbf{v}^p$. In the above equations, $m^p$ is the droplet mass, $d^p$ is the droplet diameter, and $C^p$ is the droplet specific heat. A simple drag law is used to evolve the droplets. In Eq. 2, $\dot{m}^p_i$ represents the mass source and is described below.

The current computational framework allows for the droplets to consist of multiple components. For example, in the present context each droplet is a mixture of molten $Al$ and $Al_2O_3$ in the form of an oxide cap. Even as the droplet is injected at the propellant combustion interface, part of it is in the form of aluminum oxide [13]. As the droplet moves through the core flow, $Al$ evaporates and burns. In contrast, the amount of $Al_2O_3$ in the droplet increases as the droplet collects the surrounding oxide smoke particles through a collision and agglomeration process. Thus, the droplet starts out being mostly $Al$, but ends up being mostly $Al_2O_3$. Here we track the composition of each droplet by evolving individually the mass of each of the constituents as follows: The burn rate of a droplet uses the correlation proposed by Widener and Beckstead [13].

The smoke particles are evolved according to

$$\frac{\partial}{\partial t}\int_{\Omega} \rho Y_i^s \, dV + \oint_{\partial\Omega} \rho Y_i^s \mathbf{v}_i^s \cdot \mathbf{n} \, dS = \oint_{\partial\Omega} D_i^s \nabla Y_i^s \cdot \mathbf{n} \, dS, + \int_{\Omega} s_i^s \, dV \tag{3}$$

where $Y_i^s$ is the mass fraction of the $i$th smoke species, $\mathbf{v}_i^s$ is the velocity of the $i$th smoke species, $D_i^s$ is the diffusion coefficient of the $i$th smoke species, and $s_i^s$ is the rate of generation of the $i$th smoke species. As indicated in Eq. (3), the smoke particles are advected with a velocity which may differ from that of the gas mixture. The smoke velocity can be estimated using the Equilibrium Eulerian Method of Ferry and Balachandar [3], Finally, interaction terms are present coupling the various components. The aluminum droplets influence the momentum and energy of the gas mixture. Further, due to the burning process, the droplets affect the mass of the gas mixture and the oxide smoke field. This leads to coupling of the Lagrangian droplet back onto the carrier gas mixture and oxide smoke field. The source terms arising from all droplets within a finite volume cell are assumed to contribute to that cell only. This provides an overview of the various governing equations for the gas carrier phase, the Lagrangian $Al$ droplets and the $Al_2O_3$ smoke particles. Further details are discussed in Najjar *et al.* [11] and Haselbacher *et al.* [7].

## 3. NUMERICAL METHOD AND ALGORITHMS

The solution method for the gas mixture is based on the cell-centered finite-volume method, in which the dependent variables are located at the centroids of the control volumes. The control volumes are given by the grid cells, which can be tetrahedra, hexahedra, prisms, and pyramids. The inviscid fluxes are approximated by Roe's flux function, where the face-states are obtained by a weighted essentially non-oscillatory (WENO) scheme. The viscous fluxes are computed by reconstructing the face gradients. The spatially discretized equations for the mixture are evolved using the third-order Runge-Kutta method proposed by Wray [14]. Further, the spatial discretization of the smoke equations is formally identical to that of the gas mixture in that gradients and face states of the mass

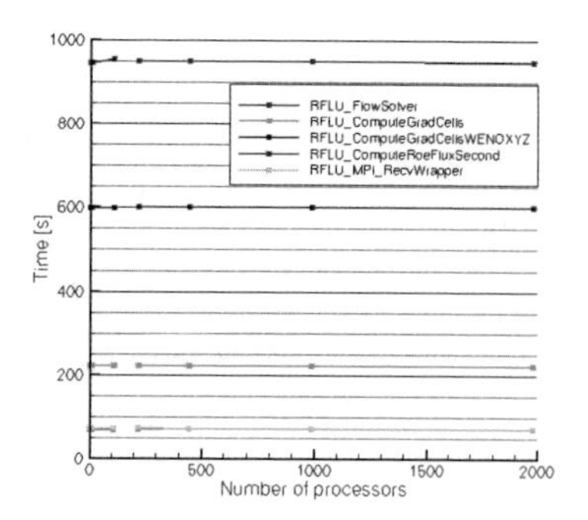


Figure 1. Results of parallel scalability study on an IBM Blue Gene machine for the mixture solver. CPU timings for various components of the CFD solver are shown. Red line corresponds to the overall CPU time of the solver; green, blue, and magenta lines are for the routines computing gradient terms; orange line shows the cpu time for MPI communication.

fraction are evaluated using the WENO scheme. The numerical method is parallelized using the domain-decomposition approach by partitioning the grid into groups of cells using METIS. For more details, see Haselbacher [4,5].

The Lagrangian droplet equations are ordinary differential equations by nature and their solution requires the evaluation of the mixture properties at the droplet locations. Hence, a robust, efficient and parallel particle-localization algorithm has been developed for unstructured mixed meshes [7]. The Lagrangian evolution equations are integrated in time using the same three-stage Runge-Kutta method [14] as for that for the mixture and smoke. The datastructure of the droplets is tightly linked to that of the mixture where each parallel region evolve its own set of droplets. Once a droplet crosses a border connecting two regions, it is communicated to the adjacent region. The adjacent region may be located on the same or a different processor. Hence, two distinct levels for communicating droplets are encountered: (a) Droplets moving between regions on the same processor, requiring only copying of data; (b) Droplets moving between regions on different processors, requiring MPI communication through buffers. One difficulty with simulating droplets in parallel is that the buffer size changes dynamically at each Runge-Kutta stage. While the buffer sizes for the mixture are constant and determined during the preprocessing stage, communicating the droplets first requires determining the buffer size. As the droplets are being evolved, the localization algorithm keeps track of the number of droplets passing through faces associated with borders. A further difficulty is that droplets have to continue to be tracked after they were received by an adjacent region and may actually have to be communicated again. The present implementation ensures that droplets can move through as many regions with differently-sized cells in a robust and consistent manner without any restriction on the time step. The parallel performance mixture solver is seen

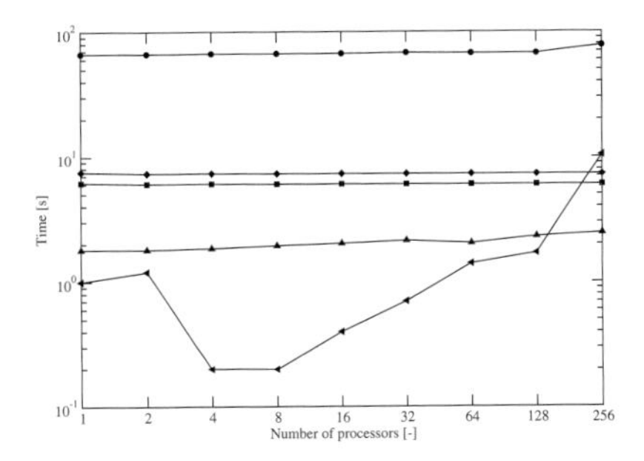


Figure 2. Results of parallel scalability study on an IBM SP5 machine. •: Run time of solver (excluding I/O), Square: Update of particle position, mass, momentum, and energy, Lozenge: Particle localization algorithm, Triangle: Communication driver (excluding MPI calls), Left Triangle: Reduction operation to determine total number of particles to be communicated.

to scale perfectly up to 1920 processors of the IBM Blue Gene machine, as shown in Fig. 1. The computational times for key components are plotted. The horizontal lines indicate the high scalability of the CFD solver as the problem size increases with processor count, the CPU count remains constant. A performance analysis is performed next on the MP coupled framework. To this end, a scaled problem with $32^3$ hexahedral cells per region is chosen. The partitions are stacked in the $x$-direction to generate a square-pencil-like geometry as the number of partitions increases. Periodic boundary conditions are employed in the $x$-direction to ensure that the ratio of communication to computation of the underlying Eulerian solver is constant across the partitions. Each partition is assigned 16384 particles at random locations within its bounding box. The random locations of the particles within a partition means that the number of particles being communicated between partitions at a given instant is only approximately constant, thus introducing a slight load imbalance. The scaling of key routines up to 256 processors on an IBM SP5 machine is shown in Fig. 2. It can be seen that the scaling is nearly perfect except for a slight increase in the run time from 128 to 256 processors. This increase is caused by the non-scalable behavior of the collective communication needed to determine the total number of particles to be communicated.

## 4. RESULTS

Various test cases have been used to perform a verification and validation (V & V) analysis for the present MP computational framework. Here, the results obtained from simulations of the Space-Shuttle SRM are discussed. The initial burn time (t=0s) is considered and for this configuration, the following complex geometrical aspects are considered: an 11-star grain is present at the head-end, slots separate the four segments, propellant fills the bucket of a submerged nozzle. Three-dimensional computations have been performed using an unstructured grid consisting of tetrahedra cells. The RSRM geometry

Figure 3. 11-star fin RSRM geometry at t=0 second. Also shown is the 256-region partitioned computational mesh.

have been discretized in 1.74 million cells and the computational mesh is partitioned into 256 regions. Figure 3 shows the configuration along with the various partitioned domains. Previous simulations on a grid of 4.5 million cells have shown that the head-end pressure has reached a value that is grid-independent for both these meshes. The following conditions are imposed along the boundaries of the computational domain: a no-slip velocity is set at the head-end and along the nozzle walls; fixed mass-rate and temperature are set along the injection surfaces of the propellant; and a mixed supersonic-subsonic outflow condition is imposed at the nozzle outlet. The mixture and droplet properties are typical of those used for these simulations and are provided elsewhere [6]. The viscosity and thermal conductivity are assumed to be constant. The mass injection rate of the system is set to a fixed value; while the droplets are injected at zero velocity and their temperature is set to the boiling point of $Al_2O_3$. The droplet distributions used are obtained from an agglomeration model discussed by Jackson *et al* [8] The $Al$ loading in the solid propellant is set to be 18%; while, the smoke particles are assumed to be generated at a fixed diameter of 1.5$\mu$m. Computations have been performed on a 256-processor IBM SP5 computer system. Single-phase flow simulation was initially carried out. This computation provides the basic dynamics of the chamber flow. It was seen that the flow is of quasi-steady nature in the main chamber. Further, the head-end pressure compares quite well with the cited measurements from the Space Shuttle Design Data Book (Thiokol Corp) and the TEM-6 test firings with a value of $\approx$ 65 atm.

Subsequently, simulations are performed for the case with burning droplets. Once the single-phase flow has reached a quasi-steady state (typically for a simulation time of $\approx$ 1 s), multiphase flow runs are performed for a simulation time of over 2.5 s: during the initial 1.0s the droplets fill the rocket chamber, after which statistics of the droplet impacting on the nozzle surface are gathered. Figures 4 and 5 presents $3-D$ plots for the contours of the mixture temperature and the droplet locations (colored by their diameter) at a representative time instant in the quasi-steady regime. It is seen in Figure 4 that the static temperature is roughly uniform in the chamber and drops quickly in the nozzle. The three-dimensionality of the flow is also captured in the temperature distribution inside the slots separating the various segments. From Figure 5, the plot of droplet locations shows

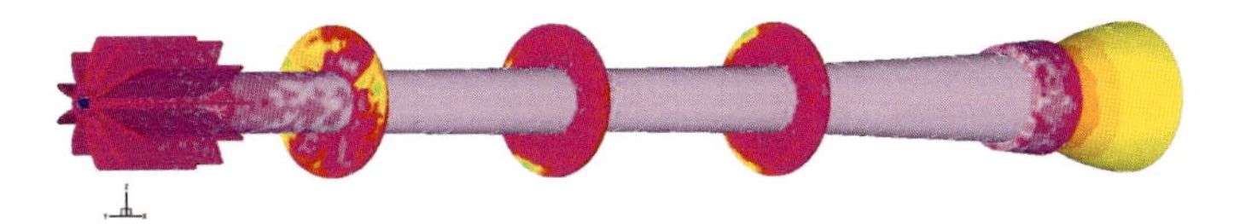

Figure 4. 3-D Multiphase flow solution for RSRM at initial burn configuration at a representative time instant. Contours of temperature.

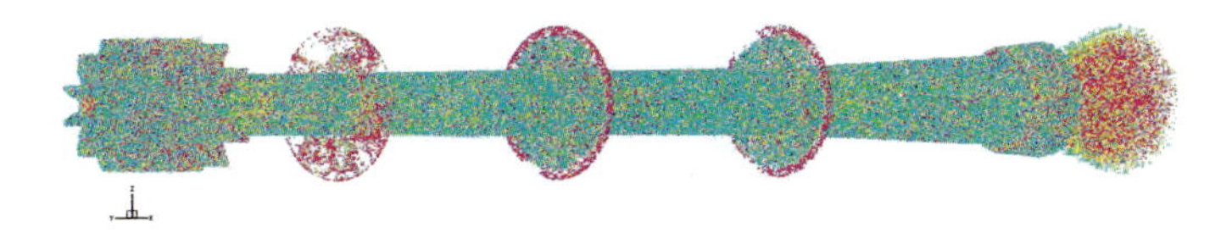

Figure 5. 3-D Multiphase flow solution for RSRM at initial burn configuration at a representative time instant. Locations of $Al$ droplets (droplets colored by their diameter).

that the chamber is filled with droplets of various sizes. Large-sized droplets are seen to be trapped in the first slot (since the propellant is inhibited); while the second and third slots are fully populated by burning droplets. Further, significant amount of droplets are seen in the nozzle bucket due to the presence of burning propellant. Preliminary statistics obtained for the droplet impact on the walls of the submerged nozzle are shown in Figure 6. Substantial number of impact is seen on the nozzle cowl as well as the exit cone plane of the nozzle (see Figure 6). Large-sized droplets ($\geq 100\,\mu m$) impact the nozzle cowl while smaller sized ones ($\leq 25\,\mu m$). The angle of impact was found to be mainly 90-degrees with smaller value (typically of skidding) near the nozzle throat. No impact is currently captured in the region close to the nozzle throat. Simulation with a longer simulation time span and a larger number of droplets are required to gather more statistically meaningful data and are currently being performed.

## 5. Conclusions

A multiphase flow computational framework was described to investigate the SRM internal chamber dynamics. Unstructured grids are used to enable the representation of geometrically complex motors. An efficient parallel droplet-localization algorithm has been developed for unstructured mixed meshes. One of its key distinguishing features is that it gathers automatically data about droplet impacts on boundaries. Hence detailed statistical data about droplet impact locations, angles, velocities, and energies can be provided. Multiphase flow simulations of the RSRM at initial time of burn (t=0 second) have been carried out and the multiphase flow dynamics including the $Al$ Lagrangian droplet evolutions in the motor chamber have been presented. Impact statistics on the submerged nozzle wall and the cowl provide further insight on the effects of the $Al$ field. Future com-

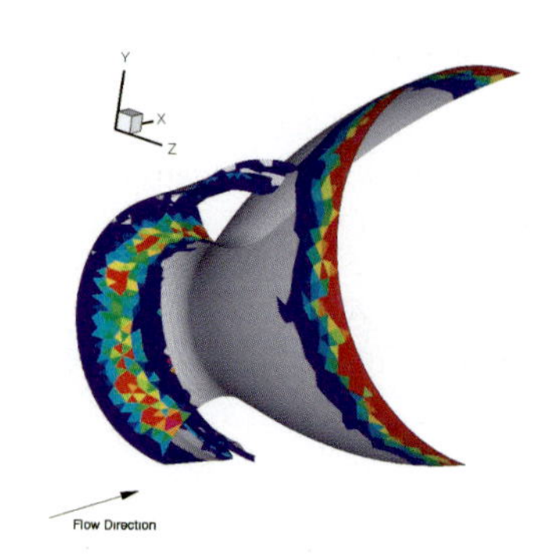


Figure 6. Droplet impact statistics on nozzle surface including nozzle cowl region. Shading represents nozzle surface. Mean diameter $d_{43}$ of impacting droplets.

putations will include LES turbulence models as well as the effects of turbulence on the droplet evolution through Langevin-based models.

## REFERENCES

1. G.P. Sutton and O. Biblarz , Rocket Propulsion Elements. 7th Edition, Wiley, 2001.
2. M. Salita, J. Prop. Power 11 (1995) 10.
3. J.P. Ferry and S. Balachandar, Power Technology 125(2002) 131-139.
4. A. Haselbacher, AIAA Paper 2005-0879 (2005).
5. A. Haselbacher, AIAA Paper 2006-1274 (2006).
6. A. Haselbacher and F.M. Najjar, AIAA Paper 2006-1292 (2006).
7. A. Haselbacher, F. Najjar, and J. Ferry, J. Comp. Phys. 225 (2007) 2198-2213.
8. T.L. Jackson, F.M. Najjar, and J. Buckmaster, J. Prop. Power 21(2005) 925.
9. B. Larrouturou, J. Comp. Phys. 95 (1991) 59.
10. F.M. Najjar, A. Haselbacher, J.P. Ferry, B. Wasistho, S. Balachandar, and R.D. Moser, AIAA Paper 2003-3700 (2003).
11. F.M. Najjar J.P. Ferry J.P., A. Haselbacher, and S. Balachandar, J. SpaceCraft & Rockets 43 (2006) 1258.
12. P.L. Roe P.L., J. Comp. Phys. 43 (1981) 357.
13. J.F. Widener and M.W. Beckstead, AIAA 98-3824 (1998).
14. A.A. Wray, unpublished manuscript (1995).

# A Parallel Aitken-Additive Schwarz Waveform Relaxation Method for Parabolic Problems

Hatem Ltaief[a] and Marc Garbey[a]*

[a]University of Houston, Department of Computer Science, 501 PGH Hall, Houston TX 77204, USA

The objective of this paper is to describe a parallel acceleration framework of the additive schwarz waveform relaxation for parabolic problems. The problem is in three space dimension and time. This new parallel domain decomposition algorithm generalizes the Aitken-like Acceleration method of the additive Schwarz algorithm for elliptic problems. Although the standard Schwarz Waveform Relaxation algorithm has a linear rate of convergence and low numerical efficiency, it is beneficent to cache use and scales with the memory. The combination with the Aitken-like Acceleration method transforms the Schwarz algorithm into a direct solver for the heat operator. This solver combines all in one the load balancing, the efficiency, the scalability and the fault tolerance features which make it suitable for grid environments.

## 1. Introduction

Nowadays, Grids offer to scientists and researchers an incredible amount of resources geographically spread but often, the poor performance of the interconnection network is a restricting factor for most of parallel applications where intensive communications are widely used. Moreover, standard processors are becoming multi-cores and there is a strong incentive to make use of all these parallel resources while avoiding conflicts in memory access. The idea here is to develop new parallel algorithms suitable for grid constraints and fitting the requirements of the next evolution in computing.

In this paper, we present a parallel three dimensional version of the Aitken-Additive Schwarz Waveform Relaxation (AASWR) for parabolic problems. This algorithm is the analogue of the Aitken-like Acceleration method of the Additive Schwarz algorithm (AS) for elliptic problems [1,2]. Further details regarding the generalization of AS can be found in [3]. The state of the art to speed up AS methods is typically based on either a coarse grid preconditioner (1) or the optimization of the transmission conditions (2). But (1) does not scale on parallel infrastructures with a slow network. (2) has been followed with success by numerous workers - see for example [4–8] and their references. Our algorithm is a different and somehow complementary approach. The core of our method is to postprocess the sequence of interfaces generated by the domain decomposition solver. Further, the AASWR method for parabolic problems is able to achieve four important

*Research reported here was partially supported by Award 0305405 from the National Science Foundation.

I.H. Tuncer et al., *Parallel Computational Fluid Dynamics 2007*,
DOI: 10.1007/978-3-540-92744-0_17, © Springer-Verlag Berlin Heidelberg 2009

features on the grid:

- *Load balancing*: adaptive number of mesh points on each subdomain in heterogeneous distributed computing environments
- *Efficiency*: having a very simple and systematic communication patterns
- *Scalability*: solving each subdomain with an embarrassing parallel fashion
- *Fault tolerance*: exchanging only an interface data at an optimal empirically defined time step interval ("naturally" fault tolerant).

The subdomain solver can be optimized independently of the overall implementation which makes the algorithm friendly to cache use and scalable with the memory in parallel environments. AASWR minimizes also the number of messages sent and is very insensitive to delays due to a high latency network. The load balancing results will be shown on a companion paper.

The paper is organized as follows. Section 2 recalls the fundamental steps of the three dimensional space and time AASWR algorithm. Section 3 describes the different approaches for the parallel implementation. Section 4 presents the results of our experiments. Finally, section 5 summarizes the paper and defines the future work plan.

## 2. The AASWR Algorithm

### 2.1. Definition of the Problem and its Discretization

First, let us assume that the domain $\Omega$ is a parallelepiped discretized by a Cartesian grid with arbitrary space steps in each direction. Let us consider the Initial Boundary Value Problem (IBVP):

$$\frac{\partial u}{\partial t} = L[u] + f(x,y,z,t),\ (x,y,z,t) \in \Omega = (0,1)^3 \times (0,T), \tag{1}$$

$$u(x,y,z,0) = u_o(x,y,z),\ (x,y,z) \in (0,1)^3, \tag{2}$$

$$u(0,y,z,t) = a(y,z,t),\ u(1,y,z,t) = b(y,z,t),\ (y,z) \in (0,1)^2,\ t \in (0,T), \tag{3}$$

$$u(x,0,z,t) = c(x,z,t),\ u(x,1,z,t) = d(x,z,t),\ (x,z) \in (0,1)^2,\ t \in (0,T), \tag{4}$$

$$u(x,y,0,t) = e(x,y,t),\ u(x,y,1,t) = f(x,y,t),\ (x,y) \in (0,1)^2,\ t \in (0,T), \tag{5}$$

where $L$ is a separable second order linear elliptic operator. We assume that the problem is well posed and has a unique solution. We introduce the following discretization in space

$$0 = x_0 < x_1 < \ldots < x_{Nx-1} < x_{Nx} = 1,\ h_{xj} = x_j - x_{j-1},$$

$$0 = y_0 < y_1 < \ldots < y_{Ny-1} < y_{Ny} = 1,\ h_{yj} = y_j - y_{j-1},$$

$$0 = z_0 < z_1 < \ldots < z_{Nz-1} < z_{Nz} = 1,\ h_{zj} = z_j - z_{j-1},$$

and time

$$t_n = n\,dt,\ n \in \{0, \ldots, M\},\ dt = \frac{T}{M}.$$

The domain $\Omega = (0,1)^3$ is decomposed into $q$ overlapping strips $\Omega_i = (X_l^i, X_r^i) \times (0,1)^2$, $i = 1..q$ with $X_l^2 < X_r^1 < X_l^3 < X_r^2, ..., X_l^q < X_{q-1}^r$ (See figure 1). The separable second order

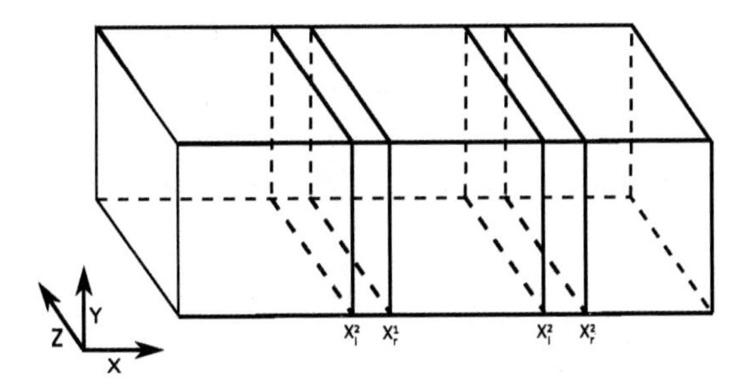


Figure 1. Domain Decomposition in 3D space with 3 overlapping subdomains.

linear elliptic operator $L$ can be written such as:

$L = L_1 + L_2 + L_3$ with

$L_1 = e_1\partial_{xx} + f_1\partial_x + g_1, \quad L_2 = e_2\partial_{yy} + f_2\partial_y + g_2, \quad L_3 = e_3\partial_{zz} + f_3\partial_z + g_3.$

$e_1, f_1, g_1$ are functions of x only, $e_2, f_2, g_2$ are functions of y only and $e_3, f_3, g_3$ are functions of z only. We write the discretized problem as follows

$$\frac{U^{n+1} - U^n}{dt} = D_{xx}[U^{n+1}] + D_{yy}[U^{n+1}] + D_{zz}[U^{n+1}] + f(X, Y, Z, t^{n+1}),\ n = 0, \ldots, M-1, \quad (6)$$

with appropriate boundary conditions corresponding to (3-5). To simplify the presentation in the next section, let us deal with homogeneous Dirichlet boundary conditions.

### 2.2. Problem to Solve

We introduce the following expansion for the discrete solution using the sine base functions:

$$U^n(X, Y, Z, t) = \sum_{j=1}^{M_y} \sum_{k=1}^{M_z} \Lambda_{j,k}^n(X, t)\ sin(jy)\ sin(kz)\ , \quad (7)$$

$$u_o(X, Y, Z) = \sum_{j=1}^{M_y} \sum_{k=1}^{M_z} \lambda_{j,k}(X) sin(jy)\ sin(kz) \quad (8)$$

$$\text{and } f(X, Y, Z, t^n) = \sum_{j=1}^{M_y} \sum_{k=1}^{M_z} f_{j,k}^n(X, t^n). \quad (9)$$

$M_y$ and $M_z$ are the number of modes in the y and z direction. Then, by plugging into the discrete solution from (7-9) to (6), we end up with the following independent one dimensional $M_y \times M_z$ problems based on the Helmotz operator:

$$\frac{\Lambda_{j,k}^{n+1} - \Lambda_{j,k}^n}{dt} = D_{xx}[\Lambda_{j,k}^{n+1}] - (\mu_j + \mu_k)\Lambda_{j,k}^{n+1} + f_{j,k}(X, t^{n+1}),\ n = 0, \ldots, M-1, \quad (10)$$

$$\Lambda_{j,k}^0 = \lambda_{j,k}(X). \quad (11)$$

$\mu_j$ and $\mu_k$ are respectively the eigenvalues of $D_{yy}$ and $D_{zz}$. This algorithm generates a sequence of vectors $W^n = (\Lambda^n_{2,l}, \Lambda^n_{1,r}, \Lambda^n_{3,l}, \Lambda^n_{2,r}, \ldots, \Lambda^n_{q,l})$ corresponding to the boundary values on the set

$$\mathcal{S} = (X^2_l, X^1_r, X^3_l, X^2_r, \ldots, X^q_l, X^{q-1}_r) \times (t^1, ..., t^M)$$

of the $\Lambda_{j,k}$ for each iterate $n$. The trace transfer operator is decomposed into $M_y \times M_z$ independent trace transfer operators as well and is defined as follows:

$$W^n_{j,k} - W^\infty_{j,k} \rightarrow W^{n+1}_{j,k} - W^\infty_{j,k}.$$

Let $P_{j,k}$ be the matrix of this linear operator. $P_{j,k}$ has a bloc diagonal structure and can be computed prior to the AASWR and once for all.
In the next section, we describe the major points of the algorithm.

### 2.3. General Algorithm

The general AASWR algorithm can then be summarized in three steps:

1. Computes the $1^{st}$ iterate of ASWR for the 3D parabolic problem (1).
2. Expands the trace of the solution in the eigenvectors basis and solve the linear problem component wise

$$(Id - P_{j,k})W^\infty_{j,k} = W^1_{j,k} - P_{j,k}\, W^0_{j,k},\ \forall j \in \{1, \ldots, M_y\},\ \forall k \in \{1, \ldots, M_z\}. \quad (12)$$

   Assemble the boundary conditions $W^\infty = \sum_{j=1}^{M_y} \sum_{k=1}^{M_z} W^\infty_{j,k}\ sin(jy)\ sin(kz)$.

3. Computes the $2^{nd}$ iterate using the exact boundary value $W^\infty$.

The computation of each subdomain in the first and the third step can be processed by any linear solvers of choice (Multigrid, Krylov, etc.). In the paper, we use a combination of Fourier transform and LU decomposition. More details on the above steps and their proof can be found in [3]. In the following section, we will go over four approaches for the parallel implementation.

## 3. Description of the Parallel Implementation

In this section, we discuss the properties of AASWR (see Section 1) that should be used in a parallel implementation. We describe the parallel algorithm using the fundamental Message Passing Interface (MPI) paradigm on a distributed network of processors and show that our new method provides a rigorous framework to optimize the use of a parallel architecture.

### 3.1. The Solver and Spare Process Concept

We recall the concept of *solver* and *spare* processes first introduced in [9]. The spare processes are not only checkpointing indirectly the application but also, they are part of the main computation by solving the interface problem (Step 2 in 2.3). Compared to [9], we do not add any overheads since it is part of the AASWR algorithm to save the interface subdomain solutions. This algorithm is therefore called *naturally* fault tolerant.

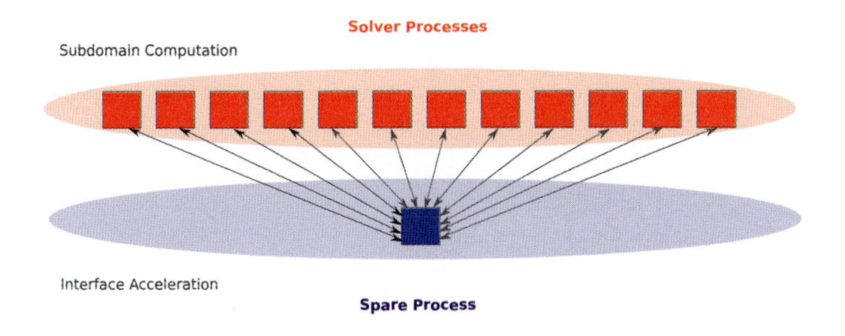


Figure 2. The Solver and Spare process configuration.

The solver processes are working on the parallel computation of the solution of the IBVP (1) in each subdomain $\Omega_i \times (0,T)$, $i = 1 \ldots q$ with $q$ the number of subdomains (Step 1 and 3 in 2.3). Figure 2 draws this framework. We have one dimensional processor topology with no local neighborhood communications implemented in each process group. The communications are only established between the solver and the spare groups.

In the next section, we describe different parallel implementation approaches which are build upon this group structure.

### 3.2. The Parallel Implementation Approaches

Four approaches have been developed to take advantage of the solver and spare process concept. Starting from the basic version to the more sophisticated one, we present the pros and cons of each method.

#### 3.2.1. The Blocking Version

The Blocking version a standard approach to start with. It is composed by five successive stages:

1. The Solver processes perform the first ASWR iteration in parallel on the entire Domain (Space and Time).

2. The Solver processes send the interface solution to the unique Spare process.

3. The Spare process solves the interface problem.

4. The Spare process sends back the new inner boundary values to the corresponding Solver processes.

5. The Solver processes compute the second iterate using the exact boundary values.

While this algorithm is very easy to implement, it does not scale when the number of Subdomains/Solver processes increases since the interface problem size becomes larger. In addition to that, the blocking *sends* in stages (2) and (4) do not allow any overlapping between computation and communication. It is also important to notice that the acceleration of the interface components at time step $t_{n_1}$ does not depend on the interface components at later time steps $t_{n_2}$, $n_2 > n_1$. This property is reflected in the fact that the blocks of the matrix $P$ are lower tridiagonal matrices. The resolution of the linear system in the acceleration step can progress with the time stepping on the interface solutions. As a matter of fact, the interface solutions should be sent as soon as they are computed.

The next approach resolves these issues by splitting the global number of time steps in equal size subsets or Time Windows (TW).

### 3.2.2. The Blocking Windowed Version

This method is the analogue of the Blocking version with 5 stages as in section 3.2.1. However, it is applied on TW, which are a subset of the total number of time steps. Figure 3(a) shows how the procedure runs over the five stages. The vertical red lines within the rectangles correspond to the interface solutions. With this technique, the size of the interface problem becomes smaller. Furthermore, the LU Decomposition needs only to be computed for the first TW. For the next TW, the update of the right hand side is sufficient to get the correct interface solutions. On the other hand, the blocking communications seem to be the bottleneck since a lot of CPU cycles are wasted on both groups in waiting the interface data to come.
In the next approach, we introduce a feature to avoid the process idle time.

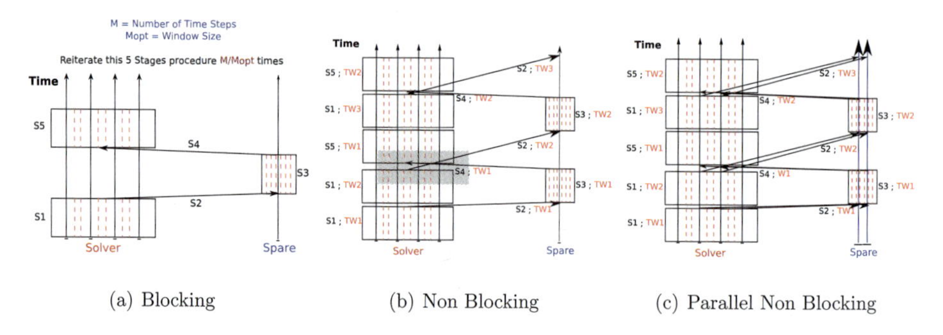


(a) Blocking (b) Non Blocking (c) Parallel Non Blocking

Figure 3. Parallel Approach Descriptions.

### 3.2.3. The Non Blocking Windowed Version

A new layer based on Non Blocking Communications is added to the earlier versions. This introduces a significant overlapping between communications of subdomain boundaries by computations of subdomain solutions. We end up with a pipelining strategy which makes this method very efficient. For example in figure 3(b), while the spare process is working on the Stage 3 of the first TW, the solver process has started the Stage 1 of the second TW. And this continues until reaching the last TW for which a special treatment is necessary.

On the spare process side, the unique process available receives lots of messages at the same time and is therefore highly loaded as the number of solver processes increases. The aim of the next version is to allocate more spare processes to handle first these simultaneous communications and second to parallelize the interface problem resolution.

### 3.2.4. The Parallel Non Blocking Windowed Version

This last technique consists on a full parallelization of the interface problem in the spare group. Figure 3(c) presents the new framework. This method is very efficient and benefits the best from the solver/spare concept. It is also very challenging to set it up since the on-going communications have to be cautiously managed. Further, the TW size has to be estimated empirically to get optimal performance. In the next section, we present experimental results of each approach depending on the TW size.

## 4. Results

The experiments have been performed on a 24 SUN $X2100$ nodes with 2.2 GHz dual core AMD Opteron processor, 2 GB main memory each and Gigabit Ethernet Network Interconnect. The elapsed time in seconds is shown for each approach in figures 4(a), 4(b), 4(c) and 4(d) compared to the sequential code. The Blocking approach does not scale when the number of subdomains/processes increases. The performance of the Blocking Windowed version is very close to the Non Blocking Windowed version since there is only one spare process taking care of the interface solutions sent by the solver processes. Finally, the Parallel Non Blocking Windowed approach is the most efficient in solving $100^3$ unknown problem with 25 time steps in 4.56 seconds with a window size of 3.

## 5. Conclusion

In this paper, we have described how AASWR can achieve efficiency, scalability and fault tolerance under grid environments. The Parallel Non Blocking Windowed approach is the most efficient compared to the other versions. Indeed, the identification of 5 successive stages in the general algorithm permits to apply a pipelining strategy and therefore, to take advantage of the solver/spare process concept. Also, the parallelization of the interface problem resolution makes AASWR scalable as the number of subdomains increases. Furthermore, AASWR is "naturally" fault tolerant and can restart the computation in case of failures, from the interface solutions located in the spare processes memory (See the forward implicit reconstruction method in [10]).

## REFERENCES

1. Nicolas Barberou, Marc Garbey, Matthias Hess, Michael M. Resch, Tuomo Rossi, Jari Toivanen and Damien Tromeur-Dervou, Efficient metacomputing of elliptic linear and non-linear problems, J. Parallel Distrib. Comput., No. 63 (5) p564–577, 2003.
2. Marc Garbey and Damien Tromeur-Dervout, On some Aitken like acceleration of the Schwarz Method, Int. J. for Numerical Methods in Fluids, No. 40 (12) p1493–1513, 2002.
3. Marc Garbey, Acceleration of a Schwarz Waveform Relaxation Method for Parabolic Problems, Domain Decomposition Methods in Science and Engineering XVII, Lecture Notes in Computational Science and Engineering, Austria, No. 60, July 3-7 2006.
4. P. L. Lions, On the Schwarz alternating method I, SIAM, Domain Decomposition Methods for Partial Differential Equations, Philadelphia, PA, USA, p1–42, 1988.

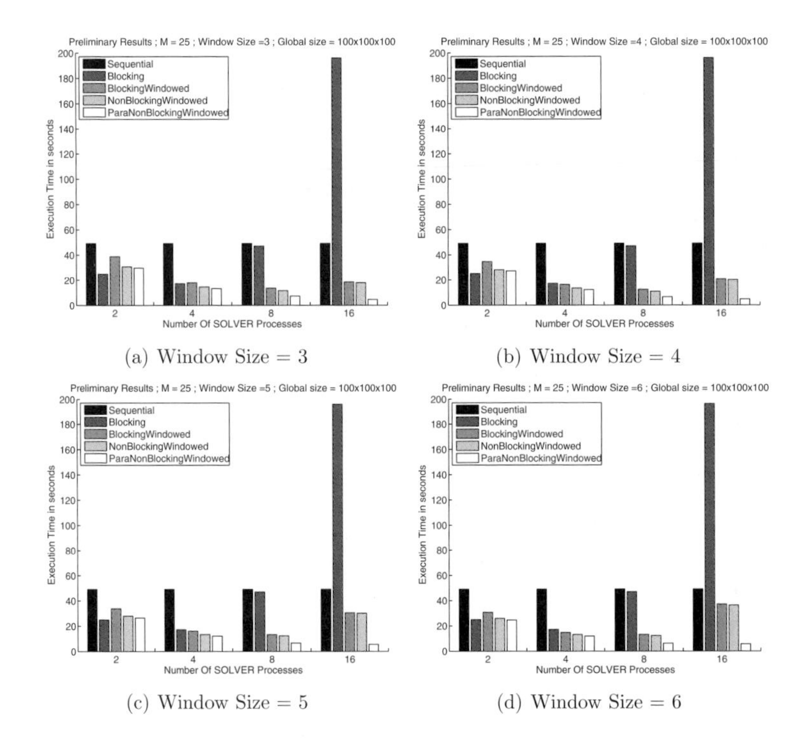


(a) Window Size = 3 (b) Window Size = 4

(c) Window Size = 5 (d) Window Size = 6

Figure 4. Experimental Results.


5. P. L. Lions, On the Schwarz alternating method II, SIAM, Domain Decomposition Methods for Partial Differential Equations, Philadelphia, PA, USA, p47–70, 1989.
6. Bruno Després, Décomposition de domaine et problème de Helmholtz, C.R. Acad. Sci. Paris, Domain Decomposition Methods for Partial Differential Equations, Philadelphia, PA, USA, No. 1 (6) p313–316, 1990.
7. Martin J. Gander and Laurence Halpern and Frédéric Nataf, Optimized Schwarz Methods, Twelfth International Conference on Domain Decomposition Methods, Domain Decomposition Press, Chiba, Japan, No. 1 (6) p15–28, 2001.
8. V.Martin, An Optimized Schwarz Waveform Relaxation Method for Unsteady Convection-Diffusion Equation, Applied Num. Math., Domain Decomposition Press, Chiba, Japan, No. 52 (4) p401–428, 2005.
9. Marc Garbey and Hatem Ltaief, On a Fault Tolerant Algorithm for a Parallel CFD Application, Parallel Computational Fluid Dynamics, Theory and Applications, A.Deane et Al edit. Elsevier, p133–140, Washington DC USA, 2006.
10. Hatem Ltaief, Marc Garbey and Edgar Gabriel, Performance Analysis of Fault Tolerant Algorithms for the Heat Equation in Three Space Dimensions, Parallel Computational Fluid Dynamics, Busan, South Korea, May 15-18 2006.

# Parallel Computation of Incompressible Flows Driven by Moving Multiple Obstacles Using a New Moving Embedded-Grid Method

Shinichi ASAO [a], Kenichi MATSUNO [b],

[a] Graduate School of Science and Technology Kyoto Institute of Technology, Matsugasaki, Sakyo-ku, Kyoto 606-8585, Japan,
asao@fe.mech.kit.ac.jp

[b] Division of Mechanical and System Engineering, Kyoto Institute of Technology, Matsugasaki, Sakyo-ku, Kyoto 606-8585, Japan,
matsuno@kit.ac.jp

In this paper, parallel computation of incompressible flows driven by moving multiple obstacles using a new moving embedded-grid method is presented. Moving embedded-grid method is the method such that the embedded grid can move arbitrarily in a main grid covering whole of the flow field. The feature of the method is to satisfy geometric and physical conservation laws. The method is applied to a flow where three square cylinders move in the stationary fluid, and computed using parallel machines.
**keyword:** Incompressible flow, Moving-Grid Finite-Volume method, OpenMP

## 1. INTRODUCTION

Today, one of the interesting problems in CFD is unsteady flow. In this paper, solutions of incompressible flows driven by multiple obstacles are presented. The target of the research is the simulations of the flow field, in which the obstacles overtake the other obstacle or multiple obstacles move independently. Such flow fields are usually seen in engineering. When we simulate such flow fields, we encounter the problems to be overcome. One of the problems is grid system. When the obstacle moves a long distance in the flow field, a body-fitted grid system is hard to adjust the motion of obstacle. For this situation, the overset-grid, or Chimera grid [1], is often used at the sacrifice of destroying conservation law due to interpolation errors. For accurate simulation, the body fitted single grid system is desirable but is not suitable for long distance movement of the obstacles because of resultant highly skewed grid. To overcome this problem, we have proposed a new moving embedded-grid method [2]. The method patches a local grid generated around an obstacle in a stationary "main grid" and shifts the embedded local grid in any direction through the main grid with eliminating front-side grids and adding back-side grid with satisfying both physical and geometric conservation laws, as illustrated in Figure 1. As a flow solver, the moving-grid finite-volume method [3] is used in the present "moving embedded-grid method".

I.H. Tuncer et al., *Parallel Computational Fluid Dynamics 2007*,
DOI: 10.1007/978-3-540-92744-0_18, © Springer-Verlag Berlin Heidelberg 2009

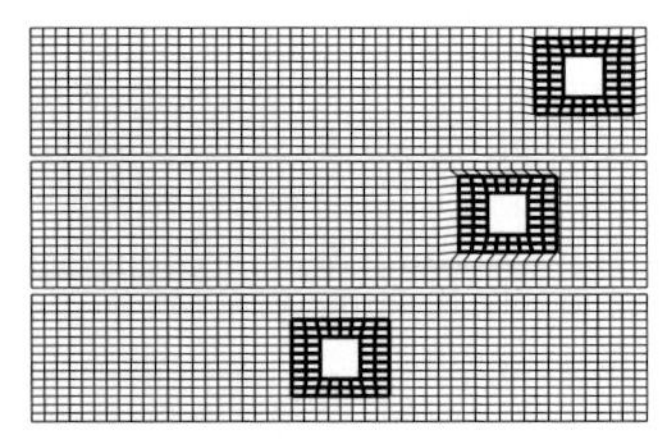

Figure 1. Moving embedded-grid method; bold line: moving embedded-grid, thin line: stationary main grid.

In this paper, simulations of the flow filed driven by multiple obstacles are presented, and further the parallel implementation of the new moving embedded grid method in a laboratory multi-core, multi CPU computer using OpenMP is challenged and the efficiency is measured.

## 2. MOVING EMBEDDED-GRID METHOD

### 2.1. Governing Equations

Governing equations are continuity equation and incompressible Navier-Stokes equations.

$$\frac{\partial u}{\partial x} + \frac{\partial v}{\partial y} = 0 \tag{1}$$

$$\frac{\partial \mathbf{q}}{\partial t} + \frac{\partial \mathbf{E}_a}{\partial x} + \frac{\partial \mathbf{F}_a}{\partial y} = -\left(\frac{\partial \mathbf{P}_1}{\partial x} + \frac{\partial \mathbf{P}_2}{\partial y}\right) + \frac{1}{Re}\left(\frac{\partial \mathbf{E}_v}{\partial x} + \frac{\partial \mathbf{F}_v}{\partial y}\right) \tag{2}$$

where, $\mathbf{q}$ is conservation amount vector, $\mathbf{E}_a$ and $\mathbf{F}_a$ are advection flux vectors in $x$ and $y$ direction, respectively, $\mathbf{E}_v$ and $\mathbf{F}_v$ are viscous-flux vectors in $x$ and $y$ direction, respectively, and $\mathbf{P}_1$ and $\mathbf{P}_2$ are pressure flux vectors of $x$ and $y$ direction respectively, The elements of the vectors are

$$\begin{aligned} &\mathbf{q} = [u, v]^T, \quad \mathbf{E}_a = [u^2, uv]^T, \quad \mathbf{F}_a = [uv, v^2]^T, \\ &\mathbf{P}_1 = [p, 0]^T, \quad \mathbf{P}_2 = [0, p]^T, \quad \mathbf{E}_v = [u_x, v_x]^T, \quad \mathbf{F}_v = [u_y, v_y]^T \end{aligned} \tag{3}$$

where, $u$ and $v$ are velocity of $x$ and $y$ direction each other, $p$ is pressure. The subscript $x$ and $y$ show derivation with respect to x and y respectively. $Re$ is Raynolds number.

### 2.2. Moving Grid Finite Volume Method

To assure the geometric conservation laws, we adopt a control volume on the space-time unified domain $(x, y, t)$, which is three-dimensional for two-dimensional flows. Now Eq.(2) can be written in divergence form as,

$$\tilde{\nabla} \cdot \tilde{\mathscr{F}} = 0 \tag{4}$$

where,

$$\tilde{\nabla} = \left[\ \frac{\partial}{\partial x}\ \ \frac{\partial}{\partial y}\ \ \frac{\partial}{\partial t}\ \right]^T, \quad \tilde{\mathscr{F}} = \left[\ \mathbf{E}\ \ \mathbf{F}\ \ \mathbf{q}\ \right]^T \tag{5}$$

$$\mathbf{E} = \mathbf{E}_a - \frac{1}{Re}\mathbf{E}_v + \mathbf{P}_1, \quad \mathbf{F} = \mathbf{F}_a - \frac{1}{Re}\mathbf{F}_v + \mathbf{P}_2 \tag{6}$$

The present method is based on a cell-centered finite-volume method and, thus, we define flow variables at the center of cell. The control volume becomes a hexahedron in the $(x, y, t)$-domain, as shown in Figure 2.

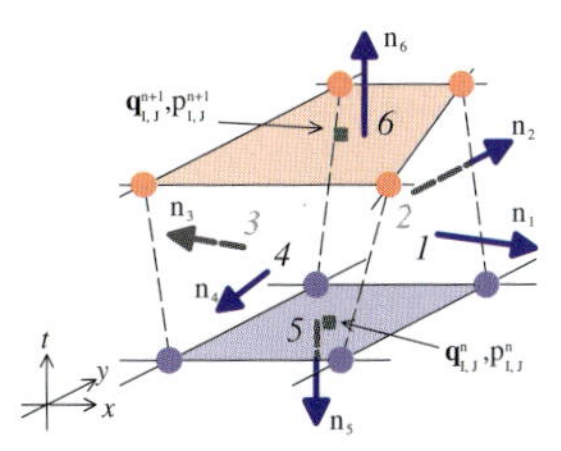

Figure 2. Control volume in space-time unified domain and its normal vectors

We apply volume integration to Eq.(2) with respect to the control volume. Then, using the Gauss theorem, Eq.(4) can be written in surface integral form as,

$$\int_{\Omega} \tilde{\nabla} \cdot \tilde{\mathscr{F}} \mathrm{d}\Omega = \oint_{S} \tilde{\mathscr{F}} \cdot \tilde{\mathbf{n}} \mathrm{d}S = \sum_{l=1}^{6} \left[ \mathbf{E} n_x + \mathbf{F} n_y + \mathbf{q} n_t \right]_l = 0 \tag{7}$$

Here, $\tilde{\mathbf{n}}$ is a outward unit normal vector of control volume surface. $\Omega$ is the hexahedron control volume. $S$ is its boundary. $\mathbf{n}_l = (n_x, n_y, n_t)_l$, $(l = 1,2,\ldots 6)$ is the normal vector of control volume surface. The length of the vector equals to the area of the surface. The upper and bottom surfaces of the control volume ($l = 5$ and 6) are perpendicular to $t$-axis, and therefore they have only $n_t$ component and its length is corresponding to the areas in the $(x, y)$-space at time $t^{n+1}$ and $t^n$ respectively. Thus, Eq.(7) can be expressed as,

$$\mathbf{q}^{n+1}(n_t)_6 + \mathbf{q}^n(n_t)_5 + \sum_{l=1}^{4} \left[ \mathbf{E}^{n+1/2} n_x + \mathbf{F}^{n+1/2} n_y + \mathbf{q}^{n+1/2} n_t \right]_l = 0 \tag{8}$$

To solve Eq.(8), we apply the SMAC method[4]. Then, Eq.(8) can be solved at three following stages. The equation to solve at the first stage contains the unknown variables at $n+1$-time step to estimate fluxes. This equation is iteratively solved using LU-SGS[5]. The equation to be solved at the second stage is Poisson equation about the pressure correction. This equation is iteratively solved using Bi-CGSTAB[6]. The flux vectors of both the moving grid and advection terms are evaluated using the QUICK method. While the flux vectors of the pressure and viscous terms are evaluated with the central difference method.

### 2.3. Formulation of Moving Embedded-Grid Method

The present moving-embedded grid method is the method such that the embedded grid moves or slides in the main grid with satisfying the physical and geometric conservation laws simultaneously. As illustrated in Figure 3, the line of grid existing in front of embedded grid is eliminated according to the movement of the embedded grid, while the grid line behind the embedded grid is added to make up for the resultant large grid spacing by passing through of the embedded grid. At the same time, the cells existing and connecting between the stationary main grid and the side of the embedded grid are skewed due to the movement of the embedded grid. The reconnection or change of the grid lines there is necessary. Hence the present method includes necessarily the following three procedures: addition and elimination of grid line as well as change of the connection relation of structured grid points.

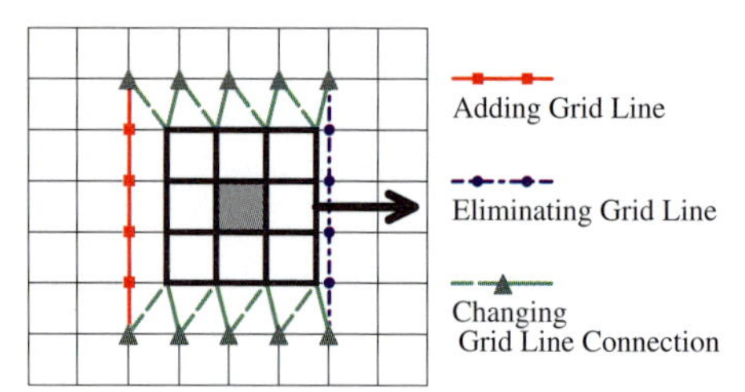


Figure 3. Detail of moving embedded grid method

The addition of the grid line is performed as follows. Suppose the $i'$-th grid line is added between $(i-1)$-th and $i$-th grid lines at $(n+1)$-time step. Then, we use the "prismatic" control volume for new $i'$-th grid line, as illustrated in Fig 4(left). The prismatic control volume is the result of zero surface area since the space area related to this $I'$-th control volume is nothing at $n$-time step. For this control volume, Eq.(8) is replaced as following equation according to no bottom surface ($(n_t)_5 = 0$) of a control volume.

$$\mathbf{q}^{n+1}(n_t)_{6'} + \sum_{l=1}^{4} \left[\mathbf{E}^{n+1/2} n_x + \mathbf{F}^{n+1\ 2} n_y + \mathbf{q}^{n+1/2} n_t\right]_l = 0 \tag{9}$$

The elimination of the grid line is accomplished through merging the two related control volumes. Suppose that the $i'$-th grid line is eliminated at $(n+1)$-time step. Then we consider that the $i'$-th line and $i$-th line of cells at $n$-time step are merged at $(n+1)$-time step and only $i$-th line of cells remains. In this situation, the control volume in space-time unified domain, $(x, y, t)$, is shown in Figure 4(center). Thus Eq.(10) replaces Eq.(8).

$$\mathbf{q}^{n+1}(n_t)_6 + \mathbf{q}^{n}(n_t)_5 + \mathbf{q}^{n}(n_t)_{5'} + \sum_{l=1}^{4} \left[\mathbf{E}^{n+1/2} n_x + \mathbf{F}^{n+1/2} n_y + \mathbf{q}^{n+1/2} n_t\right]_l = 0 \tag{10}$$

To change the grid line at $(n+1)$-time step, the control volume between $n$ and $(n+1)$-time step is shown in Figure 4(Right). The flux at $l = 5$ ($n$-time step) plane is divided according to the new connection of the grid line before the volume integral is applied to the control volume for $(n+1)$-time step calculation. This is supported because in the finite-volume formulation, the value of the variable is assumed constant in the cell, thus, the estimation of the flux at $n$-time step is straight forward. For this situation, the flux at $n$-time step is divided into two parts. Referring Figure 4(right), the following Eq.(11) replaces Eq.(8).

$$\mathbf{q}^{n+1}(n_t)_6 + \mathbf{q}^n(n_t)_{5A} + \mathbf{q}^n(n_t)_{5B} + \sum_{l=1}^{4} \left[\mathbf{E}^{n+1/2}n_x + \mathbf{F}^{n+1/2}n_y + \mathbf{q}^{n+1/2}n_t\right]_l = 0 \quad (11)$$

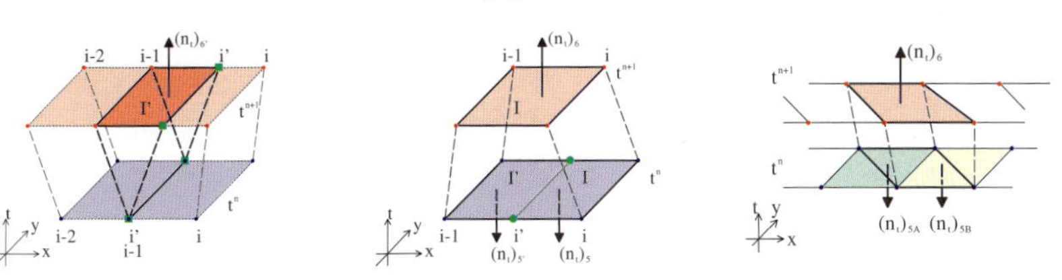

Figure 4. Control Volume for Adding Grid Line (Left), Eliminating Grid Line (Center) and Changing Grid Line Connection (Right).

In the present method, the addition, elimination and shift of grid line are performed simultaneously when the grid spacing at the boundary between the main and embedded grids exceeds a user-specified tolerance.

## 3. NUMERICAL RESULTS

### 3.1. Geometric Conservation Laws

At first it is necessary to confirm that the method satisfies the "Geometric Conservation Law" (GCL condition)[7], which means the scheme can calculate uniform flow on the moving mesh system without any error, when the embedded grid is moving in main-grid with adding, eliminating and shifting grid lines. To very that the method satisfies the GCL condition, we apply the present method to the uniform. The computational condition is that the number of grid points is $21 \times 21$, time spacing $\Delta t = 0.01$ and the Reynolds number $Re = 100$. The initial condition is assumed to be $u = u_\infty = 1.0$ , $v = v_\infty = 1.0$, and $p = p_\infty = 1.0$ as the uniform flow. The boundary condition is fixed for $u = u_\infty$ , $v = v_\infty$, and pressure at the boundary is given by linear extrapolation. Figure 6 shows the location of the embedded grid at $t = 0.2$, 0.7, 1.2 and 1.7. The calculation is performed until $t = 1000.0$. The error is defined as follow.

$$Error = \max_{I,J} \left[\left\{(u_{I,J} - u_\infty)^2 + (v_{I,J} - v_\infty)^2\right\}/2\right]^{1/2} \quad (12)$$

Figure 5 shows the history of the Error. Because the error keeps the order of $10^{-16}$, in a word, machine zero, which means the method satisfies GCL perfectly.

### 3.2. Computation of Incompressible Flows Driven by Moving Multiple Obstacles

The example is a flow where three square cylinders move in the stationary fluid illustrated in Figure 7. The length of the whole computational domain is $L_x = 20.0L$, $L_y = 11.0L$, where $L$ is the length of square cylinder. The upper and lower square cylinders move at constant speed of $v_b = 1.0$ after the constant acceleration of 0.1 from the

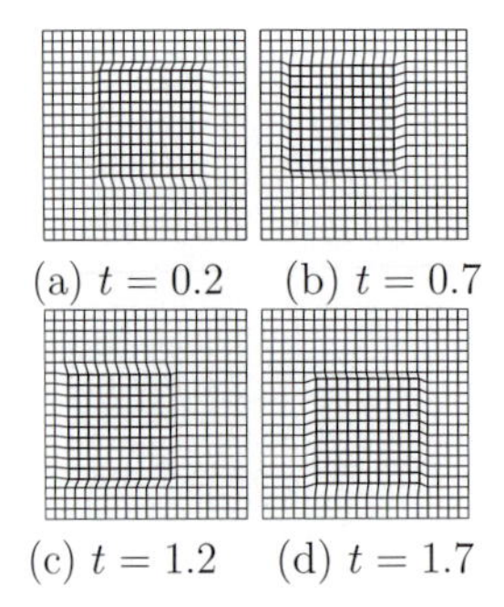


Figure 5. Grid movement

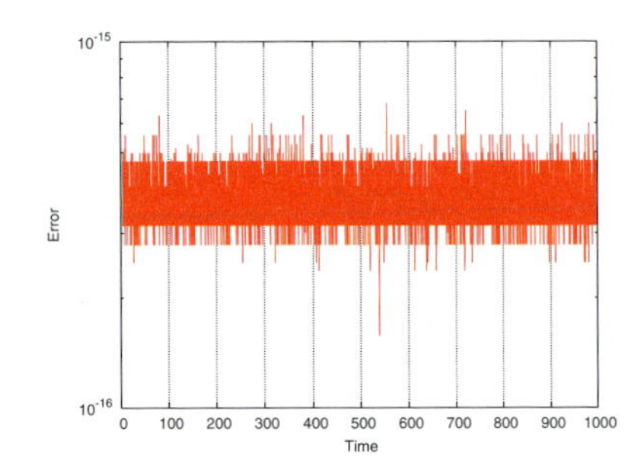


Figure 6. Error History

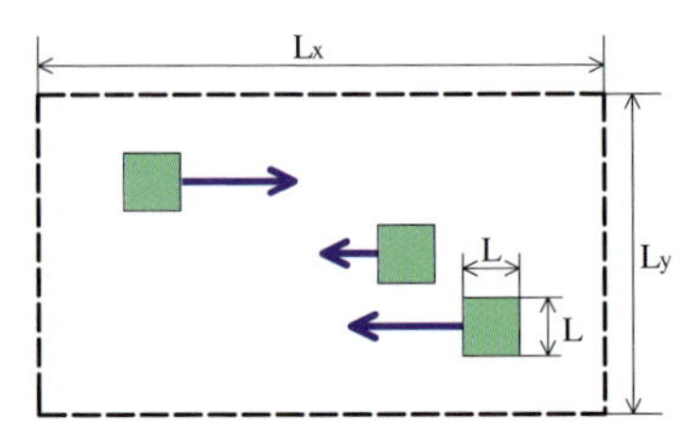


Figure 7. The flows driven by multiple cylinders.

stationary condition in the counter direction While the middle square cylinder moves at constant speed of $v_b = 0.5$ after the constant acceleration of 0.05 from the stationary condition. Figure 8 shows the vorticity and pressure contours at $t = 11.0, 13.0$ and 16.0. Figure 9 shows the time history of the lift (side force) coefficient of three square cylinders.

The lift coefficient of the upper cylinder becomes the maximum value at $t = 11.0$, which means that the lift of the upper cylinder becomes the maximum before it is overtaken by the other cylinders. The lift coefficient of the middle cylinder becomes the maximum value at $t = 13.0$. The lift of the middle cylinder also becomes the maximum before it is overtaken by the lower cylinder. As for the lift coefficient of lower cylinder, the minimum value is attained at $t = 16.0$. In this case, the lower cylinder becomes the minimum at the time when it just overtakes the middle cylinder.

## 4. PARALLEL IMPLIMENTATION

The OpenMP parallelization is implemented for the first stage of the SMAC solver, which estimates the flow velocities. In this paper, however, the second stage of the solver, in which the Bi-CGSTAB method is used for solving the Poisson equation for the pressure, has not been parallelized yet. As a result, since this Poisson solution process occupies 90 % of whole of computer time, we have not obtained the desirable parallel efficiency yet. The parallel implementation of the Bi-CGSTAB algorithm is now challenging.

The parallel efficiency of the first stage of the SMAC algorithm is as follows. The

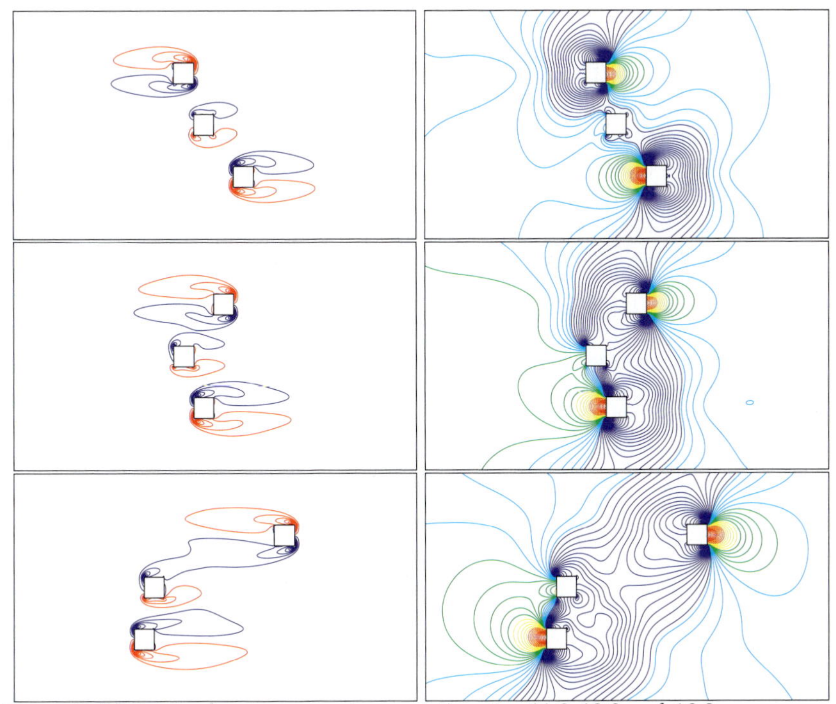

Figure 8. Vortex and pressure contours at time $t = 11.0, 13.0$ and $16.0$.

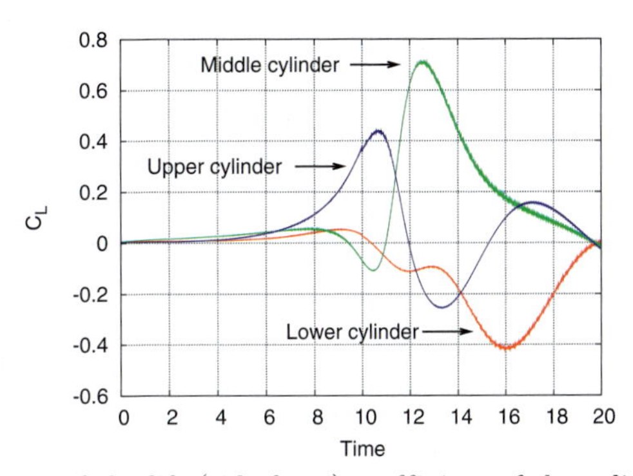


Figure 9. Time history of the lift (side force) coefficient of the cylinders.

parallel computation is carried out using OpenMP library on PC for a flow where three square cylinders move in the stationary fluid. The computer is Mac Pro, which has two Intel Xeon Dual-Core 3.0GHz processors and 4GB shared memory. The operating system is Mac OS 10.4 and the Fortran compiler is Intel Fortran 9.1. The domain is divided into four domains according to the number of cpu-core. As a result, the "Speed up" are 1.14 for the two cpu-core processors and 1.47 for the four cpu-core processors. For the fully parallelizable part of the code, such as, calculation of the right-hand-side of the solver, the "speed up" is 1.53 for the two cpu-core processors and 2.29 for the four cpu-core processors. For this case, the efficiency is not obtained in OpenMP parallel environment.

## 5. CONCLUDING REMARKS

In this paper, the moving-embedded grid method has been applied to the flows driven by multiple obstacles arbitrarily moving in the whole of the flow field. The first stage of the flow solver is parallelized using OpenMP. The present computation is parallelized in the first stage of the solver and thus the speed up is not satisfactory. The parallel speed up of the second stage is essential for incompressible solver and is under implementation.

## ACKNOWLEDGEMENT

The present research is supported by the Grant-in-Aid for Scientific Research.

## REFERENCES

1. Steger, J. L., Dougherty, F. C. and Benek, J. A., A Chimera Grid Scheme; Advances in Grid Generation, *American Society Of Mechanical Engineers Fluids Engineering Division*, Vol.5 (1983), pp.55-70.
2. Asao, S. and Matsuno, K., A Moving Embedded-Grid Method for Simulation of Unsteady Incompressible Flows with Boundaries Traveling Long Distances (in Japanese), submitted to journal.
3. Matsuno, K., Mihara, K., and Satofuka, N., A Moving-Mesh Finite-Volume Scheme for Compressible Flows, *Computational Fluid Dynamics 2000*, (2001), pp.705-710.
4. Amsden, A.A. and Harlow, F.H., A Simpli ed MAC Technique for Incompressible Fluid Flow Calculations, *Journal of Computational Physics*, Vol.6 (1970b), pp.322-325.
5. Yoon, S. and Jameson, A., Lower-Upper Symmetric-Gauss-Seidel Method for the Euler and Navier-Stokes Equations, *American Institute of Aeronautics and Astronautics Journal*, Vol.26 (1988), pp.1025-1026.
6. Murata, K., Natori, M. and Karaki, Y., Numerical Simulation (in Japanese), *Iwanami Shoten Publishers*, (1990), pp.126-131.
7. Zhang, H., Reggio, M., Trepanier, J.Y. and Camarero, R., Discrete Form of the GCL for Moving Meshes and Its Implimentation in CFD Schemes, *Computers and Fluids*, Vol.22 (1993), pp.9-23.

# Parallel Computing on Network of Windows Based PCs

S. Chien*, G. Makinabakan, A. Ecer, and H.U. Akay

Purdue School of Engineering and Technology
Indiana University-Purdue University Indianapolis
723 W. Michigan Street, Indianapolis, IN 46202, USA
E-mail: schien@iupui.edu

Key Words: parallel computing, windows based PCs, dynamic load balancing.

**Abstract:** This paper describes a dynamic load balancing tool for parallel CFD applications on Microsoft Windows (XP and Vista) based PCs. The tool encourages the PC owners to share the computing resources and provides mechanism to ensure that the parallel jobs do not run on the PCs when their owners are using the computers, while allowing the parallel jobs to use the computers when their owners are not using the computers. The tool has been successfully tested for supporting parallel CFD applications.

## 1. INTRODUCTION

Due to the rapid increase of computational power and fast decrease of costs of personal computers (PC), many PC companies, such as Compaq and Apple Computer, are promoting their PC clusters for parallel computing. Their parallel computing PC clusters are usually configured as dedicated machines and are managed in the manner similar to that of workstation clusters and supercomputers using centralized schedulers. Each PC in the cluster runs a Linux operating system. Load balancing of parallel jobs are also handled similarly as that in workstation clusters and supercomputers.

In this paper, we will discuss the use of clusters of office PCs for parallel computing. There are vast amount of PCs used in offices and laboratories in many institutions and companies. Most of these PCs use Microsoft Windows operating systems. The office PCs are usually well maintained and updated periodically. In many offices, PCs are only used several hours in a day. Therefore, the computational powers of many PCs are underused, especially in the evenings and weekends. Moreover, for companies and institutions that support PC clusters for parallel computing, the total computing power of office PCs is much higher than that of dedicated PC clusters for parallel computing. Hence it is worthwhile to explore the possibility of utilizing these readily available computation resources. We will start with the comparison of doing parallel computing between using office PC clusters and using dedicated PC

I.H. Tuncer et al., *Parallel Computational Fluid Dynamics 2007*,
DOI: 10.1007/978-3-540-92744-0_19, 

clusters, and then propose solutions to address the specific concerns for using existing office PC clusters for parallel computation.

## 2. DIFFERENCES BETWEEN OFFICE PC CLUSTERS AND DEDICATED PC CLUSTERS

Whether using office PC clusters or supercomputers, users' concern of parallel computing are the same. They want the availability of the computation power, convenience of accessing the computation power, and availability of the supporting software. However, the concerns of the office PC "owners" and the owners of dedicated parallel computer clusters are quite different. When a PC cluster is dedicated for parallel computing, the cluster is not "owned" by any PC user but "owned" by the system administrators whose goal is to maximally support all potential users and resolve the conflicts between users. Since UNIX/Linux operating systems were designed for centralized control, they are naturally the preference of operating systems of the dedicated computer clusters for parallel computing. However, the system administrator of an windows based office PC cluster needs to consider the attitude of PC "owners" to the parallel job users in addition to handle all usual issues in parallel computing clusters. In general, PC "owners" are willing to let others use their computer power for parallel computing if the parallel processes do not affect the computer performance when the "owners" are using the computer. If an owner feels inconvenience due to the execution of other people's processes, the owner will stop letting others to run program on his/her computer. Therefore, in order to use office computer cluster for parallel computing, a mechanism is needed to resolve the conflict between the computer "owner" and the other users.

The other difference is data storage and security. When a PC cluster is dedicated for parallel computing, a central file server is used and each computer node does not need local disk storage. Any needed software can be installed to the file server by the system administrator. For office PC clusters, each PC has its own disk drive to store data and may share a common file server. If the networked PCs do not have a common file server, the application program must be installed on each PC and the input data and output data of the parallel application program need to be transferred between computers. In general, keeping program and data on local disks deters the computer "owners" from sharing the computers since it opens the door for other users to access their local files. Therefore, to use office PC clusters for parallel computing, a common file server accessible by all PCs is required. Each user has an account on the file server and can only access his/her own files. When an office PC is used for parallel computing, only the CPU and local RAM of the PC can be used and local disk will not be accessed by the parallel jobs. If system software/tools are needed for parallel computing, they should be installed on each PC by the system administrator.

## 3. OUR APPROACH TO ADDRESS THE CONCERNS

In order to use office PC cluster for parallel computing, we first ported the dynamic load balancer (DLB) [1], which was originally developed for the UNIX based system, to Microsoft Windows based systems. Then we added the features to address the special concerns in the office PC cluster. The structure of DLB is shown in Figure 1. DLB is a middleware that has two major components, the Dynamic Load Balancing Agent (DLBA) and System Agent

(SA). DLBA is a user/system program that supports users to optimally distribute the parallel processes to available computers. SA is a program installed on every computer that provides the computer load information (obtained by PTrack) and communication speed information (obtained by CTrack). PTrack uses Microsoft provided free software, *PsTools*, to measure periodically the percentage of CPU time used by each and every process running on every computer. For each new measurement, PTrack considered the number of processes that uses more than 5% CPU time as the computer load count. The processes run under 5% of CPU time are usually system demons and other negligible processes. SA is also able to indicate which process is a parallel process and which process is the "owner" local process. During parallel process dispatching, DLBA requests the computer load and speed information and communication speed information from SA of each PC and then determines an optimal load distribution based on greedy algorithm [2]. During the execution of the parallel jobs, DLBA keeps monitoring the progress of the parallel job and suggests better load distribution if it can decrease the overall computation time of the parallel job. Since the original DLB was mostly written in Java, the main work for porting DLB to Windows based computers is to replace the UNIX system function calls by corresponding Windows system function calls.

In order to take the advantage of dynamic load balancing, the parallel application must be able to utilize the new load redistribution dynamically suggested by DLB. There are two ways that a parallel application can find the newly suggested load distribution. The first one is to include three simple DLB interfacing library functions *check_balance()* and *application_stopped()*into the parallel application program (see Figure 2). Before each check pointing (to be done at the end of every N timesteps), the library function *check_balance()* is called to find if there is a better load distribution. If there is a better load distribution, the parallel program stores the intermediate information, stops the program and calls the library function *application_stopped()*. Once DLB detects that the parallel program stops execution, DLB restarts the parallel program according to the proposed new load distribution. If it is not possible to include the DLB interfacing library to the application program, the application program needs to be stopped and restarted periodically.

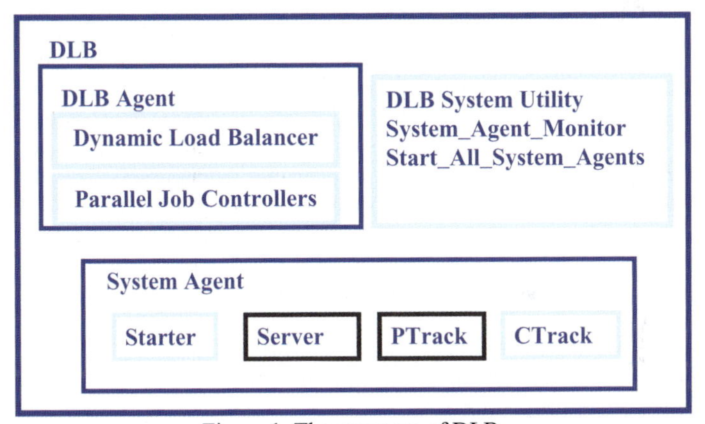


Figure 1. The structure of DLB.

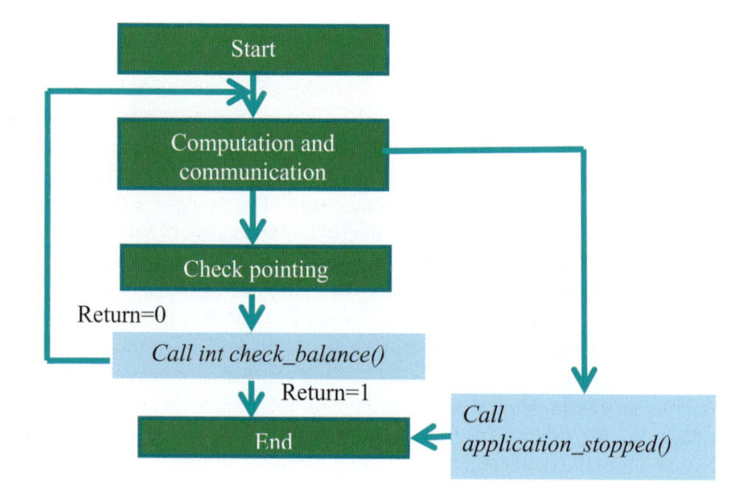


Figure 2. Inserting library function into application code.

In order to resolve the usage conflict between the PC owner and the parallel user, two mechanisms are added to each PC. The first is to enable each PC owner to set his/her PC in one of two usage modes, the Public Use Mode and Owner Use Mode, at different time periods. A PC can only be in one mode at any time instance. In the Public Use Mode (e.g., during evenings and weekends), the PC "owner" allows any parallel job to be executed on the machine and the owner's job has no priority over other jobs. In the Owner Use Mode (e.g., during weekdays), the PC owner discourages the parallel job execution on the PC. In the Public Use Mode, DLB will use the PC if it is beneficial even if PC's owner is using the computer. The PC owner specifies the usage modes at different time periods in a specific text file and the DLB reads this file to determine if the computer is allowed for parallel computing at different times. This mechanism is for avoiding potential conflicts between parallel processes and the processes of the computer owner. Since a parallel job may execute for long time, the program may be still running when the computer changes from Public Use Mode to Owner Use Mode. It is not appropriate to stop the parallel process immediately but also not appropriate to annoy the owner by sharing the computation resources. Therefore, the second mechanism is to reduce the execution priority of all parallel processes so that they are lower than the priority of the processes created by the owner when the computer is in Owner Use Mode. If the owner is using the computer, the owner will not feel any slowness since the parallel process can run only when the user is not using the computer. In this situation, the effective computer speed to the parallel job is slowed down significantly so that the dynamic load balancer will move the existing parallel processes to other low load computers. Since the whole parallel job may be slowed down as one parallel process is slowed down, it is also to the benefit of the parallel job not to submit the process to the computer that the owner is working on. Therefore, these two mechanisms ensure that the computer owner is satisfied with sharing computer resources by not feeling any negative effects.

These two mechanisms are implemented in system agent (SA) which is installed on each PC. SA has the ability to determine what processes are actively run on the computer and also has the ability to determine which processes are owner's processes. It is assumed that all parallel jobs are submitted through DLB. Before assigning a parallel job to PCs, DLB communicates with SAs on the PCs to find the effective speeds of the corresponding PCs from DLB point of view. If owner's process is running on a PC during Public Use Mode, the corresponding SA will provide a PC running slow information to DLB so that DLB will try to move the parallel process away from that computer and also not assign new parallel processes to the computer. When the computer is in Owner Use Mode, SA provides DLB the information that the computer speed is very slow. Therefore, the DLB will not try to submit new parallel process to the computer and will try to use other computers in the cluster.

This proposed approach only uses the computation resources that are volunteered by the PC "owners". The only software to be stored on each computer is (1) the MPI communication library (MPICH2) [3] and (2) a system agent (SA). MPICH2 is a demon running on the machine for assisting inter-process communication. MPICH2 uses encrypted data transfer so that the communication between parallel processes is encrypted and secure. SA is a Java based program that can communicate with other components of DLB through Java Remote Method Invocation API (RMI). SA is secure since it only provides the information about the number of single and parallel processes run on the computer and the network speed but does not allow any file being transferred to or from the host PC.

To evaluate the computer hardware utilization, *Efficiency Percentage* (EF) is defined as follows

$$EF = \frac{CP^{1} \times CS^{1} + CP^{2} \times CS^{2} + CP^{3} \times CS^{3} + .... + CP^{n} \times CS^{n}}{100\% \times CS^{1} + 100\% \times CS^{2} + 100\% \times CS^{3} + .... + 100\% \times CS^{n}}$$

where $CP^n$ = CPU utilization percentage of the nth PC, and $CS^n$ = CPU relative speed of the nth PC. EF indicates the percentage of the maximum computational power used by all processes running on the computers. The EF calculation also gives higher weight for high speed computer in efficiency calculation. CPU utilization percentage for each computer can be dynamically measured during program execution.

## 4. EXPERIMENT

The DLB tool was tested in a network of eight Windows based PCs. The computers are connected by 100Mbit Ethernet. The test aimed to show that the DLB tool can manage two different jobs running simultaneously in two set of PCs with a common PC in both sets. The application code and related data are stored in a Windows 2003 Server which is accessible by all PCs. The environment of eight heterogeneous PCs is shown in Table 1. The relative CPU speed can be derived from the benchmark execution CPU time.

Table 1. Speeds of windows based PCs used in the test

| PC NAME | BENCHMARK EXECUTION CPU TIME (seconds) |
|---|---|
| in-engr-sl11121 | 4.625 |
| in-engr-sl11122 | 4.015 |
| in-engr-sl11132 | 8.046 |
| in-engr-sl11133 | 8.764 |
| in-engr-sl11125 | 4.703 |
| in-engr-sl11126 | 4.75 |
| in-engr-sl11127 | 4.734 |
| in-engr-sl11123 | 4.25 |

The parallel application used in testing was CFDL's Parallel CFD Test Solver. This solver has a 3D transient heat equation, which is implemented on structured grids with the forward time central space differentiation method [4]. A general grid structure of the test program is shown in Figure 3. Parallel communication implementation is actualized by using MPI standards. There is a separate grid generator to generate structured grids for this application. This grid is partitioned in three spatial directions, x, y and z. The size of the grid and the total number of blocks can be chosen by the user. However, we choose to make all blocks with the same size in order to let the reader to see the load balancing results intuitively.

Two parallel applications were submitted to eight PCs. The first job had 12 even sized blocks, where 2 blocks in X direction, 2 blocks in Y direction, and 3 blocks in Z direction. The block size was approximately 8MB (15,625 nodes/block). These blocks were submitted on 5 specific machines. (in-engr-sl11125, in-engr-sl11121, in-engr-sl11122,in-engr-sl11132, and in-engr-sl11133). The second job had 20 even sized jobs, where 2 blocks in X direction, 2 blocks in Y direction, and 5 blocks in Z direction. The block size was approximately 8MB (15,625 nodes/block). The blocks were submitted to 4 specific machines (in-engr-sl11123, in-engr-sl11125, in-engr-sl11126. and in-engr-sl11127). Computer in-engr-sl11125 is the computer used by both applications. The initial load distribution is shown in Table 2.

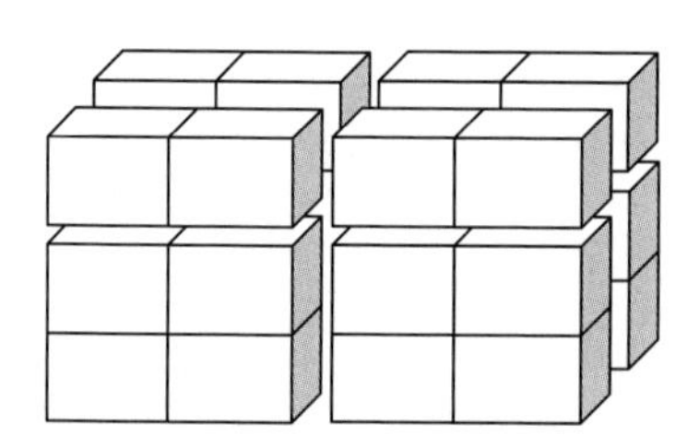

Figure 3. Grid structure of the Parallel CFD Test Application

**Table 2.** Initial load distribution

| Host Name | Total Load | %CPU Usage | Fist Job's Blocks | Second Job's Blocks |
|---|---|---|---|---|
| in-engr-sl11121 | 3 | 43.913 | 1 2 3 | |
| in-engr-sl11122 | 3 | 44.145 | 4 5 6 | |
| in-engr-sl11132 | 2 | 59.412 | 7 8 | |
| in-engr-sl11133 | 2 | 57.072 | 9 10 | |
| in-engr-sl11125 | 7 | 99.614 | 11 12 | 1 2 3 4 5 |
| in-engr-sl11126 | 5 | 62.629 | | 6 7 8 9 10 |
| in-engr-sl11127 | 5 | 61.412 | | 11 12 13 14 15 |
| in-engr-sl11123 | 5 | 62.581 | | 16 17 18 19 20 |

During the initial execution of both applications, the CPU usage percentages were measured as shown in Table 2. The %CPU usage of computer in-engr-sl11125 was 99.6% and that of other computers were less than 63%. This demonstrates that computer in-engr-sl11125 was busy and all processes running on other computers were waiting. As DLB tool allowed only one parallel job on a computer to do load balance at any time instant, Job 2 could do load balancing if Job 1 was doing load balancing. When Job 1 did load balancing, all two blocks on computer in-engr-sl11125 were moved to in-engr-sl11132 and in-engr-sl11133 (see Table 3).

Table 3. Load distribution after load balancing

| Host Name | Total Load | %CPU Usage | Job 1 Blocks | Job 2 Blocks |
|---|---|---|---|---|
| in-engr-sl11121 | 4 | 95.227 | 1 2 3 4 | |
| in-engr-sl11122 | 4 | 96.287 | 5 6 7 8 | |
| in-engr-sl11132 | 2 | 99.314 | 9 10 | |
| in-engr-sl11133 | 2 | 102.51 | 11 12 | |
| in-engr-sl11125 | 5 | 102.736 | | 1 2 3 4 5 |
| in-engr-sl11126 | 5 | 106.521 | | 6 7 8 9 10 |
| in-engr-sl11127 | 5 | 101.252 | | 11 12 13 14 15 |
| in-engr-sl11123 | 5 | 98.419 | | 16 17 18 19 20 |

By examine the %CPU Usage, we can see that the %CPU usage of all computers are over 95 percent which means that all CPUs are fully utilized. By definition, the efficiency should not exceed 1. However, the sum of the CPU usage of all processes in a computer provided by system software in both PC and Unix often slightly exceed 1 due to rounding errors. The load balancing improved the Efficiency Percentage of the whole system from initial 61.3% to about 100%.

## 5. DISCUSSION

To prevent a PC owner from only uses others' PC but not allowing others to use his/her PC, the tool makes a PC owner share the his/her computer power (set in schedule calendar).while using other PC owner's computer power. This tool requires the application code to use MPICH2 implementation of MPI. The DLB tool works only in the environment that all Microsoft Window based PCs are in the same network domain since MPICH2 requires that all the PCs to be a part of the same domain. This is not a severe restriction since all PCs in an institution can be in one domain. For example, all PCs in our institution are in one domain.

The DLB tool has also been improved to support multi core PCs. The memory size of the system has not been taken into the load balancing consideration yet. The requirement of the speed of the interconnecting network depends on the parallel applications. Based on our experience in past testing cases, we learned that if the communication cost (time) was less than 20% of the total cost for a parallel application, the load balancing tool performed well. Fast Gigabit network would definitely improve the applicability of proposed load balancing tool.

The UNIX version of the DLB tool supports fault tolerance due to hardware failure, system failure, and network failure. It assumes that the parallel application code does not create deadlock in normal operation. The fault tolerance feature is being considered in the future version of the DLB tool for the Windows based computers.

## 6. CONCLUSION

The dynamic load balancing tool is ported to be directly executed on Microsoft Windows (WindowaXP and Vista) based PCs. The tool supports dynamic load balancing on windows based PCs and ensures that the PC "owners" are not affected by the computer usages by the parallel jobs. The tool was successfully tested for supporting parallel CFD codes.

## REFERENCES

[1] Secer, Semih, "Genetic Algorithms and Communication Cost Function for Parallel CFD Problems," M.S.E.E. Thesis, Purdue University, May 1997.

[2] Y.P. Chien, A. Ecer, H.U. Akay, and F. Carpenter, "Dynamic Load Balancing on Network of Workstations for Solving Computational Fluid Dynamics Problems," *Computer Methods in Applied Mechanics and Engineering*, Vol. 119, 1994, pp. 17-33.

[3] MPICH 2, http://www-unix.mcs.anl.gov/mpi/mpich2/.

[4] Parallel CFD Test Solver, http://www.engr.iupui.edu/me/newmerl/cfdl_software.htm.

# Parallel computations of droplet oscillations

Tadashi Watanabe

*Center for Computational Science and e-Systems, Japan Atomic Energy Agency, Tokai-mura, Naka-gun, Ibaraki-ken, 319-1195, Japan*

Keywords: Droplet Oscillation, Level Set Method, MPI, Bi-CGSTAB, Preconditioning, Point Jacobi, Block Jacobi, Additive Schwarz

**Abstract**

Three-dimensional oscillations of a liquid droplet are simulated numerically using the level set method. It is shown that the oscillation frequency decreases as the amplitude increases and the frequency shift is overestimated by the theoretical prediction. Parallel computations are performed, and the effects of three preconditioners for the Bi-CGSTAB method on speed up are discussed.

## 1. Introduction

A levitated liquid droplet is used to measure material properties of molten metal at high temperature, since the levitated droplet is not in contact with a container, and the effect of the container wall is eliminated for a precise measurement. The levitation of liquid droplets, which is also used for containerless processing of materials, is controlled by using electrostatic force [1] or electromagnetic force [2] under the gravitational condition. Viscosity and surface tension are, respectively, obtained from the damping and the frequency of an oscillation of the droplet. Large-amplitude oscillations are desirable from the viewpoint of measurement. However, the relation between material properties and oscillation parameters, which is generally used to calculate material properties, is based on the linear theory given by Lamb [3], and small-amplitude oscillations are necessary for an estimation of material properties.

The effect of the amplitude on the oscillation frequency has been discussed theoretically by Azuma and Yoshihara [4]. Second order small deviations were taken into account for the linearized solution, and the oscillation frequency was shown to decrease as the amplitude increased. Although qualitative effects were shown, quantitative estimation is not enough since the theoretical approach is based on some assumptions and simplifications. In this study, numerical simulations of an oscillating

I.H. Tuncer et al., *Parallel Computational Fluid Dynamics 2007*,
DOI: 10.1007/978-3-540-92744-0_20, 

liquid droplet are performed to study the effect of the amplitude on the oscillation frequency of the droplet. Three-dimensional Navier-Stokes equations are solved using the level set method [5]. The domain decomposition technique is applied and the message passing interface (MPI) library is used for parallel computations. The Poisson equation is solved using the Bi-CGSTAB method. Three preconditioners for the Bi-CGSTAB method are compared [6]; the point Jacobi, the block Jacobi, and the overlapping additive Schwarz schemes.

## 2. Numerical Method

Governing equations for the droplet motion are the equation of continuity and the incompressible Navier-Stokes equations:

$$\nabla \cdot u = 0 \ , \tag{1}$$

and

$$\rho \frac{Du}{Dt} = -\nabla p + \nabla \cdot (2\mu \boldsymbol{D}) - F_s \ , \tag{2}$$

where $\rho$, $u$, $p$ and $\mu$, respectively, are the density, the velocity, the pressure and the viscosity, $\boldsymbol{D}$ is the viscous stress tensor, and $F_s$ is a body force due to the surface tension. The surface tension force is given by

$$F_s = \sigma \kappa \delta \nabla \phi \ , \tag{3}$$

where $\sigma$, $\kappa$, $\delta$ and $\phi$ are the surface tension, the curvature of the interface, the Dirac delta function and the level set function, respectively. The level set function is defined as $\phi=0$ at the interface, $\phi<0$ in the liquid region, and $\phi>0$ in the gas region. The curvature is expressed in terms of $\phi$:

$$\kappa = \nabla \cdot \left(\frac{\nabla \phi}{|\nabla \phi|}\right) \ . \tag{4}$$

The density and viscosity are given by

$$\rho = \rho_l + (\rho_g - \rho_l) H \ , \tag{5}$$

and

$$\mu = \mu_l + (\mu_g - \mu_l) H \ , \tag{6}$$

where $H$ is the Heaviside-like function defined by

$$H = \begin{cases} 0 & (\phi < -\varepsilon) \\ \frac{1}{2}[1 + \frac{\phi}{\varepsilon} + \frac{1}{\pi}\sin(\frac{\pi\phi}{\varepsilon})] & (-\varepsilon \le \phi \le \varepsilon) , \\ 1 & (\varepsilon < \phi), \end{cases} \tag{7}$$

where $\varepsilon$ is a small positive constant for which $|\nabla\phi| = 1$ for $|\phi| \le \varepsilon$. The evolution of $\phi$ is given by

$$\frac{\partial\phi}{\partial t} + u \cdot \nabla\phi = 0 . \tag{8}$$

All variables are nondimensionalized using liquid properties and characteristic values: $x'=x/L$, $u'=u/U$, $t'=t/(L/U)$, $p'=p/(\rho_l U^2)$, $\rho'=\rho/\rho_l$, $\mu'=\mu/\mu_l$, where the primes denote dimensionless variables, and $L$ and $U$ are representative length and velocity, respectively. The finite difference method is used to solve the governing equations. The staggered mesh is used for spatial discretization of velocities. The convection terms are discretized using the second order upwind scheme and other terms by the central difference scheme. Time integration is performed by the second order Adams-Bashforth method. The SMAC method is used to obtain pressure and velocities [7]. The Poisson equation is solved using the Bi-CGSTAB method. The domain decomposition technique is applied and the message passing interface (MPI) library is used for parallel computations. The parallel computer systems, SGI Altix 350 and 3900, which have the same processors but different communication rates, are used. Three preconditioners for the Bi-CGSTAB method are compared; the point Jacobi, the block Jacobi, and the overlapping additive Schwarz schemes. In the nondimensional form of the governing equations, the Reynolds number, $\rho_l LU/\mu_l$, and the Weber number, $\rho_l LU^2/\sigma$, are fixed at 200 and 20, respectively, in the following simulations.

In order to maintain the level set function as a distance function, an additional equation is solved:

$$\frac{\partial\phi}{\partial\tau} = (1 - |\nabla\phi|)\frac{\phi}{\sqrt{\phi^2 + \alpha^2}} , \tag{9}$$

where $\tau$ and $\alpha$ are an artificial time and a small constant, respectively. The level set function becomes a distance function in the steady-state solution of the above equation. The following equation is also solved to preserve the total mass in time [8]:

$$\frac{\partial\phi}{\partial\tau} = (A_o - A)(1 - \kappa)|\nabla\phi| , \tag{10}$$

where $A_0$ denotes the total mass for the initial condition and A denotes the total mass corresponding to the level set function. The total mass is conserved in the steady-state solution of the above equation.

## 3. Results and Discussion

Free-decay oscillations of spherical droplets are simulated. The average radius of the droplet is 1.0 in nondimensional unit, and the initial deformation is given by the Legendre polynomial of order 2. Variations of droplet shape during one oscillation period are shown in Fig. 1 as examples of simulation results. The time step is denoted by T, and the time step size is 0.0033. The initial amplitude is 0.38 and the size of the simulation region is 3.6x3.6x3.6 discretized with 120x120x120 calculation nodes.

The oscillation of droplet radius, r, in the vertical direction is shown in Fig. 2. The initial amplitude, $\Delta r$, is changed from 0.11 to 0.38 in Fig. 2. It is shown that the time history of the droplet radius shift towards the positive direction of the time axis, t, as the amplitude increases. It indicates that the frequency becomes small as the amplitude increases. The theoretical frequency is given by 0.10 based on the linear theory [3], while the calculated frequency is 0.993, 0.978, 0.954, and 0.924 for the amplitude of 0.11, 0.20, 0.29, and 0.38, respectively. The decrease in frequency is predicted theoretically [4], and this nonlinearity effect is simulated clearly in Fig. 2.

The effect of amplitude on oscillation frequency is shown in Fig. 3. The vertical axis indicates the frequency shift, which is defined as the frequency difference, $\Delta f$, normalized by the oscillation frequency, $f_0$, for the smallest amplitude, $\Delta r=0.02$. In the linear theory given by Lamb [3], the oscillation frequency is obtained in terms of the surface tension, the density, the radius and the mode of oscillation. The oscillation frequency is, however, shown to be affected much as the amplitude becomes large. Theoretical curve given by Azuma and Yoshihara [4], which was derived by taking into account a second order deviation from the linear theory, is also shown along with the simulation results in Fig. 3. The theoretical curve overestimates the frequency shift for the amplitude larger than 0.2.

The speed up of calculation is shown in Fig. 4, where three preconditioners are compared. The total calculation time for 10 time steps in 96x96x96 calculation nodes is indicated by Elapse, while the time for matrix calculations by Matrix. The domain decomposition is performed in one direction. The point Jacobi, block Jacobi, and additive Schwarz preconditioners are denoted by PJ, BJ, and AS, respectively in Fig. 4. The number of overlapping is set to 14 for the AS scheme in Fig. 4 after some sensitivity calculations. It is shown that the speed up of the PJ scheme is larger than that of the BJ and AS schemes, and the BJ scheme is almost the same as the AS scheme. The data array in each processor becomes small as the number of processors, Pe, increases, and the effect of cache becomes large. The diagonal elements are used as the preconditioner in the PJ scheme, while the block diagonal elements are used in the BJ and AS schemes. The size of the matrix is the smallest for the PJ scheme since the compressed row form is used, and the effect of cache is notable. The total calculation time using one processor is 998.0 s, 627.9 s and 627.9 s, respectively, for the PJ, BJ and AS schemes. The BJ and AS schemes are thus found to be better when the number of processors is small. It is shown that the AS scheme is not so effective in our calculations, since the matrix size is sufficiently large and the characteristics of the matrix are not much improved by the overlapping scheme.

The total number of iterations for 10 time steps is shown in Fig. 5. The number of iterations is almost the same for the BJ and AS schemes, while larger for the PJ scheme. The calculation time is larger for the PJ scheme using one processor. But, the amount of calculations is not increased as the number of processors increases for the PJ scheme. It is shown for the BJ and AS schemes that the amount of calculations is increased. This is one of the reasons why the speed up is better for the PJ scheme.

The effect of the communication rate on the speed up is shown in Fig. 6, where the BJ scheme is used. Altix 350 and 3900 are made by SGI, and Intel Itanium2 processors are used with a clock rate of 1.6 GHz. 128 processors are on one node of Altix 3900, while 2 processors are on one node of Altix 350. The inter-node communication is thus necessary for Altix 350. Although the speed up is better for Altix 3900, the effect of the inter-node communication is small if the number of processors is smaller than 8.

## 4. Conclusion

In this study, three-dimensional oscillations of a levitated liquid droplet have been simulated numerically using the level set method, and the effect of amplitude on oscillation frequency was discussed. It was shown that the oscillation frequency decreased as the amplitude increased and the frequency shift was overestimated by the theoretical prediction. Parallel computations were performed using the domain decomposition technique, and the effects of preconditioners for the Bi-CGSTAB method on speed up were discussed. The block Jacobi and additive Schwarz schemes were shown to be effective in comparison with the point Jacobi scheme when the number of processors was small.

## References

[1] Rhim, W. K., and Chung, S. K. 1990. *Methods*, **1**, 118-127
[2] Shatrov, V., Priede, J., and Gerbeth, G. 2003. *Phys. Fluid*, **15**, 668-678
[3] Lamb, H. 1932. *Hydrodynamics*. Cambridge University Press.
[4] Azuma, H., and Yoshihara, S. 1999. *J. Fluid Mech.*, **393**, 309-332.
[5] Sussman, M., and Smereka, P. 1997. *J. Fluid Mech.*, **341**, 269-294.
[6] Benzi, M. 2002. *J. Comp. Phys.*, **182**, 418-477.
[7] Amsden, A. A. and Harlow, F. H., J. Comp. Phys. **6**, 322(1970).
[8] Chang, Y. C., Hou, T. Y., Merriman, B., and Osher, S. 1996. *J. Comp. Phys.*, **124**, 449-464

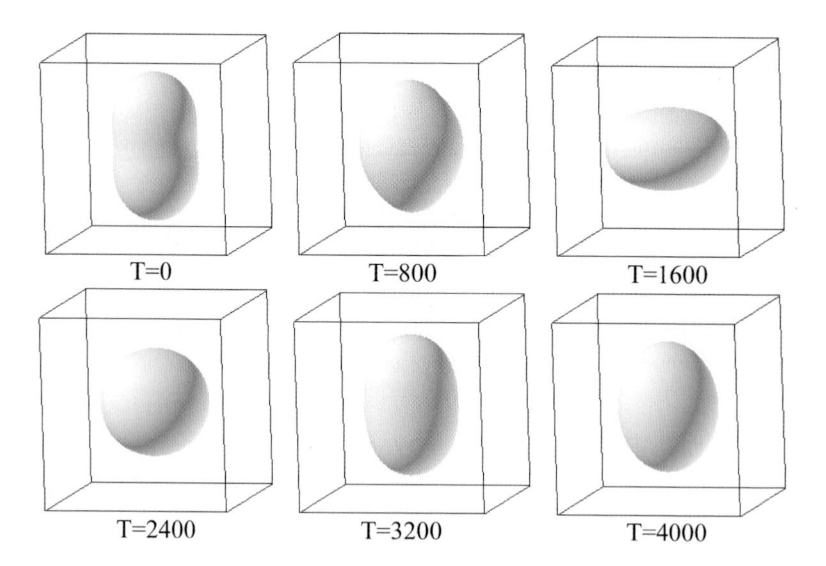


Fig. 1 Oscillation of droplet shape

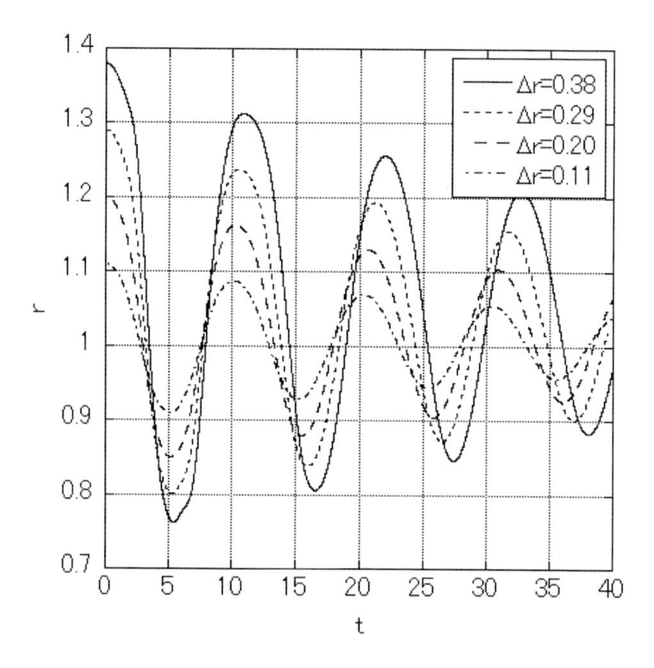


Fig. 2 Oscillation of droplet radius with different initial amplitude

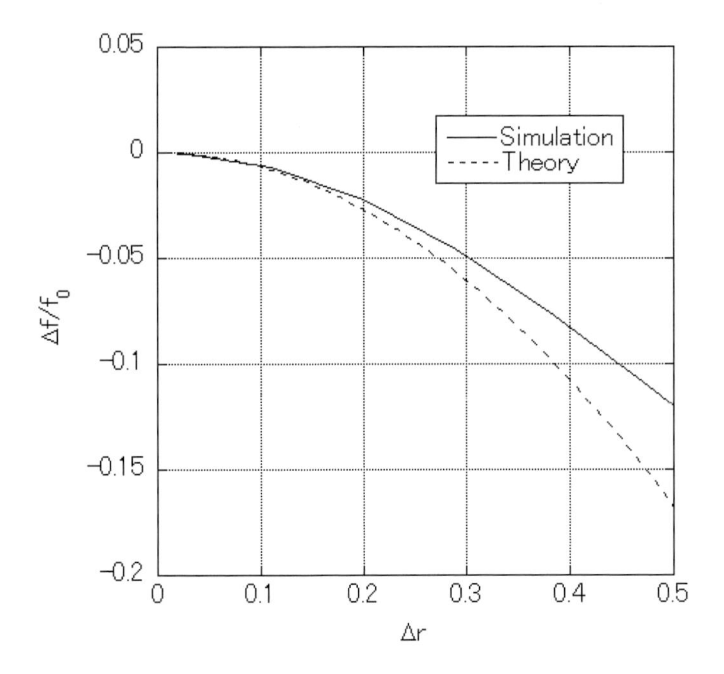


Fig. 3 Effect of amplitude on oscillation frequency

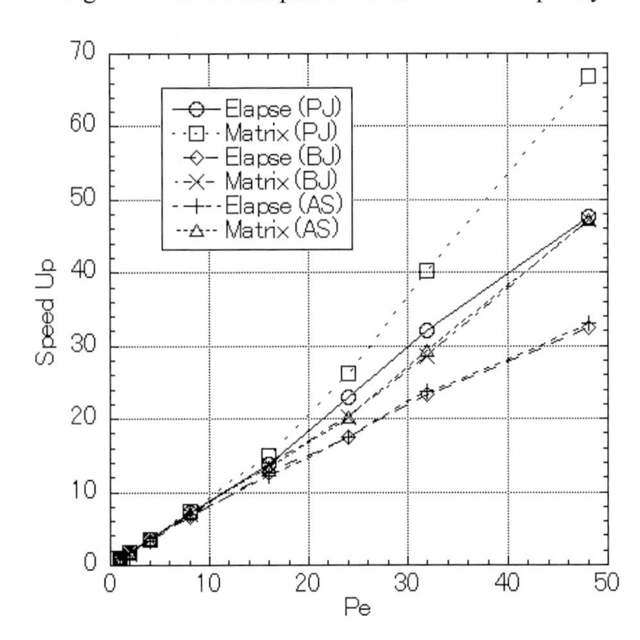


Fig. 4 Speed up of total calculation time and matrix calculation time with different preconditioners for the Bi-CGSTAB method

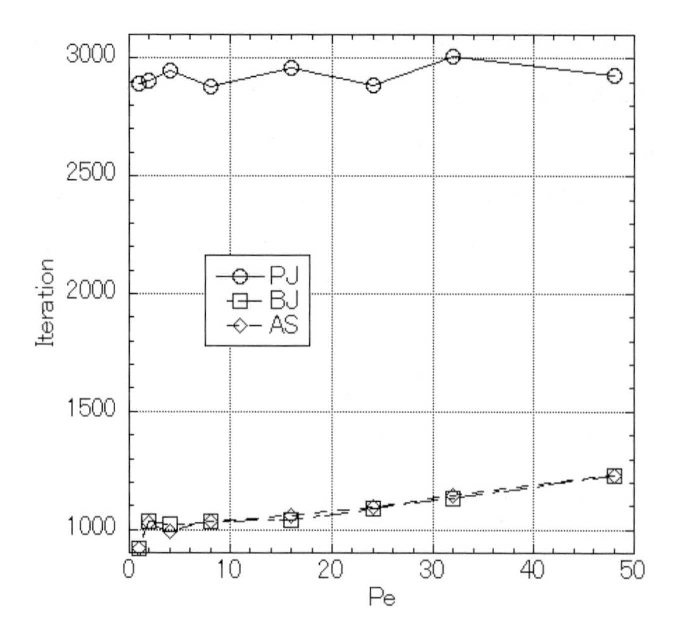


Fig. 5 Total number of iterations

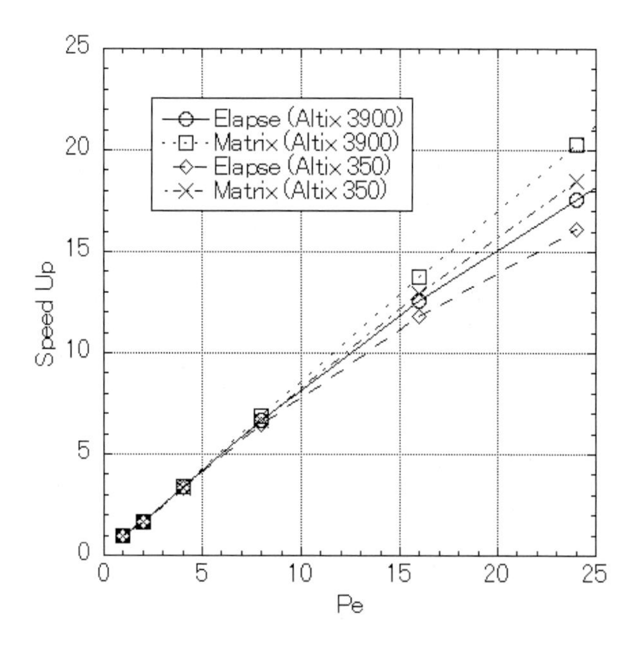


Fig. 6 Effect of communication rate

# Cyclic Distribution of Pipelined Parallel Deferred Correction method for ODE/DAE

D. Guibert [a] and D. Tromeur-Dervout[a] [b] [*]

[a]CDCSP/ICJ UMR5208-CNRS, Université Lyon 1, F-69622 Villeurbanne Cedex

[b]INRIA/IRISA/Sage Campus de Beaulieu Rennes, F-35042 Rennes Cedex, France

## Abstract

The development of large computational resources, leads to search for parallel implementation not only based on space decomposition as domain decomposition or operator splitting and other classical parallel methods in CFD. These methods have to face the lack of computational granularity when the computing resources are very large. One promising direction of parallelism can be the time domain decomposition, where the solution of problem is computed in parallel on different time slices.

In this paper, a time domain decomposition method is introduced. This method uses the spectral deferred corretion method. A cyclic distribution of the time slices is proposed to reduced the startup of the first step of the iterated correction process. Numerical results and parallel efficiency are presented on the lid driven cavity with 2D Navier-Stokes equations writen in $\omega - \psi$ formulation.

## 1. Motivation for new parallel solver for ODE/DAE

The parallelization of the systems of partial differential equations (PDE) is a field very well controlled nowadays with the distribution in space of the computing. However, domain decomposition methods (DDM) in PDE are not directly transposable for the systems of ordinary differential equations (ODE) or Differential Algebraic Equations (DAE). Because DDM in PDE are often based on the partitioning of the discretization in space of an operator which is often parabolic or elliptic. They reached very good efficiency on low order scheme in space and consequently also low order scheme in time. Nevertheless, for the evolution problems, the difficulty comes from the sequential nature of the computing. Previous time steps are required to compute the current time step. This last constraint occurs in all time integrators used in computational fluid dynamics of unsteady problems (CFD).

This difficulty occurs in parallelizing of systems of ODE or DAE. Consequently, devel-

*This work was funded by the ANR technologie logicielle 2006 PARADE and the cluster ISLES : Calcul Hautes Performances project and the MATHGRID project of the Region Rhône-Alpes

I.H. Tuncer et al., *Parallel Computational Fluid Dynamics 2007*,
DOI: 10.1007/978-3-540-92744-0_21, 

opment of efficient parallel methods for ODE/DAE will improve the parallelizing for CFD problems. For CFD problems, techniques to write systems of ODE/DAE developed in the literature are based on the method of lines and the parallel methods involved usually follow the work of K. Burrage [1]. These techniques are based on the partitioning of the stages of calculation in the implicit methods of the Runge-Kutta (RK) type. This type of parallelization is thus limited to a number of processors because of the orders of the stages of integration which are used in practice. Moreover, they do not break the sequential nature of the computing, as RK methods compute intermediary steps starting from the previous solution.

Delaying or breaking the sequential nature of the time integration schemes constitutes a challenge. Time domain decomposition methods can be a complementary approach to improve parallelism on the forthcoming parallel architectures with thousand processors. Indeed, for a given modelling problem, — e.g. through a partial differential equation with a prescribed discretization— the multiplication of the processors leads to a decrease of the computing load allocated to each processor; then the ratio of the computing time with the communication time decreases, leading to less efficiency. Several works on the parallelizing in time have been proposed during this last decade. When it is about system of ODE and DAE, the additional data-processing costs are not compensated by the resolution in space. It is in particular the case on the multiple shooting methods algorithm Parareal [7] or Pita [4] for systems of ODE, the cost of calculation and the sequential aspect of the correction of coarse grid, harms the parallel effectiveness of the method. Moreover, this correction step is an algorithm sequential. The time integrator often takes more time to solve this correction stage than it takes to solve the initial problem. They are also very sensitive to the Jabobian linearization for the correction stage on stiff non-linear problems as shown in [5]. Moreover, characteristics of the systems of DAE require the development of specific numerical software technologies. Indeed, the management of discontinuities, the stiffness of the systems of equations, the adaptivity of the steps of time, the possible junctions of the solutions and the differences in scales of time of the solutions require particular implementations like those met in [2]. These characteristics can disturb the numerical scheme of parallelization and necessarily result in reconsidering the numerical techniques of integration.

The pipelined deferred correction method that we developed in parallelCFD06 [6], combines time domain decomposition and spectral deferred correction in order to have a two level algorithm. At the first level, the time interval is split in piecewise and at the second level the spectral deferred correction iterations are pipelined onto this time slice. Some good efficiency is obtained when the pipeline is full. Nevertheless, the waiting time to fill the pipeline deteriorates the global efficiency.

In this work we propose to extend the pipelined deferred correction method with introducing a cyclic distribution of the sub-domains onto the network of processors in order to reduce the waiting time per processor in the propagation of the pipeline.

The outline of this paper is as follows. Section 2 recalls the spectral deferred correction method (SDC)while section 3 explains the technical implementation of the cyclic distribution of the pipelined SDC. Numerical and parallelism results on the lid driven cavity problem are given in 4 before to conclude in section 5.

## 2. Spectral Deferred Correction Method

The spectral deferred correction method [8] is a method to improve the accuracy of a time integrator iteratively in order to approximate solution of a functional equation with adding corrections to the defect.

Let us summarize the SDC method for the Cauchy equation:

$$y(t) \quad = \quad y_0 + \int_0^t f(\tau, y(\tau))d\tau \tag{1}$$

Let $y_m^0 \simeq y(t_m)$ be an approximation of the searched solution at time $t_m$. Then consider the polynomial $q_0$ that interpolates the function $f(t,y)$ at P points $\tilde{t}_i$ i.e:

$$\exists! q_0 \in R_P[x], q_0(\tilde{t}_i) = f(y^0(\tilde{t}_i)) \tag{2}$$

with $\{\tilde{t}_0 < \tilde{t}_1 < ... < \tilde{t}_P\}$ which are points around $t_m$. If we define the error between the approximate solution $y_m^0$ and the approximate solution using $q_0$

$$E(y^0, t_m) = y_0 + \int_0^{t_m} q^0(\tau)d\tau - y_m^0 \tag{3}$$

Then we can write the error between the exact solution and the approximation $y_m^0$:

$$y(t_m) - y_m^0 = \int_0^{t_m} \left(f(\tau, y(\tau)) - q^0(\tau)\right) d\tau + E(y^0, t_m) \tag{4}$$

The defect $\delta(t_m) = y(t_m) - y_m^0$ satisfies:

$$\delta(t_{m+1}) = \delta(t_m) + \int_{t_m}^{t_{m+1}} \left(f(\tau, y^0(\tau) + \delta(\tau)) - q^0(\tau)\right) d\tau + E(y^0, t_{m+1}) - E(y^0, t_m) \tag{5}$$

For the spectral deferred correction method an approximation $\delta_m^0$ of $\delta(t_m)$ is computed considering for example a Backward Euler scheme :

$$\delta_{m+1}^0 \quad = \quad \delta_m^0 + \Delta t(f(t_{m+1}, y_{m+1}^0 + \delta_{m+1}^0) - f(t_{m+1}, y_{m+1}^0)) - y_{m+1}^0 + y_m^0 + \int_{t_m}^{t_{m+1}} q^0(\tau)d\tau$$

Once the defect computed , the approximation $y_m^0$ is updated

$$y_m^1 = y_m^0 + \delta_m^0. \tag{6}$$

Then a new polynomial $q_1$ is computed by using the value $f(t_m, y_m^1)$ for a new iterate. This method is a sequential iterative process as follows: compute an approximation $y^0$, then iterate in parallel until convergence with the computation of $\delta$ and the update of the solution $y$ by adding the correction term $\delta$.

## 3. Pipelined SDC with cyclic distribution

The main idea to introduce parallelism in the SDC is to consider the time domain decomposition and to affect a set of time slices to each processor. Then we can pipeline the computation where each processor deals with a different iterate of the SDC. Nevertheless, a load balancing problems occurs due to two factors. Firstly, the elapsed time for the pipeline to reach the last processor can be quite long. Secondly, the problem can become physically steady, consequently the last processor, also last time slices, has to wait to do less computational work while the first ones take in charge most of the work. This lead us to introduce a cyclic distribution of the subdomains between the processors. This is illustrated in Figure 1.

The technical difficulty is to activate the time slices that have to send and receive data with synchronous communication during the pipelined propagation. We can not use asynchronous communication for data consistency in the building of the $q_0$ polynomial. Data inconsistency will lead to numerical instability in the integration scheme. To performe the data communication between processors, we introduce a vector data structure associated to the time slices managed by the processors. Then with respect to the global iteration, the vector contains the current iteration of the time slice. As all the time slices of one processors can only communicate with the time slices managed by its two neighbor processors, the communication pattern is quite simple. The size of data that one processor as to communicate increases and decreases during the iterative process following the active time slices in the vector data structure. The pipeline propagates the convergence from the left to the right of the time interval. The solution onto one slice can not be considered as converged, if it is not converged onto the left neighbor time slice managed by another processor. One convergence flag associated to the time slice is then send in the same time as the $P/2$ solution values. This rule has to be applied in order to not have dead time slice with active previous time slices in the computation that will stop the pipeline.

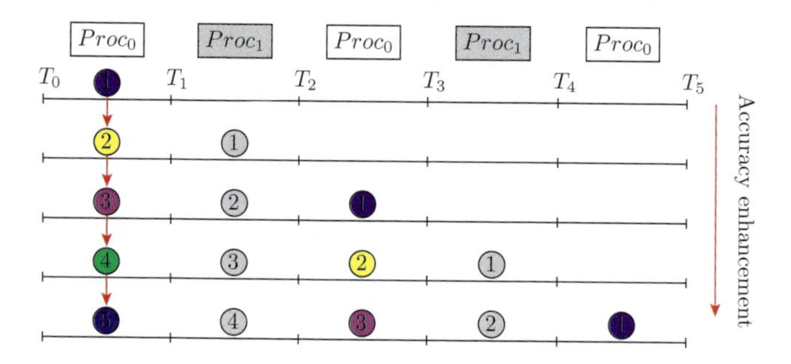


Figure 1. Pipelined Spectral DC with a cyclic distribution of the time slices onto the processors

**Algorithm 3.1** (Parallel Algorithm). *1. Initialization phase: compute an approximation $y^0$ with a very low accuracy. Then distribute overlapped time slices to processors with an overlap of $P/2$ slices. Each processor can start the next step as soon as it has been fully initialized.*

*2. Cyclic distributed pipelined SDC: Iterate until convergence*

*(a) Activate the next subdomain in the set of global subdomains.*

*(b) For each active subdomain group handled by the processor.*

*i.* Receive *the $P/2$ first overlapped $\delta$ values.*

*ii. Prepare the non blocking* receiving *of the $P/2$ last delta values plus the time slice convergence flag if the previous flag is false.*

*iii. Compute the $P/2$ first $\delta$ values and* send *them to the previous processor.*

*iv. Compute the other $\delta$ values. Be sure to have received the last values before computed the overlapped ones.*

*v.* Send *the $P/2$ last $\delta$ values plus the convergence flag.*

*vi. Update the solution $y_m^{i+1} = y_m^i + \delta_m^i$*

**Remark 3.1.** *Our approach differs a little bit from the original spectral deferred correction method because we compute a new polynomial $q_0$ for each time step $t_n$ while in the original the same $q_0$ is used for several time steps. Our experiments show that our approach seems to be more stable, and has the property to still having consistent data when the time interval is split.*

**Remark 3.2.** *The computation complexity to build the $q_0$ with Lagrange interpolation is in $O(p^2)$ per time step in our implementation. This can be replaced by an $O(p)$ Neville's algorithm that builds the $q_0$ polynomial deleting the oldest data and adding a new one.*

## 4. Numerical and parallel efficiency results

Let us consider the 2D Navier-Stokes equations in a Stream function - vorticity formulation written in a Cauchy equation form:

$$\frac{d}{dt}\begin{pmatrix}\omega\\ \psi\end{pmatrix}(t) = f(t,\omega,\psi),\ \begin{pmatrix}\omega\\ \psi\end{pmatrix}(0) = \begin{pmatrix}\omega_0\\ \psi_0\end{pmatrix} \tag{7}$$

then the discretization in finite differences with a second order scheme in space of the $f(t,\omega,\psi)$ leads to a system of ODE :

$$\frac{d}{dt}\begin{pmatrix}\omega_{i,j}\\ \psi_{i,j}\end{pmatrix} = f(t,\omega_{i[\pm 1],j[\pm 1]},\psi_{i[\pm 1],j[\pm 1]}) \tag{8}$$

This last equation can be solved by an available ODE solver such as SUNDIALS [2] SUite of Nonlinear and DIfferential/ALgebraic equation Solvers that contains the latest developments of classical ODE solvers as LSODA [2] or CVODE [3].

[2]http://www.llnl.gov/casc/sundials/

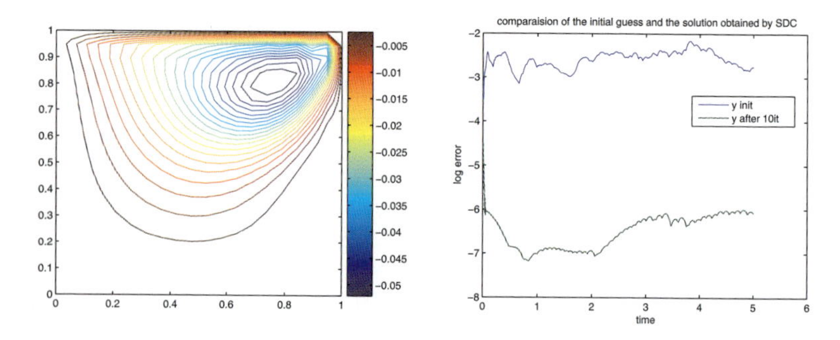


(a) stream function isovalues for the lid driven cavity with $Re = 400$ at time $T = 5$

(b) Initial guess and converged solution of the SDC method

Figure 2. Solution of 2D-NS lid driven cavity flow problem. Comparaison between a reference solution and the solution obtained by the spectral deferred correction.

Figure 2(b) gives with respect to the time, the log10 of the error in maximum norm between the iterative solution obtained by the SDC and a reference solution. This error was computed by an sequential adaptive time step ODE solver with a relative tolerance of $10^{-12}$. With 10 iterations an accuracy of $10^{-6}$ is achieved, starting from an error of $10^{-2}$ for the initial guess. The lid driven cavity problem parameters were $N_x = N_y = 20$, $Re = 400$ and $N_{timeslices} = 800$. The classical isovalues obtained at the last time step are given in 2(a).

Figure 3 shows the execution time and the waiting time of the current implementation of the pipelined SDC with 800 time slices where each processor has only one cycle in charge (e.g. two processors take in charge 400 time slices each). The execution time decreases with the number of processors but we see that the waiting time to fill the pipe — the bottom curve — increases when the number of processors increases. For 16 processors $\frac{700s}{500s} \approx 71\%$ of the time on the last processor is devoted to wait for the first incoming data from the previous processor in order to start the computation. It's called the waiting time to fill the pipe. This is the problem of load balancing that motivates our cyclic distribution of the time slices onto the processors.

Figure 4 shows the execution time and the waiting time of the current implementation on 12 processors of the pipelined SDC with 800 time slices where each processor can have from 1 to 8 cycles in charge for 10 iterates. A better execution time is achieved when the number of cycles increases. As the computation complexity remains the same, the same behavior of the execution time and the waiting time are expected. But the waiting time decreases faster than the execution time with respect to the number of cycles. Consequently, the waiting time reduction is compensated by the data volume communication that is increasing with the number of active cycles managed by the processor. We deduce that the granularity of the computation is not sufficient to overlap the non blocking communications between active slices.

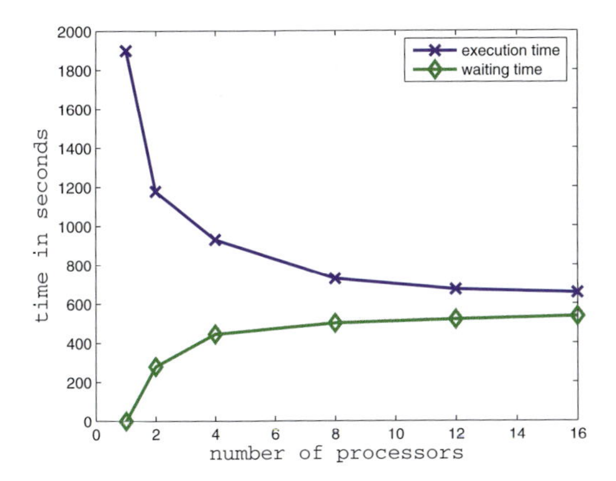


Figure 3. Efficiency of the Pipelined SDC

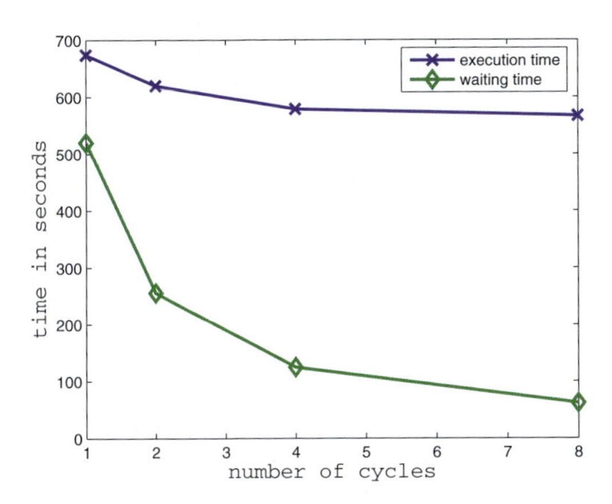


Figure 4. Execution/Waiting time of the Pipelined SDC increasing the number of cycles per processor for 12 processors

## 5. Conclusions

In this paper we developed a cyclic distribution of the spectral deferred correction. The communication pattern in this method still also available for the classical deferred correction method. Some improvement in term of elapsed time have been achieved. Nevertheless, a balance between the granularity of the computation on the time slice and the communication/waiting times have to be reached. Especially, as the time slices do not have the same computational load, it will be necessary to adapt the size of the time slices based on the initial guess computational load.

## REFERENCES


1. K Burrage and H. Suhartanto. Parallel iterated methods based on multistep runge-kutta methods of radau type for stiff problems. Advances in Computational Mathematics, 7 :59-77, 1997.
2. L.R. Petzold. DASSL: A Differential/Algebraic System Solver, June 24, 1991. Available at http://www.netlib.org/ode/ddassl.f.
3. S. D. Cohen and A. C. Hindmarsh. *CVODE, a Stiff/Nonstiff ODE Solver in C.* Computers in Physics, 10(2):138-143, 1996.
4. C. Farhat and M. Chandesris. *Time-decomposed parallel time-integrators: theory and feasibility studies for fluid, structure, and fluid-structure applications.* Int. J. Numer. Meth. Engng., **58**(9):1397–1434, 2003.
5. D. Guibert and D. Tromeur-Dervout, *Parallel Adaptive Time Domain Decomposition for Stiff Systems of ODE/DAE*, Computer and Structure, 85(9):553-562, 2007
6. D. Guibert and D. Tromeur-Dervout, *Parallel Deferred Correction method for CFD Problems* , Proc. of international Parallel CFD 06, J. Kwon, A. Ecer, J. Periaux, N. Satofuka, and P. Fox editors, elsevier, ISBN-13: 978-0-444-53035-6, 2007
7. J.-L. Lions, Y. Maday, and G. Turinici. *Résolution d'EDP par un schéma en temps "pararéel"*, C.R.A.S. Sér. I Math., **332**(7):661–668, 2000.
8. Minion, Michael L., *Semi-implicit spectral deferred correction methods for ordinary differential equations*, Commun. Math. Sci.,1(3):471–500,2003.

# Hybrid Parallelization Techniques for Lattice Boltzmann Free Surface Flows

Nils Thürey, T. Pohl and U. Rüde

University of Erlangen-Nuremberg, Institute for System-Simulation,
Cauerstr. 6, D-91054 Erlangen, Germany

In the following, we will present an algorithm to perform adaptive free surface simulations with the lattice Boltzmann method (LBM) on machines with shared and distributed memory architectures. Performance results for different test cases and architectures will be given. The algorithm for parallelization yields a high performance, and can be combined with the adaptive LBM simulations. Moreover, the effects of the adaptive simulation on the parallel performance will be evaluated.

## 1. Introduction

When a two phase flow involving a liquid and a gas, such as air and water, is simplified by only simulating the liquid phase with appropriate boundary conditions, this is known as a *free surface flow*. These flows a interesting for a variety of applications, from engineering and material science to special effects in movies. Our fluid simulator to solve these free surface problems uses the lattice Boltzmann method and a *Volume-of-Fluid* (VOF) model for the liquid-gas interface. Together with a turbulence model, adaptive time steps and adaptive grids it can efficiently solve complex free surface flows with a high stability. A central component is its ability to adaptively coarsen the grid for large volumes of fluid, so that these can be computed on a coarser grid and require less computations.

The free surface LBM model we apply is based on a D3Q19 lattice, with an LBGK collision operator, and a Smagorinsky turbulence model. For the free surface, we have to distinguish gas, fluid, and interface cells. While fluid cells can be handled normally, interface cells require

Figure 1. Example of an adaptive free surface simulation with the VOF LBM algorithm.

I.H. Tuncer et al., *Parallel Computational Fluid Dynamics 2007*,
DOI: 10.1007/978-3-540-92744-0_22, © Springer-Verlag Berlin Heidelberg 2009

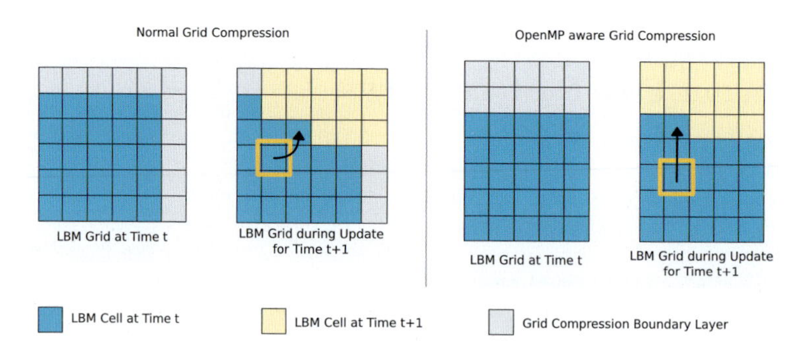


Figure 2. Comparison of the normal grid compression, and the grid compression for OpenMP. Instead of copying an updated cell to its diagonal neighbor, the cell is copied to a target cell in two cells distance (along the y direction for this 2D example, along the z direction for an actual 3D implementation).

additional computations to track the fluid-gas interface. As there is no information from the gas phase, we have to reconstruct the DFs that are streamed from gas cells with $f_I'(\mathbf{x}, t + \Delta t) = f_i^{eq}(\rho_A, \mathbf{u}) + f_{\tilde{i}}^{eq}(\rho_A, \mathbf{u}) - f_i(\mathbf{x}, t)$, where $f_i$ is a distribution function (DF) with opposing velocity vector to $f_I$, $f_i^{eq}$ is the equilibrium DF, and $\rho$ and $\mathbf{u}$ are fluid density and velocity, respectively. In addition, we compute a mass $m$ for interface cells, to determine where the interface is located within the cell. The mass is updated by the difference of incoming and outgoing DFs as $\Delta m_i(\mathbf{x}, t + \Delta t) = f_{\tilde{i}}(\mathbf{x} + \Delta t \ \mathbf{e}_i, t) - f_i(\mathbf{x}, t)$ . Whenever the mass of a cell is larger than its density, or less than zero, the cell type has to be converted correspondingly from interface to fluid or empty. The adaptive coarsening explicitly targets large volumes of fluid cells away from the interface to speed up the computations there with larger cells. The full algorithm is described in [7]. In the following, we will focus on its paralellization and the resulting performance.

## 2. OpenMP Parallelization

*OpenMP* is a programming model for parallel shared-memory architectures, and has become a commonly used standard for multi-platform development. An overview of the whole API is given in, e.g., [1].

The main part of the computations of the solver (ca. 73%) need to be performed for the computations of the finest grid. Thus, the parallelization aims to speed up this central loop over the finest grid of the solver. A natural way to do this would be to let the OpenMP compiler parallelize the outermost loop over the grid, usually the z direction. However, as we make use of the grid compression technique [5], this would violate the data dependencies for a cell update. With grid compression, the updated DFs of a cell at position $(i, j, k)$ in the grid are written to the position $(i-1, j-1, k-1)$. This only works for a linear update of all cells in the grid. Instead, to use grid compression with OpenMP, the DFs of cell $(i, j, k)$ are written back to the position $(i, j, k-2)$, as shown in Figure 2. This allows the update of all cells of an xy plane in arbitrary order. Note that this modified grid compression only requires slightly more memory than the original version (a single line of cells along the z direction, assuming the same x and y

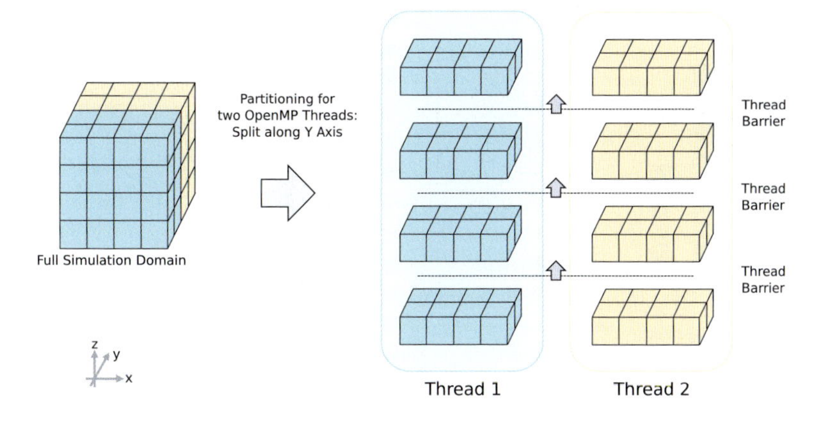


Figure 3. The OpenMP parallelization splits the y component of the loop over the grid.

grid resolution).

Hence, the loop over the y component is parallelized, as shown in Figure 3. This is advantageous over shortening the loops along the x direction, as cells on a line along the x axis lie successively in memory. Long loops in this direction can thus fully exploit spatial coherence of the data, and prefetching techniques of the CPU if available. After each update of a plane the threads have to be synchronized before continuing with the next plane. This can be done with the OpenMP *barrier* instruction. Afterwards, the cells of the next xy plane can again be updated in any order. In the following a gravitational force along the z axis will be assumed. This usually causes the fluid to spread in the xy plane, which justifies a domain partitioning along the x and y axes.

**OpenMP Performance Measurements**

Performance measurements of the OpenMP parallelized solver can be seen in Figure 4. The graphs show absolute time measurements for a fixed number of LB steps, as *mega lattice site updates per second* (MLSUPS) measurements are not suitable for simulations with the adaptive coarsening of [7]. For these measurements, a test case of a drop falling into a basin of fluid was used. The left graph was measured on a 2.2GHz dual Opteron workstation[A], with a grid resolution of $304^3$. The graph to the right of Figure 4 was measured on a 2.2GHz quad Opteron workstation[B], using a resolution of $480^3$. For the dual nodes, as well as the quad nodes, the CPUs are connected by HyperTransport links with a bandwidth of 6.4 GB/s. Each graph shows timing measurements for different numbers of CPUs, and with or without the use of the adaptive coarsening algorithm.

The results without the adaptive coarsening show the full effect of the parallelization, as in this case almost 100% of the computational work is performed on the finest grid. It is apparent that the speedup is directly proportional to the number of CPUs used in this case. The four

[A]CPU: 2 x AMD Opteron 248, 2.2 GHz, 1MB L2-cache; 4GB DDR333 RAM.
[B]CPU: 4 x AMD Opteron 848, 2.2 GHz, 1MB L2-cache; 16GB DDR333 RAM.

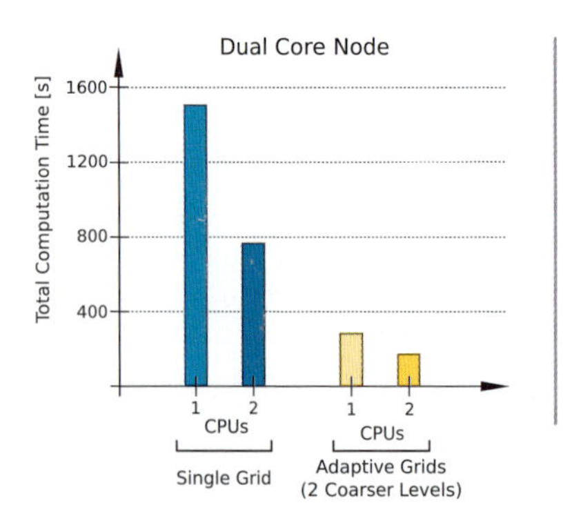


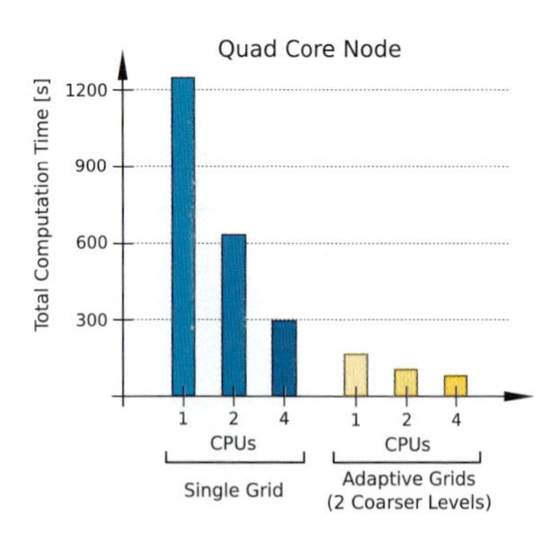


Figure 4. Time measurements for the OpenMP version of the solver: the runs to the left were measured on a dual Opteron workstation, while those to the right were measured on a workstation with four Opteron CPUs. The blue bars represent runs without the adaptive coarsening algorithm, while the orange bars use two coarse levels in addition to the finest one.

blue bars of the right graph from Figure 4 even show a speedup of 4.11 for four CPUs. This can be explained with the architecture of the quad node – the simulation setup uses most of the available memory of the machine, but each CPU has one fourth of the memory with a fast local connection, while memory accesses to the remaining memory have to performed using the HyperTransport interconnect. Thus, with four OpenMP threads the full memory bandwidth can be used, while a single thread, solving the same problem, frequently has to access memory from the other CPUs.

The timing results with adaptive coarsening show less evident speedups, as in this case only roughly 70% of the overall runtime are affected by the parallelization. As expected, the runtime for two CPUs is 65% of the runtime for a single CPU. The OpenMP parallelization thus yields the full speedup for the finest grid. It can be seen in the right graph of Figure 4 that there is a speedup factor of more than 15 between the version without adaptive coarsening running on a single CPU and the version with adaptive grids running on four CPUs. To further parallelize the adaptive coarsening algorithm, a parallelization of the grid reinitialization would be required, which is complicated due to the complex dependencies of the flag checks.

## 3. MPI Parallelization

For the development of applications for distributed memory machines, the *Message Passing Interface* (MPI) is the most widely used approach. In contrast to OpenMP, MPI requires more low level work from a developer, as most of its functions only deal with the actual sending and receiving of messages over the network. Details of the introductory and more advanced functions of MPI can be found, e.g., in [2] and [3].

For the MPI parallelization the domain is split along the x axis, as shown in Figure 5. In this figure two nodes are used, the domain is thus halved along the x axis, and a ghost layer is added at the interface of the two halves. Before each actual LB step, the boundary planes are exchanged via MPI, to assure valid boundary conditions for all nodes. As indicated in Figure 5,

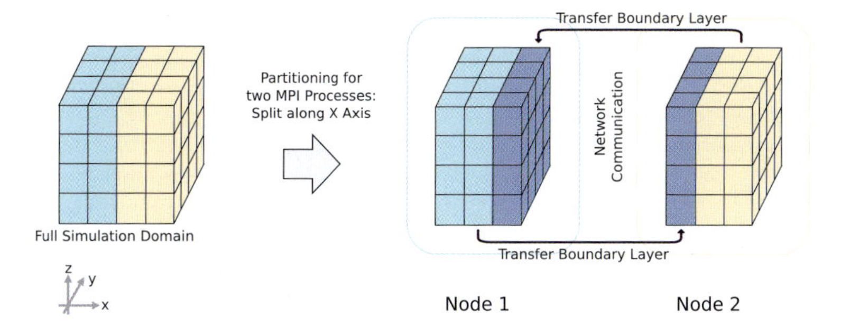


Figure 5. The MPI parallelization splits the x component of the loop over the grid.

the boundary layer contains the full information from a single plane of the neighboring node. For a normal LB solver, this would be enough to perform a stream-collide-step. However, the free surface handling can require changes in the neighborhoods of filled and emptied interface cells from a previous step. All fluid cells in the layer next to the boundary layer thus have to be validated again. If one of them has an empty cell as a neighboring node, it is converted to an interface cell. This simple handling ensures a valid computation, but causes slight errors in the mass conservation, as the converted cell might have received excess mass from the former neighboring interface cell. We have found that this is unproblematic, especially for physically based animations, as the error is small enough to be negligible. For engineering applications, an additional transfer between the nodes could ensure a correct exchange of the excess mass, similar to the algorithm proposed in [6,4]. The error in mass conservation is less than 1% for 1000 LB steps.

If this scheme is used in combination with the adaptively coarsened grids, it has to be ensured that there is no coarsening of the ghost and transfer layers. Therefore, the transfer is only required for the finest grid level. A coarsening of the ghost layers would require information from a wider neighborhood of the cells, and result in the exchange of several layers near the node boundary, in addition to the ghost layers of the different grid levels. As the bandwidth of the network is a bottleneck, such an increase of the exchanged data would result in a reduced performance. Hence, only the fine layers are connected with a ghost layer, and are treated similar to the free surface or obstacle boundaries to prevent coarsening in this region. As the node boundary has to be represented on the finest grid, this means that large volumes of fluid spanning across this boundary can not be fully coarsened. This parallelization scheme thus modifies the actual layout of the coarse and fine grids in comparison to the serial version of the solver. Further care has to be taken an adaptive resizing of the time step. This is based on the maximum velocity in the simulation, and requires the global maximum for all nodes to be computed.

## 4. MPI Performance Measurements and Discussion

To measure the performance of this MPI parallelized version of the solver, the test case is again that of a drop falling into a basin of fluid. The timing measurements of Figure 6 were measured on multiple quad Opteron nodes[C]. Details of these quad nodes can be found in Section 2. The x axis of each graph shows the number of nodes used for the corresponding measurement. For each node, the OpenMP parallelization of the previous section was used to execute four OpenMP threads on each node. As the parallelization changes the adaptive coarsening, Figure 6 again shows timing measurements for a fixed number of LB steps, instead of MLSUPS or MFLOPS rates. The figure shows two graphs in each row: the graph to the left was measured on a grid without adaptive coarsening, while the one to the right was measured from runs solving the same problem with two levels of adaptive coarsening. The two rows show the effect of the overhead due to MPI communication: the upper row shows results for a cubic domain of $480^3$, denoted as test case $Q$ in the following, while the lower row was measured with a wider channel and a resolution of $704 \cdot 352 \cdot 352$ (test case $W$). The grid resolution remains constant for any number of CPUs involved (strong scaling). As the domain is equally split for the number of participating MPI nodes, test case Q results in thinner slices with a larger amount of boundary layer cells to be transferred. For 8 nodes and test case Q, this means that each node has a grid resolution of $60 \cdot 480 \cdot 480$ with $480^2$ boundary cells to be exchanged. Splitting the domain of test case W, on the other hand, results in slices of size $88 \cdot 352 \cdot 352$ with $352^2$ boundary cells to be exchanged.

Overall, the graphs without adaptive coarsening show a speedup of around 1.8 for the strong scaling. While the speedup for test case Q, from four to eight nodes, is around 1.62, it is 1.75 for test case W, due to the better ratio between computations and communication in the latter case. This effect can also be seen for the graphs with adaptive coarsening (the right column of Figure 6). While the curve flattens out for test case Q, there is a larger speedup for test case W. For test case Q with adaptive coarsening, the speedup factor is ca. 1.3 – 1.35, while it is between 1.45 and 1.55 for test case W. This lower speedup factor, in comparison to the test cases with only a single fine grid, is caused by the increased overhead due to MPI communication, compared to the amount of computations required for each LB step with the adaptive coarsening. Moreover, the amount of coarsening that can be performed for each slice of the grid is reduced with the increasing number of MPI processes. Practical test cases will, however, usually exhibit a behavior that is a mixture of the four test cases of Figure 6. An example of a large scale test case that was computed with four MPI processes, and required almost 40GB of memory, can be seen in Figure 8.

To evaluate the overall performance of the solver, varying rectangular grid sizes without adaptive coarsening were used to simulate problems requiring the whole memory of all participating nodes (weak scaling). While the MLSUPS rate for a single quad Opteron is 5.43 with a grid resolution of $704 \cdot 352 \cdot 352$, eight quad nodes with a resolution of $1040 \cdot 520 \cdot 520$ achieve a performance of 37.3 MLSUPS, as shown in Figure 7. This represents a total speedup factor of 6.87 for the eight nodes.

## 5. Conclusions

We have demonstrated that the parallel algorithm presented here is suitable to perform efficient large scale computations. Both algorithms for OpenMP and MPI parallelization can be combined to solve large problems on hybrid shared- and distributed-memory systems. However, the algorithm does not yield the full performance when the only goal is to reduce the computational time for small problems with MPI. For large problems, the speedup will effectively depend on the setup – for large volumes of fluid, the speedup can be around 1.3 – 1.5, while fluids with many interfaces and fine structures can almost yield the full speedup of a factor two for each doubling of the CPUs or nodes used for the computation.

[C] The nodes are connected by an InfiniBand interconnect with a bandwidth of 10GBit/s

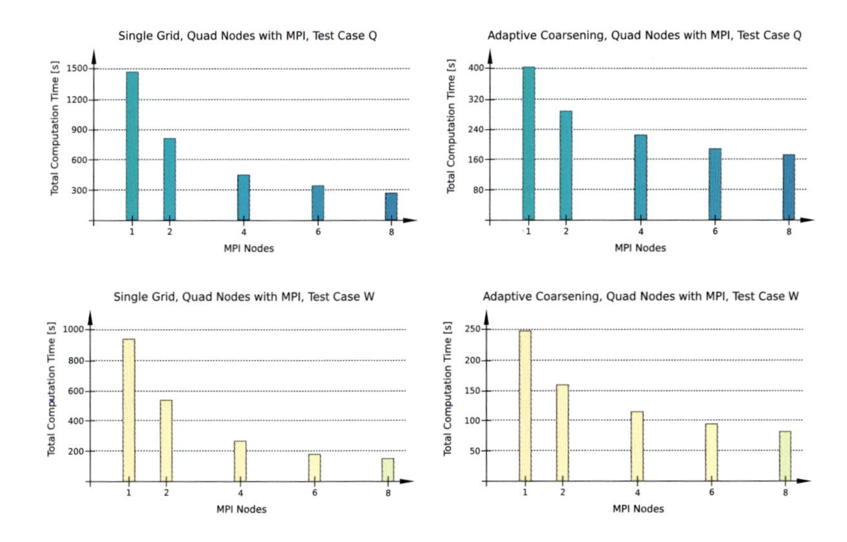


Figure 6. Time measurements for the MPI version of the solver running a problem with $480^3$ (test case Q) in the upper row, and for a problem with $704 \cdot 352 \cdot 352$ (test case W) in the lower row of graphs.

For dynamic flows, an interesting topic of future research will be the inclusion of algorithms for load balancing, e.g., those described in [4]. The algorithm currently assumes an distribution of fluid in the xy plane due to a gravity along the z direction. If this is not the case, the static and equidistant domain partitioning along the x and y axes will not yield a high performance.

## REFERENCES

1. R. Chandra, L. Dagum, D. Kohr, D. Maydan, J. McDonald, and R. Menon. *Parallel Programming in OpenMP.* Academic Press, 2001.
2. W. Gropp, E. Lusk, and A. Skjellum. *Using MPI, Portable Parallel Programming with the Mesage-Passing Interface.* MIT Press, second edition, 1999.
3. W. Gropp, E. Lusk, and R. Thakur. *Using MPI-2, Advances Features of the Message-Passing Interface.* MIT Press, 1999.
4. C. Körner, T. Pohl, U. Rüde, N. Thürey, and T. Zeiser. Parallel Lattice Boltzmann Methods for CFD Applications. In A.M. Bruaset and A. Tveito, editors, *Numerical Solution of Partial Differential Equations on Parallel Computers*, volume 51 of *LNCSE*, pages 439–465. Springer, 2005.
5. T. Pohl, M. Kowarschik, J. Wilke, K. Iglberger, and U. Rüde. Optimization and Profiling of the Cache Performance of Parallel Lattice Boltzmann Codes in 2D and 3D. Technical Report 03–8, Germany, 2003.
6. T. Pohl, N. Thürey, F. Deserno, U. Rüde, P. Lammers, G. Wellein, and T. Zeiser. Performance Evaluation of Parallel Large-Scale Lattice Boltzmann Applications on Three Supercomputing Architectures. In *Proc. of Supercomputing Conference 2004*, 2004.
7. N. Thürey and U. Rüde. Stable Free Surface Flows with the Lattice Boltzmann Method on adaptively coarsened Grids. *to appear in Computing and Visualization in Science*, 2008.

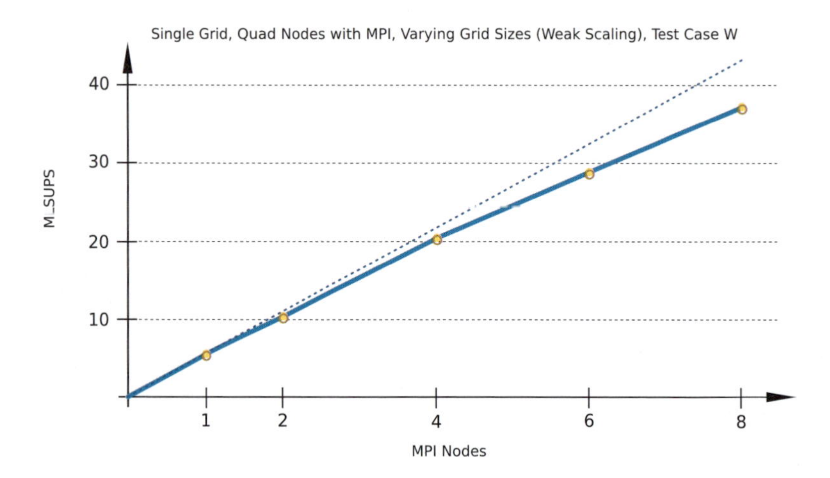


Figure 7. MLSUPS measurements for runs with varying grid resolutions (weak scaling) of test case W and without adaptive coarsening. The dotted line represents the projected ideal speedup according to the performance of a single quad node.

Figure 8. Pictures of a wave test case with a grid resolution of $880 \cdot 880 \cdot 336$. On average, only 6.5 million grid cells were simulated during each time step due to two levels of adaptive coarsening.

# Flow-structure interaction and flow analysis of hydraulic machinery on a computational grid

F. Lippold[a*], I. Buntić Ogor[a] and A. Ruprecht[a]

[a]Universität Stuttgart, Institute of Fluid Mechanics and Hydraulic Machinery (IHS) Pfaffenwaldring 10, 70550 Stuttgart, Germany

In order to simulate the interaction between flexible structures and the flow in hydraulic machinery (flow-structure interaction, FSI), the parallel CFD code FENFLOSS is enhanced with moving-mesh algorithms. Parallel versions of these algorithms are implemented and discussed. Furthermore, an interface to couple the CFD code with structural codes is required. A third code is used for this purpose. It needs an interface to exchange data with FENFLOSS during the simulation. Additionally, the opportunity offered by the usage of computational grids rather than single clusters is evaluated for the application in the field of optimisation and analysis, including FSI.

## 1. INTRODUCTION

The simulation of flow-structure interactions requires special features of the involved codes. On the CFD side this is the capability to account for the movement of the flexible structure, which is represented by the boundaries of the flow domain. Usually, this includes an Arbitrary-Lagrange-Euler (ALE) formulation of the Navier-Stokes Equations and a moving mesh algorithm. This algorithm transfers the changes of the domain boundaries to the whole fluid grid. Several moving mesh methods have been developed in the last decade. In this paper an interpolation method and a pseudo structural approach are presented and discussed. Furthermore, the CFD code FENFLOSS[2], developed at the IHS, is coupled with the commercial structural analysis code ABAQUS©[3]. The exchange of the coupling data is done via the coupling manager MpCCI[4]. For this purpose an interface that allows FENFLOSS and MpCCI to exchange data is developed. All these enhancements in FENFLOSS have to be implemented to run on parallel and serial computers.

In the process of numerical design phase of hydraulic machinery there are two basic steps. The first one is the optimisation of the geometry, using a simplified representation. The second one is the detailed analysis of a small set of geometries found to be optimal, including FSI simulations. Both steps require a certain amount of computational power

*Results were obtained within the projects InGrid (www.ingrid-info.de) funded by the German Ministry of Education and Research - BMBF, and HPC-Europa (www.hpc-europa.org) funded by the European Union

[2]**F**inite **E**lement based **N**umerical **FL**ow **S**imulation **S**ystem

[3]www.abaqus.com

[4]**M**esh based **p**arallel **C**ode **C**oupling **I**nterface©, www.scai.fraunhofer.de/mpcci.html

I.H. Tuncer et al., *Parallel Computational Fluid Dynamics 2007*,
DOI: 10.1007/978-3-540-92744-0_23, © Springer-Verlag Berlin Heidelberg 2009

and, especially in the second case, storage. Whereas the optimisation needs many CFD runs the demands for a FSI analysis are different, i.e. a few runs but on a powerful computer and with different codes exchanging data. A computational grid offers all the necessary hard and software. It provides a variety of different architectures and software, fitting to many applications.

Figure 1 shows the typical engineering flow chart in hydraulic machinery development.

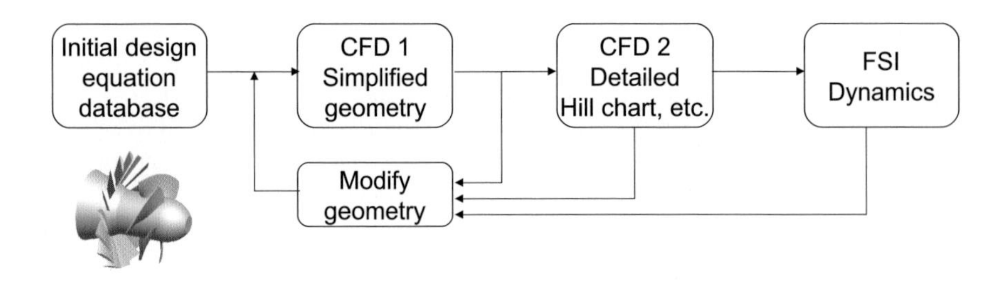


Figure 1. Engineering flow chart in hydraulic machinery development.

## 2. FLUID-STRUCTURE INTERACTION MODELLING

Seen from the physical point of view, fluid-structure interaction is a two field problem. But numerically, the problem has three fields, where the third field is the fluid grid that has to be updated after every deformation step of the structure, to propagate the movement of the fluid boundaries, the wetted structure, into the flow domain, see figure 2. The choice of the appropriate mesh update algorithm depends on the requirements of the applications regarding performance and deflection of the boundaries.

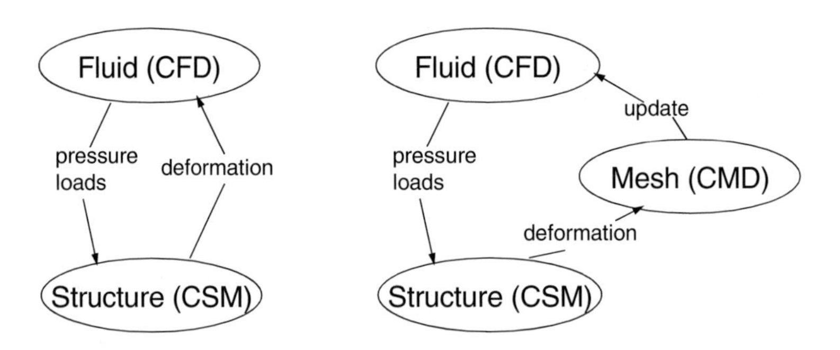


Figure 2. Physical and numerical coupling model

The solution of the numerical problem can be arranged in a monolithic scheme, which solves the structural and flow equations in one step. This method is applied for strongly coupled problems, e.g. for modal analyses. Another scheme, suitable for loosely and intermediately coupled simulations, is the separated solution of flow and structural equations with two independent codes. In the separated scheme well proven codes and models can be used. This is the reason why for engineering problems usually the second approach is employed. However, a data exchange between the codes has to be arranged, including the interpolation between the two computational meshes.

### 2.1. Moving mesh methods

The first mesh update method, discussed here, is an interpolation scheme using the nodal distance $\mathbf{r}$ and $\mathbf{s}$ between moving and fixed boundaries, respectively, to compute the new nodal positions after a displacement step of the moving boundary. The most simple approach is to use a linear interpolation value $0 \leq \kappa \leq 1$. Here we are using a modification of the parameter $\kappa = \frac{|\mathbf{s}|}{|\mathbf{r}|+|\mathbf{s}|}$ proposed by Kjellgren and Hyvärinen [1].

$$\tilde{\kappa} = \begin{cases} 0, & \kappa < \delta \\ \frac{1}{2}(\cos(1-(\frac{\kappa-\delta}{1-2\delta})\cdot\pi)+1, & \delta \leq \kappa \leq 1-\delta \\ 1, & 1-\delta < \kappa \leq 1 \end{cases} \quad (1)$$

This parameter is found from the nearest distance to the moving boundary and the distance to the fixed boundary in the opposite direction. The parameter $\delta$ is used to obtain a rigid boundary layer mesh. Typically it will be $0.0 \leq \delta \leq 0.1$.

To use this approach for parallel computations the boundaries have to be available on all processors. Since a graph based domain decomposition is used here this is not implicitly given. Hence, the boundary exchange has to be implemented additionally. Usually, the number of boundary nodes is considerably small, i.e. the additional communication time and overhead are negligible.

Another approach is based on a pseudo-structural approach using the lineal springs introduced by Batina [4]. In order to stabilise the element angles this method is combined with the torsional springs approach by Farhat et. al. [5,6] for two- and three-dimensional problems. Since we use quadrilateral and hexahedral elements the given formulation for the torsional springs is enhanced for these element types.

Here, a complete matrix is built to solve the elasto-static problem of the dynamic mesh. The stiffness matrix built for the grid smoothing has the same graph as the matrix of the implicit CFD-problem. Hence, the whole cfd-solver structure including the memory can be reused and the matrix graph has to be computed only once for both fields, the fluid and the structure. This means that there is almost no extra memory needed for the moving mesh. Furthermore, the parallelised and optimised, preconditioned BICGStab(2) performs good on cache and vector CPUs for the solution of the flow problem, which brings a good performance for the solution of the dynamic mesh equations, as well. Nevertheless, the stiffness matrix has to be computed and solved for every mesh update step, i.e. usually every time step. The overall computation time for a three-dimensional grid shows that the percentage of computing time needed for the mesh update compared to the total time is independent of the machine and number of CPUs used, see Lippold [7]. This means that the parallel performance is of the same quality as the one of the flow solver.

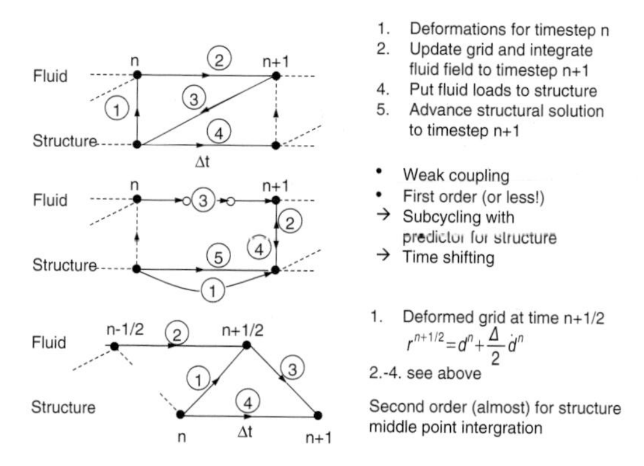


Figure 3. Exchange schemes for loosely coupled problems

Regarding the usage of the torsional springs two issues have to be addressed. Computing the torsional part of the element matrices includes a bunch of nested loops costing quite a considerable amount of computational time. Because of that, the respective routine requires approximately twice the time than the lineal part does. Due to the fact that the lineal springs use the edge length to determine the stiffness but the torsional springs use the element area or volume, respectively, it is important to notice that the entries contributed to the matrix may differ within some scales. Hence, using the additional torsional springs the numbers shown above are changing for these two reasons. Furthermore, it yields higher values on the off-diagonal elements of the matrix which reduces its condition number. Hence, the smoother needs more iterations to reduce the residual to a given value. This leads to additional computational time. Additionally, the element quality might be unsatisfying if the matrix entries, coming from the torsional springs are dominant. Meaning that the smoothed grid fulfills the angle criterion but not the nodal distance. In order to reduce this disadvantage, the matrix entries of the torsional springs have to be scaled to the same size as the contribution of the lineal springs. For these reasons the interpolation scheme is used for subsequent computations.

## 2.2. Coupling schemes

One-way coupling means to put the pressure values obtained by a steady or unsteady flow analysis as loads to the structure and to neglect the effect of the structural deflection on the flow.

In order to account for the coupling and to avoid unstable simulations, some coupling schemes for two-way interaction have been developed [2]. Figure 3 shows three of the mostly used schemes.

Using the simple coupling scheme results in the flow chart shown in figure 4. The left part shows the solution of the flow problem as it is implemented in FENFLOSS. On the right side the structural solver and the grid deformation is sketched. The moving grid algorithm may be implemented either directly in FENFLOSS or as a stand alone

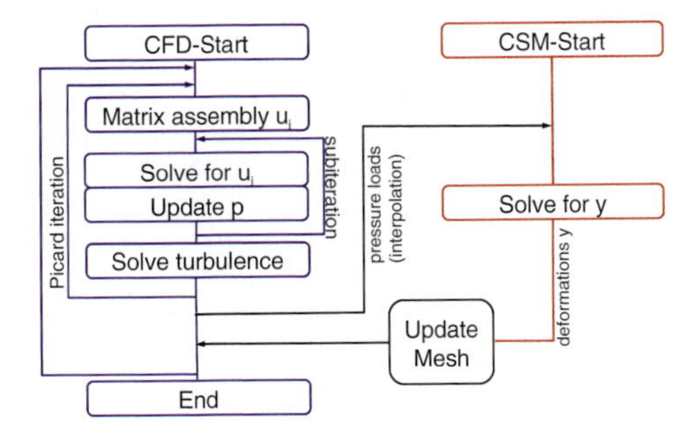


Figure 4. Coupling scheme

software called during the coupling step. Here, an extended interpolation method, which is implemented in FENFLOSS, see Lippold [7], is used.

### 2.3. Coupling FENFLOSS and ABAQUS

In order to couple the IHS CFD code FENFLOSS with the commercial software ABAQUS to perform fluid-structure simulations, an interface between FENFLOSS and MpCCI has to be available. The old MpCCI 2.0 SDK interface in FENFLOSS is replaced with a new one based on the adapter-driver connection required by the current MpCCI versions. In this case it means mixing the MpCCI-API written in C with classical FORTRAN code. The implementation of some special function calls in FENFLOSS is possible, creating an inflexible interface, which introduces new code into the program. Furthermore, the code then can only be used for special analysis types, i.e. fluid-structure simulations. Since, it is recommended to link the adapter and driver via an external library, the best and most flexible way is to use the shared object interface that is provided by FENFLOSS. Once written, all algorithms needed for coupling the two codes, updating the grid, as well as coupling schemes can be implemented by user-subroutines without changing the simulation code itself. This yields a better code quality and a faster development of the interface. The set-up and configuration for the graphical MpCCI interface is done with simple perl scripts. Figure 5 shows the architecture of the coupled codes.

## 3. FSI APPLICATION

One application is the two-way coupled solution of an inclined elastic 3D NACA0012 wing using FENFLOSS and ABAQUS coupled with MpCCI. It is clamped at one end and free at the other one. All surrounding walls have slip conditions. The fluid is water and the mean flow velocity is 10 m/s. The length of the wing is chosen as $2 \cdot$ c.

The wing's chord length is c=1m and the Young's modulus E$= 5.0 \cdot 10^9$N/m$^2$. Performing a regular static analysis with the steady state pressure loads produced by the undeflected wing yields a deflection at the wing tip of $\Delta y = 0.14$m. Running a two-way coupled analysis ends up in an equilibrium state with $\Delta y = 0.09$m. This shows the need for the choice of a proper coupling. Figure 6 shows the original and the deflected wing

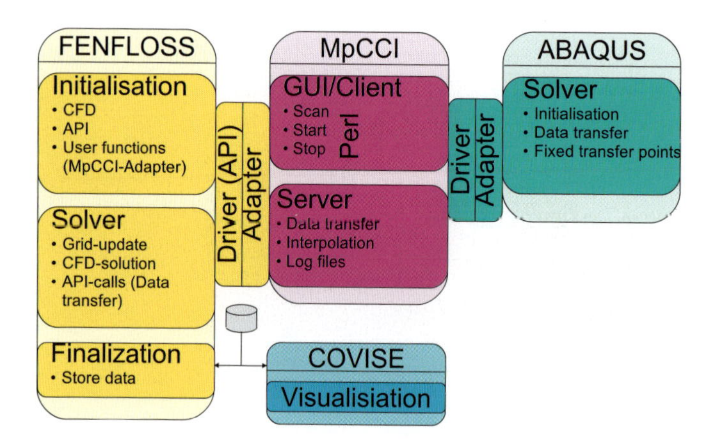


Figure 5. Architecture for coupling FENFLOSS with the shared object adapter-driver library and ABAQUS

including the surface pressure distribution.

## 4. CFD and FSI ON A GRID

In the last years there were several attempts to use computational grids and distributed systems [8–10]. However, there are difficulties to overcome for CFD in a grid if the code uses lots of communication for parallel computing. So does FENFLOSS, which uses a parallel version of the Krylov solver BICGStab(2) with communication in the most inner iteration loop. This means rather small packages but high frequent communication. Hence, the latency and not the network bandwidth has a considerable impact on the overall computational time.

However, grid environments are used for parameter studies as it is done during the geometry optimisation. Another type of analysis is the determination of the dynamic behaviour of structures excited by unsteady flow. In our application the structural simulation requires less computational effort than the fluid simulation. Furthermore, structural and fluid codes usually require different hardware architectures. The former usually run very well on regular CPUs, whereas the latter ones, especially FENFLOSS, provide a good performance on vector computers.

For the use on a grid FENFLOSS is started by GLOBUS or alternatively by a simulation grid-service. Since it is capable of discovering running FENFLOSS simulations on a grid resource the **CO**llaborative **VI**sualisation and **SI**mulation **E**nvironment, COVISE©[5] is used for online visualisation of the simulation results.

One drawback that became clear when using this set-up is the large amount of ports required by GLOBUS. In the most cases this is in contrast to security issues and not

[5]www.visenso.de and www.hlrs.de/organization/vis/covise/

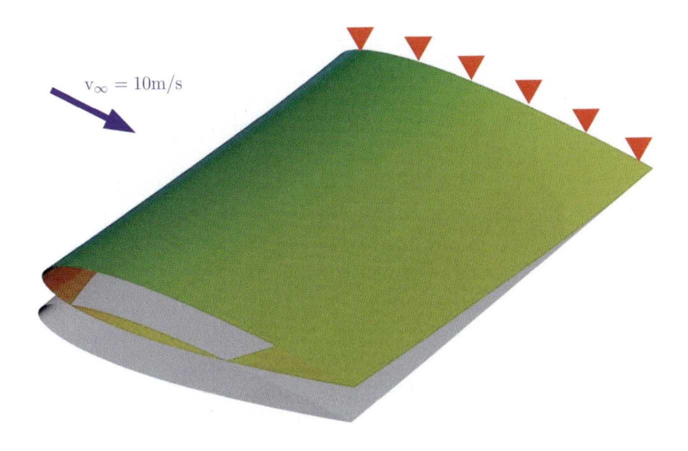


Figure 6. 3D NACA0012, inclined by 10°, undeflected and deflected with pressure distribution, red 50 kPa, blue -150 kPa

supported by most system administrators. Here, the grid resource has no connection to other non-grid resources and the client is running on a separate computer. Furthermore, it is recommended to suppress the standard output of the simulation codes since this is quite costly over the grid.

## 5. CONCLUSION and OUTLOOK

A procedure for fluid-structure interaction (FSI) problems is introduced. Furthermore, the CFD-code FENFLOSS and the necessary implementations for an FSI-analysis on computational grids as well as first results are discussed.

Further developments and examinations will focus on the performance of the moving grid algorithm and the fluid structure coupling. The other part will be the optimisation of the FENFLOSS grid-service and the coupling to the visualisation environment COVISE.

## REFERENCES

1. P. Kjellgren, J. Hyvärinen, An Arbitrary Langrangian-Eulerian Finite Element Method. Computational Mechanics, No. 21, 81, 1998.
2. C. Farhat, M. Lesoinne, P. le Tallec, Load and motion transfer algorithms for fluid/structure interaction problems with non-matching interfaces, Computer Methods in Applied Mechanics and Engineering, No. 157, 95, 1998.
3. S. Piperno, Explicit/implicit fluid/structure staggered procedures with a structural predictor and fluid subcycling for 2D inviscid aeroelastic simulations, International Journal for Numerical Methods in Fluids, No. 25, 1207, 1997.
4. J.T. Batina, Unsteady Euler airfoil solutions using unstructured dynamic meshes,

AIAA Paper No. 89-0115, AIAA 27th Aerospace Sciences Meeting, Reno, Nevada (1989).
5. C. Farhat, C. Degand, B. Koobus, M. Lesoinne, Torsional springs for two-dimensional dynamic unstructred fluid meshes, Comput. Methdos in Applied Mechanics and Engineering No. 163, 231, 1998.
6. C. Degand, C. Farhat, A three-dimensional torsional spring analogy method for unstructured dynamic meshes, Computers and Structures, No. 80, 305, 2002.
7. F. Lippold, Fluid-Structure-Interaction in an Axial Fan, HPC-Europa report, 2006.
8. P. Nivet, A computational grid devoted to fluid mechanics. INRIA, Rapport de Recherche, 2006
9. S. Wornom, Dynamic Load balancing and CFD Simulations on the MecaGRID and GRID5000. INRIA, Rapport de Recherche No. 5884, 2006
10. S.-H. Ko, C. Kim, O.-H. Rho, S. Lee, A Grid-based Flow Analysis and Investigation of Load Balance in Heterogeneous Computing Environment, Parallel Computational Fluid Dynamics, 277, Elsevier, 2004.

# Parallel Computation of Incompressible Flow Using Building-Cube Method

**Shun Takahashi** [a], **Takashi Ishida** [b], **Kazuhiro Nakahashi** [c]

[a] *Ph.D. Student, Department of Aerospace Engineering, Tohoku University.*

[b] *Graduate Student, Department of Aerospace Engineering, Tohoku University*

[c] *Professor, Department of Aerospace Engineering, Tohoku University*

Keywords: Cartesian mesh; Incompressible flow; Adaptive Mesh Refinement (AMR); Building-Cube Method (BCM); Dynamic load-balancing;

## 1. INTRODUCTION

The past CFD progress has been highly supported by the improvement of computer performance. Moor's Law tells us that the degree of integration of computer chips has been doubled in 18 months. It means that the computer performance increases by a factor of 100 every 10 years. The latest Top500 Supercomputers Site [1] on the other hand, tells us that the performance improved to 1,000 times in the last 10 years. Increase in the number of CPUs in a system also contributes to this rapid progress. With the progress of the computer performance, a pre-processing becomes more important in a series of numerical analysis. For instance, structured mesh or unstructured mesh generation demands human skill for complex geometries in particular. Quick and easy grid generation is one of the tasks for next generation CFD. In solving complex flow-field, a higher-order scheme is significant as same as the mesh resolution. The algorithm has to be simple, and that leads to high scalability. It becomes considerable to maintain and update the software, too. Thus, there are many requests for next generation numerical method.

One of the authors developed a new high-resolution numerical method based on a block-structured Cartesian mesh approach [2-4]. We call it Building-Cube Method (BCM). The method equips both characteristics of nonuniform and uniform Cartesian mesh. Entire flow field is covered with many quadrangles like nonuniform Cartesian mesh they are called as "cube" in this method. Furthermore each cube includes same number of computational cells as uniform Cartesian mesh. The data structure is simplified owing to this grid system that the same number of uniform computational cell is involved in all cubes. And the characteristic is useful to achieve an efficient parallel computation since the load balance of each cube is perfectly equivalent

All cubes have information of geometrical location and all computational cells are appointed the indices (I, J, K) and an index of cube they are belonging. In the two phases of nonuniform and uniform Cartesian mesh configuration, adaptive mesh refinement can be easily performed. Nonuniform Cartesian mesh refinement method has been studied for a long time [5-6]. But our approach adds it new concept of the nonuniform and uniform mixed Cartesian mesh to it. In this paper, the applicability of adaptive cube refinement in BCM is discussed with dynamic load-balancing.

## 2. NUMERICAL METHOD

### 2. 1 Computational Grid in Building-Cube Method

The BCM was proposed by present author aimed for high-resolution flow computations around real geometries using high-density mesh [2-4]. Entire flow field is covered with many quadrangles like nonuniform

I.H. Tuncer et al., *Parallel Computational Fluid Dynamics 2007*,
DOI: 10.1007/978-3-540-92744-0_24, 

Cartesian mesh which are called as "cube" in this method. Furthermore each cube includes same number of computational cell as uniform Cartesian mesh. The data structure is simplified owing to the grid system. All cubes have information of geometrical location and all computational cells are appointed the indices (I, J, K) and the cube index they are belonging. Figure 1 shows cube and cell distribution in two-dimensional BCM grid. The size of a cube becomes small with getting closer to the object. The cube which is a sub-domain is useful to change the distribution of computational cell. The mesh distribution is determined by the minimum cell size and the number of division of a cube.

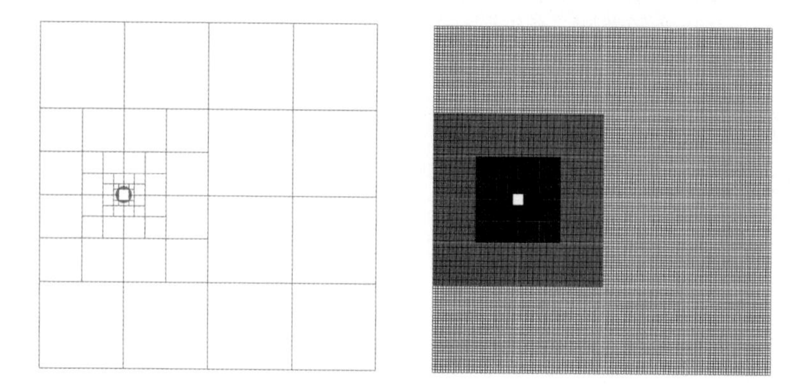

Fig. 1 Cube (left) and cell (right) distribution in whole BCM grid

### 2. 2 Incompressible Flow Solver

The incompressible Navier-Stokes equations are solved by finite-difference scheme in this study. The pressure term is calculated by solving Poisson equation with successive over relaxation (SOR) method. And the velocity terms are updated by Adams-Bashforth second-order explicit method. Convective terms are computed using third-order upwind scheme [7]. In the present study, the wall boundary is described as simplest staircase pattern. The higher-order boundary treatment like the immersed boundary method has to be performed as one of future works.

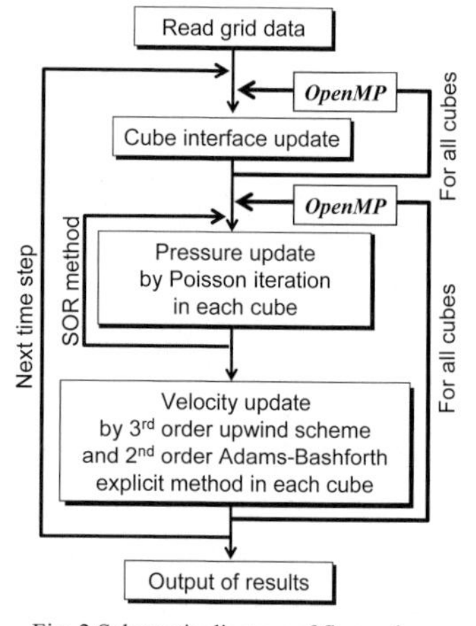


Fig. 2 Schematic diagram of flow solver

The computational grid of BCM consists of assemblage of cubes as many sub-domains. All numerical procedures are performed in each cube. Therefore, the exchange of flow information is necessary during each time-step. For solving Poisson equation of the pressure with second-order central difference scheme, one stencil is overlapped with next one at the boundary cell in a cube. For solving third-order upwind scheme, two stencils are overlapped. To prevent the numerical error, the size of adjacent cubes is limited within twice in this study. Though the first-order interpolation is used for the exchanging information when the size of adjacent cubes is different, more higher-order interpolation has to be used to keep whole solution accuracy.

### 2. 3 Parallelization

Parallelization methods are classified into two categories. These are SMP which consists of shared memory architecture and Cluster/MPP which consists of distributed memory architecture. OpenMP and MPI/PVM are used to perform parallel computation by them. OpenMP parallelization is easier than MPI parallelization, that is performed by only few directive lines insertion. Though significant program modification is needed in the case of MPI parallelization, the distributed memory system can treat large-scale data. Moreover, the latency for memory-access increases in the shared memory system if the scale of numerical simulation becomes large. In short, OpenMP parallelization is suitable for small or medium size computation, and MPI/PVM parallelization is suitable for enormous numerical simulation utilizing a number of processors and distributed memory.

In present study, OpenMP parallelization is selected to confirm the parallel efficiency in BCM at first. After inserting few directive lines, the parallelization is applied to iterative loops automatically by the architecture if only a user has specified the number of threads. Even if the loop-size changed, it is possible to parallelize efficiently owing to the characteristic. One of subjects in this research is to keep dynamic load-balancing and it is realized by the automatic parallelization.

## 3. NUMERICAL RESULTS

### 3. 1 Flow Simulation around Circular Cylinder

The flow solver is validated by a circular cylinder at *Re*=20 and 200. The number of cubes is 60 and each cube has 32[I]x32[J] computational cells, so that the total number of computational cells is 61,440. The minimum computational cell size near the wall boundary is 0.01 to the cylinder diameter.

Table. 1 Validation at *Re*=20

| | Vortex Length |
|---|---|
| Fornberg | 0.91 |
| Dennis et al. | 0.94 |
| Linnick et al. | 0.93 |
| Present | 0.96 |

Table. 2 Validation at *Re*=200

| | Strouhal number |
|---|---|
| Miyake et al. | 0.196 |
| Liu et al. | 0.192 |
| Linnick et al. | 0.197 |
| Present | 0.200 |

Twin vortices are observed at *Re*=20 as shown in Fig.3. The vortex length is compared with other studies [8-12] in Table.1. The color map shows the distribution of the vorticity magnitude. In the case of *Re*=200, Karman's vortex street is observed as shown in Fig.4. And Strouhal number is compared with other studies [12-14] in Table.2. The color map shows the vorticity distribution. Both results are in agreement with other studies.

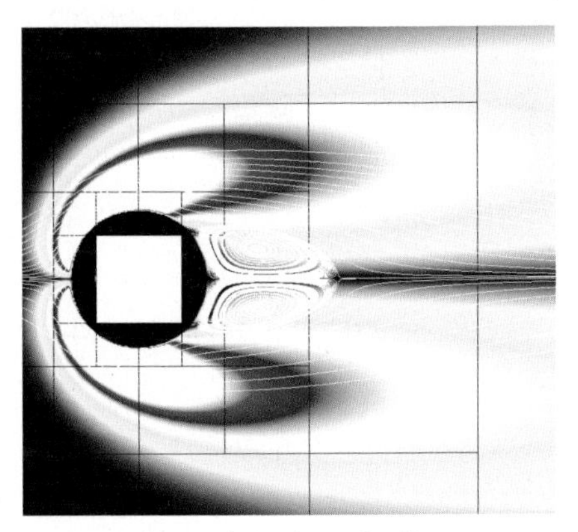

Fig. 3 Twin Vortices at *Re*=20

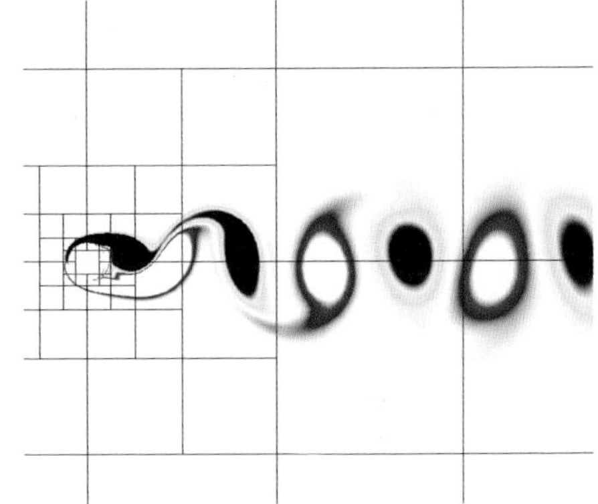

Fig. 4 Karman's vortex street at *Re*=200

### 3. 3 Parallel Efficiency

Present parallel computations are performed by NEC TX7 which is a scalar parallel computer with shared memory architecture. One node includes 64 Intel Itanium 2 (IA-64) processors, 512 GB main memory and about 410 GFLOPS performance.

The structure of the flow solver is shown in Fig.2. The loops for all cubes are parallelized by OpenMP. The grain size is important in OpenMP parallelization, that is, the parallelization of long iterative loops including many procedures is appropriate. Second loop for all cubes in the flow solver consists of some procedures to realize the coarse grain parallelization. In present OpenMP parallelization, these loops for all cubes are divided and allocated automatically for specified number of threads.

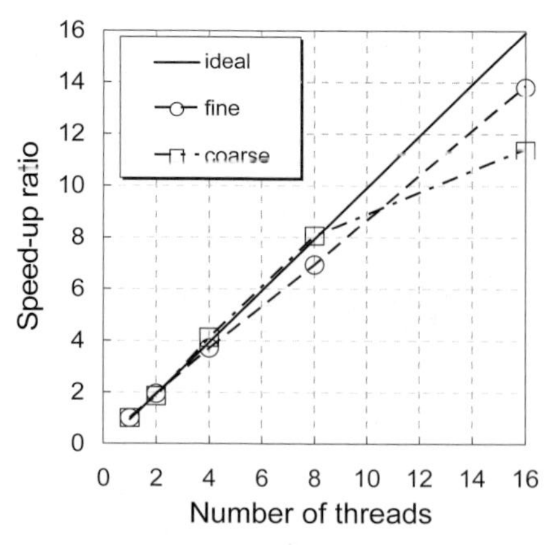


Fig. 5 Comparison of Speed-up ratio

The speed-up ratio is estimated with fine and coarse grid to verify the scalability. The fine grid and coarse grid consist of 1,453 and 286 cubes, respectively. The same number 32[I]x32[J] of computational cells are included in all cubes in both cases. Total numbers of computational cell are 1,487,872 and 292,864 respectively. In the case of fine grid, good scalability is achieved up to 16 CPUs parallel computation. In the case of coarse grid on the other hand, the speed-up ratio becomes worse at 16 CPUs parallel computation. It is caused by inequity of the load balance, that is, sufficient number of cubes is needed for efficient parallelization.

### 3. 4 Dynamic Load-Balancing

Cartesian mesh approach has an advantage of easy mesh adaptation. The procedure is applied to the computational cells directly in usual Cartesian mesh cases. In BCM however, it is applied to cubes constructing sub-domains. In applying the cube adaptation, only the size of cubes changes though the number of computational cells in each cube doesn't change. The cube adaptation is implemented by two procedures that are refinement and coarsening. These procedures are sequentially implemented during flow simulation in this study.

Dynamic load-balancing can be discussed by speed-up ratio using cube adaptation. In Fig.6, in the case of not using the adaptation, the total number of cubes is 1,453 constantly. In the case of using the adaptation, the total number of cubes changes between 286 and about 1,450. OpenMP can efficiently partition the number of cubes even with the change of the number of cubes during the flow computation.

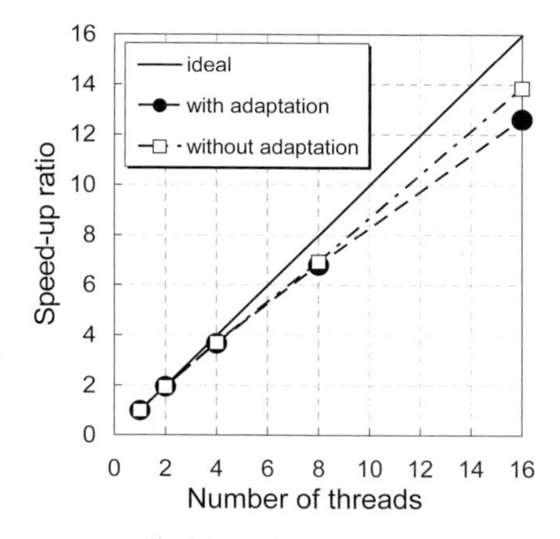


Fig. 6 Comparison of Speed-up ratio

As shown in Fig.6, both cases show good scalability, that is to say, dynamic load-balancing is practically kept even with the change of the number of cubes with the adaptation. A little deterioration is caused by several sequential loops in the adaptation procedure which is needed to remake computational grid. The distributions of vorticity magnitude and cubes in Fig.7 are made by the adaptation using the vorticity magnitude sensor in Karman's vortex street at *Re*=200. It is observed that refined region moves with the vortices displacement.

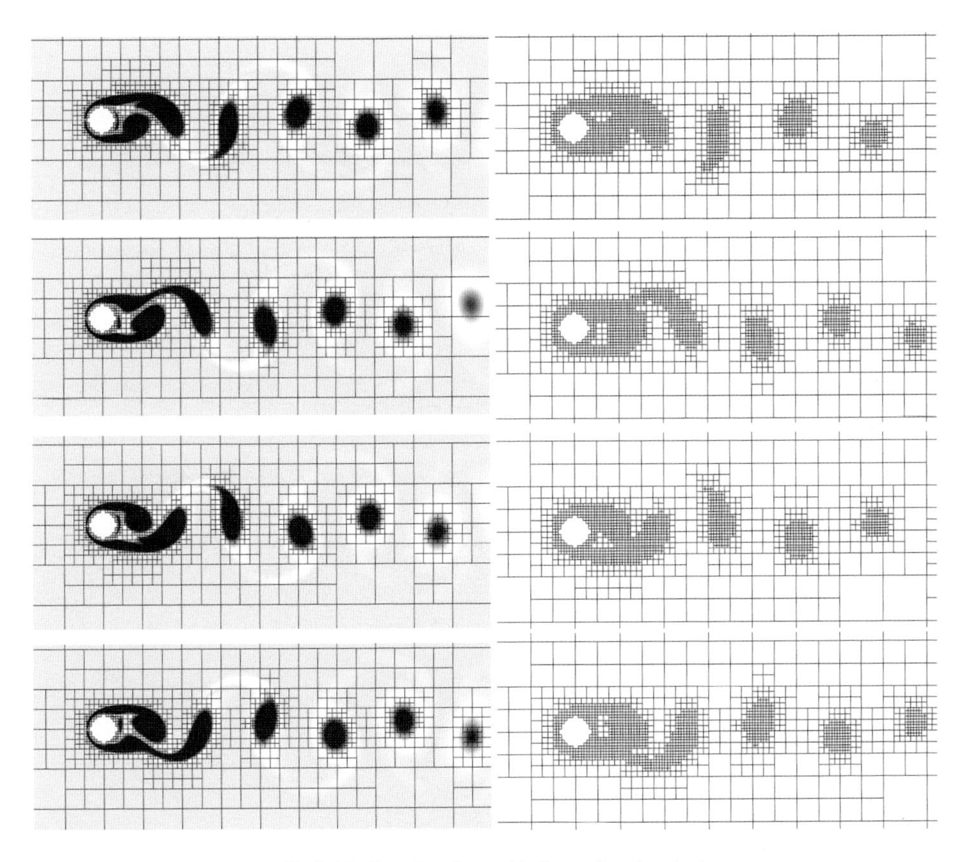

Fig. 7 Adaptive cube refinement in Karman's vortex street

## 4. CONCLUSIONS

A parallel incompressible flow solver was developed in the framework of the Building-Cube Method. Good scalability was confirmed in the parallelized simulation using OpenMP. An efficient adaptive cube refinement and coarsening methods for enhancing the local resolution was developed for accurate computation of unsteady flows. At the same time, the performance of the dynamic load-balancing for changing the number of cubes was demonstrated. Though MPI parallelization has to be used if the simulation scale becomes larger like three-dimensional complex geometry problem at high Reynolds number, the simplicity and scalability of BCM was investigated in incompressible flow simulation in this study.

## ACKNOWLEDGEMENT

Present computation was implemented by the use of NEC TX7 in Super-Computing System at Information Synergy Center of Tohoku University. We sincerely thank all the staffs.

**REFERENCES**


1. http://www.top500.org/
2. Nakahashi K., "High-Density Mesh Flow Computations with Pre-/Post-Data Compressions," AIAA paper, 2005-4876, 2005
3. Nakahashi K., Kitoh A., Sakurai Y., "Three-Dimensional Flow Computations around an Airfoil by Building-Cube Method," AIAA paper, 2006-1104, 2006
4. Kamatsuchi T., "Turbulent Flow Simulation around Complex Geometries with Cartesian Grid Method," AIAA paper, 2007-1459
5. Aftosmis M.J., "Solution Adaptive Cartesian Grid Methods for Aerodynamic Flows with Complex Geometries," 28th computational fluid dynamics von Karman Institute for Fluid Dynamics Lecture Series 1997-02, 1997
6. Ogawa, T., "Development of a flow solver using the adaptive Cartesian mesh algorithm for wind environment assessment," *J Industrial Aerodynamics*, 81, 377-389, 1999
7. Kawamura T., Kuwahara K., "Computation of high Reynolds number flow around circular cylinder with surface roughness," AIAA paper, 84-0340, 1984
8. Coutanceau M., Bouard R., "Experimental determination of the main features of the viscous flow in the wake of a circular cylinder in uniform translation. Part 1," *J Fluid Mech* 79, 231–256., 1977
9. Tritton D.J., "Experiments on the flow past a circular cylinder at low Reynolds number," *J Fluid Mech* 6, 547–567, 1959
10. Fornberg B., "A numerical study of the steady viscous flow past a circular cylinder," *J Fluid Mech* 98, 819–855, 1980
11. Dennis S.C.R., Chang G., "Numerical solutions for steady flow past a circular cylinder at Reynolds number up to 100," *J Fluid Mech* 42, 471–489, 1970
12. Linnick M.N., Fasel H.F., "A high-order immersed boundary method for unsteady incompressible flow calculations," AIAA paper, 2003-1124. 2003
13. Liu C., Zheng X., Sung C.H., "Preconditioned multigrid methods for unsteady incompressible flows," *J Comp Phys* 139, 35–57, 1998
14. Belov A., Martinelli L., Jameson A., "A new implicit algorithm with multigrid for unsteady incompressible flows calculations," AIAA paper, 95-0049, 1995

# 3D Model of pollution distribution in city air and its parallel realization

**Alexander I. Sukhinov,[a] Valeriy K. Gadelshin,[a] Denis S. Lyubomischenko[a]**

[a]*Institute of Technology of Southern Federal University, Taganrog, 347928, Russia*

**Abstract**

The monitoring of atmospheric air pollution in urban conditions is large-scale problem. It is necessary to take into account many environmental parameters. As a result numerical algorithms are complicated and need many system resources. It was decided to use object – oriented technologies in association with parallel development methods to provide operative estimation of environmental conditions.

Keywords: Air pollution, 3D advection - diffusion models, modelling on distributed computer clusters;

The purpose of the work is to construct algorithms and appropriate set of programs for atmospheric air pollution modeling in urban area. Coupled 3D models of aerodynamics near the earth surface and pollution model in urban area have been presented in this report. For authentic construction of the velocity distributions it is necessary to use high-resolution grids in discrete models; great amount of computations must be fulfilled in time interval no greater than tens - hundreds of seconds. It requires in turn fast parallel realization of 3D models. Another using of described models – for prediction of air conditions in emergency situations when in atmosphere of city harmful or dangerous pollutants have been emitted suddenly. Numerical and program realization of these models have been fulfilled for the Taganrog city which is situated on the beach of the Azov sea in the South of Russia.

The equations of model of near-ground aerodynamics are considered in area 13 km on 10 km on 100 m in Cartesian coordinate system using geophysics data about Taganrog city with step in horizontal coordinate directions of 5 m. Characteristic


DOI: 10.1007/978-3-540-92744-0_25, 

thickness of a ground layer of atmosphere in which pollution substances transport is of the order 100 meters.

The basic program project is divided on several blocks: main processor is managed by MPI, database, pre-processor, post-processor. The interactions between blocks are presented on fig. 1:

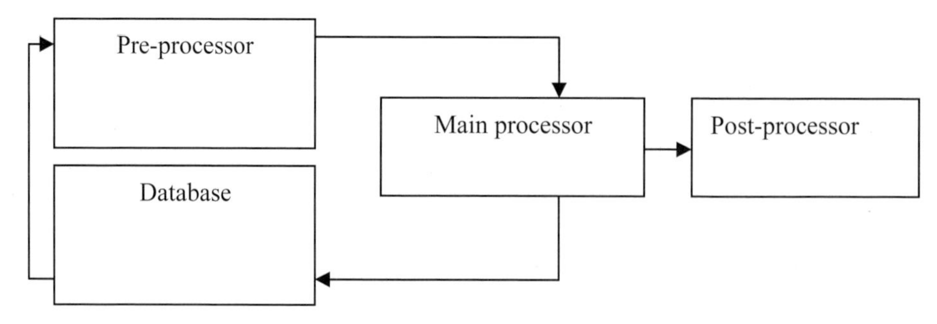


Fig.1 The scheme of project

It is possible to use data of previous experiments including the information of weather conditions extracting data from database or setting up the condition data directly in pre-processor. Also the adaptation of the information for multiprocessor system has been performed in pre-processor. After these preparing procedures data has been sent to main processor. Visualization has been performed by post-processor.

The project supports the fast estimation of pollution using previously modeling results stored in the database.

Main processor contains meteorological and emission classes. The problem of boundary conditions defining in such complicated area is nontrivial. It caused by complicated grounding surface, great stratification and numerical scheme inaccuracy.

## The algorithm of computations

It was decided to consider the problem of velocity field computation in two steps. On the first step the boundary and initial conditions have been obtained from the database. The region is of rather great value area including Taganrog and neighborhoods. Second step is local problem treating with boundary conditions from first step. On first step to extend the considering area in horizontal and vertical direction is required. Grid step has been increased too. The area 30 km x 20 km x 1 km has been considered on first step. On the upper border geostrophy wind conditions have been used, on the bottom border the surface is considered like a plane with attached roughness parameters. On second step the geometry of local area has been considered explicitly.

The system of equations of ground aerodynamics includes three-dimensional equations of motion, conservation mass equation, the state equation for ideal gas and heat transport equation.

The system of equations for aerothermodynamics is [1, 4].

The equations of motions (Navier - Stokes):

$$u'_t + (uu)'_x + (vu)'_y + (wu)'_z = -\frac{1}{\rho} p'_x + \\ + \frac{\eta_h}{\rho}(u''_{xx} + u''_{yy}) + \frac{1}{\rho}(\eta_v u'_z)'_z ,$$

$$v'_t + (uv)'_x + (vv)'_y + (wv)'_z = -\frac{1}{\rho} p'_y + \\ + \frac{\eta_h}{\rho}(v''_{xx} + v''_{yy}) + \frac{1}{\rho}(\eta_v v'_z)'_z , \quad (1)$$

$$w'_t + (uw)'_x + (vw)'_y + (ww)'_z = -\frac{1}{\rho} p'_z + \\ + \frac{\eta_h}{\rho}(w''_{xx} + w''_{yy}) + \frac{1}{\rho}(\eta_v w'_z)'_z .$$

The equation of indissolubility (mass conservation law):

$$\rho'_t + (\rho u)'_x + (\rho v)'_y + (\rho w)'_z = 0 . \quad (2)$$

The equation of state for ideal gas:

$$\rho = \frac{P}{RT} . \quad (3)$$

The equation of heat transport:

$$T'_t + (uT)'_x + (vT)'_y + (wT)'_z = \eta_h(T''_{xx} + T''_{yy}) + (\eta_v T'_z)'_z + f . \quad (4)$$

In system (1) - (4) $u(x,y,z,t)$, $v(x,y,z,t)$, $w(x,y,z,t)$ - components of velocity vector in point with coordinates $(x,y,z)$ at the moment of time $t$, $\rho(x,y,z,t)$ - density, $p(x,y,z,t)$ - pressure, $T(x,y,z,t)$ - temperature, $\eta_h$ - turbulent mixing coefficients in horizontal directions, $\eta_v$ - turbulent mixing coefficient in vertical direction, $f(x,y,z,t)$ - function of thermal sources. For the system (1)-(4) correspondent boundary conditions should be added. For initial conditions previously obtained weather conditions may be taken.

Using property quasi-stationary of ground air layer [3], it is possible to consider humidity as practically not dependent on height and accepts it as a constant value.

In given work the variant of MAC- method has been used. It based on additive schemes of splitting on physical processes. This method guarantees fulfillment of the balance mass law and it is computationally stable.

The scheme which approximates differential problem of aerodynamics is:

$$\frac{\tilde{u} - u}{\tau} + uu_{0\,x} + vu_{0\,y} + wu_{0\,z} = \frac{\eta_h}{\rho}(u_{\bar{x}x} + u_{\bar{y}y}) + \frac{1}{\rho}(\eta_v u_{\bar{z}})_z , \quad (5)$$

$$\frac{\widetilde{v}-v}{\tau}+u\underset{x}{v_0}+v\underset{y}{v_0}+w\underset{z}{v_0}=\frac{\eta_h}{\rho}(v_{\bar{x}x}+v_{\bar{y}y})+\frac{1}{\rho}(\eta_v v_{\bar{z}})_z, \quad (6)$$

$$\frac{\widetilde{w}-w}{\tau}+u\underset{x}{w_0}+v\underset{y}{w_0}+w\underset{z}{w_0}=\frac{\eta_h}{\rho}(w_{\bar{x}x}+w_{\bar{y}y})+\frac{1}{\rho}(\eta_v w_{\bar{z}})_z, \quad (7)$$

$$\widetilde{P}_{\bar{x}x}+\widetilde{P}_{\bar{y}y}+\widetilde{P}_{\bar{z}z}=\frac{1}{\tau}((\rho\widetilde{u})_{\bar{x}}+(\rho\widetilde{v})_{\bar{y}}+(\rho\widetilde{w})_{\bar{z}}), \quad (8)$$

$$\widetilde{\rho}=\frac{M\widetilde{P}}{RT}, \quad (9)$$

$$\hat{P}_{\bar{x}x}+\hat{P}_{\bar{y}y}+\hat{P}_{\bar{z}z}=\frac{\widetilde{\rho}-\rho}{\tau^2}+\frac{1}{\tau}((\widetilde{\rho}\widetilde{u})_{\bar{x}}+(\widetilde{\rho}\widetilde{v})_{\bar{y}}+(\widetilde{\rho}\widetilde{w})_{\bar{z}}), \quad (10)$$

$$\frac{\hat{u}-\widetilde{u}}{\tau}=-\frac{1}{\rho}(\hat{P})_x, \quad (11)$$

$$\frac{\hat{v}-\widetilde{v}}{\tau}=-\frac{1}{\rho}(\hat{P})_y, \quad (12)$$

$$\frac{\hat{w}-\widetilde{w}}{\tau}=-\frac{1}{\rho}(\hat{P})_z, \quad (13)$$

$$\frac{\hat{T}-T}{\tau}+\hat{u}\underset{x}{\hat{T}_0}+\hat{v}\underset{y}{\hat{T}_0}+\hat{w}\underset{z}{\hat{T}_0}=\eta_h(\hat{T}_{\bar{x}x}+\hat{T}_{\bar{y}y})+(\eta_v\hat{T}_{\bar{z}})_z+f, \quad (14)$$

$$\hat{\rho}=\frac{M\hat{P}}{R\hat{T}}. \quad (15)$$

Computational algorithm consists of following steps.

Equations (5) - (7), (9), (11) – (13) and (15) have been realized explicitly. To solve (8), (10) and (14) Seidel type iteration algorithms is used.

The transport equation for polluting substance transportation and transformation has been formulated as three-dimensional parabolic equation with variable coefficients of turbulent mixing and with constant value of destruction parameter.

$$\varphi_t'+u\varphi_x'+v\varphi_y'+(w-w_g)\varphi_z'+\sigma\varphi= \\ =(\eta_h\varphi_x')'_x+(\eta_h\varphi_y')'_y+(\eta_v\varphi_z')'_z+f_\varphi, \quad (16)$$

where $\varphi(x,y,z,t)$ - concentration polluting substance, $u$ $v$, $w$ - components of a vector of air medium velocity, $w_g$ -polluting substance velocity of falling down due to gravity, $\sigma$ - coefficient of substance destruction intensity, $f_\varphi(x,y,z,t)$ - function of sources of polluting substances, $\eta_h$ and $\eta_v$ - coefficients of turbulent mixing in horizontal and vertical directions respectively.

Discrete transport equations of polluting substances have been solved by means of known iterative Seidel method. Due to appropriate choice of time step the dominant diagonal feature for the matrix of equation system should be fulfilled. It guarantees the convergence of iteration with the rate of geometrical progression.

The difference scheme approximating the equation (5) is following:

$$\frac{\varphi_{i,j,k}^{t+1}-\varphi_{i,j,k}^{t}}{\tau}+$$

$$+\frac{1}{h_x}\left(\frac{\varphi_{i,j,k}^{t}+\varphi_{i+1,j,k}^{t}}{2}\cdot\frac{u_{i,j,k}^{t}+u_{i+1,j,k}^{t}}{2}-\frac{\varphi_{i,j,k}^{t}+\varphi_{i-1,j,k}^{t}}{2}\cdot\frac{u_{i,j,k}^{t}+u_{i-1,j,k}^{t}}{2}\right)+$$

$$+\frac{1}{h_y}\left(\frac{\varphi_{i,j,k}^{t}+\varphi_{i,j+1,k}^{t}}{2}\cdot\frac{v_{i,j,k}^{t}+v_{i,j+1,k}^{t}}{2}-\frac{\varphi_{i,j,k}^{t}+\varphi_{i,j-1,k}^{t}}{2}\cdot\frac{v_{i,j,k}^{t}+v_{i,j-1,k}^{t}}{2}\right)+$$

$$+\frac{1}{h_z}\left(\frac{\varphi_{i,j,k}^{t}+\varphi_{i+1,j,k}^{t}}{2}\cdot\frac{\overline{w}_{i,j,k}^{t}+\overline{w}_{i+1,j,k}^{t}}{2}-\frac{\varphi_{i,j,k}^{t}+\varphi_{i-1,j,k}^{t}}{2}\cdot\frac{\overline{w}_{i,j,k}^{t}+\overline{w}_{i-1,j,k}^{t}}{2}\right)=$$

$$=\eta_h\left(\frac{\varphi_{i-1,j1,k}^{t+1}+\varphi_{i,j,k}^{t+1}+\varphi_{i-1,j,k}^{t+1}}{h_x^{\,2}}+\frac{\varphi_{i,j-11,k}^{t+1}+\varphi_{i,j,k}^{t+1}+\varphi_{i,j+1,k}^{t+1}}{h_y^{\,2}}\right)+$$

$$+\eta_v\left(\frac{\varphi_{i,j1,k-1}^{t+1}+\varphi_{i,j,k}^{t+1}+\varphi_{i1,j,k+1}^{t+1}}{h_z^{\,2}}\right)+f_{i,j,k}\ ,$$

where $\varphi_{i,j,k}^{t+1}$ - value of concentration on $t+1$ time layer, $\varphi_{i,j,k}^{t}$ - value of concentration on $t$ time layer $u_{i,j,k}^{t}$ $v_{i,j,k}^{t}$, $w_{i,j,k}^{t}$ - components of a vector of velocity in corresponding directions $\overline{w}_{i,j,k}^{t}=w_{i,j,k}^{t}-w_g$, $\eta_h$ and $\eta_v$ - turbulent coefficients (see above), $f_{i,j,k}$ - function of polluting source.

The contribution of car traffic on an environment in models of transport of a harmful substance may be simulated by ground linear source, which corresponds to the contour of a transport system [1]. Function of a mobile transport source may be presented thus:

$$f=\sum_{i=1}^{N}E_n^i\delta(\bar{r}-\bar{r}_i), \tag{17}$$

where $E_n^i=E_n(\bar{r}_i,t)|\Delta\bar{r}_i|$ - quantity of emissions of an substance from a mobile source in unit of a discrete grid $\bar{r}_i=(x_i,y_i,z_i)$ at the time moment $t$.

**Parallel realization**

Parallel realization of the scheme [2, 5] is carried out by means of domain decomposition method, using 8-processor cluster of distributed computations under the government of MPI.

Computing experiment has been lead for two different grid sizes and numbers of the involved processors (np). Fig. 2 shows efficiency of algorithm for various numbers of processors and for two 3D grids:

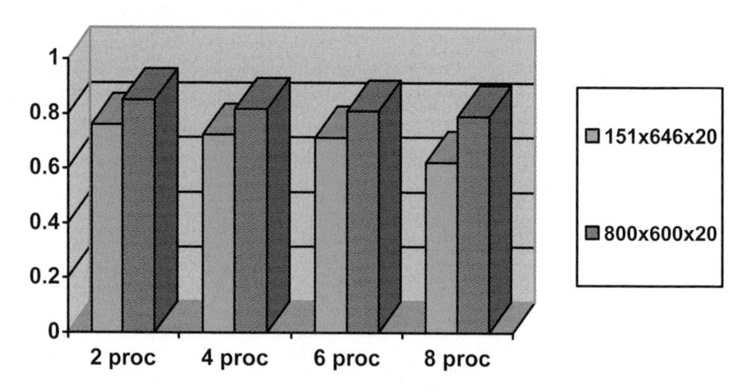


Fig. 2 The efficiency of parallel algorithm

On fig.3 results of modeling have been presented for following conditions: intensity of a motor transportation stream of 1400 units/hours, a polluting impurity – one-component gas CO, averaged car's velocity is 50 km/h, capacity of emissions 5.5 g / (sec/km), a wind of east direction, the velocity value is 5 m/s, temperature of air medium is 20 $^{0}$ C, pressure is $10^5$ Pa. Concentration plot is corresponding to concentrations at height 1 m as follows:

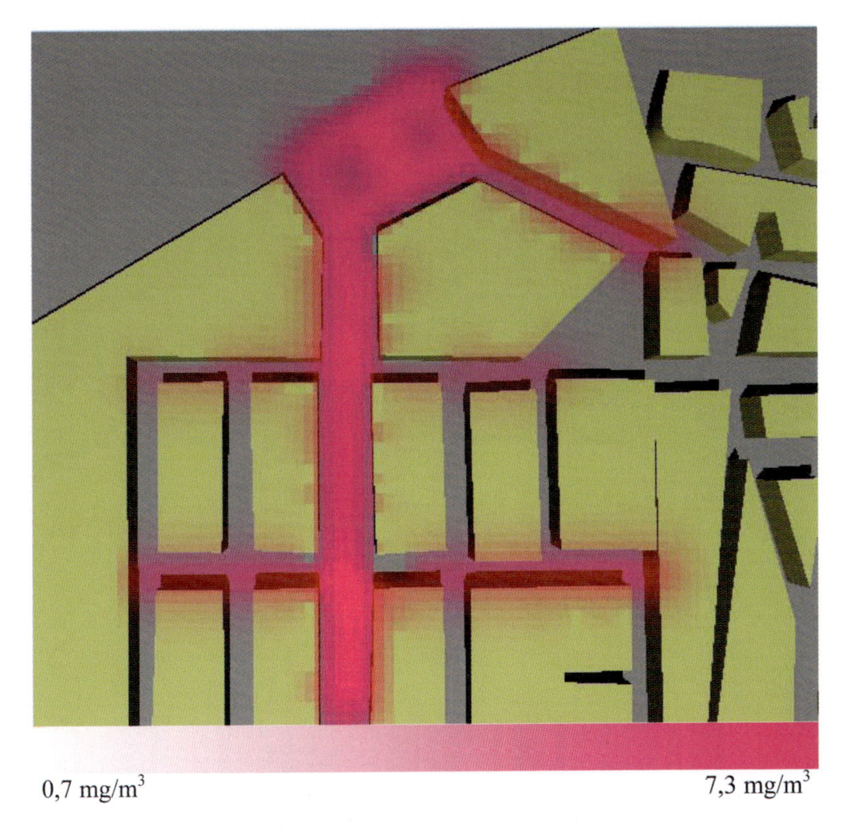


Fig. 3 Fragment of modeling results

## References

[1] Marchuk G. I. (1982) " Mathematical modeling in a problem of an environment " Moscow: Nauka, 1982,319p., (in Russian).

[2] Suhinov A. I., Suhinov A.A. (2004) "3D Model of Diffusion-Advection-Aggregation Suspenshions in Water Basins and Its Parallel Realization". International Conference Parallel CFD 2004, pp 189 – 192.

[3] Berlyand M. "Forecast and regulation of pollution of an atmosphere " Leningrad: Gidrometeoizdat, 1985, 271p., (in Russian).

[4] Koldoba A. V., Poveshchenko J. A., Samarskaya E. A., Tishkin V. F. "Methods of mathematical modelling of an environment " Moscow: Nauka, 2000, 254p., (in Russian).

[5] Korneev V. V. "Parallel calculation systems" Moscow: Noledzh, 1999,320p., (in Russian).

# PARALLEL NAVIER-STOKES SOLUTIONS OF A WING-FLAP CONFIGURATION ON STRUCTURED MULTI-BLOCK OVERSETTING GRIDS

**Erhan Tarhan, Yüksel Ortakaya, Emre Gürdamar, Bülent Korkem**

*Design and Engineering Dept., TAI, Technopolis M.E.T.U. Ankara, TURKEY*

Keywords: Fowler flap, lift force, drag force, Navier-Stokes Equations, point matching multi-block, oversetting grids, pc-cluster.

## 1. INTRODUCTION

In order to improve the landing and take-off performance characteristics, most of the airplanes use high lift devices. Among several types of high lift devices, Fowler flap [1] is preferred due to its advantages. Fowler flaps can be described by three main features: variable area, variable camber, a gap and overlap between the wing and flap.

A Fowler flap has been designed [2] to meet the prescribed lift requirements for an intermediate trainer wing with emphasis on drag and pitching moment. The flap configuration has been designed by using a two-dimensional unstructured Reynolds averaged Navier-Stokes solver implementing Spalart-Allmaras turbulence model. Clean, 20 and 30 degrees flap configurations and streamlines at 10 degrees angle of attack are shown in Figure 1.

Due to complex geometry generated by extension of the flap and existence of the slot between the main wing and the flap, the mesh generation becomes difficult especially for the structured grid applications. Moreover, the number of grid points increases dramatically near the complex regions which directly affects the computational time. To overcome such difficulties, there are several methods to apply such as multi-block and overset grid techniques with utilization of parallel processing of each sub-divided components. In the overset grid methodology, several sub-domains are created to

I.H. Tuncer et al., *Parallel Computational Fluid Dynamics 2007*,
DOI: 10.1007/978-3-540-92744-0_26, © Springer-Verlag Berlin Heidelberg 2009

resemble the whole computational domain where each individual domain is meshed separately and with no requirement for grid line continuity.

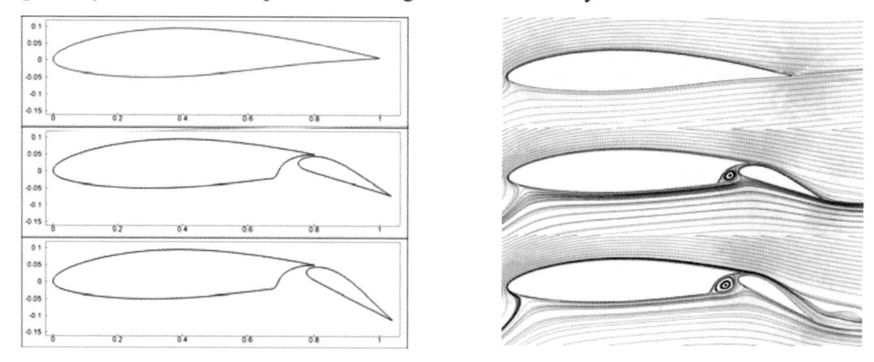

Figure 1. Clean, 20 and 30 degrees flap configurations and streamlines at 10 degrees angle of attack.

In this work, a three-dimensional parallel point-matching multi-block and overset structured grid Navier-Stokes flow solver (xFLOWsv-pmbc) is used to solve the flowfield around the trainer aircraft's wing-flap configuration. The base line flow solver (xFLOWsv) [3] uses an explicit finite-volume method with multi-stage Runge-Kutta time integration. Point-matching multi-block and overset grid techniques with the parallelization directives of MPI have been added to the core program in order to solve the complex geometries and decrease the CPU time. In the point-matching multi-block grid technique, a computational domain is decomposed to a number of sub-domains which are faced with another in a point-point way. Even the subdomains are placed one-cell or more overlap on another, grid points must be coincided.

## 2. Geometry and Mesh Generation

The model geometry is a wing with a 30 degrees deflected, partial span Fowler flap. For the entire computational domain, initially three structured mesh domains are generated. The first of these is of C-H type and generated around the clean wing enclosing the slot. The second one is the structured mesh domain created for the slot which is located inside the wing. The interfaces between the slot and the wing mesh are the point-matching structured grid planes. The end planes of slot domain are composed of singular lines lying along the spanwise direction. The last domain is another C-H type of structured mesh generated around the Fowler flap which is overset onto the slot and wing mesh. A side and an overall view of the generated mesh are shown in the Figure 2.

The surface mesh of the wing and the volume mesh for the overset flap and the slot domains are generated by a commercial grid generator tool GRIDGEN. Whereas, to obtain the volume mesh for the wing, a two-dimensional hyperbolic grid generator is used for each spanwise surface grid layers. The outcome two-dimensional planes are stacked together to unify the volume mesh around the wing. Mesh size for each block is given in Table 1.

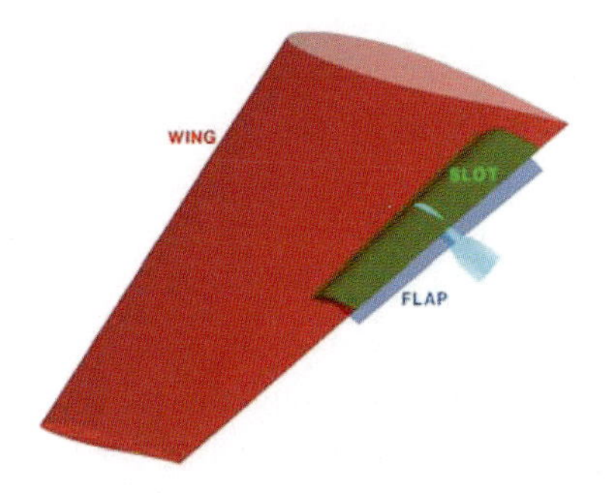


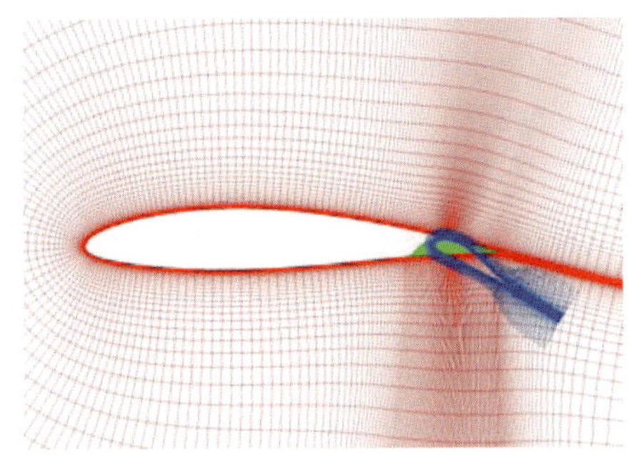

Figure 2. Wing-Flap Configuration

Table 1 Mesh size breakdown

| Block id | Region | imax | jmax | kmax | Total |
|---|---|---|---|---|---|
| 1 | Wing (Root) | 421 | 65 | 22 | 602,030 |
| 2 | Wing (Middle, lower side) | 201 | 65 | 99 | 1,293,435 |
| 3 | Wing (Middle, upper side) | 201 | 65 | 99 | 1,293,435 |
| 4 | Wing (Tip, lower side) | 201 | 65 | 91 | 1,188,915 |
| 5 | Wing (Tip upper side) | 201 | 65 | 91 | 1,188,915 |
| 6 | Slot | 81 | 33 | 99 | 264,627 |
| 7 | Flap | 205 | 56 | 99 | 1,136,520 |

The total number of mesh points is 6,967,877.

## 3. Overset Grid Methodology

Overset grid technique has been more pronounced since 1980's. Steger, Dougherty and Benek [4] first applied the technique to two-dimensional flow equations which utilize stream functions. Later on Steger, Benek, Dougherty and Buning [5,6,7,8] used the technique for three-dimensional flow solvers for both inviscid and viscous flows. Starting from the 1990's, moving body problems and the flow solution around complex geometries have become popular to solve in conjunction with the overset grid technique.

In the overset grid methodology, individual separate body conforming grids are overlapped on one another, without forcing the grid points to be aligned with neighboring components [9]. For a typical domain with two overset grid blocks, the method is applied in a way that the overset mesh produces a "hole" in the donor domain. The donor grid points in hole regions are flagged as "iblank = 0". Next, the hole boundaries are searched in the overset mesh so that the flow properties are passed to the donor. On the other hand, the overset mesh collects the information from its outermost planes. Hole boundaries of the donor domain and the outermost planes of the overset mesh are "searched and localized" in each other by using suitable algorithms [9]. At the localized overset grid points, the flow properties can be interpolated by linear interpolation methods such as weighting coefficients based on the inverse of the distances or the trilinear interpolation method [9].

### 3.1. Search and Localization

In this work overset mesh points are checked in each donor cell by defining the sub-volumes. The total volume of the sub-volumes must be equal to the volume of the donor cell. In order not to search through the entire donor domain, it is coarsened as much as possible by taking the alternate points at sequenced grid levels but without losing the boundary planes. The coarsest donor grid has the smallest number of elements. For the wing-flap case, block 2 has 8x3x5 nodes at this level. Each overset mesh is first checked at this level and the localized cell indices are transferred to a finer level by doubling the coarse level index value. Some extra number of cells may be added to the finer level not to miss the donor cell due to grid inaccuracies. The advantage of the model is that it is not sensitive to the cell skewness which may require a special way of treatment in three dimensions for directional based algorithms. Also, the model does not need a special treatment at the wake or split planes. The hole boundary created by the flap domain on the slot and wing mesh is shown in Figure 3.

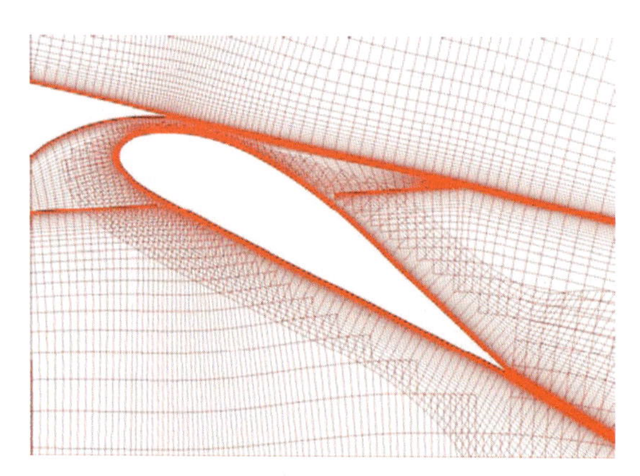

Figure 3. Hole boundary created by the flap domain.

### 3.2. Hole Cutting

Hole cutting is the essential part of the overset grid methodology where the overlapping domain creates an artificial hole in the donor. A common way of hole cutting is based on the algorithm which checks the dot product of two vectors which are the face normals at the cutting plane and relative position vectors defined between the cell centers and the mesh points to be checked [4]. According to the sign of the dot product in/out cells are identified. To speed up the algorithm, an initial routine is added for the closed walls. According to the algorithm, a temporary geometric center is calculated by using the wall nodes. Next, a minimum distance is calculated between the closest wall node and the center. The donor mesh points are first checked to see whether they fall into the circular region defined by the minimum distance. A similar version can also be applied by using the circular area defined by the maximum radius.

In this work classical treatment is applied for the hole cutting process. For the wall boundaries which are not closed, the accelerator modules are not used.

## 4. Flow Solver

The base line flow solver (xFLOWsv) [3, 10] is a three-dimensional Navier-Stokes solver using the multi-stage Runge-Kutta time integration and an artificial dissipation model [12] with a cell-vertex finite volume formulation, implemented on the structured grids. The governing Euler/Navier-Stokes equations for flow of a compressible gas in integral form are written as in the following:

$$\frac{\partial}{\partial t}\iiint_{\Omega} w\,d\Omega + \iint_{S} \vec{F} \bullet d\vec{S} - \iint_{S} \vec{F}_v \bullet d\vec{S} = 0 \qquad (1)$$

where, **w** is the vector for the flow variables in the conservative form, $\Omega$ is the control volume element with boundary defined by S. F and $F_v$ are the inviscid and viscous flux vectors: $\mathbf{w} = [\rho, \rho u, \rho v, \rho w\ \rho E]^T$

An algebraic turbulence model (Baldwin-Lomax) [13] is implemented on the flow solver which calculates the turbulent eddy viscosity.

For the boundary conditions, at inflow and outflow planes one-dimensional Riemann invariants are used as the characteristic type of boundary conditions. No-slip boundary condition is used at the solid walls where the velocity vectors are set to zero, pressure and density are determined by using the ghost planes. For the symmetry plane, flow tangency is applied as in the case of inviscid walls. To evaluate the properties at the point matching interfaces, a linear interpolation is applied by using the adjacent planes. For this purpose, flow properties and fluxes are passed to the neighboring domain by the use of classical message send and receive operations of parallel processing.

To apply a boundary condition for the overset grid nodes, donor cells of neighboring domain are used for a linear interpolation which is based on the inverse distances. In the hole regions, flow field calculations are skipped by the use of "iblank" arrays which are previously assigned during the search and localization process.

## 5. Parallelization

For parallel processing, MPI libraries are used via mpich-1.2.7 [14]. For point-matching multi-block solutions, the flow properties and the fluxes are passed to the adjacent domain at the matching interfaces. On the other hand, to interpolate overset grid points, flow properties and fluxes calculated at corresponding donor cells are passed by using MPI directives. For message passing, standard send and receive commands are used. Each processor is assigned to solve its corresponding domain and to communicate with the neighbours, before updating the boundaries. Communication sequence is known before the solution starts and kept in a global variable "italk". The total number of "talking" is determined by each process after the matching interface and the overset grid data are read. For every value of "italk", each process knows the data sizes to be sent or received and the corresponding blocks and hence the processes. The parallel environment is composed of 32 computers having 64 bit 3.2 GHz Pentium processors and 2GB of RAM. Communication between computers is made with a switch which allows 10/100 megabit frequency. The cluster has Linux based operating system and in order to allow parallel processing, mpich-1.2.7 is used.

## 6. Results

Wing-flap configuration is solved at 0, 2, 10, 12 and 14 degrees angles of attack with a freestream Mach number 0.25 and 6 million Reynolds number. Flow solutions around the wing-flap configuration are presented in Figures 4 and 5. In Figure 4, contours of pressure coefficient are presented at 10 degrees angle of attack. In the next figure, streamlines are shown in the slot and in side passage between wing and flap. The additional lift gained by the flap is plotted against the clean wing Figure 6. Figure 7 shows the comparison of pitching moment coefficients between the wing-flap and clean wing. In Figure 8, convergence history is shown in terms of normal force coefficients. A typical run-time with 7 parallel processors which is the same as the number of mesh blocks and for 13000 time iterations is around 4 days.

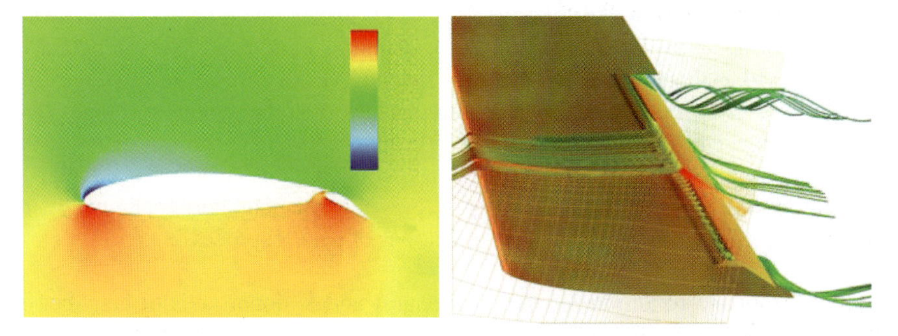

Figure 4, 5. Pressure coefficient contours and streamlines in vortical formation through the slot.

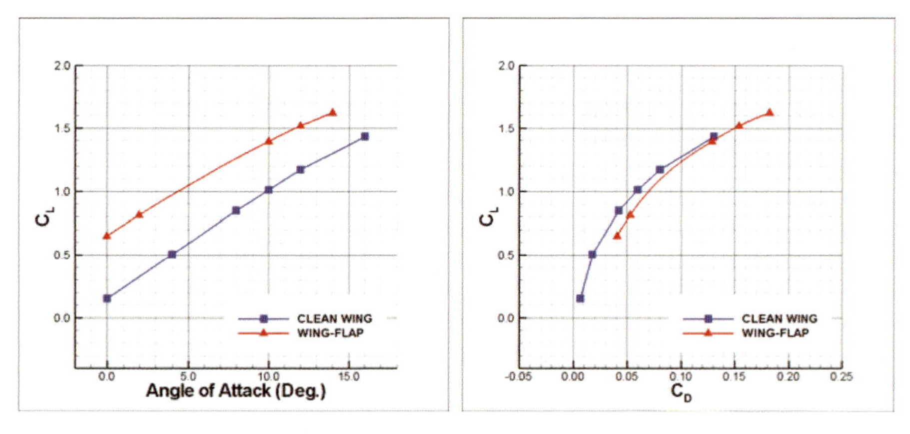


Figure 6. Comparison of the force coefficients of wing-flap configuration and clean wing.

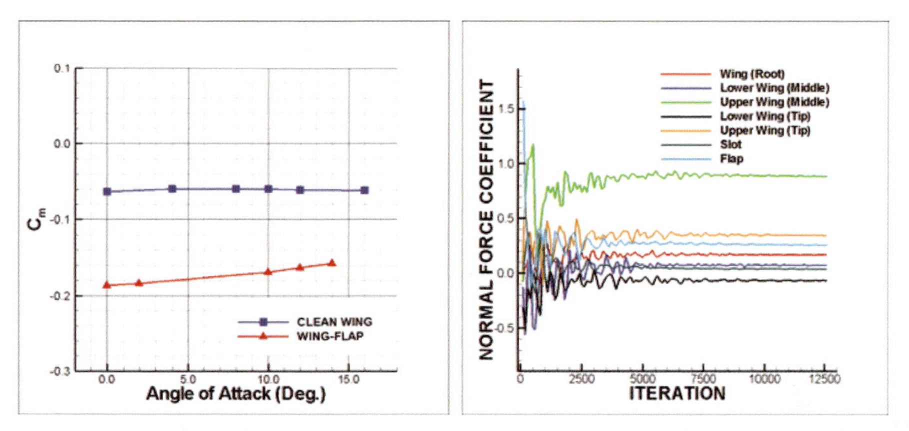


Figure 7. Pitching moment coefficients

Figure 8. Convergence history.

## 7. Conclusion and Further Work

The overset grid and point-matching multi-block techniques are added to a base line flow solver (xFLOWsv). Parallel computations are performed by using MPI in a linux pc-cluster with 32 processors. A Fowler type wing-flap configuration is solved by utilizing the parallel overset and point-matching multi-block grids. The results are as expected when drag polar and lift curve are compared with results for the clean geometry.

The studies show, although overset grid approach eliminates some difficulties in mesh generation, mesh generation is still the main difficulty in flow solution problems.

To improve the overset methodology, "search and localization" algorithm needs to be robust. Besides, for the flow solver part non-matching mesh handling capability will be implemented and to realize a more general description of turbulence through the flowfield, a one-equation Spalart-Allmaras turbulence model will be applied with "Detached-Eddy Simulation".

To realize the slipstream effects induced by a propeller there are ongoing studies about implementation of a propeller model, based on a non-linear actuator disk theory. The propeller model will be implemented as a boundary condition by utilizing the overset grid techniques.

**References**

1. Harlan, F., *The Fowler Flap*, Published by Aircraft Designs Inc., 1948.
2. Wenzinger C.J., Harris T.A., "Wind-Tunnel Investigation of an N.A.C.A 23012 Airfoil with Various Arrangments of Slotted Flaps", NACA Report No. 664, 1939
3. Şahin, P., *Navier-Stokes Calculations over Swept Wings*, M.S., Aerospace Engineering Department, 2006, M.E.T.U., Ankara, Turkey
4. Steger, J. L., Dougherty, F. C., and Benek, J. A., *A Chimera Grid Scheme*, Advances in Grid Generation, ASME FED-5, Vol.5, American Society of Mechanical Engineers, New York, 1983, pp. 59-69.
5. Steger, J.L., and Benek, J.A., *On the Use of Composite Grid Schemes in Computational Aerodynamics*, Computer Methods in Applied Mechanics and Engineering, Vol. 64, Nos. 1-3,1987, pp. 301-320.
6. Dougherty, F.C., Benek, J.A., and Steger, J.L., *On Application of Chimera Grid Schemes*, NASA TM-88193, Oct. 1985.
7. Benek, J.A., Buning, P.G. and Steger J. L., *A 3-D Grid Embedding Technique*, AIAA Paper 85-1523, July 1985.
8. Benek, J.A., Steger, J.L., and Dougherty, F.C., *A Flexible Grid Embedding Technique With Application to the Euler Equations*, AIAA Paper 83-1944, July 1983
9. Meakin, R., *Composite Overset Structured Grids*, Handbook of Grid Generation, Vol. 1, CRC Press 1999.
10. Tarhan, E., Korkem B., Şahin P., Ortakaya Y., Tezok F., *Comparison of Panel Methods and a Navier-Stokes Flow Solver on ARA-M100 Wing-Body Configuration,* AIAC 2005, M.E.T.U. Ankara, Turkey.
11. Tarhan, E., Ortakaya, Y., *TAI – AIAA CFD Drag Prediction Workshop III presentation* http://aaac.larc.nasa.gov/tsab/cfdlarc/aiaa-dpw/, last access on 15/02/2007.
12. Jameson, A., *Analysis and Design of Numerical Schemes for Gas Dynamics 1: Artificial Diffusion, Upwind Biasing, Limiters and their Effect on Accuracy and Multi-grid Convergence,* International Journal of Computational Fluid Dynamics, Vol. 4, 1995, pp. 171–218.
13. Baldwin B. and Lomax, H., *Thin Layer Approximation and Algebraic Turbulence Model for Separated Flows*, AIAA Paper 78-257, 1978.
14. http://www-unix.mcs.anl.gov/mpi/mpich1/, *MPICH Home Page,* last access on 15/02/2007.

# PARALLEL NAVIER-STOKES SOLUTIONS OF NASA 65° DELTA-WING

**Emre Gürdamar, Erhan Tarhan, Yüksel Ortakaya, Bülent Korkem**

*Design and Engineering Dept., TAI, Technopolis M.E.T.U., Ankara, TURKEY*

Keywords: Delta wing, vortical flow, turbulence modelling, multi-block, Navier-Stokes equations

## 1. Introduction

xFLOWg -an in-house developed, three dimensional structured Navier-Stokes solver- is used in computations of NASA 65° delta wing. The case study is a part of NATO-RTA AVT-113 working group, named as Vortex Flow Experiment 2 (VFE2) [1] which investigates the vortical flow formation in order to improve the technology readiness level for military aircraft.

Single block computational mesh for the test geometry is decomposed into 8 domains to be solved on a parallel cluster of PC's. Since turbulence modeling appears to be a vital aspect for resolving the vortical motion accurately, one-equation Spalart-Allmaras [2] is selected as a model in computations.

## 2. Test Geometry

Research in VFE-2 includes vortical flows over 65° swept delta wings having various types of leading edges mounted. The leading edges are sharp, small range, medium range and long range radius leading edges. A sting is mounted to the rear part of the wing root for the wind tunnel experiments and it is blended to the wing surface by a fairing. The schematic view of the test geometry and locations of the surface pressure tabs for the wind tunnel experiments are shown in Figure 1 [3].

I.H. Tuncer et al., *Parallel Computational Fluid Dynamics 2007*,
DOI: 10.1007/978-3-540-92744-0_27, © Springer-Verlag Berlin Heidelberg 2009

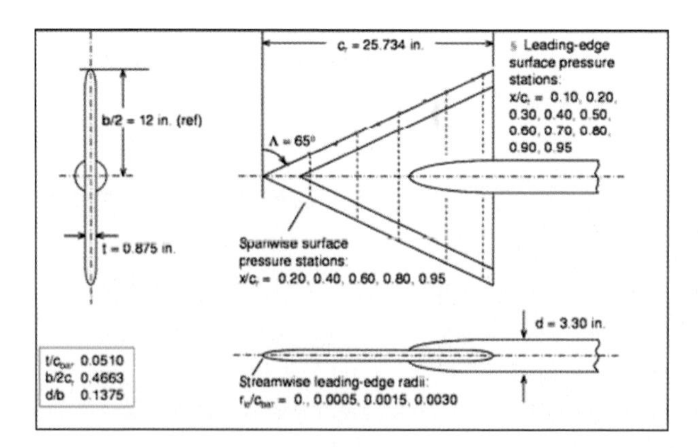


Figure 1. Description of the test model and the pressure stations.

Present computations are done for the medium and sharp range radius leading edge configurations.

## 3. Grid Generation

Accurate resolution of highly vortical regions with precise determination of the vortices' core locations is the main aim of VFE-2 study. Accuracy is directly related to the fineness level of mesh spacing in locations where vortical flow phenomena is relatively dominant. Current grid is generated by taking these facts into consideration.

Structured mesh with a 2,070,024 number of total nodes is decomposed into 8 sub domains with point matching interfaces and used in current computations. The breakdown of mesh sizes is given in Table 1.

Table 1. Breakdown of mesh sizes for each block.

| Block id | $i_{max}$ | $j_{max}$ | $k_{max}$ | total |
|---|---|---|---|---|
| 1 | 66 | 80 | 48 | 253,440 |
| 2 | 66 | 82 | 48 | 259,776 |
| 3 | 66 | 82 | 50 | 270,600 |
| 4 | 66 | 80 | 50 | 264,000 |
| 5 | 64 | 80 | 50 | 256,000 |
| 6 | 64 | 82 | 50 | 262,400 |
| 7 | 64 | 82 | 48 | 251,904 |
| 8 | 64 | 82 | 48 | 251,904 |

Multi-blocks are obtained by dividing the single block mesh in several directions. Block numbers 1, 2, 7 and 8 enclose wall boundaries where blocks 2 and 7 surround the wing part. Block numbers 1 and 8 has both the wing and sting boundaries. The rest of the blocks mentioned above and extend to farfield. The multi-block structure is shown in Figure 2.

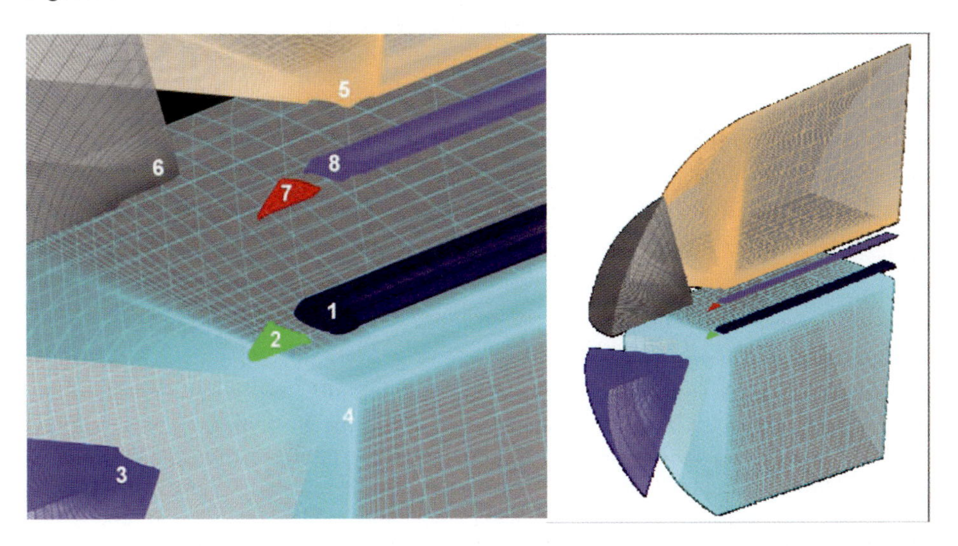


Figure 2. Multi-block mesh structure of medium range radius leading edge VFE-2 delta wing

## 4. Flow Solver

The base line flow solver (xFLOWg) is a three-dimensional Navier-Stokes solver using the multi-stage Runge-Kutta time integration and an artificial dissipation model. The solver works on structured grids and has a cell-vertex finite volume formulation implemented. The governing Euler/Navier-Stokes equations for flow of a compressible gas in integral form are written as in the following:

$$\frac{\partial}{\partial t}\iiint_{\Omega} w\, d\Omega + \iint_{S} \vec{F} \bullet d\vec{S} - \iint_{S} \vec{F}_v \bullet d\vec{S} = 0 \qquad (1)$$

where, **w** is the vector for the flow variables in the conservative form, $\Omega$ is the control volume element with boundary defined by S. F and $F_v$ are the inviscid and viscous flux vectors: $\mathbf{w} = [\rho, \rho u, \rho v, \rho w\ \rho E]^T$

Several turbulence models are implemented on the flow solver which calculates the turbulent eddy viscosity. Besides the algebraic Baldwin-Lomax, [8] and half-equation Johnson-King turbulence models, currently, one-equation Spalart-Allmaras [2] and its derivative, namely Detached Eddy Simulation (DES) [4], are used in the solver. Parallel multi-block VFE-2 computations of sharp edge swept wings are done by Spalart-Allmaras turbulence model.

For the boundary conditions, at inflow and outflow planes, one-dimensional Riemann invariants are used as the characteristic type of boundary conditions. No-slip boundary condition is used at the solid walls where the velocity vectors are set to zero; pressure and density are determined by using the ghost planes. For the symmetry plane, flow tangency is applied as in the case of inviscid walls.

To evaluate the properties at the point matching interfaces, a linear interpolation is applied by using the adjacent planes. For this purpose, flow properties and fluxes are passed to the neighboring domain by the use of classical message send and receive operations of parallel processing.

### 4.1. Turbulence Modeling

Due to the fact that discrete domains are present in multi-block applications, all flow quantities must be properly "transported" through the interfaces between the neighboring domains. The inner-outer layer definitions of Baldwin-Lomax turbulence model appears to be inefficient. Besides the previously implemented Baldwin-Lomax algebraic turbulence model, one equation turbulence model Spalart-Allmaras is added to the solver. The transport equation of Spalart-Allmaras model working variable $\widetilde{\nu}$ is given in Equation (2).

$$\begin{aligned}\frac{\partial(\rho\widetilde{\upsilon})}{\partial t}+\frac{\partial(\rho u_j\widetilde{\upsilon})}{\partial x_j} &= \rho C_{b1}(1-f_{t2})\widetilde{S}\widetilde{\upsilon} \\ &+\rho\frac{1}{\sigma}\left[\nabla((\upsilon+\widetilde{\upsilon})\nabla\widetilde{\upsilon})+C_{b2}(\nabla\widetilde{\upsilon})^2\right] \\ &-\rho\left(C_{w1}f_w-\frac{C_{b1}}{\kappa^2}f_{t2}\right)\left(\frac{\widetilde{\upsilon}^2}{d}\right)+\rho f_{t1}\Delta U^2\end{aligned} \quad (2)$$

This equation is composed of convective and diffusive terms with source parameters that dominate the turbulent viscosity development. These transport properties of the one or two-equation models, allow them to work properly through multi-blocks by having introduced proper boundary conditions at interfaces. After the turbulence model equation is solved for $\widetilde{\nu}$, eddy viscosity $\nu_t$ is obtained by a correlation given in Equation (3).

$$\upsilon_t=\widetilde{\upsilon}f_{\upsilon1} \quad (3)$$

The function $f_{\upsilon1}$ is defined as $f_{\upsilon1}=\dfrac{\chi^3}{\chi^3+c_{\upsilon1}^3}$ where $\chi\equiv\dfrac{\widetilde{\upsilon}}{\upsilon}$.

## 5. Parallel Environment

Parallel processing procedure includes the usage of MPI libraries via mpich-1.2.7 [10]. In point-matching multi-block solutions, the flow properties and the fluxes are passed to the adjacent domain at the matching interfaces with standard send and receive commands. Each processor is assigned to solve its corresponding domain and to communicate with the neighbours, before updating the boundaries. Communication sequence is known before the solution starts and kept in a global variable "italk". The total number of "talking" is determined by each process after the matching interface is read. For every value of "italk", each process knows the data sizes to be sent or received and the corresponding blocks and hence the processes.

The parallel environment is composed of 32 computers having 64 bit 3.2 GHz Pentium processors and 2GB of RAM. Communication between computers is made with a switch which allows 10/100 megabit frequency. The cluster has Linux based operating system and in order to allow parallel processing, mpich-1.2.7 is used.

For the delta wing configuration with a mesh of 2,070,024 nodes, a typical wall clock time is 20 hours with a single computer. This run time reduces to 3 hours when the mesh is decomposed into 8 sub domains and run on 8 computers.

## 6. Results

This study concentrates on computations of six different cases. The flow solutions of medium range radius leading edge configuration are obtained at 0.85 Mach number at various angles of attack. Sharp range radius leading edge delta wing is exposed to 0.85 Mach number at similar angle of attacks. Spalart-Allmaras turbulence model is used in the solutions and results are compared with experimental results and Baldwin-Lomax solutions that are presented in the previous meetings of this working group. The Reynolds number is same for all of these 6 cases that is 6 millions. The flow solution sets are tabulated and given in Table 2.

Table 2. Computation cases

| Case | Leading Edge | AoA | Mach | Re |
|---|---|---|---|---|
| 1 | Medium | 13.4° | 0.851 | $6x10^6$ |
| 2 | Medium | 18.5° | 0.849 | $6x10^6$ |
| 3 | Medium | 24.6° | 0.850 | $6x10^6$ |
| 4 | Sharp | 13.5° | 0.851 | $6x10^6$ |
| 5 | Sharp | 18.6° | 0.851 | $6x10^6$ |
| 6 | Sharp | 24.6° | 0.851 | $6x10^6$ |

The phenomena to be investigated in these computations are mainly related to vorticity driven flow events. Detection of secondary vortex formations with accurate vortex core locations is one of the foremost objectives of this study. At high angles of attacks, vortex breakdown is expected to be seen. As a primary comparison of the computational results, surface pressure coefficient values are compared with experimental results.

First three test cases are the medium range radius leading edge (MLE) type wing solutions at 0.85 Mach number. The first one is at 13.4 degree angle of attack. One main vortex is resolved in this solution. The surface pressure coefficient distributions, obtained at 0.2, 0.4, 0.6, 0.8, 0.95 sections of the root chord of 1, are plotted against experimental results are given in Figure 3.

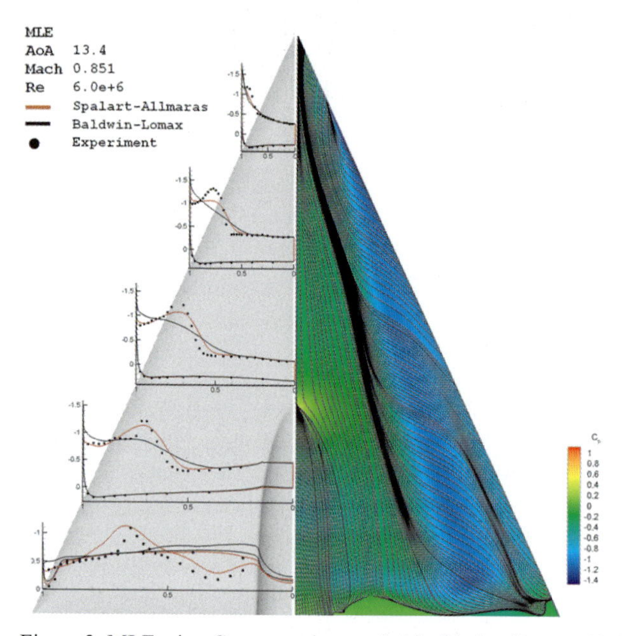


Figure 3. MLE wing Cp comparison and skin friction lines at 13.4 deg. angle of attack.

Pressure jump at the 0.4 section of the wing could not be detected properly, however the general trend in further sections are suited well to the experimental results. Baldwin-Lomax turbulence model showed a diffusive effect on the solutions. The leading edge separation is reattached and observed by the concentrated skin friction lines.

Vortex core location is given in Figure 4 using tangent vectors on the sections of experimental comparisons.

A similar behavior is examined in the second test case which is at 18.5 angle of attack. However, a secondary inner vortex is observed in the aft sections of the wing. The pressure coefficient distributions and skin friction lines are given in Figure 4. The surface tangent vectors and primary/secondary vortex formations are plotted in the same figure.

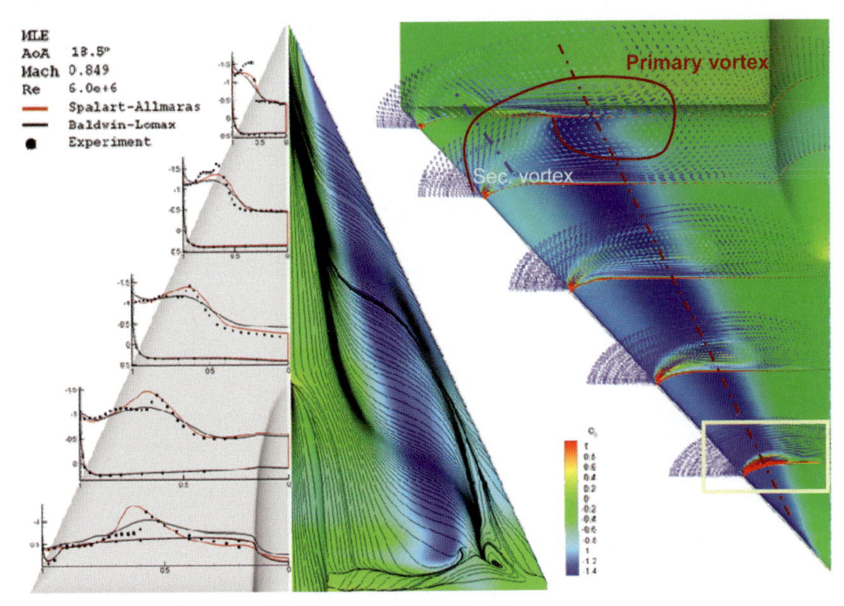


Figure 4. MLE solution at 18.5 degrees angle of attack

In the last test case of MLE set, the start of the vortex breakdown is expected to be seen. Figure 5 shows the iso-surface of vorticity magnitude and a cut through the aft part of the wing is analyzed.

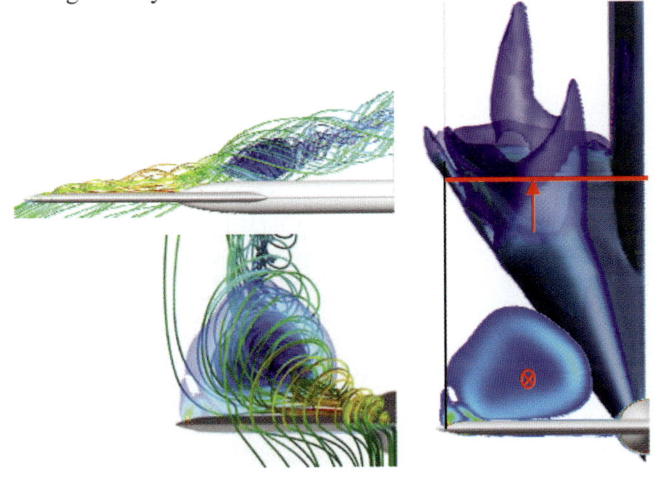

Figure 5. MLE solution at 24.6 degrees angles of attack

The results of the sharp leading edge (SLE) test cases are given in Figure 6.

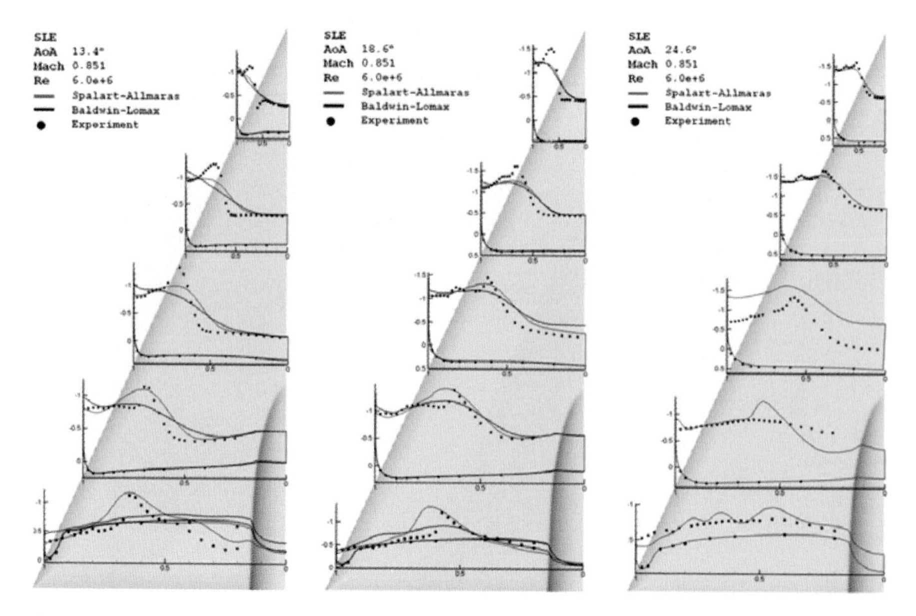


Figure 6. SLE solutions

The results of these computations show that the unsteady effects could not be resolved properly. However, improvements with the use of Spalart-Allmaras turbulence model are possible.

## References

1. http://www.rta.nato.int/ACTIVITY_META.asp?ACT=AVT-113, AVT-113 *Understanding and Modelling Vortical Flows to Improve the Technology Readiness Level for Military Aircraft*, last access on: 24/04/2007
2. Spalart, P. A., Allmaras, S. R., *A One-Equation Turbulence Model for Aerodynamic Flows,* AIAA-92-0439, 30th Aerospace Sciences Meeting and Exhibit, Jan. 6-9, 1992, Reno, NV.
3. Chu, J., Luckring, J. M., *Experimental Surface Pressure Data Obtained on 65° Delta Wing Across Reynolds Number and Mach Number Ranges, Volume 2 – Small Radius Leading Edge*, NASA Technical Memorandum 4645, Feb 1996.
4. Spalart, P. R., Jou, W.-H., Stretlets, M., and Allmaras, S. R., *Comments on the Feasibility of LES for Wings and on the Hybrid RANS/LES Approach,* Advances in DNS/LES, Proceedings of the First AFOSR International Conference on DNS/LES, 1997.
5. Tarhan, E., Korkem B., Şahin P., Ortakaya Y., Tezok F., *Comparison of Panel Methods and a Navier-Stokes Flow Solver on ARA-M100 Wing-Body Configuration,* AIAC 2005, M.E.T.U. Ankara, Turkey.

6. Şahin, P., *Navier-Stokes Calculations over Swept Wings*, M.S., Aerospace Engineering Department, 2006, M.E.T.U., Ankara, Turkey
7. Jameson, A., *Analysis and Design of Numerical Schemes for Gas Dynamics 1: Artificial Diffusion, Upwind Biasing, Limiters and their Effect on Accuracy and Multi-grid Convergence,* International Journal of Computational Fluid Dynamics, Vol. 4, 1995, pp. 171–218.
8. Baldwin B. and Lomax, H., *Thin Layer Approximation and Algebraic Turbulence Model for Separated Flows*, AIAA Paper 78-257, 1978.
9. P. R. Spalart , S. Deck, M. L. Shur, K. D. Squires, M. Kh. Strelets and A. Travin, *A New Version of Detached-eddy Simulation, Resistant to Ambiguous Grid Densities,* Theoretical and Computational Fluid Dynamics, Volume 20, Number 3 / July, 2006
10. http://www-unix.mcs.anl.gov/mpi/mpich1/, *MPICH Home Page,* last access on 15/02/2007.

# PARALLEL TURBULENT NAVIER-STOKES SOLUTIONS OF WING ALONE GEOMETRIES FOR DRAG PREDICTION

**Pınar Şahin, Emre Gürdamar, Erhan Tarhan, Yüksel Ortakaya, Bülent Korkem**

*Design and Engineering Dept., TAI, Technopolis M.E.T.U., Ankara, TURKEY*

Keywords: Drag Prediction Workshop, multi-block, transonic wing, drag force, turbulence modelling

## 1. Introduction

An in-house-developed, three dimensional structured Navier-Stokes solver (xFLOWg) is used in the transonic wing computations. Wing alone geometries are taken from Drag Prediction Workshop III (DPW3) [1], which was held on June 3-4, 2006, in San Francisco, CA. Although experimental results are not available for the test cases of DPW3, drag and lift comparison with the well-known CFD codes appears to be a satisfactory validation study.

The test problems, which are two wing-only configurations, one of which is the optimized version of the other, are solved at a transonic speed and at various angles of attack. A finite volume based solver which operates the multi-stage Runge-Kutta method for time integration and applies a flux limiter is used in computations. The solver utilizes point-matching multi-block grids with parallel processing using the MPI libraries. For the turbulent eddy viscosity computation, one-equation Spalart-Allmaras [2] turbulence model is used.

## 2. Test Geometries and Grid Generation

Transonic wing-only configurations are the test geometries of DPW3. Among the two alternative wing-only geometries, wing 2 (DPW-W2), is the aerodynamically optimized

I.H. Tuncer et al., *Parallel Computational Fluid Dynamics 2007*,
DOI: 10.1007/978-3-540-92744-0_28, © Springer-Verlag Berlin Heidelberg 2009

version of the base geometry, wing 1 (DPW-W1). Geometrical comparisons of DPW-W1 and DPW-W2 are presented in Figure 1.

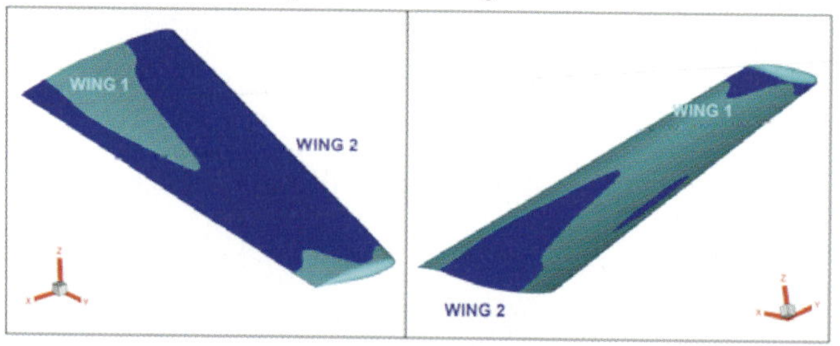


Figure 1. Geometrical comparisons of DPW-W1 and DPW-W2.

Two meshes of coarse and fine levels are investigated in this work. Coarse mesh is generated with algebraic grid generation. Sufficient number of grid points is embedded into boundary layer. Orthogonality of grid lines with the wing surface is taken into account to resolve boundary layer phenomena properly. Point matching multi-blocks are generated by splitting the single block mesh into sub-domains in the spanwise direction. The mesh size for the coarse level grid is about 700 thousand nodes with 8 multi-blocks for both of the geometries. A breakdown for the mesh sizes is given in Table 1. Figure 2 shows the point matching multi-block coarse grid generated for DPW-W1.

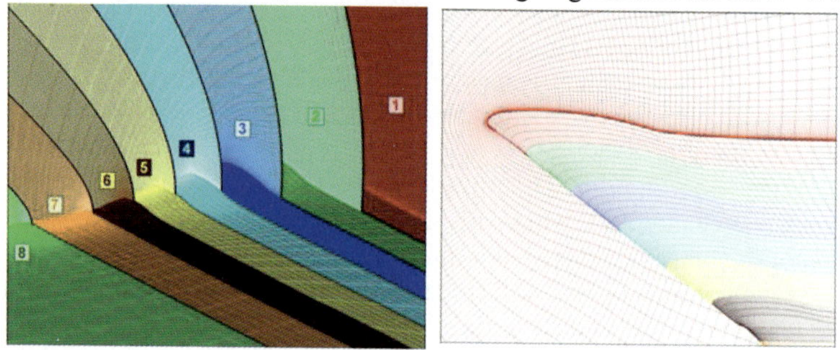


Figure 2. Point matching multi-block grids of coarse mesh.

Table 1. Breakdown of mesh sizes for each block in coarse mesh configuration.

| Block id | imax | jmax | kmax | total |
|---|---|---|---|---|
| 1 | 193 | 65 | 6 | 75,270 |
| 2 | 193 | 65 | 7 | 87,815 |
| 3 | 193 | 65 | 7 | 87,815 |
| 4 | 193 | 65 | 7 | 87,815 |
| 5 | 193 | 65 | 7 | 87,815 |
| 6 | 193 | 65 | 7 | 87,815 |
| 7 | 193 | 65 | 7 | 87,815 |
| 8 | 193 | 65 | 8 | 100,360 |

Fine multi-block mesh is originally coded as "medium mesh" in DPW3 computations, generated by Boeing Company with ICEMCFD [3] grid generation tool. The mesh has around 4.2 million nodes and comprised of 8 multi-blocks, which is used in parallel computations. The multi-blocks with wing surface mesh is shown in Figure 3. The breakdown of mesh sizes for fine mesh is given in Table 2.

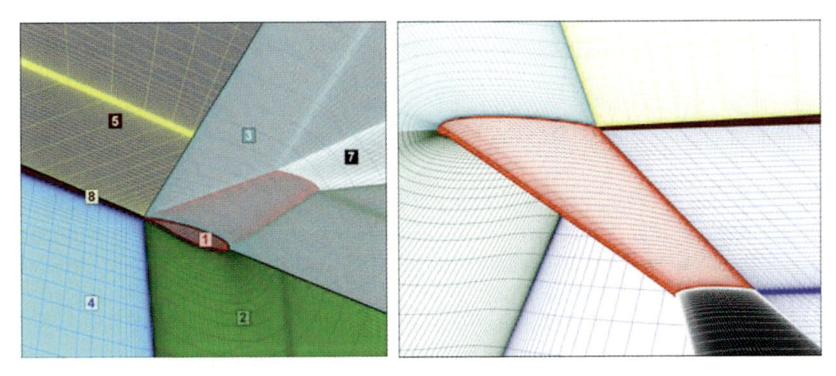


Figure 3. Point matching multi-block grids of fine mesh.

Table 2. Breakdown of mesh sizes for each block in the fine mesh configuration.

| Block id | imax | jmax | kmax | total |
|---|---|---|---|---|
| 1 | 193 | 49 | 73 | 690,361 |
| 2 | 97 | 73 | 121 | 856,801 |
| 3 | 97 | 73 | 121 | 856,801 |
| 4 | 65 | 73 | 121 | 574,145 |
| 5 | 65 | 73 | 121 | 574,145 |
| 6 | 81 | 97 | 33 | 259,281 |
| 7 | 49 | 33 | 73 | 118,041 |
| 8 | 65 | 33 | 121 | 259,545 |

## 3. Flow Solver

The base line flow solver (xFLOWg) [4, 5] is a three-dimensional, Navier-Stokes solver using the multi-stage Runge-Kutta time integration and an artificial dissipation model [6]. The solver works on structured grids and has a cell-vertex finite volume formulation implemented. The governing Euler/Navier-Stokes equations for flow of a compressible gas in integral form are written as in the following:

$$\frac{\partial}{\partial t}\iiint_{\Omega} w \, d\Omega + \iint_{S} \vec{F} \bullet d\vec{S} - \iint_{S} \vec{F}_v \bullet d\vec{S} = 0 \qquad (1)$$

where, **w** is the vector for the flow variables in the conservative form, Ω is the control volume element with boundary defined by S. F and $F_v$ are the inviscid and viscous flux vectors: $\mathbf{w} = [\rho, \rho u, \rho v, \rho w\ \rho E]^T$

Several turbulence models are implemented on the flow solver which calculates the turbulent eddy viscosity. Besides the algebraic Baldwin-Lomax, [7] and half-equation Johnson-King turbulence models, currently, one-equation Spalart-Allmaras [8] and its derivative, namely Detached Eddy Simulation (DES) [8], are used in the solver. Parallel multi-block DPW3 computations of wing 1 and wing 2 are done by Spalart-Allmaras turbulence model.

For the boundary conditions, at inflow and outflow planes, one-dimensional Riemann invariants are used as the characteristic type of boundary conditions. No-slip boundary condition is used at the solid walls where the velocity vectors are set to zero; pressure and density are determined by using the ghost planes. For the symmetry plane, flow tangency is applied as in the case of inviscid walls.

To evaluate the properties at the point matching interfaces, a linear interpolation is applied by using the adjacent planes. For this purpose, flow properties and fluxes are passed to the neighboring domain by the use of message send and receive operations of parallel processing.

### 3.1. Turbulence Modeling

Due to the fact that discrete domains are present in multi-block applications, all flow quantities must be properly "transported" through the interfaces between the neighboring domains. The inner-outer layer definitions of Baldwin-Lomax turbulence model appears to be inefficient. Besides the previously implemented Baldwin-Lomax algebraic turbulence model, one equation turbulence model Spalart-Allmaras is added to the solver. The transport equation of Spalart-Allmaras model working variable $\widetilde{\nu}$ is given in Equation (2).

$$\begin{aligned}\frac{\partial(\rho\widetilde{\upsilon})}{\partial t}+\frac{\partial(\rho u_j\widetilde{\upsilon})}{\partial x_j} &= \rho C_{b1}(1-f_{t2})\widetilde{S}\widetilde{\upsilon} \\ &+\rho\frac{1}{\sigma}\left[\nabla((\upsilon+\widetilde{\upsilon})\nabla\widetilde{\upsilon})+C_{b2}(\nabla\widetilde{\upsilon})^2\right] \\ &-\rho\left(C_{w1}f_w-\frac{C_{b1}}{\kappa^2}f_{t2}\right)\left(\frac{\widetilde{\upsilon}^2}{d}\right)+\rho f_{t1}\Delta U^2\end{aligned} \tag{2}$$

This equation is composed of convective and diffusive terms with source parameters that dominate the turbulent viscosity development. These transport properties of the one or two-equation models, allow them to work properly through multi-blocks by having introduced proper boundary conditions at interfaces. After the turbulence model equation is solved for $\widetilde{\nu}$, eddy viscosity $\nu_t$ is obtained by a correlation given in Equation (3).

$$\upsilon_t = \widetilde{\upsilon} f_{\upsilon 1} \qquad (3)$$

The function $f_{\upsilon 1}$ is defined as $f_{\upsilon 1} = \dfrac{\chi^3}{\chi^3 + c_{\upsilon 1}^3}$ where $\chi \equiv \dfrac{\widetilde{\upsilon}}{\upsilon}$.

## 4. Parallel Environment

Parallel processing procedure includes the usage of MPI libraries via mpich-1.2.7 [9]. In point-matching multi-block solutions, the flow properties and the fluxes are passed to the adjacent domain at the matching interfaces with send and receive commands. Each processor is assigned to solve its corresponding domain and to communicate with the neighbours, before updating the boundaries. Communication sequence is known before the solution starts and kept in a global variable "italk". The total number of "talking" is determined by each process after the matching interface is read. For every value of "italk", each process knows the data sizes to be sent or received and the corresponding blocks and hence the processes.

The parallel environment is composed of 32 computers having 64 bit 3.2 GHz Pentium processors and 2GB of RAM. Communication between computers is made with a switch which allows 10/100 megabit frequency. The cluster has Linux based operating system and in order to allow parallel processing, mpich-1.2.7 is used.

## 5. Results

TAI contribution in DPW3 was carried out using xFLOWg solver with Baldwin-Lomax and Johnson-King turbulence models. Up to date results of the computations for DPW3 wings are accomplished by using xFLOWg with Spalart-Allmaras turbulence model using both coarse and fine meshes. In order to relax and accelerate the convergence, viscosity production is scaled from higher values to its normal value gradually. Total residual drop with the force coefficient convergence is shown in Figure 4 for each of the multi-block grids in coarse mesh.

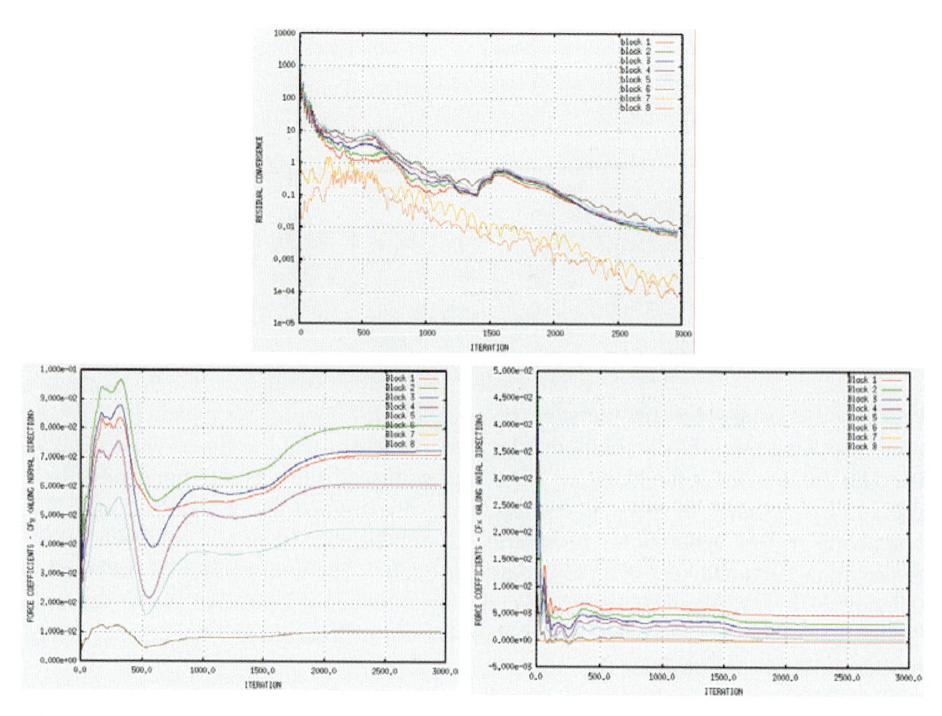


Figure 4. Total, $C_L$ and $C_D$ residuals

Force coefficients are compared with other solvers for a range of angle-of-attack values from -1 to 3 degrees. Mach number for this study is specified as 0.76 and a Reynolds number of 5 million is used. The force coefficients are calculated using coarse and fine multi-block meshes. Previously calculated Baldwin-Lomax and Johnson-King results are also included in the Figures 5 and 6. Lift coefficients for both of the wing configurations are in a good agreement with other solvers. In both cases, close to the linear region, lift curve obtained from coarse mesh shows similar characteristics with FLUENT-RKE solution, where on the point of the stall point xFLOWg is comparable with FLOWer-SSG/LLR. Fine mesh solutions exhibit relatively lower lift coefficients than the coarse mesh solutions for both wing geometries. On the basis of comparison with other CFD codes, lift and drag trends are well correlated with Boeing Mesh.

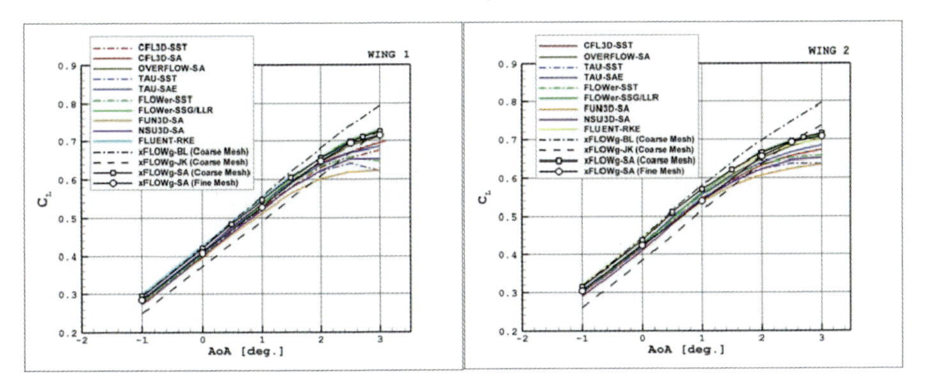


Figure 5. Lift coefficients for DPW-W1 and DPW-W2

From the drag curves, it can be seen that the previous computations of xFLOWg are inadequate to resolve the drag characteristics. Current computations with xFLOWg SA, appears to produce a similar drag curve with the other well-known solvers. As for drag value comparison, highest drag values are obtained from the coarse mesh computations. Fine mesh results are appeared similar to FLUENT-RKE solutions. OVERFLOW, FUN3D and NSU3D presents similar drag coefficient values at small angles of attack. As a matter of fact that, good accuracy in lift and drag values has been achieved when compared to the other computational solutions.

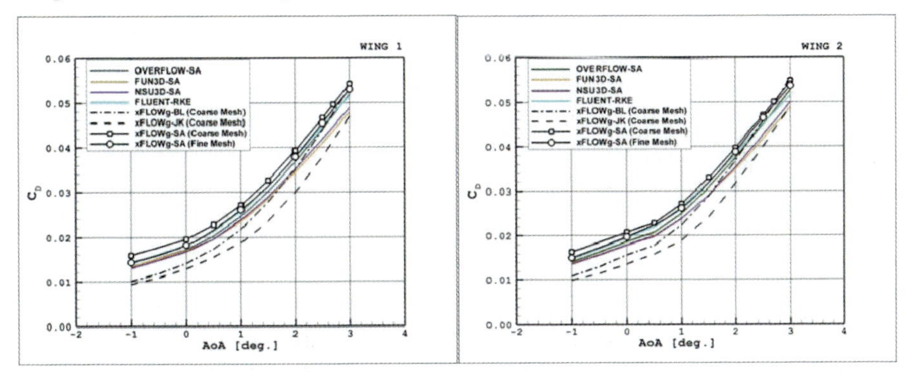


Figure 6. Drag coefficients for DPW-W1 and DPW-W2

Pressure coefficient contours for the wing 1 is shown in Figure 7. A shock formation through the mid of the sections is observed along the wing. The continuity of the contour lines is important to examine the sensation of the interface communication between the blocks. Interfaces are investigated in the left graph of Figure 7 and no discontinuity between the boundaries is noticed. A three-dimensional visualization of pressure coefficient on the wing surface is given in the second figure. The change in shock strength as traveled from root to tip is seen clearly.

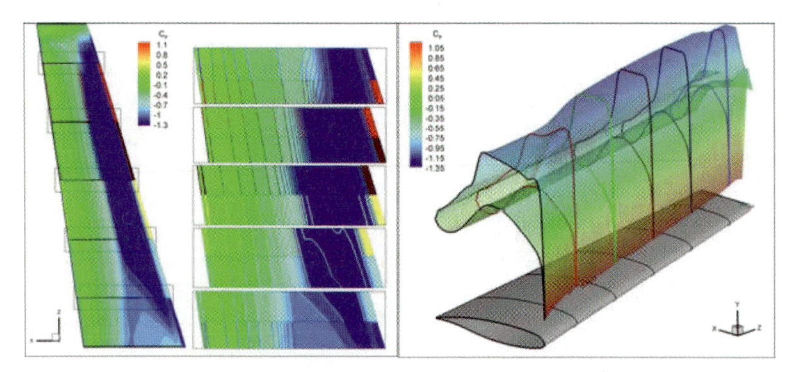


Figure 7. Upper surface Cp contours wing 1 (coarse mesh)

Further investigation with the fine mesh is done for certain angles-of-attack. Figure 8 shows 0° angle-of-attack solution. The first figure represents the turbulent viscosity distribution for spanwise sections. A sudden increase in turbulent viscosity is observed after the shock formation which is mainly due to the shock-boundary layer interaction. The second figure shows the pressure coefficient distribution on the upper surface of the wing. The wing tip vortices are seen in the sub graph of the figure. The vortex core location is visualized in the first figure by plotting the turbulent viscosity contours at the wing tip. A secondary flow is monitored that is generated from the leading edge of the wing tip.

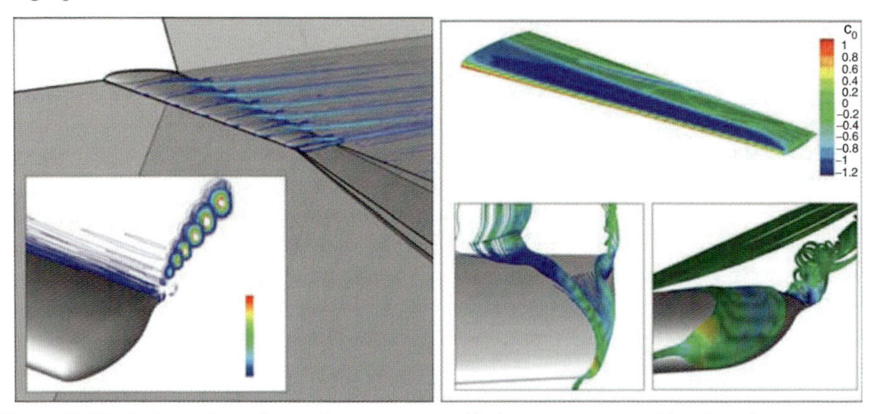


Figure 8. Turbulent viscosity and pressure coefficient contours with wing tip visualization for the fine mesh

The comparison of coarse and fine mesh solutions against these well-known CFD codes shows that the results are comparable with this validated software. The fine mesh solutions appear close to the other results of the flow solvers, but the increase in mesh dimensions ends up with longer computational times. xFLOWg SA results are robust and give reasonable results even if the grid sizes are distinctive.

**References**

1. Tarhan, E., Ortakaya, Y., *TAI – AIAA CFD Drag Prediction Workshop III presentation* http://aaac.larc.nasa.gov/tsab/cfdlarc/aiaa-dpw/, last access on: 15/02/2007
2. Spalart, P. A., Allmaras, S. R., *A One-Equation Turbulence Model for Aerodynamic Flows,* AIAA-92-0439, 30th Aerospace Sciences Meeting and Exhibit, Jan. 6-9, 1992, Reno, NV.
3. http://www.ansys.com/products/icemcfd.asp, *ANSYS, ICEM CFD,* last access on 15/02/2007.
4. Tarhan, E., Korkem B., Şahin P., Ortakaya Y., Tezok F., *Comparison of Panel Methods and a Navier-Stokes Flow Solver on ARA-M100 Wing-Body Configuration,* AIAC 2005, M.E.T.U. Ankara, Turkey.
5. Şahin, P., *Navier-Stokes Calculations over Swept Wings*, M.S., Aerospace Engineering Department, 2006, M.E.T.U., Ankara, Turkey
6. Jameson, A., *Analysis and Design of Numerical Schemes for Gas Dynamics 1: Artificial Diffusion, Upwind Biasing, Limiters and their Effect on Accuracy and Multi-grid Convergence,* International Journal of Computational Fluid Dynamics, Vol. 4, 1995, pp. 171–218.
7. Baldwin B. and Lomax, H., *Thin Layer Approximation and Algebraic Turbulence Model for Separated Flows*, AIAA Paper 78-257, 1978.
8. Spalart, P. R., Jou, W.-H., Stretlets, M., and Allmaras, S. R., *Comments on the Feasibility of LES for Wings and on the Hybrid RANS/LES Approach,* Advances in DNS/LES, Proceedings of the First AFOSR International Conference on DNS/LES, 1997
9. http://www-unix.mcs.anl.gov/mpi/mpich1/, *MPICH Home Page,* last access on 15/02/2007.

Adaptive Aitken-Schwarz for Darcy 3D flow on heterogeneous media

A. Frullone [a], P. Linel[a] and D. Tromeur-Dervout[a b *]

[a]CDCSP/ICJ UMR5208-CNRS, Université Lyon 1, F-69622 Villeurbanne Cedex

[b]INRIA/IRISA/Sage Campus de Beaulieu Rennes, F-35042 Rennes Cedex, France

## 1. Introduction

An adaptive Aitken-Schwarz method is developed to solve Darcy flow problem that follows:

$$\nabla.(k(x)\nabla h(x)) = f(x),\ x \in \Omega \subset R^d, d = \{2,3\}\,,\ Bh = g \text{ on } \partial\Omega \quad (1)$$

where $h(x)$ represents the hydraulic head and $k(x)$ the hydraulic conductivity of the media. $B$ represents formally the boundary conditions. The domain is discretized with Q1-finite elements.

In this paper we develop a method to target a 3D benchmark proposed by ANDRA (the french national agency for the nuclear waste management), where the hydraulic conductivity exhibits strong variations in the geological layers as shown in Table 1.

Even the linear Darcy flow model of the heterogeneous media leads to several types of difficulties as the treatment of various scales leading to solve sparse linear systems of very big size with bad condition number due to the heterogeneous permeability.

Direct solver as MUMPS can be used but their efficiency decreases quite strongly for non regular data structure. Table 2 gives the efficiency for the use of MUMPS direct solver running onto an Altix350 with 16 processors ia64 and 32 Go memory to perform the Andra Benchmark leading to a matrix of size 39000 with 9841790 Q1-elements.

The main idea here is to develop a domain decomposition methodology to benefit from the natural decomposition of the domain in layers. In each layer, the domain is similar to a Poisson problem with constant hydraulic conductivity, leading to a problem better conditioned than the full problem.

The Aitken-Schwarz domain decomposition method (DDM) uses the purely linear convergence property of Schwarz methods for accelerating the solution convergence at the artificial interfaces using the Aitken acceleration technique [11] [12] [10] .

The explicit construction of the error operator at the artificial interfaces with respect to a nonuniform Fourier orthogonal basis associated to the interface discretisation is developed [8][7]. We developed an adaptive construction of the acceleration matrix according to numerical *a posteriori* estimate of the Fourier mode behavior of the solution at artificial interface [6] for non separable operators. Preliminary result on the robustness of the method has been validated on the Darcy problem with large contrasts in the permeability coefficients and on some test problems characterized by typical difficulties in Domain Decomposition.

Let us consider the multidimensional case with the discretized version of the problem (1). We restrict ourselves for simplicity to the two overlapping subdomains case and the Gauss-Seidel Generalized Schwarz Alternated Method algorithm (GSAM). In the multidimensional framework

*This work was partially funded by the GDR MOMAS and the cluster ISLES : Calcul Hautes Performances project of the Region Rhône-Alpes

I.H. Tuncer et al., *Parallel Computational Fluid Dynamics 2007*,
DOI: 10.1007/978-3-540-92744-0_29, © Springer-Verlag Berlin Heidelberg 2009

| Hydrogeologic layers | Thickness [m] | Permeability [m/s] | |
|---|---|---|---|
| | | Regional | Local |
| Tithonian | Variable | 3.10e-5 | 3.10e-5 |
| Kimmeridgian when it outcrops | Variable | 3.10e-4 | 3.10e-4 |
| Kimmeridgian under cover | | 10e-11 | 10e-12 |
| Oxfordian L2a-L2b | 165 | 2.10e-7 | 10e-9 |
| Oxfordian Hp1 Hp4 | 50 | 6.10e-7 | 8.10e-9 |
| Oxfordian C3a-C3b | 60 | 10e-10 | 10e-12 |
| Callovo-Oxfordian Cox | 135 | Kv=10e-14 Kh=10e-12 | |

Table 1
Hydraulic conductivity in the 3D Andra Benchmark

| MUMPS: MUltifrontal Massively Parallel sparse direct Solver | | | | | |
|---|---|---|---|---|---|
| proc(s) | 1 | 2 | 4 | 8 | 16 |
| Elapsed time (s) | 550 | 405 | 211 | 204 | 132 |
| Efficiency (%) | 100 | 68 | 65 | 33 | 26 |

Table 2
Efficiency of the MUMPS direct solver to solve the Andra 3D Benchmark

the main difficulty consists to define a representation of the interface solutions that can produce a cheap approximation of the matrix $P$, the error transfer operator.

For a linear operator one can write the sequel of Schwarz iterates $U^0_{\Gamma^1_h}, U^0_{\Gamma^2_h}, U^1_{\Gamma^1_h}, U^1_{\Gamma^2_h}, ...,$ then the error between two Schwarz iterates behaves linearly. Errors $e^i_{\Gamma^j_h} = U^{i+1}_{\Gamma^j_h} - U^i_{\Gamma^j_h}$ satisfy $\left( e^{i+1}_{\Gamma^1_h} \;\; e^{i+1}_{\Gamma^2_h} \right)^t = P \left( e^i_{\Gamma^1_h} \;\; e^i_{\Gamma^2_h}, \right)^t$ where $P$ represents the iteration error operator at artificial interfaces. The explicit construction of $P$ needs as many iterates as the cumulative size of the degrees of freedom of the artificial interfaces. A band approximation of $P$ can be performed in order to reduce the number of Schwarz iterates. Nevertheless, there is no *a priori* estimate to determine the band width to take into consideration. It is possible to apply a diagonalization of $P$ and to accelerate the solution modes with respect to the eigenvalue basis $V_P$ (see Baranger, Garbey and Oudin [1]).

## 2. Explicit building of $P_{[[.,.]]}$ and Adaptive Aitken-Schwarz

Here, we use a discretisation of the interfaces $\Gamma_h$ with a set $V$ of orthogonal vectors $\Phi_k$ with respect to a discrete Hermitian form $[[.,.]]$. One can consider without loss of generality that these vectors satisfy $[[\Phi_k, \Phi_k]] = 1, \forall k \in \{0, ..., N\}$.

Consider the decomposition of the trace of the Schwarz solution $U_{\Gamma_h}$ with respect to the orthogonal basis $V$: $U_{\Gamma_h} = \sum_{k=0}^{N} \alpha_k \Phi_k$. with $\alpha_k = [[U_\Gamma, \Phi_k]]$. Then the nature of the convergence does not change if we consider the error coefficients in the basis $V$ instead of the error in the physical space. One can write the error components iterations equation but in the coefficient space. Let $\beta^i_{\Gamma^j_h}$ be the coefficients of the error $e^i_{\Gamma^j_h}$. Then, we can write in the coefficient space: $\left( \beta^{i+1}_{\Gamma^1_h} \;\; \beta^{i+1}_{\Gamma^2_h} \right)^t = P_{[[.,.]]} \left( \beta^i_{\Gamma^1_h} \;\; \beta^i_{\Gamma^2_h} \right)^t$. This matrix $P_{[[.,.]]}$ has the same size as the matrix $P$. Nevertheless, we have more flexibility to define some consistent approximation of this matrix, since we have access to *a posteriori* estimate based on the module value of the Fourier coefficients.

The explicit building of $P_{[[.,.]]}$ consists to compute how the basis functions $\Phi_k$ are modified by the Schwarz iterate. Figure 1 describes the steps for constructing the matrix $P_{[[.,.]]}$. Step (a) starts from the the basis function $\Phi_k$ and gets its value on the interface in the physical space. Then step (b) performs two Schwarz iterates with zero local right hand sides and homogeneous boundary conditions on $\partial\Omega = \partial(\Omega_1 \cap \Omega_2)$. Step (c) decomposes the trace solution on the interface in the basis $V$. Thus, we obtains the column $k$ of the matrix $P_{[[.,.]]}$. The full computation of

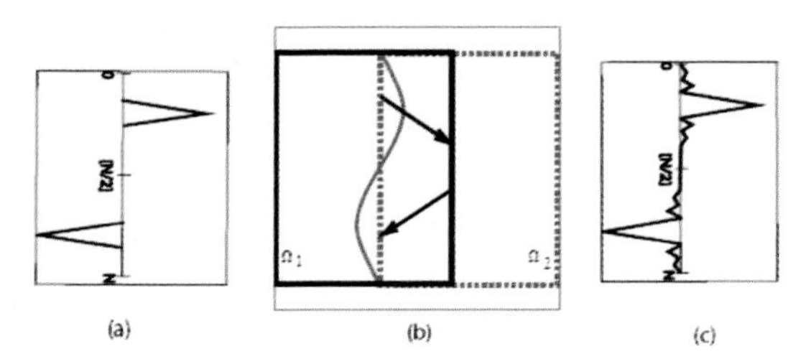


Figure 1. Steps to build the $P_{[[.,.]]}$ matrix

$P_{[[.,.]]}$ can be done in parallel, but it needs as much local subdomain solves as the number of interface points (i.e the size of the matrix $P_{[[.,.]]}$). Its adaptive computation is required to save computing. The Fourier mode convergence gives a tool to select the Fourier modes that slow the convergence and have to be accelerated, contrary to the approach of [9], which builds an approximation of the operator in the physical space relatively to the trace solution on a coarse grid and then compute the Schur eigenvalue decomposition of this approximation. The adaptive Aitken-Schwarz algorithm can be described as follows:

**Algorithm 2.0.1.**

1. *Perform two GSAM iterates.*
2. *Write the difference between two successive iterates in the $V$ basis and select the component modes higher than a fixed tolerance and with the worst convergence slope. Set Index the array of selected modes.*
3. *Take the subset $\tilde{v}$ of $m$ Fourier modes from 1 to max(Index).*
4. *Compute the $m \times m$ ${P^*}_{[[.,.]]}$ matrix associated to the $m$ modes selected which is an approximation of $P_{[[.,.]]}$.*
5. *Accelerate the $m$ modes with the Aitken formula:*$\tilde{v}^{\infty} = (Id - P^*_{[[.,.]]})^{-1}(\tilde{v}^{n+1} - P^*_{[[.,.]]}\tilde{v}^n)$.
6. *Recompose the solution with the $m$ mode accelerated and the $N - m$ other modes of the last iterate solution.*

The arithmetical computing cost is greatly reduced if $m \ll N$. Because the local problems do not change, the use of direct solver or of Krylov methods with projection techniques on Krylov spaces, generated during previous Schwarz iterations as shown in [14], reduces the complexity to $2m \times O(N^2)$.

### 3. Numerical results for nonseparable operator in 2D

Let us consider the elliptic problem (1) with $\Omega = [0,\pi]^2$, $B = I, g = 0$, with a domain decomposition in the $x$-direction into two overlapping subdomains. $k(x,y) = a_0 + (1-a_0)(1 + tanh((x - (3 * h_x * y + 1/2 - h_{ref}))/\epsilon_a))/2$, and $a_0 = 10$,
$\epsilon_a = 10^{-2}$, taking into account a jump not parallel to the interface between subdomains (see figure 2 (left)). The right hand side $f$ is such that the exact solution is:
$u(x,y) = 150x(x-\pi)y(y-\pi)(y-\pi/2)/\pi^5$. We use a Cartesian grid with $90 \times 90$ elements, uniform in $x$ and nonuniform in $y$ (the maximum distance between the grid points in the $y$-direction is of the order of $2h_{ref}$ , ( $\epsilon = h_{ref}/2$)). Figure2 (right) exhibits the convergence of the

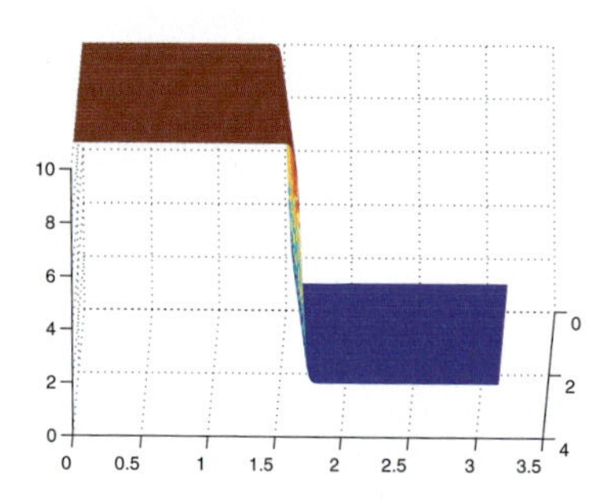

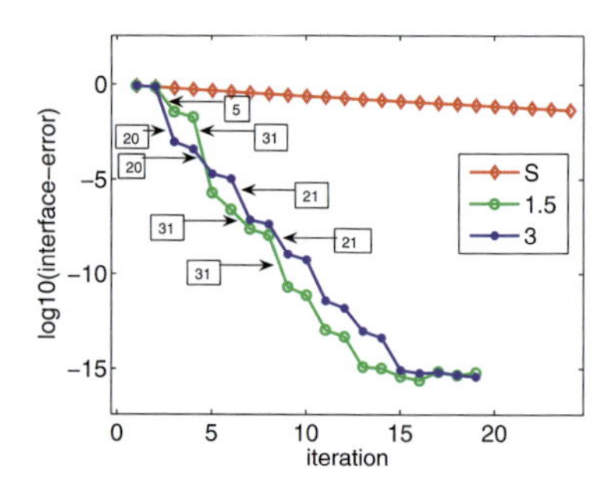


Figure 2. The function $k(x,y)$ of the test problem, showing the jump not parallel to the interfaces (left).Adaptive acceleration using sub-blocks of $P_{[[.,.]]}$, with 100 points on the interface, overlap= 1, $\epsilon = h_{ref}/8$ and Fourier modes tolerance $= ||\hat{u}^k||_\infty/10^i$ for $i = 1.5$ and 3 for 1st iteration and $i = 4$ for successive iterations (right).

Aitken-Schwarz method with respect to the number of modes kept to build the reduced matrix $\tilde{P}_{[[.,.]]}$ in accordance to algorithm (2.0.1).

Next result shows the potentiality of the Aitken-Schwarz domain decomposition method to simulate the transport of fluid solutions in heterogeneous underground media. Figure 3 represents the results obtained for $k(x,y) \in [0.0091, 242.66]$, following a lognormal data distribution. It exhibits the convergence with respect to the Fourier modes kept to build adaptively the $P_{[[.,.]]}$ matrix.

### 4. Numerical analysis for the Neumann-Dirichlet algorithm with 3 subdomains

This section illustrates the concept of linear convergence of the Schwarz Neumann-Dirichlet algorithm and the choice of the interface representation that should be accelerate. Consider the multiplicative Schwarz algorithm with Neumann-Dirichlet boundary conditions to solve the 1D Poisson problem with 3 non overlapping subdomains splitting the domain $[\alpha,\Gamma_1] \cup [\Gamma_1,\Gamma_2] \cup [\Gamma_2,\beta]$, where $\Gamma_1 < \Gamma_2$.

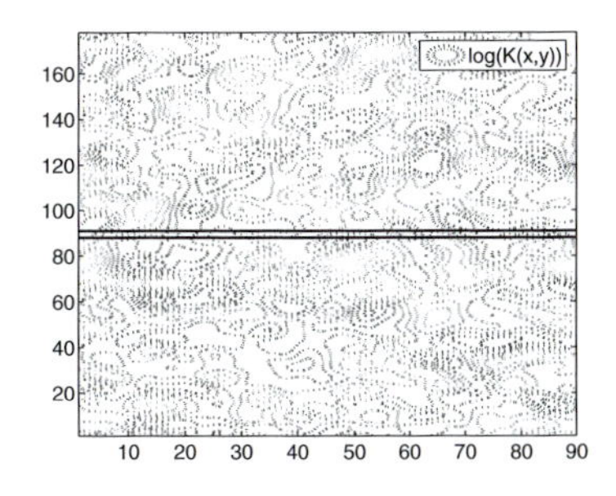


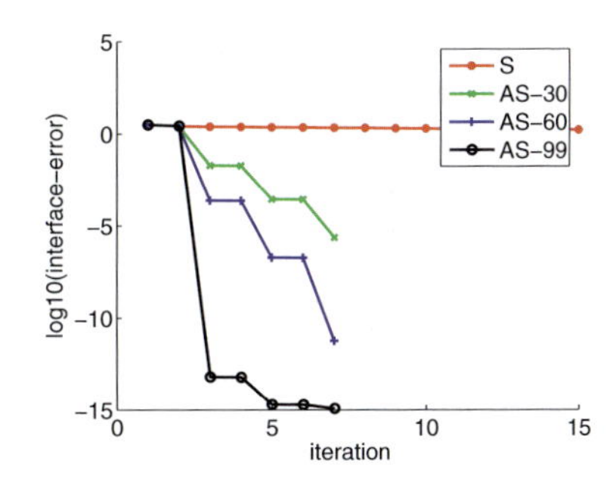


Figure 3. Data distribution for the 2D Darcy problem with $k(x,y) \in [0.0091, 242.66]$ (left) and convergence of the Aitken-Schwarz domain decomposition with respect to the Fourier modes kept to build adaptively $P_{[[,]]}$ (right).

Then the Schwarz algorithm writes:

$$\left\{\begin{array}{l}\Delta u_1^{(j)} = f \text{ on } [\alpha, \Gamma_1]\\ u_1^{(j)}(\alpha) = 0\\ u_1^{(j)}(\Gamma_1) = u_1^{(j-1/2)}(\Gamma_1)\end{array}\right., \quad \left\{\begin{array}{l}\Delta u_2^{(j+1/2)} = f \text{ on } [\Gamma_1, \Gamma_2]\\ \dfrac{\partial u_2^{(j+1/2)}(\Gamma_1)}{\partial n} = \dfrac{\partial u_1^{(j)}(\Gamma_1)}{\partial n}\\ u_2^{(j+1/2)}(\Gamma_2) = u_3^{(j)}(\Gamma_2)\end{array}\right., \quad \left\{\begin{array}{l}\Delta u_3^{(j)} = f \text{ on } [\Gamma_2, \beta]\\ \dfrac{\partial u_3^{(j)}(\Gamma_2)}{\partial n} = \dfrac{\partial u_2^{(j-1/2)}(\Gamma_2)}{\partial n}\\ u_3^{(j)}(\beta) = 0\end{array}\right..$$

The error on subdomain $i$ writes $e_i(x) = c_i x + d_i$. For subdomains 1 and 3 we have

$$\begin{pmatrix}\alpha & 1 & 0 & 0\\ \Gamma_1 & 1 & 0 & 0\\ 0 & 0 & \beta & 1\\ 0 & 0 & 1 & 0\end{pmatrix}\begin{pmatrix}c_1\\ d_1\\ c_3\\ d_3\end{pmatrix} = \begin{pmatrix}0\\ e_2^{(j-1/2)}(\Gamma_1)\\ 0\\ \frac{\partial e_2^{(j-1/2)}(\Gamma_2)}{\partial n}\end{pmatrix} = \begin{pmatrix}0\\ e_2 g_1^{(j-1/2)}\\ 0\\ de_2 g_2^{(j-1/2)}\end{pmatrix} \tag{2}$$

This equation gives

$$e_1^{(j)}(x) = -\frac{e_2 g_1^{(j-1/2)}}{\alpha - \Gamma_1} x + \frac{\alpha e_2 g_1^{(j-1/2)}}{\alpha - \Gamma_1} \tag{3}$$

$$e_3^{(j)}(x) = de_2 g_2^{(j-1/2)} x - \beta de_2 g_2^{(j-1/2)} \tag{4}$$

Error on the second subdomain satisfies

$$\begin{pmatrix}1 & 0\\ \Gamma_2 & 1\end{pmatrix}\begin{pmatrix}c_2\\ d_2\end{pmatrix} = \begin{pmatrix}\frac{\partial e_1^{(j)}(\Gamma_1)}{\partial n}\\ e_3^{(j)}(\Gamma_2)\end{pmatrix} = \begin{pmatrix}de_1 g_1^{(j)}\\ e_3 g_2^{(j)}\end{pmatrix} \tag{5}$$

$$e_2^{(j+1/2)}(x) = de_1 g_1^{(j)} x - de_1 g_1^{(j)} \Gamma_2 + e_3 g_2^{(j)} \tag{6}$$

Replacing $e_3 g_2^{(j)}$ and $de_1 g_1^{(j)}$, $e_2^{(j+1/2)}(x)$ writes:

$$e_2^{(j+1/2)}(x) = -\frac{x - \Gamma_2}{\alpha - \Gamma_1} e_2 g_1^{(j-1/2)} + (\Gamma_2 - \beta) de_2 g_2^{(j-1/2)} \tag{7}$$

Consequently, the following identity holds:

$$\begin{pmatrix} e_2 g_1^{(j)} \\ de_2 g_2^{(j)} \end{pmatrix} = \begin{pmatrix} \dfrac{\Gamma_2 - \Gamma_1}{\alpha - \Gamma_1} & \Gamma_2 - \beta \\ \dfrac{-1}{\alpha - \Gamma_1} & 0 \end{pmatrix} \begin{pmatrix} e_2 g_1^{(j-1)} \\ de_2 g_2^{(j-1)} \end{pmatrix} \tag{8}$$

Consequently the matrix do not depends of the solution, neither of the iteration, but only of the operator and the shape of the domain. This why the convergence (eventually the divergence is purely linear). More generally speaking, this property is related to the Neumann to Dirichlet map associated to the problem (see [13].

Figure 4 shows the linear convergence of the Schwarz algorithm on the second subdomain with $\alpha = 0$, $\beta = 1$, $\Gamma_1 = 0.44$ ,$\Gamma_2 = 0.7$.

Similar results are obtained in 3D with a finite element Q1 discretisation of the domain. Following the 1D results, the interface is constituted by the vector where the first components are the Dirichlet boundary condition and the last components the Neumann boundary conditions onto the second domain in the Fourier space. The matrix $P$ is approximatively build with considering only the lowest Fourier modes of each condition.

Figure 5 exhibits the convergence on the second subdomain and the Aitken acceleration.

## 5. Conclusions

We present an adaptive Aitken-Schwarz domain decomposition strategy that can be applied to solve Darcy flow in heterogenous media. Some numerical experiment in 2D validate our approach. It perform also well on 3D problems with regular mesh interfaces between the layer. Some extension of the methodology to handle non regular meshing onto the interface are under progress.

## REFERENCES


1. J. Baranger, M. Garbey, and F. Oudin-Dardun. Generalized aitken-like acceleration of the schwarz method. In *Lecture Notes in Computational Science and Engineering*, volume 40, pages 505–512, 2004.
2. J. Baranger, M. Garbey, and F. Oudin-Dardun. The Aitken-like Acceleration of the Schwarz Method on Non-Uniform Cartesian Grids. Technical Report UH-CS-05-18, University of Houston, July 21, 2005.
3. Nicolas Barberou, Marc Garbey, Mathias Hess, Mickael Resch, Tuomo Rossi, Jari Toivanen, and Damien Tromeur-Dervout. Efficient meta-computing of elliptic linear and non linear problems. *J. of Parallel and Distributed Computing*, (63(5)):564–577, 2003.
4. Isabelle Boursier, Damien Tromeur-Dervout, and Yuri Vassilevsky. Aitken-schwarz methods with non matching finite elements andspectral elements grids for the parallel simulation of anunderground waste disposal site modelized upscaling. In G. Winter, E. Ecer, J. Periaux, N. Satofuka, and P. Fox, editors, *Proc. Int. Conf. PCFD04*, pages 69–76. Elsevier, 2005. ISBN 0-444-52024-4.
5. Dan Gabriel Calugaru and Damien Tromeur-Dervout. Non-overlapping DDMs of Schwarz type to solve flow in discontinuous porous media. In R. Kornhuber, R. Hoppe, J. Périaux, o. Pironneau, O. Widlund, and J. Xu, editors, *Domain Decomposition Methods in Science and Engineering*, volume 40 of *Lecture Notes in Computational Science and Engineering*, pages 529–536. Springer, 2004. ISBN: 3-540-22523-4.
6. A. Frullone and D. Tromeur-Dervout. Adaptive acceleration of the Aitken-Schwarz Domain Decomposition on nonuniform nonmatching grids. *preprint cdcsp-0700*, 2007. submitted.

7. A. Frullone and D. Tromeur-Dervout. Non uniform discrete fourier transform for the adaptive acceleration of the aitken-schwarz ddm. In U. Langer, M. Discacciati, D. Keyes, O. Widlund, and W. Zulehner, editors, *Domain Decomposition Methods in Science and Engineering XVII*, volume 60 of *Lecture Notes in Computational Science and Engineering*. springer, 2008.
8. Andrea Frullone and Damien Tromeur-Dervout. A new formulation of NUDFT applied to Aitken-Schwarz DDM on Nonuniform Meshes. In A. Deane and al, editors, *Proc. Int. Conf. PCFD05*, pages 493–500. Elsevier, 2006. ISBN-10: 0-444-52206-9.
9. M. Garbey. Acceleration of the Schwarz Method for Elliptic Problems. *SIAM J. Sci. Comput.*, 26:1871–1893, 2005.
10. M. Garbey and D. Tromeur-Dervout. On some Aitken-like acceleration of the Schwarz method. *Internat. J. Numer. Methods Fluids*, 40(12):1493–1513, 2002. LMS Workshop on Domain Decomposition Methods in Fluid Mechanics (London, 2001).
11. Marc Garbey and Damien Tromeur-Dervout. Operator Splitting and Domain Decomposition for Multicluster. In D. Keyes, A. Ecer, N. Satofuka, P. Fox, and J. Periaux, editors, *Proc. Int. Conf. Parallel CFD99*, pages 27–36, williamsburg, 2000. North-Holland. (Invited Lecture) ISBN 0-444-82851-6.
12. Marc Garbey and Damien Tromeur-Dervout. Two Level Domain Decomposition for Multicluster. In H. Kawarada T. Chan, T. Kako and O. Pironneau, editors, *Proc. Int. Conf. on Domain Decomposition Methods DD12*, pages 325–340. DDM org, 2001.
13. D. Tromeur-Dervout. *Domain decomposition methods: theory and applications, F. Magoules and T. Kako editors*, volume 25 of *GAKUTO Internat. Ser. Math. Sci. Appl.*, chapter Aitken-Schwarz method: acceleration of the convergence of the Schwarz method, pages 37–64. Gakkotosho, Tokyo, 2006.
14. Damien Tromeur-Dervout and Yuri Vassilevski. Choice of initial guess in iterative solution of series of systems arising in fluid flow simulations. *J. Comput. Phys.*, 219(1):210–227, 2006.

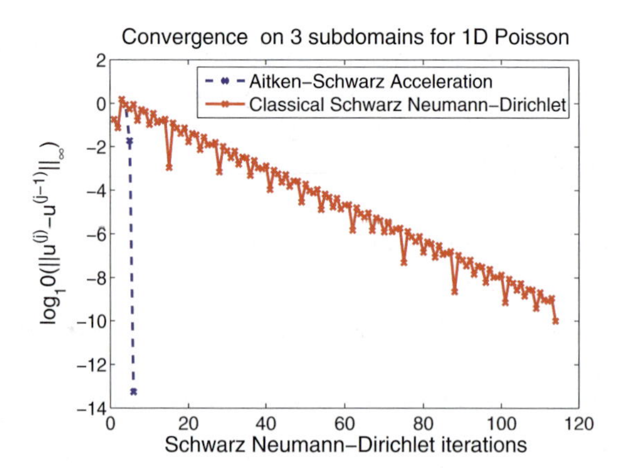


Figure 4. Convergence of the non overlapping Schwarz algorithm with Neumann-Dirichlet boundary condition for 1D Poisson problem with 3 subdomains

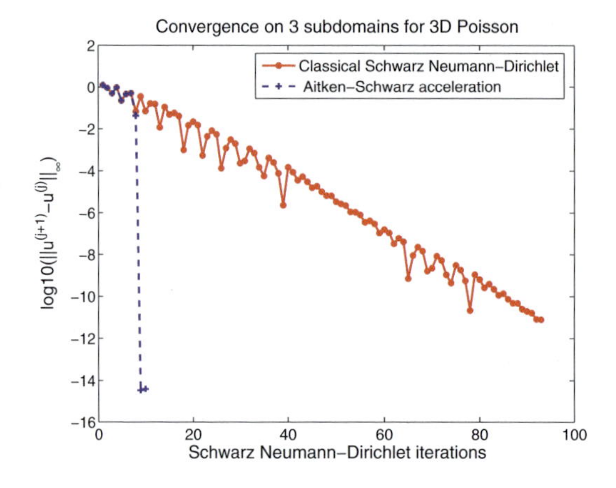


Figure 5. Convergence on the subdomain 2 of the non overlapping Schwarz algorithm with Neumann-Dirichlet boundary condition for 3D Poisson problem with 3 subdomains

# Numerical Simulation of Compressible Flow using Three-Dimensional Unstructured Added/Eliminated Grid Method

**Masashi Yamakawa[a] and Kenichi Matsuno[a]**

[a]*Division of Mechanical and System Engineering, Kyoto Institute of Technology, Matsugasaki, Sakyo-ku, Kyoto 606-8585, Japan*

A geometric conservation law should be considered on moving boundary problems in body-fitted coordinated grid system. In this paper, as moving grid problem, three-dimensional unstructured mesh with add and eliminated grid system is dealt with. In using add and eliminated grid method, the unstructured moving grid finite volume method is adopted. In this case, a control volume is treated as four-dimensional space-time unified domain. So, a procedure of calculation is relatively complicatedly, espacially, in the case of parallel computation. In this paper, parallelization with OpenMP of the computation is estimated.

Keywords: Moving-Grid; Unstructured Grid; OpenMP Parallelization

## 1. INTRODUCTION

Our interest is a numerical simulation of flows around moving and deforming bodies. Especially, in the case that a body moves long-distance and in the case of that a shape of a body is large deformed, they attracted attention also in a industrial field. To simulate such a flow field in the body-fitted coordinated system, great skewness of gridline may occur according to the motion of the body. In this case, there are two problems for calculated mesh. One is a conservation of flow parameter on moving grid, so it is how to satisfy a geometric conservation law[1] in addition to satisfy a physical conservation law. Another is instability or breakdown of calculation by skewness and overlap of gridline. For the former issue, we have already proposed the scheme which satisfy both conservation laws on moving unstructured mesh, it is the Unstructured moving-grid finite-volume method[2]. The method adopts a control volume in a space-time unified domain on the unstructured grid system, then it is implicit and is solved iteratively at every time-step. The latter issue is how to prevent generating skew and overlapped gridline when grid moves. Especially, if it is going to maintain the topology of grid,


DOI: 10.1007/978-3-540-92744-0_30, © Springer-Verlag Berlin Heidelberg 2009

solution of problem will become very difficult. So, it should be adding and eliminating the grid-points according to a bias of distribution of grid-points. Then the skewness and overlap of grid will be avoidable simply. In this case, it is necessary to satisfy the conservation laws for the moving grid also for the added/eliminated grid. Although the combination of two problems is very difficult, in two-dimensional system, the solution for added/eliminated grid already has been shown. In this paper, there arc two objectives. First, in a three-dimensional system, we propose the method that physical and geometric conservation laws are satisfied in added/eliminated grid system. Second, A numerical simulation of flow around a body is computed using the added/eliminated grid method in OpenMP parallel environment. Then, it is applied for three-dimensional compressible flow in an Unstructured grid system.

## 2. THREE-DIMENSIONAL UNSTRUCTURED MOVING-GRID FINITE-VOLUME METHOD WITH ADDED/ELIMINATED GRID

### 2.1. Governing Equation

The Euler equation in three-dimensional coordinate system is dealt with, in this paper. Then the equation can be written in the conservation law form as follow:

$$\frac{\partial q}{\partial t}+\frac{\partial E}{\partial x}+\frac{\partial F}{\partial y}+\frac{\partial G}{\partial z}=0, \tag{1}$$

where

$$q=\begin{pmatrix}\rho\\ \rho u\\ \rho v\\ \rho w\\ e\end{pmatrix},\quad E=\begin{pmatrix}\rho u\\ \rho u^2+p\\ \rho uv\\ \rho uw\\ u(e+p)\end{pmatrix},\quad F=\begin{pmatrix}\rho v\\ \rho uv\\ \rho v^2+p\\ \rho vw\\ v(e+p)\end{pmatrix},\quad G=\begin{pmatrix}\rho w\\ \rho uw\\ \rho vw\\ \rho w^2+p\\ w(e+p)\end{pmatrix}. \tag{2}$$

The unknown variables $\rho$, $u$, $v$, $w$, $p$ and $e$ represent the gas density, velocity components in the $x$, $y$ and $z$ directions, pressure, and total energy per unit volume, respectively. The working fluid is assumed to be perfect, and the total energy $e$ per unit volume is defined by

$$e=\frac{p}{\gamma-1}+\frac{1}{2}\rho\left(u^2+v^2+w^2\right), \tag{3}$$

where the ratio of specific heats $\gamma$ is typically taken as being 1.4.

**2.2. Description of the scheme**

In the case of dealing with moving boundary problem on body fitted coordinated system, a grid system must dynamically change and deform its shape according to the movement of boundary. Then, one of the most important issue in the grid system is how to satisfy a geometric conservation law on the moving grid. To solve this problem, in this section, the unstructured moving-grid finite-volume method in three-dimensional coordinate system is explained. The method is able to satisfy the geometric conservation law as well as a physical conservation law.

At first, the Euler equation in conservation form Eq.(1) can be written in divergence form:

$$\widetilde{\nabla}\widetilde{\mathbf{F}} = 0 , \tag{4}$$

where

$$\widetilde{\mathbf{F}} = E\vec{\mathbf{e}}_x + F\vec{\mathbf{e}}_y + G\vec{\mathbf{e}}_z + q\vec{\mathbf{e}}_t , \tag{5}$$

$$\widetilde{\nabla} = \vec{\mathbf{e}}_x \frac{\partial}{\partial x} + \vec{\mathbf{e}}_y \frac{\partial}{\partial y} + \vec{\mathbf{e}}_z \frac{\partial}{\partial z} + \vec{\mathbf{e}}_t \frac{\partial}{\partial t} . \tag{6}$$

Here $\vec{\mathbf{e}}_x$, $\vec{\mathbf{e}}_y$, $\vec{\mathbf{e}}_z$ and $\vec{\mathbf{e}}_t$ are unit vectors in $x$, $y$, $z$ and $t$ directions respectively.

Now, we consider a change of shape of element by moving boundary. When time step advances from $n$ to $n$+1, the element is moved as shown in Fig.1.

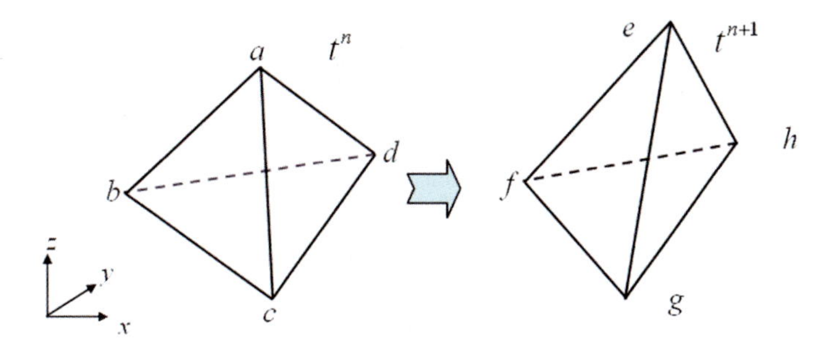


Figure 1. Movement of element

In the three-dimensional coordinate system, we deal with the control-volume as four-dimensional space-time unified domain ($x$, $y$, $z$, $t$). Then, the method is based on a cell-centered finite-volume method, thus, flow variables are defined at the center of cell in unstructured mesh. The unified domain is shown in Fig.2.

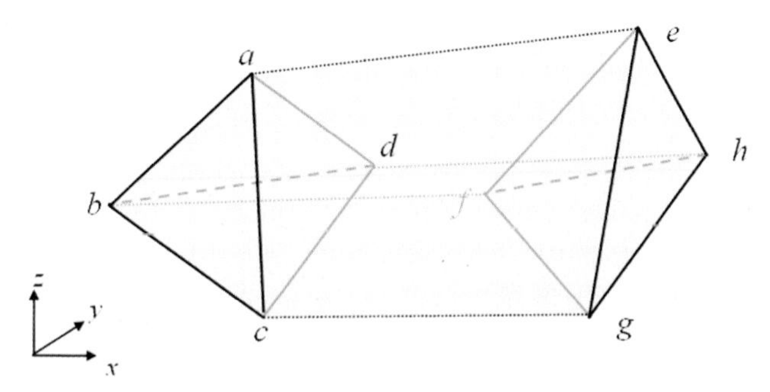


Figure 2. Control-volume in space-time unified domain

$n$-time step tetrahedron is made by four-grid points( $a, b, c, d$ ). And $n$+1-time step tetrahedron is made by other grid points( $e, f, g, h$ ). Then, the space-time unified control-volume is formed by moving of tetrahedron from n-time step to n+1 time step. So, the control-volume is constructed by eight grid points( $a, b, c, d, e, f, g, h$ ).

We apply volume integration to Eq.(4) with respect to this control volume. Then Eq.(4) can be written in surface integral form, using the Gauss theorem:

$$\int_V \tilde{\nabla}\tilde{\mathbf{F}} dV = \oint_S \tilde{\mathbf{F}} \cdot \mathbf{n} dS = \sum_{l=1}^{6} \left( En_x + Fn_y + Gn_z + qn_t \right)_l = 0 \tag{7}$$

where $\mathbf{n}$ is a outward unit normal vector of control volume boundary. $V$ is the polyhedron control volume and $S$ is its boundary. $\mathbf{n}_l = ( \mathbf{n}_x, \mathbf{n}_y, \mathbf{n}_z, \mathbf{n}_t )_l$ ( $l = 1,2,\ldots 6$ ) is the normal vector of control volume boundary, and the length of the vector equals to the volume of the boundary. For example at $l = 1$, the volume of the boundary is constructed by six grid points ( $a, b, c, e, f, g$ ). This is formed by sweeping a triangle ( $a, b, c$ ) to next time-step triangle ( $e, f, g$ ). The components $\mathbf{n}_x$, $\mathbf{n}_y$, $\mathbf{n}_z$, and $\mathbf{n}_t$ are unit normal vectors in $x$, $y$, $z$ and $t$ directions respectively, and these components are expressed by the combination of a suitable inner product and a suitable outer product.

Next, the case of added grid is described. In this case, addition of elements is expressed as division of elements with added grid-points. When a new grid-point is placed on the center of the gridline, one calculated element is divided to two elements as shown in Fig.3.

Then for all the elements which share the gridline (a-$c$: Fig.3.) by which a new point has been placed, the divisions are carried out simultaneously. For one of divided elements, the unstructured moving-grid finite-volume method is applied to satisfy the geometric conservation law. In three-dimensional system, the method is featured treatment of a control volume on the space-time unified domain ($x$, $y$, $z$, $t$), which is four-dimensional system, too. Thus, a change of number of grid-points for time-axis must be taken into consideration as well as a change of positions of the grid-points.

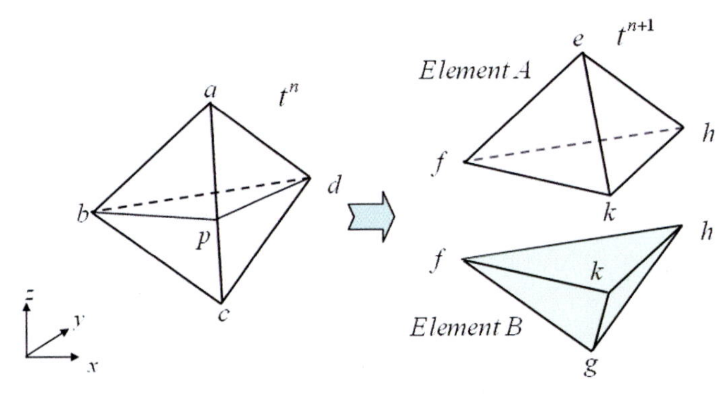


Figure 3. Division of element

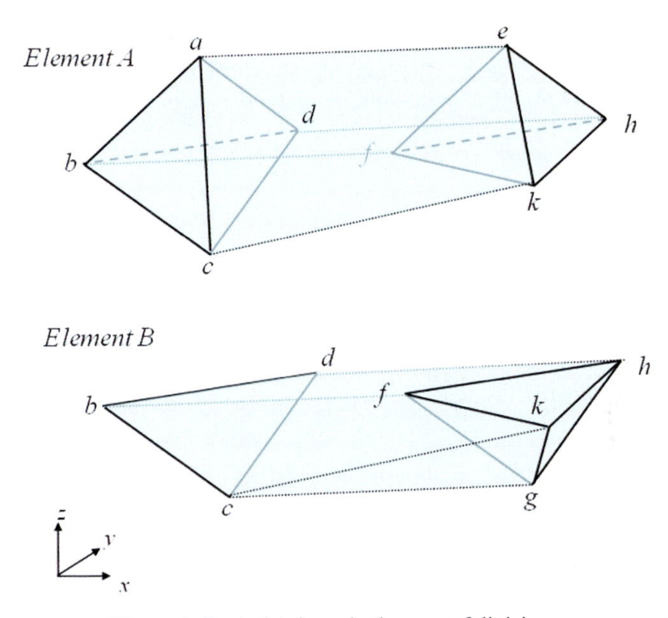


Figure 4. Control volume in the case of division
( Upper: for the element A, Lower: for the element B )

As element A in Fig.3., in this case, the change for time-axis is position of grid-points only. So the cell is tetrahedron in three-dimensional $(x, y, z)$-domain. When grid moves, the control volume becomes a complicated polyhedron in the $(x, y, z, t)$-domain as shown in Fig.3. It is formed when the tetrahedron sweeps space which is built by grid points *a, b, c, d, e, f, h, k* in Fig.4(upper). Then, descritization is same to equation(7). On the other hand, as element B is considered the element which appeared suddenly at n+1 time-step by placing the new grid-point. In this case, for satisfaction of conservation, the control volume is defined as a space which is formed by sweeping of the element from n time-step to n+1 time step. Here, the element changes its shape from triangle-plate to tetrahedron-volume. So, the control volume is built by grid points *b, c, d, f, g, h, k* in Fig.4(lower). In this case, the volume at n-step is zero, thus the descritization become to equation(8).

$$q^{n+1}(n_t)_5 + \sum_{l=1}^{4} \left\{ \left( E^{n+1/2}, F^{n+1/2}, G^{n+1/2}, q^{n+1/2} \right) \cdot \mathbf{n} \right\}_l = 0 \qquad (8)$$

Next, the case of eliminated grid is described. In this case, elimination of element is expressed as union of elements with eliminated grid-point. Now, disappeared element is defined as element A in Fig.5. It is eliminated by moving grid-point from position *p* to position *c* in Fig.5, so it like to be merged by adjoining element B. Thus, the disappearance of element occurs by union of elements. Then for all the elements which share the gridline( *p-c* : Fig.5. ) the merging are carried out simultaneously, too. Also the unstructured moving-grid finite-volume method is applied to satisfy the geometric conservation law.

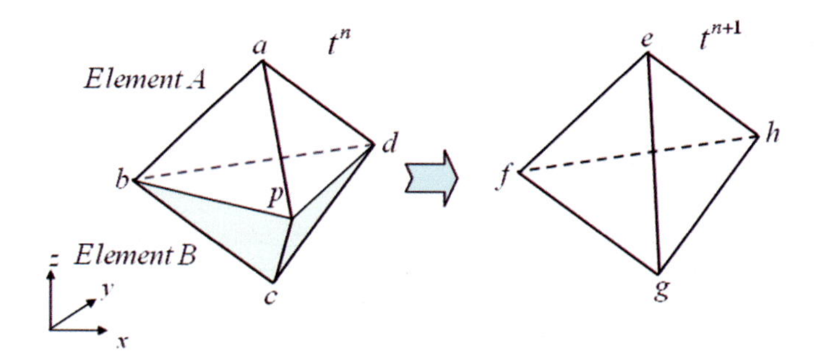


Figure 5. Union of element

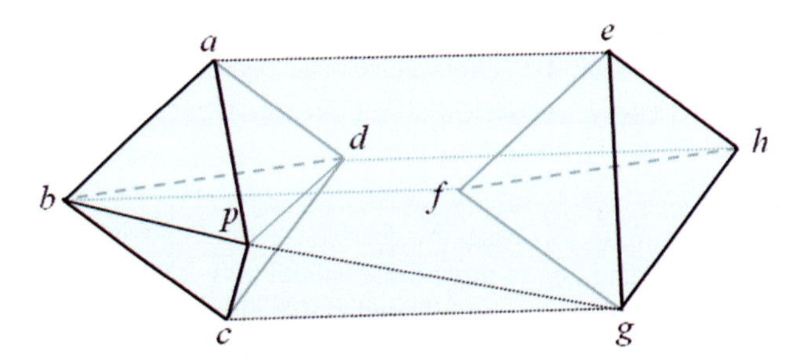


Figure 6. Control volume in the case of union

When time-step advances from $n$ step to $n$+1 step, the volume of the element A become to zero. Thus, the element A and B must be considered as union. When the pentahedron which build by $a$, $b$, $c$, $e$, $p$ sweeps space and changes a shape of itself to tetrahedron, the control volume is formed. Then the control volume is built by grid points $a$, $b$, $c$, $e$, $p$, $e$, $f$, $g$, $h$ in Fig.6. And the descritization become to equation(9).

$$q^{n+1}(n_t)_8 + q^n(n_t)_7 + \sum_{l=1}^{6}\{(E^{n+1/2}, F^{n+1/2}, G^{n+1/2}, q^{n+1/2})\cdot \mathbf{n}\}_l = 0\ , \tag{9}$$

where $q^n(n_t)_7$ is build by $q_A{}^n(n_t)$ and $q_B{}^n(n_t)$.
Here, at $l = 4$ and 5, the control volumes have only $\mathbf{n}_t$ component and correspond to the volumes in the ( $x$, $y$, $z$ )-space at time $t^{n+1}$ and $t^n$ themselves respectively. Thus, Eq.(7) can be expressed as,

$$q^{n+1}(n_t)_6 + q^n(n_t)_5 + \sum_{l=1}^{4}\{(E^{n+1/2}, F^{n+1/2}, G^{n+1/2}, q^{n+1/2})\cdot \mathbf{n}\}_l = 0\ . \tag{8}$$

Here, the conservative variable vector and flux vector at ($n$+1/2)-time step are estimated by the average between $n$-time and ($n$+1)-time steps. Thus, for example, $E^{n+1/2}$ can be expressed as,

$$E^{n+1/2} = (E^n + E^{n+1})/2\,. \tag{9}$$

The flux vectors are evaluated using the Roe flux difference splitting scheme [3] with MUSCL approach, as well as the Venkatakrishnan limiter [4].
The method uses the variable at (n+1)-time step, and thus the method is completely implicit. We introduce sub-iteration strategy with a pseudo-time approach [5] to solve the implicit algorithm. Now by defining that the operating function L($q^{n+1}$) as Eq.(10), the pseudo-time sub-iteration is represented as Eq.(11).

$$L(q^{n+1}) = \frac{1}{\Delta t(n_t)_6}[q^{n+1}(n_t)_6 + q^n(n_t)_5 + \sum_{l=1}^{4}\{(E^{n+1/2}, F^{n+1/2}, G^{n+1/2}, q^{n+1/2})\cdot n\}_l\,], \tag{10}$$

$$\frac{dq^{n+1\langle\nu\rangle}}{d\tau} = -L(q^{n+1\langle\nu\rangle})\,, \tag{11}$$

where $\nu$ is index for iteration, and $\tau$ is pseudo-time step. To solve Eq.(11), we adopted the Runge-Kutta scheme to advance $\Delta\tau$ pseudo-time step. When $\nu$ inner iteration is converged, we can get ($n$+1)-time step solution, $q^{n+1}$.

## 3. NUMERICAL RESULTS OF PARALLELIZATION

### 3.1. Parallel Environment

The parallel computation of the present method is carried on for test problem using OpenMP library on PC. The computer has two Intel Xeon E5160 processors and 4GB shared memory. The operating system is WinXP, fortran compiler is Intel Fortran 8.1.

### 3.2. Parallel Efficiency

On the test problem, an error for an uniform flow is estimated on three-dimensional added and eliminated grid system. Then, a results of the estimating is shown in Table 1. However, we cannot get enough speed up in these environment.

Table 1. Parallel computing performance

| Number of elements | 5,000 | 50,000 | 100,000 | 200,000 |
|---|---|---|---|---|
| Speed up | 0.990 | 1.042 | 1.030 | 1.014 |

## 4. CONCLUSIONS

In this paper, three-dimensional unstructured moving-grid finite-volume method with added and eliminated grid system is formulated and estimated in OpenMP parallel environment. The numerical computation using OpenMP parallelization is performed for some deferent number of elements. However, the efficiency of parallelization was not enough obtained in any case. Although at initial step, it was considered that the problem was related to shortage of the number of elements, the algorithm should be improved in the calculated code. In future work, a sufficient efficiency algorithm also including MPI parallelization should be studied.

## REFERENCES


1. S. Obayashi, "Freestream Capturing for Moving Coordinates in Three Dimensions" , *AIAA Journal*, Vol.30, pp.1125-1128, (1992)
2. M. Yamakawa and K. Matsuno, Unstructured Moving-Grid Finite-Volume Method for Unsteady Shocked Flows, *Journal of Computational Fluids Engineering*, 10-1 (2005) pp.24-30.
3. P. L. Roe, Approximate Riemann Solvers Parameter Vectors and Difference Schemes, *Journal of Computational Physics*, 43 (1981) pp.357-372.
4. V. Venkatakrishnan, On the Accuracy of Limiters and Convergence to Steady State Solutions, *AIAA Paper*, 93-0880, (1993).
5. C. L. Rumsey, M. D. Sanetrik, R. T. Biedron, N. D. Melson, and E. B. Parlette, Efficiency and Accuracy of Time-Accurate Turbulent Navier-Stokes Computations, *Computers and Fluids*, 25 (1996) pp. 217-236.

# Technology of parallelization for 2D and 3D CFD/CAA codes based on high-accuracy explicit methods on unstructured meshes

**Andrey V. Gorobets, Ilya V. Abalakin, Tatiana K. Kozubskaya**

*Institute for Mathematical Modeling of Russian Academy of Science, 4A, Miusskaya Sq., Moscow, 125047, Russia*

Keywords: Parallel; Aeroacoustics; Unstructured; High-order; Liners; Resonator;

**Summary**

The paper is devoted to the parallelization of existing high-accuracy sequential CFD/CAA codes which become inapplicable to majority of the modern engineering problems due to a single CPU performance limitation. An efficient parallel upgrade gets especially complicated in case of extensive irregular approximation stencils used in explicit algorithms of high accuracy order. The parallelization technology developed is applied to the code NOISEtte for solving 2D and 3D gas dynamics and aeroacoustics problems using unstructured meshes and high-order (up to 6) explicit algorithms. The parallel version of NOISEtte is used for modeling resonator-type problems considered as typical problems of nonlinear aeroacoustics. In particular, the paper presents 2D and 3D computational experiments on simulating acoustic liners which are used for sound suppression in jet engines.

## 1. Introduction

It is well known that many of modern CFD/CAA (computational aeroacoustics) applications require high accuracy numerical algorithms and fine space and time discretization. This leads to such an amount of computations that can only be carried out using parallel computer systems.

Parallel CFD codes can be designed in different ways. Parallelism can be incorporated from the very beginning, so the code is originally designed for parallel systems. Another option is parallelization of a sequential code that was originally designed for computations on a single CPU. The second approach is important because

I.H. Tuncer et al., *Parallel Computational Fluid Dynamics 2007*,
DOI: 10.1007/978-3-540-92744-0_31, © Springer-Verlag Berlin Heidelberg 2009

it allows sequential codes (which are robust and well developed but obsolete due to single CPU limitations) to be adapted for parallel computations. The paper is devoted to the second approach and describes specifics of the parallelization technology for codes based on explicit algorithms.

A specificity of the high-order explicit algorithms needed in modern engineering applications is an extensive space stencil that consists of a large number of nodes and has an irregular structure in case of non-structured meshes. It makes the parallelization more complicated because a lot of nodes are included in data exchange, which strongly depends on a stencil structure. In spite of these difficulties, the parallelization must satisfy the main requirements like minimization of man-hour needed, maximization of parallel efficiency and minimization of differences between original and parallel code.

The technology of parallelization developed under these requirements has been applied to the high-order explicit DNS code NOISEtte intended for solving aeroacoustic problems using non-structured triangular meshes for 2D cases and tetrahedral meshes for 3D cases [1]. The code was originally created by IMM RAS (Russia) and INRIA (France). It provides up to 6$^{th}$ order of accuracy that can be reached on "Cartesian" areas of the mesh.

A parallel version of the code has been used for solving a set of model resonator-type problems considered as basic in numerical simulation of acoustic liners [3]. The problems of that kind are typical for nonlinear aeroacoustics. First, they are characterized by a great difference in determining scales that explains a necessity of highly refined meshes and, consequently, high computational costs. Second, as all the aeroacoustics problems these problems require high accuracy algorithms to provide a proper resolution of acoustic waves. Both the first and the second properties make the parallel implementation vitally needed and motivate a choice of the resonator-type problems, in order to demonstrate the capacities of the parallel NOISEtte 2D&3D.

## 2. Overview of numerical algorithm

The following three basic models are used in computational experiments considered:
1. Navier-Stokes equations;
2. Euler equations;
3. Linearized Euler equations;

The system of equations for any of these models can be represented in a generalized form:

$$\frac{\partial \mathbf{Q}}{\partial t}+\frac{\partial \mathbf{F}}{\partial x}+\frac{\partial \mathbf{G}}{\partial y}+\frac{\partial \mathbf{H}}{\partial y}=\frac{1}{\mathrm{Re}}\left(\frac{\partial \mathbf{F}_{NS}}{\partial x}+\frac{\partial \mathbf{G}_{NS}}{\partial y}+\frac{\partial \mathbf{H}_{NS}}{\partial y}\right), \tag{1}$$

where $\mathbf{Q}$ – is a vector of full or linearized conservative variables, $\mathbf{F},\mathbf{G},\mathbf{H}$ – vectors of full or linearized conservative fluxes, $\mathbf{F}_{NS},\mathbf{G}_{NS},\mathbf{H}_{NS}$ – vectors of full or linearized dissipative fluxes, Re – Reynolds number. A method of high order approximation for nonlinear aeroacoustics problems on unstructured meshes [4-6] is used for numerical solution of (1) regardless of model. The method was developed on a basis of a finite-volume approach and a 2$^{nd}$ order approximation on arbitrary unstructured triangular or tetrahedral meshes. Viscosity terms in the Navier-Stokes equations are approximated using finite-element method of 2$^{nd}$ order of accuracy.

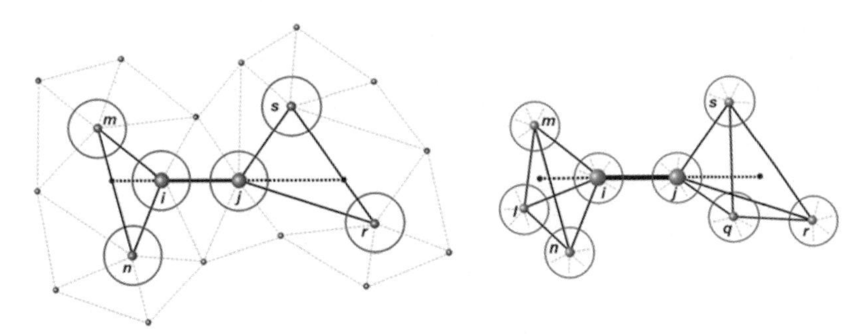


Figure 1: Stencil structure for 2D triangular (left) and 3D tetrahedral (right) unstructured meshes.

A specificity of the method is a type of space approximation. It is designed in a way that when being applied to the “Cartesian” regions of the mesh it coincides with a high order (up to the 6$^{th}$ order depending on the scheme parameters) finite-difference approximation. Here the “Cartesian” mesh means a mesh obtained by decomposing the rectangles (or cubes) into rectangular triangles (or tetrahedrons). It is clear that as larger part of computational domain is covered with the “Cartesian” mesh as higher total accuracy is reached. For the flux reconstruction between the nodes $i$ and $j$ an extended stencil of irregular structure (see Fig. 1) is used to provide that scheme property. The flux calculation is of the highest computational cost in the algorithm, so an efficient parallel treatment of data on those stencils mainly determines parallelization efficiency.

## 3. Technology of parallelization

The parallelization of a sequential code can be divided into following stages:

1. infrastructure development that includes additional modules for automatic building of data exchange, parallel output, etc.;
2. modification of the original code including additional data structures, upgrade of computational loops, placing data transmission calls;
3. verification and performance testing of the parallel version.

Infrastructure functionality can be implemented as separate program modules which allows first to minimize changes in the original code and second allows to use the same modules for other codes. External programs can be used as well, in particular Metis software, etc.

A few specific ideas to provide a substantial increase in parallel efficiency are proposed [2]. One of the main techniques is reducing the data exchange by means of overlapping the computations. First, it is overlap of mesh elements: edge, triangle of tetrahedron belongs to a subdomain if at least one its node belongs to the subdomain. This way some elements may belong to more than one subdomain. But this saves from a lot of data exchange leading to substantial parallel performance improvement. Second, it is nodes overlap: most of the computational loops over nodes are extended by including halo nodes from neighbour subdomains. In this case corresponding data exchange is replaced where it is possible by computations that are in most cases much faster. The set of nodes which include all nodes of subdomain and adjacent halo nodes from neighbors will be denoted as extended subdomain.

A use of a special set of index arrays and concept of extended subdomain allows to minimize modifications and to make parallelization easier. Three numerations are used: 1 - global, 2 - local and 3 - extended local. The first corresponds to the whole domain, the second - to subdomain and the third - to extended subdomain. Index arrays shown on fig. 2 are used to switch from one numeration to another. As all node arrays are allocated in extended local numeration most of the computational loops in the code don't require any changes at all.

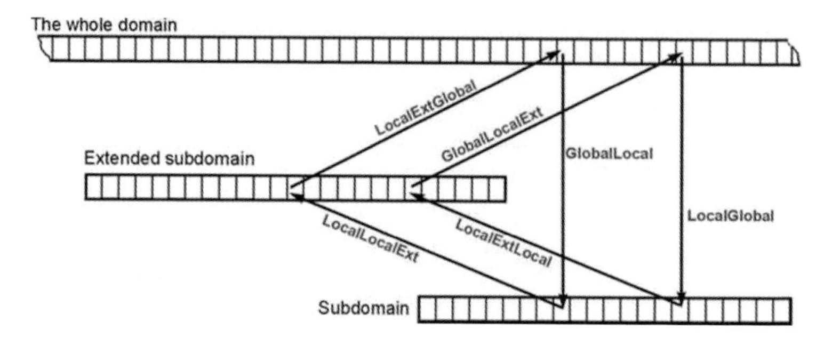


Figure 2: Additional index arrays, indicated by arrows, are used to switch from one numeration to another.

The parallel version of NOISEtte shows parallel efficiency of 85% in performance test on a typical low-cost system (so-called Beowulf cluster) with common Ethernet network even using relatively small mesh with $8*10^4$ nodes on 30 CPU. Initially optimized for low-cost clusters the code demonstrates efficiency of 99.5% in the same test on a cluster with high performance Myrinet network. The comparison of the performance of the code on these two systems shows clearly that parallel performance loss is only due to the data exchange. That fact confirms a high efficiency of the parallelization approach that consists in replacing where it is possible the data exchange with overlapped computations. A speedup plot is represented on fig. 3.

The verification stage performed includes comparison of the results obtained with parallel version of the code and the original sequential version. Full coincidence of the results verifies correctness of the parallelization. The parallel code has been carefully verified on a set of known benchmark problems.

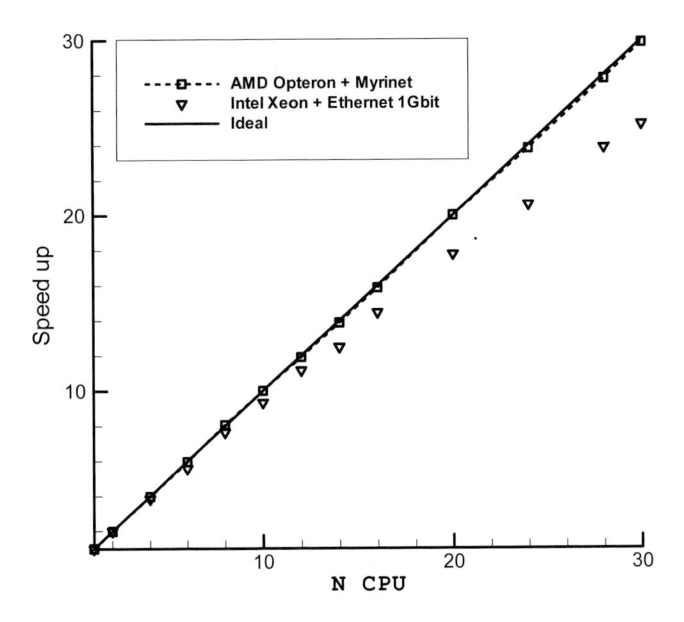


Figure 3: Speedup on different parallel systems in case of mesh with $8*10^4$ nodes

## 4. Numerical results

### 4.1. Impedance tube

The first group of 2D and 3D model problems simulates the conditions of the physical experiment, namely, experiment in impedance tubes. These problems are intended to study sound-absorbing properties of the resonator and reveal the mechanisms of acoustic energy losses. Figure 4 left demonstrates a scheme of computational domain. A plane acoustic wave is given at the input boundary. The problem parameters are following: tube diameter D=2.35cm, mouth diameter d=0.75cm, width of the perforated screen L=0.6cm. The monochromatic acoustic wave of power 147.1 Db and frequency 273 Hz enters tube from the left.

An effect of transition of acoustic energy into the energy of the turbulence was obtained and investigated. The wave breaks into vorticity structures in the resonator mouth area. Figure 4 right shows a 3D flow pattern with density iso-surfaces.

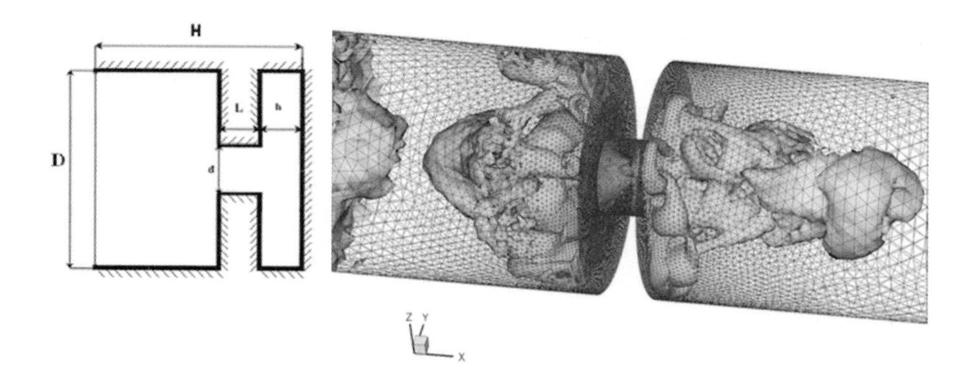


Figure 4: Impedance tube: scheme (left) and density iso-surface near the hole (right)

**4.2. Channel with resonators**

The second model problems in simplified 2D configurations correspond to the physical experiment on acoustic liners in channels covered with acoustic absorbing panels. To study an influence of collectivity, the acoustic liners are modeled by a system of one, five and eleven resonators. A top picture in Fig. 5 shows a scheme of computational domain in case of 5 resonators. The experiment has the following input conditions: subsonic inflow Mach 0.4 from left boundary combined with monochromatic acoustic wave 3431 Hz, 150Db.

A field of density disturbances corresponding to the acoustic waves radiated from the resonator mouths is represented in the bottom picture (Fig. 5). Following effects have been studied within a set of computational experiments:

1. Resonator whistle
2. Turbulence: mixing layer development, vorticity structures inside of the resonator camera
3. Influence of the developed boundary layer
4. Interaction of the whistle with the input acoustic wave
5. Sound suppression effect for different number of resonators

The sound suppression was estimated by a comparison of the output acoustic energy between three cases: channel without resonators, channel with 5 resonators and channel with 11 resonators. Results are represented in table 1.

Table 1. Sound suppression effect

| Number of resonators | Acoustic power, Db | Sound suppression, Db |
|---|---|---|
| 0 | 148.8 | 0 |
| 5 | 147.1 | 1.7 |
| 11 | 144.3 | 4.5 |

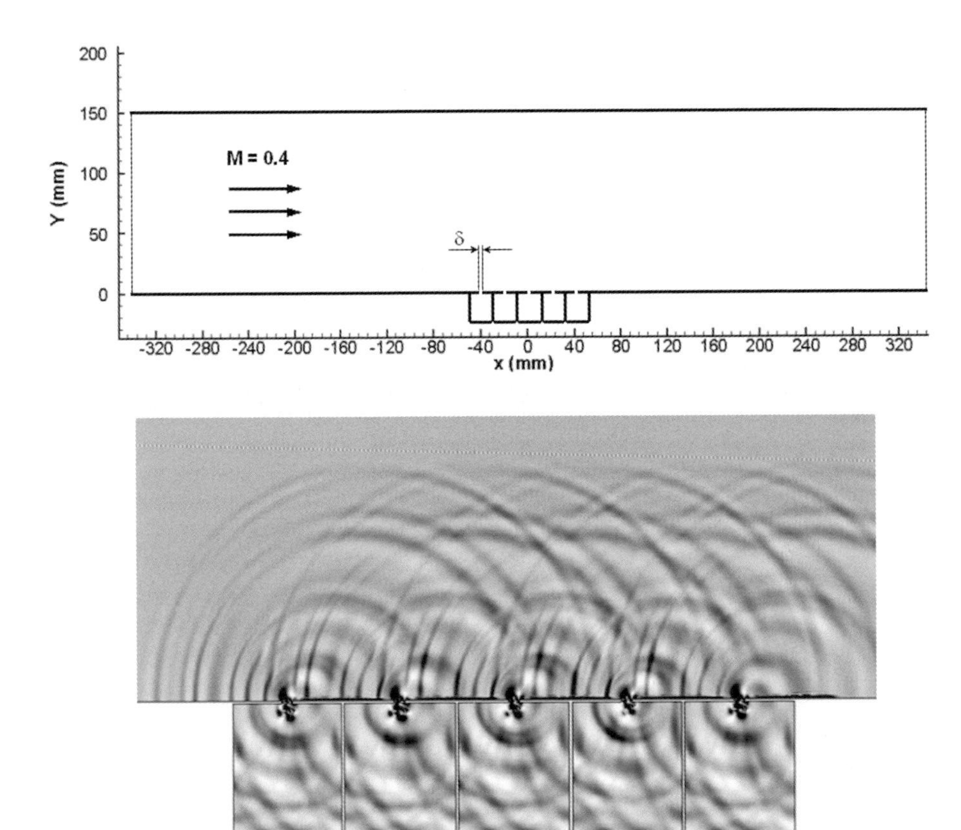


Figure 5: Channel with resonators: scheme (top) and density disturbances produced by the acoustic waves near resonators (bottom).

## 5. Conclusions

The main result of the present work is the development and demonstration of parallelization technology for explicit sequential codes. The parallelization technology described in the paper was applied to the code NOISEtte for solving 2D and 3D gas dynamics and aeroacoustics problems using unstructured meshes and high-order (up to 6) explicit algorithms.

Parallel efficiency of the parallel version and its applicability for typical CAA problems was demonstrated on a set of 2D and 3D computational experiments on resonator-type problems, which can be considered as first important steps in development of robust and efficient computational tools for modeling and optimization of acoustic liners.

**Literature**


1. *Ilya Abalakin, Alain Dervieux, Tatiana Kozubskaya*, High Accuracy Finite Volume Method for Solving Nonlinear Aeroacoustics Problems on Unstructured Meshes, *Chinese Journal of Aeroanautics*, 19 (2), (2006), pp. 97-104.
2. *A.Gorobets, T.Kozubskaya*, Technology of parallelization of explicit high-accuracy algorithms on unstructured meshes in computational fluid dynamics and aeroacoustics, *Matematicheskoe modelirovanie*, 19 (2), (2007), pp. 68-86
3. *Tatiana Kozubskaya., Ilya Abalakin, Andrey Gorobets, and Alexey Mironov*, Simulation of Acoustic Fields in Resonator-Type Problems Using Unstructured Meshes, AIAA Paper 2519-2006, (2006)
4. *Abalakin, I.V., Dervieux, A., and Kozubskaya T.K.* A vertex centered high order MUSCL scheme applying to linearised Euler acoustics, INRIA report RR4459, April 2002.
5. *Debiez, C., and Dervieux*, A. Mixed element volume MUSCL methods with weak viscosity for steady and unsteady flow calculation, *Computer and Fluids*, Vol. 29, 1999, pp. 89-118.
6. *Gourvitch N., Rogé G., Abalakin I., Dervieux A., and Kozubskaya T.* A tetrahedral–based superconvergent scheme for aeroacoustics, INRIA report RR5212, May 2004.

# Separate treatment of momentum and heat flows in parallel environment

**Aydin Misirlioglu and Ulgen Gulcat**

*Istanbul Technical University,Faculty of Aeronautics & Astronautics, Maslak 34469, Istanbul/TURKEY*

In this study, already developed two separate solvers, one for the velocity field and one for the temperature field are run concurrently on two different machines. Each machine here also has multiprocessors which solve each equation in parallel using the domain decomposition technique. The parallel processing is performed via PVM on SGI Origin 2000 and Origin 3000 connected through 10 Mbit Ethernet. Since the machines are not dedicated the communication times may vary for different runs. The runs are first performed for a single domain on each machine handling the mass and the heat flow separately. Afterwards, two overlapping domains are used on each machine. Overall elapsed computational times indicate satisfactory speed up.

Keywords: separate treatment, heat and mass flow

## 1. INTRODUCTION

Modeling of some physical phenomena requires simultaneous numerical solution of separate differential equations. The mass and the heat flow past solid bodies is a good example for such flows. Furthermore, the numerical solvers for the mass flow and for the heat flow may also, for any reason, be developed separately by different group of people on different machines. If the solution of the total problem requires marching, at each step the velocity field for the flow obtained via solution of the momentum and the continuity equations is transferred to the energy equation to be used in convective terms and, similarly the temperature field is transferred to the momentum equation to calculate the heat dissipation and the density variations in continuity equation. The exchanged information is used in linearization of nonlinear terms as coefficients lagging one step in marching process for the solution.
The mass and heat flow case considered in this study is compressible flow past a flat plate immersed in a free stream Mach number of 8. The surface temperature to free

I.H. Tuncer et al., *Parallel Computational Fluid Dynamics 2007*,
DOI: 10.1007/978-3-540-92744-0_32, © Springer-Verlag Berlin Heidelberg 2009

stream temperature ratio is taken as 4. First, the classical boundary layer equations are solved with finite difference discretization for a steady flow case while marching in the flow direction. Then, the effect of the additional terms appearing because of high temperature and velocity gradients are studied.
The domain decomposition technique is implemented with matching and overlapping interfaces for multi domain discretizations while the PVM [1] is employed for parallel processing since the other tools were not even able for some reason to start the process.

## 2. FORMULATIONS

Formulations for the governing equations and the numerical formulations after the discretization via the finite differencing will be given in this section.

### 2.1. Governing equations

The continuity, momentum and the energy equations are, for the sake of convenience, cast into the following form with the Stewartson transformation, [2, 3].

$$\frac{\partial u}{\partial x}+\frac{\partial v}{\partial y}=-\frac{1}{2\rho_e}\frac{\partial}{\partial y}\left(\frac{\bar{\mu}}{h}\frac{\partial h}{\partial y}\right) \tag{1}$$

$$u\frac{\partial u}{\partial x}+v\frac{\partial v}{\partial y}=\frac{1}{\rho_e}\frac{\partial}{\partial y}\left(\bar{\mu}\frac{\partial u}{\partial y}\right)-\frac{1}{2\rho_e}\frac{\partial}{\partial y}\left(\frac{\bar{\mu}u}{h}\frac{\partial h}{\partial y}\right) \tag{2}$$

$$u\frac{\partial h}{\partial x}+v\frac{\partial h}{\partial y}=\frac{1}{\rho_e}\frac{\partial}{\partial y}\left(\frac{\bar{\mu}}{\Pr}\frac{\partial h}{\partial y}\right)+\frac{\bar{\mu}}{\rho_e}\left(\frac{\partial u}{\partial y}\right)^2+\frac{1}{2\rho_e}\frac{\bar{\mu}u}{h}\frac{\partial u}{\partial y}\frac{\partial h}{\partial y} \tag{3}$$

Here, u, v, h and $\rho_e$ denote the velocity components, enthalpy and the free stream density respectively. The Stewartson transformation makes the equations appear in incompressible form which makes the numerical handling of the resulting matrix equations simple. In addition, Pr denotes the Prandtl number and the $\bar{\mu}=\mu\rho/\rho_e$ denotes the modified viscosity. The additional enthalpy related terms appearing at the right hand side of (1-3) represent the molecular mass diffusion rate and its effects because of the presence of high velocity and enthalpy gradients, [4]. The governing equations without those terms give the classical boundary layer equations.

### 2.2. Numerical formulation

The finite difference discretization of equations (2-3) results in two separate tri-diagonal matrix equations to be solved on two different machines. The continuity equation, on the other hand, is used to obtain explicitly the v component of the velocity. Therefore,

the momentum and continuity equations are evaluated on one machine and the energy equation and equation of state are evaluated on the other machine to obtain the temperature and the density values.
The tri-diagonal system of equations takes the following form

$$A_i u_{i-1} + D_i u_i + C_i u_{i+1} = B_i \tag{4}$$

Where u is the unknown vector and A,B,C an D are the coefficients in terms of known values. The inversion of (4) can be found in any numerical analysis book or in [5]. The additional terms in (1-3) are also discretized with finite differencing and placed in proper places in algebraic equations, ie., either as unknowns or as known quantities as forcing terms at the right hand side of the equations.

## 3. DOMAIN DECOMPOSITION

During the separate solution of the equations, if more than a single domain is used then the domain decomposition technique is employed. At the interfaces, matching and overlapping grids are used as shown in Figure 1.

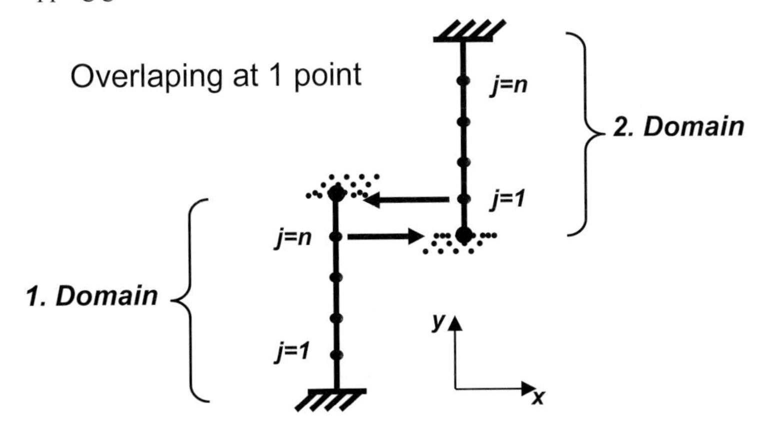


Figure 1. Matching and overlapping domains.

During iterations, the problem in each domain is handled as a Dirichlet problem where the unknown at the $n$th node of the first domain is imposed as the boundary condition on the second domain and, similarly, the unknown value obtained at the first node of the second domain is imposed as the boundary condition on the first domain. Obviously, since the momentum and the energy equations are solved separately on each machine, the message passing for the boundary conditions of the domains is handled separately, as an internal affair of the machine under consideration. Therefore, for each machine the internal communication speed plays an important role for single step calculations which are being performed almost in a dedicated manner. Once the velocity field and the temperature field are computed on different machines, they are sent to corresponding nodes of the corresponding domains. The communication between the separate machines, however, is performed through general network channels which are shared with the other users, which in turn, causes the communication time to vary for different steps. Figure 2 shows Origin 2000 and Origin 3000 as two different machines connected via Ethernet.

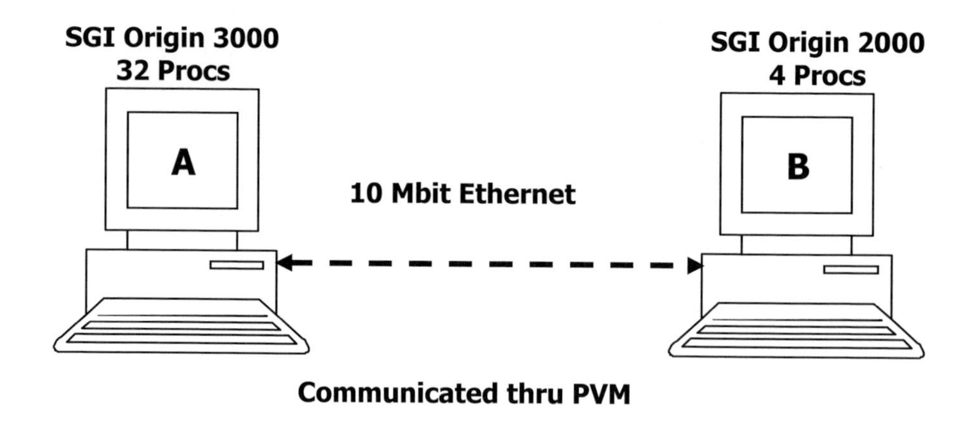


Figure 2. The computational environment

## 4. RESULTS and DISCUSSION

Laminar flow over a flat plate immersed in a free stream Mach number of 8 wherein the ratio of the plate temperature to free stream temperature equals 4 is studied. Two separate machines are employed in parallel manner. The momentum and the continuity are solved on SGI 3000 and the energy equation is solved on SGI 2000.
A very fine grid is used in discretizing the flow field. For the sake of simplicity, equally spaced grid with $\Delta x = 0.005$ and $\Delta y = 0.0001$ having 1400 grid points in y direction and 60 points in x direction is used. Figure 3 shows the comparisons of the boundary layer profiles obtained at x=0.3 with this grid resolution, where 10-15 iteration is sufficient for the solution of Equation 4 for each domain. The classical and the modified solutions of the boundary layers are shown in Figure 3, where, in case of the modified equations the boundary layer thickness is almost the double of the classical boundary layer thickness.
Figure 4. shows the skin friction comparisons for the boundary layer flow. As depicted in the figure, the skin friction obtained with additional terms is higher near the leading edge and it gets smaller afterwards compared to the classical solution.

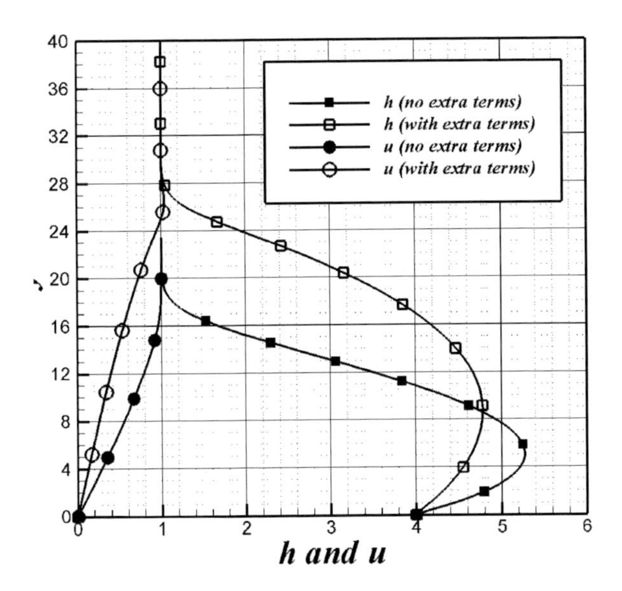


Figure 3. Comparison of the boundary layer solution for M=8 and $T_w/T_e$=4 cases.

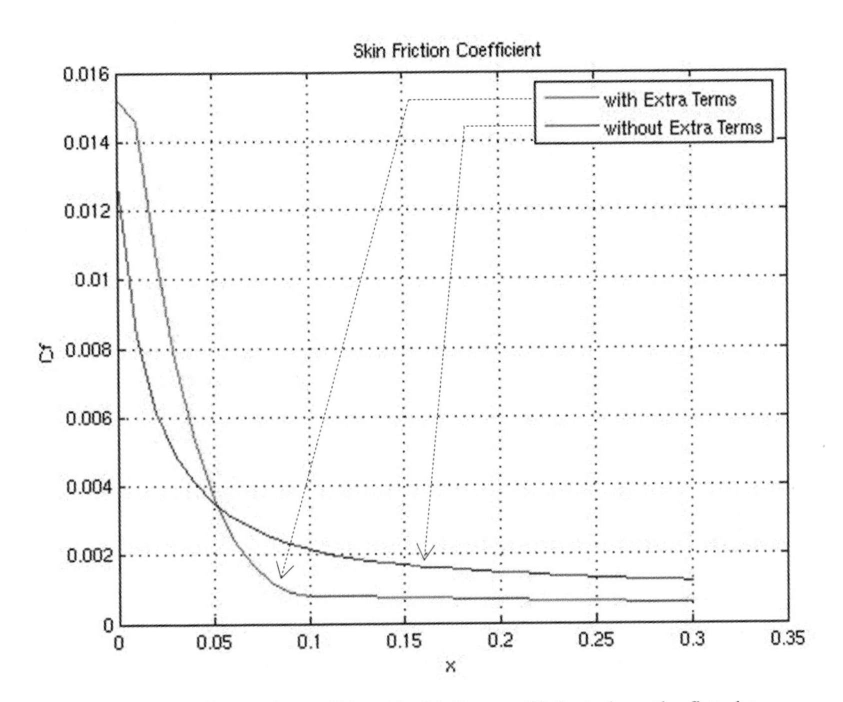


Figure 4. Comparison of the skin friction coefficient along the flat plate

There are 7 different runs performed for the same problem. These runs and the elapsed times are as given in Table 1. Since the problem size is small a few processors are used. The contents of the Table 1 give enough information about the speed ups.

Table 1. Run cases and elapsed times

| Case | Runs | Elapsed time (s) |
|---|---|---|
| 1 | Sequential on A | $6.303 \times 10^{-3}$ |
| 2 | 1 procs for momentum and 1 procs for energy on A | $2.887 \times 10^{-3}$ |
| 3 | 1 procs for momentum and 1 procs for energy on B | $5.764 \times 10^{-3}$ |
| 4 | 1 procs for momentum on A and 1 procs for energy on B | $3.179 \times 10^{-3}$ |
| 5 | 2 procs for momentum and 2 procs for energy on A | $2.895 \times 10^{-3}$ |
| 6 | 2 procs for momentum and 2 procs for energy on B | $5.378 \times 10^{-3}$ |
| 7 | 2 procs for momentum on A and 2 procs for energy on B | $2.928 \times 10^{-3}$ |

According to Table 1, the timings indicate that i) in any event Origin 3000 is almost twice as fast as Origin 2000, ii) comparison of case 1 and case 2 shows that super linear speed up is achieved when the two equations are solved concurrently as opposed to the sequential solution of the same machine, ie, Origin 3000, iii) comparison of case1 and case 4 indicates that even if the two equations are solved on different machines, the speed up is almost doubled compared to the sequential solution obtained on the faster of the two, iv) similarly, comparison of case 5 and 7 shows that the total time elapsed is almost the same for computations made on two different machines with two processors on each and the computations on a faster one with 4 processors.

## 5. CONCLUSION

The parallel separate solution of the momentum and the energy equation on two separate machines show that although there is a large data communication at each time step there still is a considerable speed up.
When two different machines are involved in any parallel processing, PVM, still is the most handy tool to resort to.

### REFERENCES

1. PVM 3 User Guide and Reference Manual, 1994
2. Schlichting, H., Gersten, K., Krause, E. and Oertel, H., Boundary-Layer Theory, Springer; 8th ed. 2000. Corr. 2nd printing edition, 2004
3. White, F.M, Viscous Fluid Flow, McGraw-Hill Science/Engineering/Math; 3 edition 2005
4. Durst, F., Transport Equations for Compressible Fluid Flows with Heat and Mass Transfer Private communucation, 2006.
5. Hoffman, J.D., Numerical Methods for Engineers and Scientists, 2nd Edition, CRC, 2001

# DNS of Turbulent Natural Convection Flows on the MareNostrum supercomputer

F. X. Trias[a], A. Gorobets[a], M. Soria[a] and A. Oliva[a]

[a]Centre Tecnològic de Transferència de Calor (CTTC)
ETSEIAT, c/ Colom 11, 08222 Terrassa, Spain
e-mail: oliva@cttc.upc.edu, web page: http://www.cttc.upc.edu

A code for the direct numerical simulation (DNS) of incompressible turbulent flows that provides a fairly good scalability for a wide range of computer architectures has been developed. The spatial discretization of the incompressible Navier-Stokes equations is carried out using a fourth-order symmetry-preserving discretization. Since the code is fully explicit, from a parallel point of view, the main bottleneck is the Poisson equation. In the previous version of the code, that was conceived for low cost PC clusters with poor network performance, a Direct Schur-Fourier Decomposition (DSFD) algorithm [1–6] was used to solve the Poisson equation. Such method, that was very efficient for PC clusters, can not be efficiently used with an arbitrarily large number of processors, mainly due to the RAM requirements [5] (that grows with the number of processors). To do so, a new version of the solver, named Krylov-Schur-Fourier Decomposition (KSFD), is presented here. Basically, it is based on the Direct Schur Decomposition (DSD) algorithm [7] that is used as a preconditioner for a Krylov method (CG) after Fourier decomposition. Benchmark results illustrating the robustness and scalability of the method on the MareNostrum supercomputer are presented and discussed. Finally, illustrative DNS simulations of wall-bounded turbulent flows are also presented.

**Keywords:** Direct Numerical Simulation; MareNostrum supercomputer; parallel Poisson solver; Schur complement method; Conjugate gradient; Natural convection

## 1. INTRODUCTION

Direct numerical simulation (DNS) and large-eddy simulation (LES) of transition and turbulent flows are applications with huge computing demands, that needs parallel computers to be feasible. In this context, efficient and scalable algorithms for the solution of the Poisson equation that arises from the incompressibility constraint are of high interest.

The initial version of our DNS code [1–6] was conceived for low-cost PC clusters with poor network performances (low bandwith/high latency). To do so, a Direct Schur-Fourier Decomposition (DSFD) algorithm [2–5] was used to solve each Poisson equation to machine accuracy with a single all-to-all message. It allowed to perform DNS simulations of three-dimensional natural convection flows in a cavity bounded in two directions [1,6,8] on a PC cluster.

I.H. Tuncer et al., *Parallel Computational Fluid Dynamics 2007*,
DOI: 10.1007/978-3-540-92744-0_33, © Springer-Verlag Berlin Heidelberg 2009

However, the DSFD algorithm can not be efficiently used with an arbitrarily large number of processors, mainly due to the RAM [5] (that grows with the number of processors). To do so, a Krylov-Fourier-Schur Decomposition (KFSD) algorithm has been developed. Again, a Fourier decomposition is used to uncouple the original 3D Poisson equation into a set of 2D planes. Then, each 2D problem is solved using a combination of the DSD algorithm [7] and a Conjugated Gradient (CG) method [9].

The KSFD algorithm provides a fairly good scalability up to hundreds of CPUs. Moreover, the flexibility of the solver allows to use it efficiently both on high-performance networks and low-cost high-latency networks. Benchmark results illustrating the robustness and scalability of method on the MareNostrum supercomputer are presented and discussed. The new version of the code has been used to carry out a new DNS simulation of an air-filled differentially heated cavity with $Ra = 10^{11}$, $Pr = 0.71$ and height aspect ratio 4 on the MareNostrum supercomputer. Illustrative results of this simulation are also presented.

## 2. NUMERICAL METHODS FOR DNS

The non-dimensional incompressible Navier-Stokes (NS) equations coupled with the thermal transport equation in a bounded parallelepipedic domain of height $L_z$, width $L_y$ and depth $L_x$ are considered

$$\nabla \cdot \mathbf{u} = 0 \tag{1a}$$

$$\frac{\partial \mathbf{u}}{\partial t} + (\mathbf{u} \cdot \nabla)\mathbf{u} = Pr\Delta\mathbf{u} - \nabla p + \mathbf{f} \tag{1b}$$

$$\frac{\partial T}{\partial t} + (\mathbf{u} \cdot \nabla) T = \Delta T \tag{1c}$$

where $Pr$ is the Prandtl number and $\mathbf{f}$ is an externally applied body force (e.g. gravity). Periodic boundary conditions are prescribed in the $x$-direction,

$$\mathbf{u}(\mathbf{x}, t) = \mathbf{u}(\mathbf{x} + L_x\mathbf{e}_x, t) \tag{2a}$$

$$T(\mathbf{x}, t) = T(\mathbf{x} + L_x\mathbf{e}_x, t) \tag{2b}$$

because it allows to study the 3D effects due to intrinsic instability of the main flow and not due to the boundary conditions [1,6].

### 2.1. Spatial and temporal discretizations

Governing equations (1a-1c) are discretized on a staggered grid in space by fourth-order symmetry-preserving schemes [10]. Such family of discretizations preserves the underlying differential operators symmetry properties. These global discrete operator properties ensure both stability and that the global kinetic-energy balance is exactly satisfied even for coarse meshes if incompressibility constraint is accomplished. More details about these schemes can be found in [10].

For the temporal discretization, a fully explicit dynamic second-order one-leg scheme [6, 10] is used for both convective and diffusive terms. In practice, this method allows to increase the time step to a value that is about twice greater, that is twice cheaper, than the standard second-order Adams-Bashforth scheme. Finally, to solve the pressure-velocity coupling we use a classical fractional step projection method [11]. Further details about the type-integration method can be found in [6].

## 3. PARALLELISATION

Spatial decomposition is the typical approach to parallelise CFD codes. As the momentum and energy equations (1b,1c) are discretized in a fully explicit method, their parallelisation is straightforward. Therefore, the main problem is the Poisson equation arising from the incompressibility constraint (1a). Moreover, the use of the fourth-order symmetry-preserving schemes [10] for the convective and diffusive terms of the equations implies that the Laplacian operator in the Poisson equation must be also fourth-order accurate (otherwise the symmetry-preserving property is lost). This is the most problematic aspect from a parallel computing point of view.

### 3.1. Review of the Direct Schur-Fourier Decomposition

In the previous version of our code a parallel direct solver, named Direct Schur-Fourier Decomposition (DSFD) [2–6], was used to solve the Poisson equation. It was based on a combination of FFT and a Direct Schur Decomposition (DSD) algorithm [7]. It has a very good performance on loosely coupled parallel systems with relatively small number of CPUs (up to about 30-50 CPUs). The use of FFT uncouples the domain on one direction and leads to a set of 2D problems that can be solved independently. Then, a direct DSD method was used to solve each plane. This method requires only one data exchange to obtain the exact solution but it has some specific bottlenecks such as RAM requirements and size of communications that grows with the number of CPUs and the mesh size [5]. All these problems limit the solver scalability specially for high-order schemes. Moreover, it must be noted that the weight of all these scalability limitations would depend on the mesh size, the number of CPUs, the computer architecture and the total time-integration period demanded for our applications[1]. In conclusion, all these scalability limitations motivated the necessity to investigate a different approach for really large-scale problems on supercomputers.

### 3.2. Alternative: Krylov-Schur-Fourier Decomposition algorithm

Therefore, if due to the aforementioned scalability limitations the DSD method cannot be applied efficiently (or cannot be applied at all) to couple all the planes, then it can be used to couple only some planes of the set. In this case, the rest of planes can be solved using an iterative method. Moreover, DSD itself can be used as an auxiliary direct solver or preconditioner inside of the iterative method. In the later case, the DSD algorithm is used to couple some parts of the plane and the iterative method provides coupling of the whole plane. Henceforth, we will refer those parts of the plane that are coupled with the DSD solver as *blocks*.

Since all the 2D systems planes are fully decoupled after FFT [3,5], only the solution

[1]Based on our experience, for real CFD applications the use of the DSD is valid for problems with mesh sizes up to $\sim 6 \times 10^6$ points and $30 \sim 50$ processors

of one arbitrary plane is considered. Therefore, the linear systems to be solved are of the form

$$Ax = b \tag{3}$$

where the matrix $A$ is, by construction (see reference [5], for instance), a symmetric positive-definite non-singular[2] matrix the Conjugate Gradient (CG) method [9] is chosen as the most appropriate. The plane is decomposed into blocks and each block is coupled with the DSD solver [7]. Thus, the matrix $A$ is splitted as follows

$$A = A_D + A_I \tag{4}$$

where matrix $A_D$ is obtained from matrix $A$ after discarding elements that correspond to the coupling between different blocks and consequently matrix $A_I$ is obtained from $A$ after discarding elements that correspond to the coupling inside the blocks. An example of the decomposition of one plane into 4 blocks is shown in figure 1.

Then, iterative CG solver uses a direct solver, $DSD\left(A_D, b\right)$, as a preconditioner which provides exact solution for the system $A_D x = b$. Therefore, the preconditioned CG algorithm results,

First iteration:

1. $r^0 = b - Ax^0$
2. $z^0 = DSD\left(A_D, r^0\right)$
3. $p^1 = z^0$

$i$-th iteration:

1. $z^{i-1} = DSD\left(A_D, r^{i-1}\right)$
2. $^*$ $\rho_{i-1} = r^{i-1^T} \cdot z^{i-1}$
3. $\beta_{i-1} = \rho_{i-1}/\rho_{i-2}$
4. $p^i = z^{i-1} + \beta_{i-1} p^{i-1}$
5. Update halos for $p^i$
6. $q^i = Ap^i$
7. $^*$ $\alpha_i = \rho_{i-1}/(p^{i^T} \cdot q^i)$
8. $x^i = x^{i-1} + \alpha_i p^i$
9. $r^i = r^{i-1} - \alpha_i q^i$
10. $^*$ Calculation of residual norm

where $^*$ operations imply collective communication for global reduction operations:

- Summation of a single value: steps 2 and 7
- Maximum of a single value: step 10

Therefore, one of the main features of the algorithm presented, named Krylov-Fourier-Schur Decomposition (KFSD), is its flexibility. The number of blocks governs the convergence properties of the algorithm: the smaller is the number of blocks, the CG algorithm

[2]Originally there is one plane whose corresponding $A$ matrix is singular. In this case, singularity is removed preserving the symmetry by changing one element on main diagonal that corresponds to an arbitrary inner node. This modification fixes the value of this particular node.

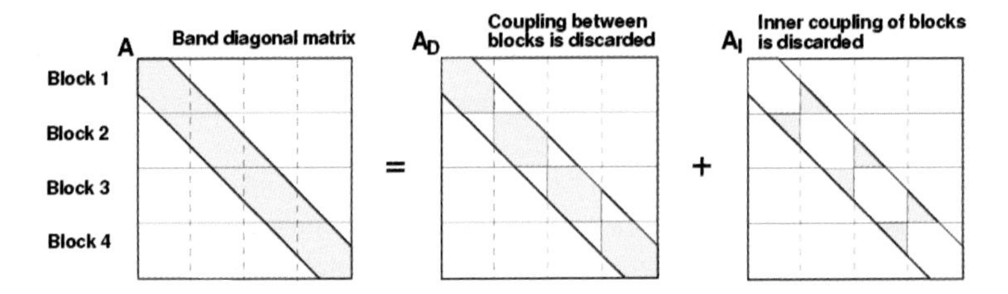


Figure 1. Example of the decomposition of one plane into 4 blocks. Structure of the matrices $A_D$ and $A_I$

is better preconditioned[3] and consequently the number of iterations reduced. However, bigger blocks implies more RAM requirements. Hence, the block size of each plane must be chosen in order to minimise the global computational cost accomplishing the memory limitations imposed by our computer architecture. The general idea behind is demonstrated on a simplified example in the following section.

## 4. PARALLEL PERFORMANCE TEST

The computing times have been measured on the MareNostrum supercomputer of the Barcelona Supercomputer Center. It is a IBM BladeCenter JS21 cluster with 10240 CPU Power PC 970MP processors at 2.3GHz. Dual CPU nodes with 8Gb of RAM are coupled with a high-performance Myrinet network. Auxiliary Ethernet Gigabit network is used for the shared file system.

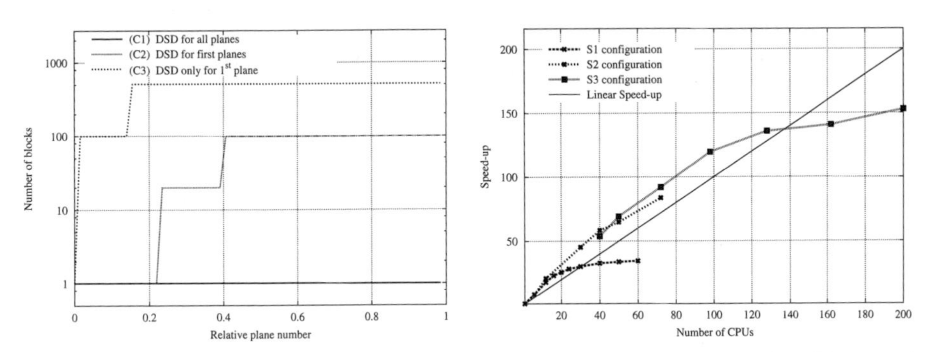


Figure 2. Left: Example of different KSFD solver configurations. Right: Estimated speed-ups for different configurations.

Illustrative speed-up results for a $32 \times 170 \times 320$ mesh are displayed in Fig. 2 (right). It shows adaptation of the solver for different number of CPUs to provide maximal per-

[3]Actually, number of blocks equal to one means that the direct DSD algorithm is being used.

formance and scalability. To do so, the following solver configurations (see figure 2, left) have been used:

- Configuration 1: Direct solution using the DSD algorithm for all planes (only one block).
- Configuration 2: From planes 1 to 5 DSD algorithm is used[4]. The rest of planes (6-32) have been solved using an iterative CG+DSD method.
- Configuration 3: Only the first plane is solved directly using the DSD algorithm. The rest of planes (2-32) have been solved using an iterative CG+DSD method.

Configurations are switched sequentially with the growth of CPU number when one configuration becomes slower than the others. Measurement starts with the first configuration from 1 CPU where DSD method is reduced to band-LU (it should be noted that band-LU for all planes cannot fit in memory of 1 CPU, so computing time of only one band-LU solution was measured and then multiplied by 32). Then, the DSD method begin losing efficiency and it is replaced by configuration 2. Finally, configuration 2 is replaced by configuration 3 which reaches speedup of 136 with 128 CPUs and then starts to lose efficiency reaching finally speedup of at least 153.

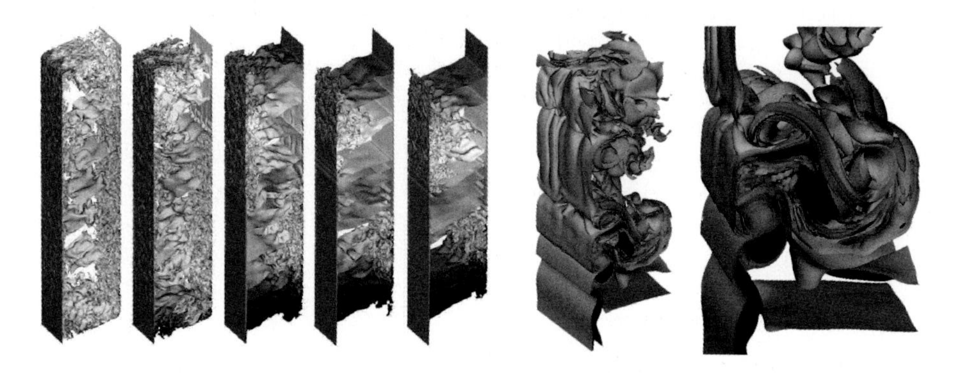

Figure 3. Left: several instantaneous temperature fields at $Ra = 10^{11}$ and $Pr = 0.71$ (air). Right: zoom around the vertical boundary layer becomes totally disrupted and large eddies are ejected to the cavity core.

## 5. ILLUSTRATIVE DNS RESULTS

The direct numerical simulation (DNS) of a buoyancy-driven turbulent natural convection flow in an enclosed cavity has been used as a problem model. Several illustrative DNS results of instantaneous temperature maps obtained for a differentially heated cavity at $Ra = 10^{11}$, with $Pr = 0.71$ (air) and height aspect ratio 4 are shown in Fig. 3. The mesh

[4]Since the first planes have worst convergence properties, that is for large values of the condition number, the direct DSD algorithm is used.

size chosen for this simulation is $128 \times 682 \times 1278$. It has been carried out on the MareNostrum supercomputer using up to 512 processors. To the best of authors' knowledge, this is the largest DNS simulation with two wall-normal directions ever performed.

This case has been chosen as an extension of the work initiated by Soria *et al.* [1] and followed by Trias *et al.* [8]. Despite the geometric simplicity of this configuration turbulent natural convection is a very complex phenomenon, that is not still well understood [12]. This new DNS results shall give new insights into the physics of turbulence and will provide indispensable data for future progresses on turbulence modelling [13,14]. Further details about this configuration can be found in [1,4,6].

## 6. CONCLUSIONS

A code for the direct numerical simulation of incompressible turbulent flows that provides a fairly good scalability for a wide range of computer architectures has been developed. In the previous version of the code, that was originally conceived for low cost PC clusters, a DSFD algorithm [7,2,5,4] was used to solve the Poisson equation. However, such method can not be used for an arbitrarily large number of processors and mesh size [5], mainly due to RAM limitations. To do so, a new version of the solver, named Krylov-Schur-Fourier Decomposition (KSFD) has been developed. It basically consist on using the DSD algorithm as a preconditioner for a CG method. The scalable solver model and the criteria used have been briefly described and the main idea behind have been shown in a simplified example. The good efficiency and accuracy of the proposed method have been shown by performing several numerical experiments on the MareNostrum supercomputer.

Finally, to benchmark the KSFD algorithm, a DNS simulation of a differentially heated cavity with $Ra = 10^{11}$, $Pr = 0.71$ (air) and height aspect ratio 4 has been carried out using up to 512 processors on the MareNostrum supercomputer.

## ACKNOWLEDGEMENTS

This work has been financially supported by the *Ministerio de Educación y Ciencia*, Spain; Project: "Development of high performance parallel codes for the optimal design of thermal equipments". Contract/grant number ENE2007-67185.

Calculations have been performed on MareNostrum supercomputer at the Barcelona Supercomputing Center. The authors thankfully acknowledges this institution.

## REFERENCES

1. M. Soria, F. X. Trias, C. D. Pérez-Segarra, and A. Oliva. Direct numerical simulation of a three-dimensional natural-convection flow in a differentially heated cavity of aspect ratio 4. *Numerical Heat Transfer, part A*, 45:649–673, April 2004.
2. M. Soria, C. D. Pérez-Segarra, and A.Oliva. A Direct Schur-Fourier Decomposition for the Solution of the Three-Dimensional Poisson Equation of Incompressible Flow Problems Using Loosely Parallel Computers. *Numerical Heat Transfer, Part B*, 43:467–488, 2003.

3. F. X. Trias, M. Soria, O. Lehmkuhl, and C. D. Pérez-Segarra. An Efficient Direct Algorithm for the Solution of the Fourth-Order Poisson Equation on Loosely Coupled Parallel Computers. In *Parallel Computational Fluid Dynamics*, Gran Canaria, Spain, May 2004. Elsevier.
4. F. X. Trias, M. Soria, C. D. Pérez-Segarra, and A. Oliva. Direct Numerical Simulation of Turbulent Flows on a low cost PC Cluster. In *Parallel Computational Fluid Dynamics*, Washington DC, USA, May 2005. Elsevier.
5. F. X. Trias, M. Soria, C. D. Pérez-Segarra, and A. Oliva. A Direct Schur-Fourier Decomposition for the Efficient Solution of High-Order Poisson Equations on Loosely Coupled Parallel Computers. *Numerical Linear Algebra with Applications*, 13:303–326, 2006.
6. F. X. Trias, M. Soria, A. Oliva, and C. D. Pérez-Segarra. Direct numerical simulations of two- and three-dimensional turbulent natural convection flows in a differentially heated cavity of aspect ratio 4. *Journal of Fluid Mechanics*, 586:259–293, 2007.
7. M. Soria, C. D. Pérez-Segarra, and A. Oliva. A Direct Parallel Algorithm for the Efficient Solution of the Pressure-Correction Equation of Incompressible Flow Problems Using Loosely Coupled Computers. *Numerical Heat Transfer, Part B*, 41:117–138, 2002.
8. F. X. Trias, M. Soria, O. Lehmkuhl, and A. Oliva. Direct Numerical Simulations of Turbulent Natural Convection in a Differentially Heated Cavity of Aspect Ratio 4 at Rayleigh numbers $6.4 \times 10^8$, $2 \times 10^9$ and $10^{10}$. In K. Hanjalić, Y. Nagano, and S. Jakirlić, editors, *International Symposium on Turbulence Heat and Mass Transfer*, Dubrovnik, Croatia, September 2006.
9. Yousef Saad. *Iterative Methods for Sparse Linear Systems.* PWS, 1996.
10. R. W. C. P. Verstappen and A. E. P. Veldman. Symmetry-Preserving Discretization of Turbulent Flow. *Journal of Computational Physics*, 187:343–368, May 2003.
11. N. N. Yanenko. *The Method of Fractional Steps.* Springer-Verlag, 1971.
12. J. Salat, S. Xin, P. Joubert, A. Sergent, F. Penot, and P. Le Quéré. Experimental and numerical investigation of turbulent natural convection in large air-filled cavity. *International Journal of Heat and Fluid Flow*, 25:824–832, 2004.
13. F. X. Trias, M. Soria, A. Oliva, and R. W. C. P. Verstappen. Regularization models for the simulation of turbulence in a differentially heated cavity. In *Proceedings of the European Computational Fluid Dynamics Conference (ECCOMAS CFD 2006)*, Egmond aan Zee, The Netherlands, September 2006.
14. F. X. Trias, R. W. C. P. Verstappen, M. Soria, A. Gorobets, and A. Oliva. Regularization modelling of a turbulent differentially heated cavity at $Ra = 10^{11}$. In *5th European Thermal-Sciences Conference, EUROTHERM 2008*, Eindhoven, The Netherlands, May 2008.

# TermoFluids: A new Parallel unstructured CFD code for the simulation of turbulent industrial problems on low cost PC Cluster

O.Lehmkuhl [a], C.D. Perez-Segarra,[b] R.Borrell[a], M.Soria[b] and A.Oliva[b]

[a]TERMO FLUIDS S.L.
c/ Magi Colet 8, 08204 Sabadell, Spain
email: termofluids@yahoo.es

[b]Centre Tecnològic de Transferència de Calor(CTTC)
ETSEIAT,c/Colom 11, 08222 Terrassa, Spain
email: oliva@cttc.upc.edu, web page: htpp://www.cttc.upc.edu

The main features of *TermoFluids* are presented. It is a new unstructured and parallel object-oriented CFD code for accurate and reliable solving of industrial flows. The more relevant aspects from a parallel computing point of view, such as communication between CPUs and parallel direct and iterative algebraic solvers that allow *TermoFluids* to run efficiently on loosely-coupled parallel computers are presented. Also, the different approaches for turbulence modelling implemented in *TermoFluids* (RANS, LES and hybrid LES/RANS models) are pointed out. Illustrative results of numerical simulation of industrial problems, as the thermal optimisation of the nacelle of a wind turbine, are also presented.

**Keywords:** loosely coupled parallel computers; linear solvers; turbulence modelling; unstructured meshes

## 1. INTRODUCTION

The increase in the computational power and the improvement in the numerical methods have been significant over the last decades. This fact, together with the emergence of low-cost parallel computers, has made possible the application of numerical methods to the study of complex phenomena and geometries, such as the simulation of turbulent industrial flows.

Parallel computing of turbulent flows using DNS [1], LES [2] or hybrid LES/RANS [3] models are currently being used. However, most of these techniques are commonly applied on structured Cartesian or body-fitted multi-block codes.

Taking into account the current state-of-the-art of parallel techniques, and the ability of unstructured meshes to create grids around complex geometries, an unstructured and parallel object-oriented code has been developed. This code uses efficient algorithms, that work well on slow networks of PC clusters. The code implements turbulence models such as LES, hybrid LES/RANS and RANS models. The use of such techniques have made possible the accurate simulation of industrial flows with few simplifications in the geometry and the fluid dynamic and heat and mass transfer phenomena.

I.H. Tuncer et al., *Parallel Computational Fluid Dynamics 2007*,
DOI: 10.1007/978-3-540-92744-0_34, 

The use of state-of-the-art in object oriented based software engineering techniques and the implementation in C++ has allowed a quick development of the code.

In this paper, the main features of this unstructured parallel code for solving industrial CFD problems with low-cost PC clusters are presented. This contribution also covers the recent experience in LES/RANS computations of such kind of problems, presenting numerical results of different cases.

## 2. NUMERICAL METHODS AND PARALLEL ALGORITHMS

In this section the main features of a general purpose CFD code called *TermoFluids* are presented. The code has been designed for the simulation of complex fluid dynamics and heat and mass transfer problems of industrial interest. Its most revelant technical specifications are:

- Finite volume method on 3D or 2D unstructured collocated meshes (a mixture of tetrahedral, hexahedral, prism and/or pyramid elements, see Figure 1).
- Steady state or time integration flows with fully implicit (SIMPLE-like) and explicit (fractional step method) algorithm.
- Direct and iterative sparse linear solvers.
- Laminar and turbulent flow modelling.
- 2nd order space and time discretisation.
- Multi-component fluids.
- Conjugate heat transfer with a loosely coupled algorithm (possibility of different time steps for the fluid and the solid).
- Radiative heat transfer using Discrete Ordinates Method for transparent and non-transparent medium.
- Local refinement and Laplacian smoother post-processing tools.

Figure 1. Illustrative example of *TermoFluids* features. Left: External aerodynamics simulation of a car at $Re = 10^5$ with 5 CPUs, Right: Computational mesh (over $5 \cdot 10^5$ CV)

### 2.1. Sparse linear solvers

Once the system of equations is discretised, the resulting system of linear equations (all the dependent variables except the pressure which receive an special treatment), are solved by means of a GMRES ($M = 5$) [4] solver preconditioned with the inverse of the diagonal of the system. This method does not lose performance with the increase of the number of CPU and has low memory requirements.

For incompressible flows, the above solution does not work well with the pressure equation. For this situation, the pressure equation is reduced to the Poisson equation, which has an infinite speed propagation of the information in the spatial domain, i.e. the problem becomes fully elliptic. For the treatment of this particularity, especial solvers have been developed such as:

*Direct solvers*: Direct Schur Decomposition [5] using sparse Cholesky for the local variables with iterative or direct solver for the interface system.
*Iterative solvers*: CG with a PFSAI preconditioner.

In a companion paper [6] the main features of the aforementioned solvers are explained in detail.

## 2.2. Parallel algorithm implementation

The typical approach for the parallelisation of the non-linear system of equations is the spatial domain decomposition. The computational domain to be solved is divided into a number of blocks, that are assigned to different CPUs.

On parallel implementation, the non uniform load of the processes can affect the overall efficiency of the code (this aspect is more important in low-cost PC clusters). Special techniques to balance the load of each CPU are used. Partitioning of the computational domain is carried out by means of METIS [7] software. This code divides the domain into different groups of nodes and minimises the number of graph edges having nodes from different groups. Each of these groups has a similar number of nodes that represent the control volumes of the mesh to be solved in the associated CPU.

An important feature of the code is the communication algorithm developed to be used on high latency networks. This algorithm implements an optimisation method to minimise the synchronic point-to-point communication between CPUs using a constructive proof of Vizing's theorem [8].

The parallel efficiency of the algorithm on loosely coupled parallel computers is shown in Figure 2. The computing time have been measured using a PC cluster with 10 standard PCs

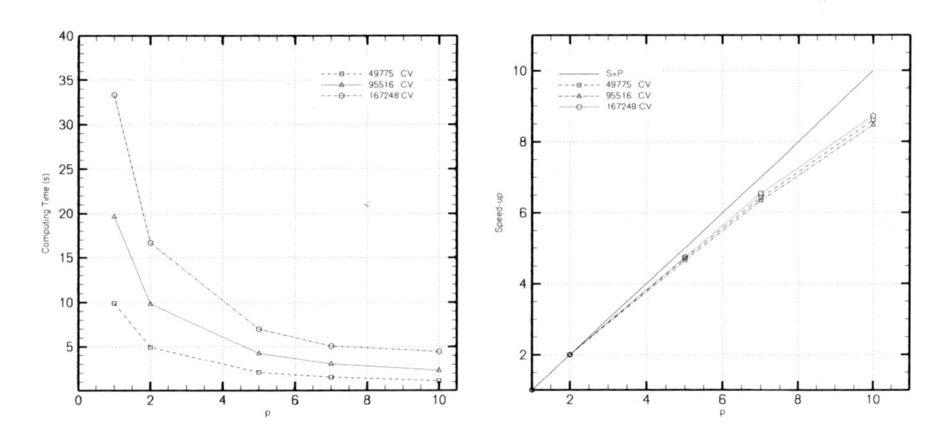


Figure 2. Left: Computing time on a conventional network versus the number of processors for different meshes, Right: Estimated speed-ups on a conventional network for different meshes.

(AMD k7 CPU 2600 MHz and 1024 Mbytes of RAM) and, a conventional network (100 Mbits/s 3COM cards and a CISCO switch), running Debian Linux 3.0 kernel version 2.6.16. MPI has been used as message-passing protocol (LAM 7.2.1).

### 2.3. Turbulence Modelling

Traditionally, in most industrial unstructured CFD codes turbulence is treated by means of RANS models. *TermoFluids* code employs a wide variety of RANS models usually used for industrial applications [9], such as: $k-\omega$ models: Wilcox; Peng, Davidson and Holmberg; Bredberg, Davidson and Peng; $k-\epsilon$ models: Ince and Lauder with standard Yap correction; Abe, Kondoh and Nagano; Goldberg, Peroomian and Chakravarthy; RNG-Based $k-\epsilon$ ; hybrid $k-\omega/k-\epsilon$ models: Menter Standard SST Model.

To deal with the coupling between the turbulent variables, an efficient locally coupled algorithm has been implemented, combined with BiCGSTAB to solve the spatial couples.

Wall boundary conditions are treated by means of different approaches. The first one, integrates the boundary layer but requires a high mesh density at this zone (low-Reynolds turbulent number version of the RANS model). The second one uses the two-layer approach. That means, wall layer is calculated with a one-equation model (i.e. a differential equation for turbulent kinetic energy and algebraic expressions for viscosity and turbulent dissipation) while the external zone is treated with the selected RANS model. In this case, mesh requirement is lower than the first approach. The third implements the standard wall functions, being the less restrictive of the three. The code implementation allows also the use of the different approaches in the same computational domain, depending on the boundary conditions.

Numerical simulations using traditional RANS models appear to be unable to accurately account for the time-depending and three-dimensional motions governing flows with massive separation or to predict flow transitions. This unsteady three-dimensional behaviour of the flow can be better solved by means of LES modelling [10]. However, due to the need of solving the small scales motion in the boundary layer, they require a high amount of computational resources.

This drawback can be circumvented if hybrid LES/RANS models are used. The main idea of such modelling is to solve the whole wall layer with a RANS model, while the external zone is treated with LES. These kind of hybrid models take advantage of the good performance of RANS models in the region near the walls and the capability of LES to resolve the time-depending and three-dimensional behaviour that dominate the separated regions. Among the main LES and hybrid LES/RANS models implemented in the code *TermoFluids* can be cited: Smagorinsky SGS model; Dynamic Smagorinsky SGS model; one-Equation SGS Model Yoshizawa with possibility of hybrid LES/RANS using all linear RANS models; Spalart Detached Eddy Simulation Model; Menter Detached Eddy Simulation Model.

When LES or RANS/LES models are used, some important properties of the discretisation must be conserved in order to achieve numerical stability. The first one, is the conservation of the turbulent kinetic energy. In the unstructured mesh background, due to collocated arrangement of the variables, this property can only be numerically conserved by using a special Least-Squares method [11] to evaluate the pressure gradient at the cell center (see figure 3) and a skew symmetric interpolation scheme for the convective term:

$$\sum_{faces}\left(\frac{\partial p}{\partial x_i}|_p n_i^{face} - \frac{\partial p}{\partial n}|_{face}\right)A_{face} \Longrightarrow \frac{\partial p}{\partial x_i} \tag{1}$$

$$\phi_{face} = \frac{\phi_p + \phi_{nb}}{2} \tag{2}$$

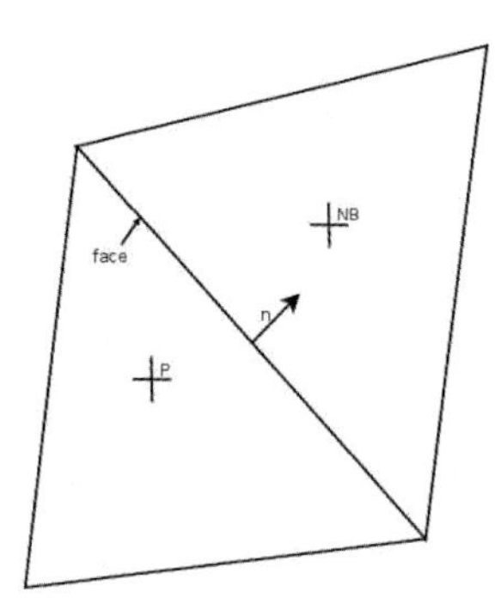


Figure 3. Typical control volume used in the discretisation. $P$ represent the cell center, $NB$ the neighbour cell center and $n$ the normal to the face

Another important factor to achieve numerical stability in RANS/LES methods is the transition function that switch the RANS model (in the boundary layer) and the LES model (in the detached zone). Two different functions can be used in the *TermoFluids* code. The first one, is the traditional transfer function used by Spalart in DES model [3]. This function assume that the model used to solve the smaller turbulent length ($l$) in a control volume is the one that dominate (see equation 6). That is, for a generic two-equation turbulence model:

$$\frac{\partial k}{\partial t} + \frac{\partial}{\partial x_i}(\rho u_i k) = \frac{\partial}{\partial x_i}(\mu_{eff}\frac{\partial k}{\partial x_i}) + P_k - \rho\beta_k \frac{k^{\frac{3}{2}}}{l} \tag{3}$$

$$\frac{\partial \omega}{\partial t} + \frac{\partial}{\partial x_i}(\rho u_i \omega) = \frac{\partial}{\partial x_i}(\mu_{eff}\frac{\partial \omega}{\partial x_i}) + P_\omega - \rho\beta_\omega \omega^2 + C_D \tag{4}$$

$$\mu_t = \rho C_\mu l \sqrt{k} \tag{5}$$

two different turbulence lengths are defined,

- **RANS mode:** $l = \sqrt{\frac{k}{\omega}}$
- **LES mode:** $l = C_{LES}\triangle$

where $C$ is a constant (usually 0.65, see [3]) and $\triangle = (\int_{cv} \delta V)^{\frac{1}{3}}$. So with the Spalart criteria the correct turbulence length at each control volume is:

$$l = min(\sqrt{\frac{k}{\omega}}, C_{LES}\triangle) \tag{6}$$

One of the main drawback of this formulation is that the RANS and LES zone can be mesh dependent. If the boundary layer is meshed with a regular mesh (i.e. with local refinement methods), the LES model can be switched on inside the boundary layer (see figure 4). On the other hand, Menter [12] in his SST model, has used a transfer function that can be adapted to this situation,

$$l = \sqrt{\frac{k}{\omega}} \cdot (F_2) - (1 - F_2) \cdot (C_{LES}\triangle) \tag{7}$$

$$F_2 = tanh(pow(max(2\frac{\sqrt{k}}{0.09\omega y}, \frac{400\nu}{y^2\omega}), 2)) \tag{8}$$

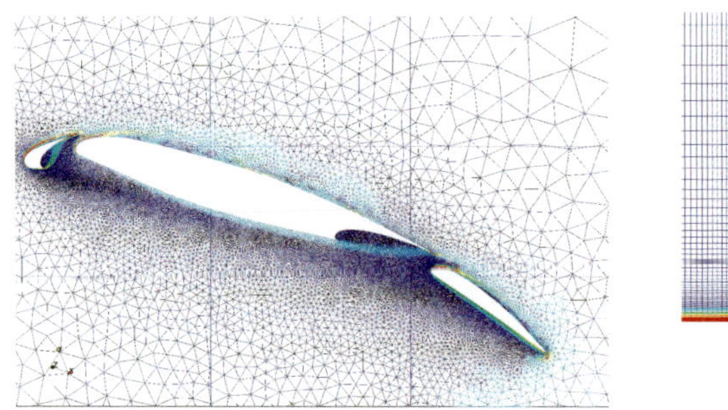

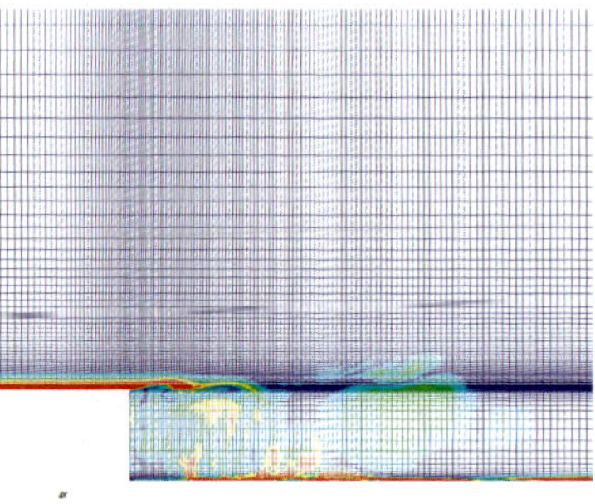

Figure 4. Left: Optimal mesh arrangement for a full LES boundary layer or the second transition function, Right: Optimal mesh arrangement for the traditional DES boundary layer transition function

where $\nu$ is the cinematic viscosity and $y$ is the normal wall distance.

This function is mesh independent (works equal in irregular or regular meshes) but uses normal distance to the wall, so in dynamic mesh methods a huge computational effort is required.

## 3. ILLUSTRATIVE EXAMPLES

The *TermoFluids* CFD code, have been tested in different industrial situations. For example, external aerodynamics optimisation of a car, simulation of industrial food storage chambers and also in the thermal optimisation of the nacelle of wind turbines.

The nacelle contains the key components of the wind turbine, including the gearbox, the electrical generator, the electronic controller and the electrical transformer. Each of these components contain internal heat sources so a heat transfer optimisation is needed to avoid overheating in the gearbox, the generator or/and the transformer. The fluid motion is governed by mixted convection and fluid flow is turbulent (being turbulence modelling important to achieve accurate results).

The boundary conditions of the simulation are:

- Gearbox: solid wall temperature fixed at $65^oC$.
- Electrical generator: solid wall temperature fixed at $50^oC$.
- Electrical transformer: solid wall temperature fixed at $105^oC$.
- Electronic controller: solid wall temperature fixed at $100^oC$.
- Output zones: pressure condition,

$$\gamma_b \frac{\rho_b v_b^2}{2} = p_o - p_b \tag{9}$$

- Input zones: input velocity at $1\frac{m}{s}$

The turbulence model used is the Dynamic Smagorinsky SGS model with a mesh of $1.5 \cdot 10^6$ of control volumes. Six days of computations were necessary in order to achieve the statistically stationary motion using 20 CPU's.

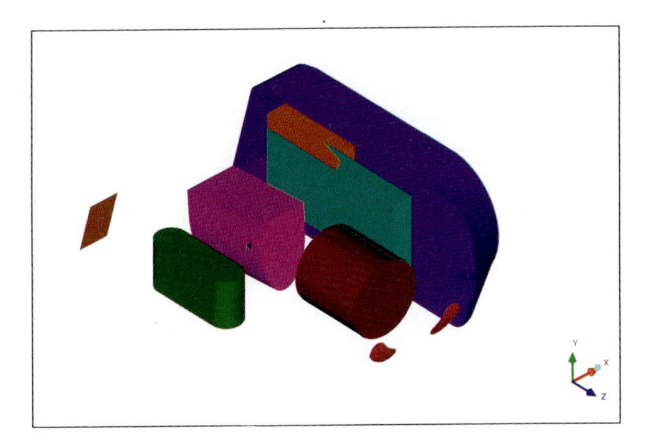


Figure 5. Internal elements of the nacelle of the wind turbine. In green the transformer, in red the gearbox, in pink the generator and the other elements are the electronics controllers

In figure 6, the average stream traces can bee seen. The flow is governed by the mixed convection. So there are some zones where the governing motion is a the cellular flow (being natural convection dominant) and other zones where the forced convection is the principal source of heat transfer. In this kind of flow, the traditional models like RANS, fail to predict the flow motion.

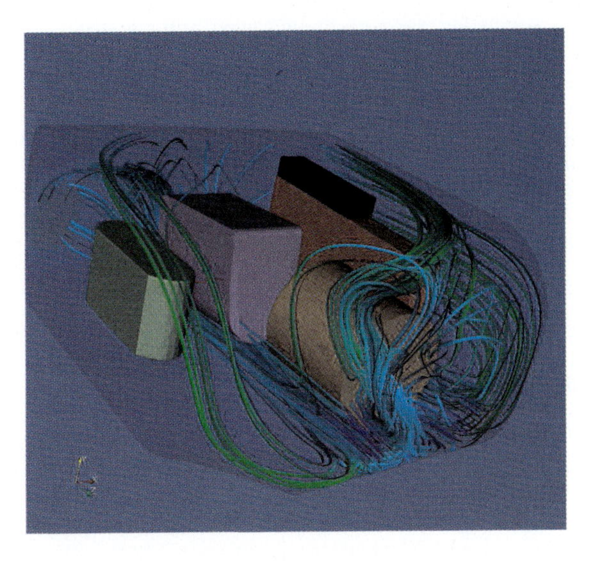

Figure 6. Averaged stream traces of the nacelle of the wind turbine simulation are presented. Dynamic Smagorinsky SGS model with a mesh of $1.5 \cdot 10^6$ of c.v.

## 4. CONCLUSIONS

The main features of *TermoFluids* have been presented. The most relevant aspects from a parallel computing point of view, such as the communication between CPUs and parallel direct and iterative algebraic solvers that allow *TermoFluids* to run efficiently on loosely-coupled parallel computers have been explained in detail. Furthermore, the different approaches for turbulence modelling implemented in *TermoFluids* (RANS, LES and hybrid LES/RANS models) have been pointed out. Illustrative results of numerical simulation of industrial problems have showed the potential of *TermoFluids* to accurate simulation of industrial flows as the flow inside the nacelle of a wind turbine.

## ACKNOWLEDGEMENTS

This work has been partially supported financially through the Ministerio de Educacion y Ciencia Español with the Research project ENE2007-67185 (Development of high performance parallel codes for optimal design of thermal systems) and the Research project C-06619 (Company: Ecotecnia, Analysis of the thermal and fluid-dynamic behaviour of the nacelle of a w.t.).

## REFERENCES

1. M.Soria, F.X. Trias, C.D. Perez-Segarra and A.Oliva; Direct numerical simulation of a three-dimensional natural convection flow in a differentially heated cavity of aspect ratio 4; Numerical Heat Transfer, Part A, 45 (2004); 649-673.
2. A.G.Kravchenko and P.Moin; Numerical studies of flow over a circular cylinder at $Re_D = 3900$; Physics of Fluids 12 (2000); 403-417.
3. P.R.Spalart, S.Deck, M.L.Shur, K.D.Squires, M.Kh.Strelets and A.Travin; A new version of detached-eddy simulation, resistant to ambiguous grid densities; Theor. Comput. Fluid Dyn. 20 (2006).
4. Y.Saad, M.H.Schultz; GMRES: A generalized minimal residual algorithm for solving non-symmetric linear systems; SIAM J.Sci.Stat.Comput. (1986).
5. F.X.Trias, M.Soria, C.D. Perez-Segarra and A.Oliva; A Direct Schur-Fourier Decomposition for the Efficient Solution of High-Order Poisson Equations on Loosely Coupled Parallel Computers; Numerical Linear Algebra with Applications 13 (2006); 303-326.
6. R.Borrell, O.Lehmkuhl, M.Soria and A.Oliva; A Direct Schur method for the solution of Poisson equation with unstructured meshes; to be presented at Parallel CFD Conference (2007).
7. G.Karypis and V.Kumar; MeTIS: A software package for partitioning unstructured graphs, partitioning meshes and computing fill-reducing ordering of sparse matrixes, version 4; 1998.
8. J.Misra and D.Gries; A constructive proof of Vizing's theorem; Inform. Process. Lett. 41 (1992); 131-133.
9. M.A.Leschziner; Modelling turbulent separated flow in the context of aerodynamics applications; Fluid Dynamics Research 38 (2006); 174-210.
10. U.Piomelli; Large-eddy simulation: achievements and challenges; Progress in Aerospace Sciences 35 (1999); 335-362.
11. K.Mahesh, G.Constantinescu, P.Moin; A numerical method for large-eddy simulation in complex geometries;Journal of Computational Physics 197 (2004) 215-240.
12. F.R.Menter; Improved Two-Equation k-w Turbulence Models for Aerodynamic Flows; NASA TM-103975.

# Schur Complement Methods for the solution of Poisson equation with unstructured meshes

R.Borrell[a], O.Lehmkuhl[a], M.Soria[b] and A.Oliva[b]

[a]TERMO FLUIDS S.L.
c/ Magi Colet 8, 08204 Sabadell, Spain

[b]Centre Tecnològic de Transferència de Calor(CTTC)
ETSEIAT,c/Colom 11, 08222 Terrassa, Spain
email: oliva@cttc.upc.edu, web page: htpp://www.cttc.upc.edu

Some variants of the Schur Decomposition algorithm [1] to solve the Poisson equation on complex geometries with unstructured meshes are presented. This algorithms have been designed to be applied to Large Eddy Simulations with low cost parallel computers, in the context of the new CFD code *TermoFluids* [2]. Numerical experiments carried out in order to test the robustness and efficiency of the algorithms are presented. Preliminary tests show that in general Schur Complement techniques accelerate the convergence and are significantly faster than others approaches such as Krylov methods with sparse approximate inverse preconditioners [7].

**Keywords:** loosely coupled parallel computers; linear solvers; Schur complement methods; Poisson equation; unstructured meshes

## 1. INTRODUCTION

Many important applications in the computational fluid dynamics (CFD) field demand huge computing power and need parallel computers to be feasible. The Poisson equation, which arises from the incompressibility constraint and has to be solved at least once at each time step, is usually the main bottleneck from a parallel point of view. In this context, efficient and scalable algorithms for the solution of the Poisson equation on a wide range of parallel systems are of high interest.

There are different factors that can affect the efficiency of a parallel algorithm. However, considering the limited communication capacity of PC clusters, one of the major factors that contributes to the degradation of the efficiency of the parallel implementation is the overhead due to the exchange of data among the different processors.

In order to overcome this problem, in this work some variants of the Schur complement algorithm have been developed. Based on previous experiences of the application of this technique for the solution of the Poisson equation on *Cartesian grids* with a single message

I.H. Tuncer et al., *Parallel Computational Fluid Dynamics 2007*,
DOI: 10.1007/978-3-540-92744-0_35, © Springer-Verlag Berlin Heidelberg 2009

[3], an adaptation to unstructured meshes and general sparse matrices is here proposed.

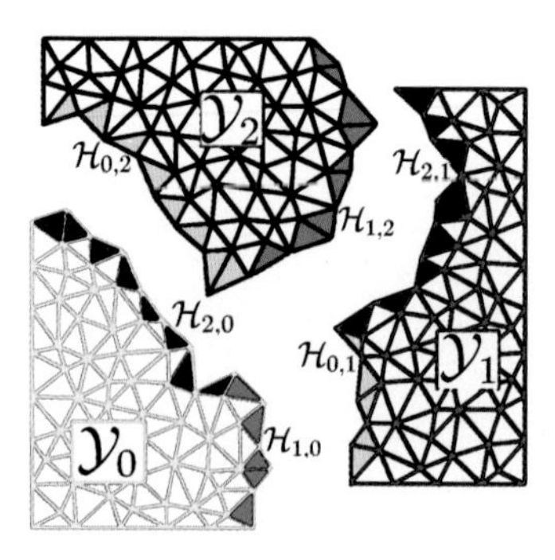


Figure 1. Sparse matrix distribution.

## 2. LINEAR SYSTEM DISTRIBUTION

The Schur Decomposition Methods presented on this paper have been implemented to solve the discretized Poisson Equation derived from the pressure correction schemes of pressure based algorithms used by *TermoFluids* [2] code. As a result of the discretization, the linear system of equations obtained is:

$$Ax = b \tag{1}$$

where $A$ is an n x n sparse matrix, symmetric and positive definite.

It is necessary to subdivide the problem into $p$ parts, being $p$ the number of processors used on the iterations. A partition of the unknowns set $\mathcal{Y}$ is carried out. After this process an unknowns subset $\mathcal{Y}_i$, called *local* unknowns, and the corresponding equations are assigned tho each processor. Some equations may link variables of different processors making exchange of data between them necessary.

Given two subsets of variables $\mathcal{J}_1$, $\mathcal{J}_2 \subset \mathcal{Y}$ , the *halo* of $\mathcal{J}_1$ in $\mathcal{J}_2$ defined as

$$\mathcal{H}(\mathcal{J}_1, \mathcal{J}_2) := \{k \in \mathcal{J}_2 : \exists s \in \mathcal{J}_1 \ with \ a_{s,k} \neq 0, a_{s,k} \in A\} \tag{2}$$

is the set of unknowns of $\mathcal{J}_2$ coupled with unknowns of $\mathcal{J}_1$. Some abbreviations are defined (see figure 1):

$$\mathcal{H}_{i,j} := \mathcal{H}(\mathcal{Y}_i, \mathcal{Y}_j) \qquad \mathcal{H}_i := \bigcup_{j \neq i} \mathcal{H}_{i,j} \tag{3}$$

A good distribution of data between processors is a crucial point for the efficiency of the parallel algorithms, some characteristics that are required to the partition are: 1) similar size of the partition elements $\mathcal{Y}_0,...,\mathcal{Y}_{p-1}$ to ensure load balance; 2)minimum size of $\mathcal{H}_0,...,\mathcal{H}_{p-1}$ to reduce communication costs. The partition is carried out by means of a graph partitioner such as METIS [9] applied to the adjacency graph of A. METIS minimizes the communications costs (edge-cuts) and balance the computational load (number of vertices for subgraphs).

Although the methods presented in this work are purely algebraic, the associated nodes or control volumes of the mesh are used to clarify the arguments. Since the discretization schemas of *TermoFluids* [2] code only use the adjacent nodes, the adjacency graph of A is strictly equivalent to the adjacency graph of mesh nodes, see figure 1.

## 3. SCHUR COMPLEMENT TECHNIQUES

The main idea of Schur complement methods is decompose the initial system of equations into the *internal* and *interface* systems, verifying that the internal equations do not link variables of different processors. As a consequence, the internal equations are separated from the distributed system and can be solved by each processor independently.

### 3.1. Internal and Interface unknowns

The unknowns set of one processor $\mathcal{Y}_i$ is partitioned into the *internal* unknowns $\mathcal{U}_i$, and *inferface* unknowns $\mathcal{S}_i$. The internal unknowns of one processor are not coupled with internal unknowns of other processors, this is equivalent to the condition

$$\mathcal{H}(\mathcal{U}_i, \mathcal{Y}_j) \cap \mathcal{U}_j = \varnothing \quad \forall j \neq i, \;\; i,j \in \{0,...p-1\} \tag{4}$$

.

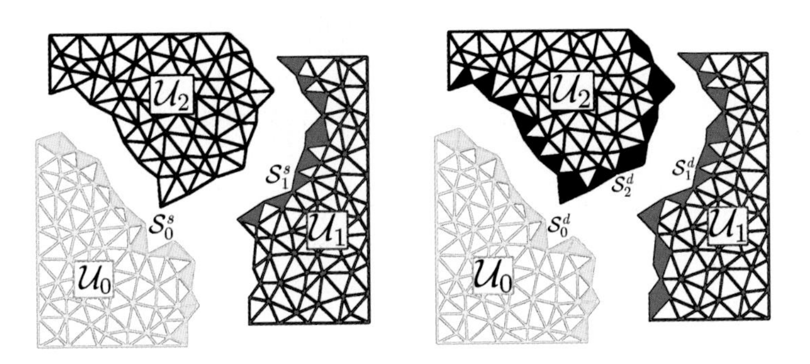


Figure 2. Left: simple interface. Right: double interface.

The global interface $\mathcal{S}$ and internal $\mathcal{U}$ unknowns sets are defined as

$$\mathcal{S} := \bigcup_{i\in\{0,..p-1\}} \mathcal{S}_i \qquad\qquad \mathcal{U} := \bigcup_{i\in\{0,..p-1\}} \mathcal{U}_i \tag{5}$$

There are different possibilities to fix $\mathcal{S}$. One approach, here called *double* interface $\mathcal{S}^d$ , is:

$$\mathcal{S}^d := \bigcup_{i \in \{0,..p-1\}} \mathcal{S}_i^d, \qquad where \quad \mathcal{S}_i^d := \bigcup_{j \neq i} \mathcal{H}_{j,i} \tag{6}$$

the interface of each processor is determined as the local variables coupled with variables of other processors (right part of Figure 2) [4]. $\mathcal{H}(\mathcal{U}_i, \mathcal{S}^d) \subseteq \mathcal{S}_i^d$ and the balance of $\mathcal{S}^d$ depends on the initial balance of $\mathcal{H}_0,...,\mathcal{H}_{p-1}$.

An other possibility, here called *simple* interface $\mathcal{S}^s$, is:

$$\mathcal{S}^s := \bigcup_{i \in \{0,..p-1\}} \mathcal{S}_i^s, \qquad where \quad \mathcal{S}_i^s := \bigcup_{j > i} \mathcal{H}_{j,i} \tag{7}$$

the interface of one processor is determined as the variables in $\mathcal{Y}_i$ coupled with variables of $\mathcal{Y}_j$ with $j > i$ (left part of Figure 2) . It is easy to see that condition (4) is verified ($A$ is symmetric). $|\mathcal{S}^s| < |\mathcal{S}^d|$ , but the imbalance of $\mathcal{S}^s$ is worse in general, for example, $\mathcal{S}_{p-1}^s = \varnothing$. On the other hand $\mathcal{H}(\mathcal{U}_i, \mathcal{S}^s) \nsubseteq \mathcal{S}_i^s$ in general.

In order to obtain a well balanced distribution of $\mathcal{S}^s$ an iterative algorithm has been developed on this paper, see figure 3. The main step of the algorithm is:

Interface balance algorithm

1. $\forall\, i, j \in \{0, ..p-1\}$
2. $If(\ |\mathcal{S}_i^s| > |\mathcal{S}_j^s|\ and\ \exists k \in \mathcal{H}_{j,i} \cap \mathcal{S}_i^s\ )$
3. $\{$
4. $\quad \mathcal{S}_i^s := \mathcal{S}_i^s - \{k\}$
5. $\quad \forall l \neq i\ \ \mathcal{S}_l^s := \mathcal{S}_l^s \cup \mathcal{H}(\{k\}, \mathcal{Y}_l)$
6. $\}$

Figure 3. Interface balance algorithm

A few explanations are in order. In line 4 the size of $\mathcal{S}_i^s$ is reduced, while in line 5 the interface size of processors with variables coupled with $k$ is increased. It is obvious that processor $j$ is one of these because $k \in \mathcal{H}_{j,i}$ and $A$ is symmetric. This step is repeated until the imbalance reduction gets stalled.

### 3.2. Schur complement system derivation

In order to clarify the exposition the unknowns are re-ordered listing first the internal unknowns of all processors $\mathcal{U}_0, ..., \mathcal{U}_{p-1}$ followed by the interface unknowns $\mathcal{S}_0, .., \mathcal{S}_{p-1}$. The block system derived is:

$$\begin{bmatrix} A_0 & 0 & \cdots & 0 & F_0 \\ 0 & A_1 & \cdots & 0 & F_1 \\ \vdots & & & \vdots & \\ 0 & 0 & \cdots & A_{p-1} & F_{p-1} \\ E_0 & E_1 & \cdots & E_{p-1} & C \end{bmatrix} \begin{bmatrix} x_{\mathcal{U}_0} \\ x_{\mathcal{U}_1} \\ \vdots \\ x_{\mathcal{U}_{p-1}} \\ x_{\mathcal{S}} \end{bmatrix} = \begin{bmatrix} b_{\mathcal{U}_0} \\ b_{\mathcal{U}_1} \\ \vdots \\ b_{\mathcal{U}_{p-1}} \\ b_{\mathcal{S}} \end{bmatrix} \tag{8}$$

where $A_i \in \mathcal{U}_i \times \mathcal{U}_i$ represents the linear dependences between one processor internal variables, $F_i \in \mathcal{U}_i \times \mathcal{S}$ the linear dependences between internal and interface variables, $E_i = F_i^T$ and $C \in \mathcal{S} \times \mathcal{S}$ are the linear dependences between the interface variables. Gaussian elimination is applied to (8) and the next triangular block system is obtained:

$$\begin{bmatrix} A_0 & 0 & \cdots & 0 & F_0 \\ 0 & A_1 & \cdots & 0 & F_1 \\ \vdots & & & \vdots & \\ 0 & 0 & \cdots & A_{p-1} & F_{p-1} \\ 0 & 0 & \cdots & 0 & S \end{bmatrix} \begin{bmatrix} x_{\mathcal{U}_0} \\ x_{\mathcal{U}_1} \\ \vdots \\ x_{\mathcal{U}_{p-1}} \\ x_{\mathcal{S}} \end{bmatrix} = \begin{bmatrix} b_{\mathcal{U}_0} \\ b_{\mathcal{U}_1} \\ \vdots \\ b_{\mathcal{U}_{p-1}} \\ z \end{bmatrix} \tag{9}$$

where S is the Schur complement matrix and z the modified rhs for the interface variables

$$S = C - \sum_{i=0}^{p-1} E_i A_i^{-1} F_i \qquad \textit{and} \qquad z = b_{\mathcal{S}} - \sum_{i=0}^{p-1} E_i A_i^{-1} b_{\mathcal{U}_i} \tag{10}$$

in system (9) the linear dependences between variables of different processor are all located on the Schur complement submatrix $S$, the internal equations can be separated and its corresponding $x$ values evaluated by each processor independently:

$$x_{\mathcal{U}_i} = A_i^{-1}(b_{\mathcal{U}_i} - F_i x_{\mathcal{S}}) \tag{11}$$

## 4. SCHUR COMPLEMENT ALGORITHM

The different proposes of algorithms based on Schur complement decomposition methods depend on the solvers used to solve the interface and internal systems. On this paper complete sparse Cholesky factorization is used for the internal systems and preconditioned CG for Schur complement system. The general algorithm can be divided into two stages: *preprocessing* and *solution*. On the preprocessing stage the interface is evaluated, the Schur complement decomposition is carried out, and the internal factorizations and Schur complement matrix preconditioner are evaluated. The algorithm for the solution stage is described in detail hereafter.

**Internal systems via sparse Cholesky factorization**. The solution of internal system is required twice on the solution stage, first to evaluate the interface rhs $z$ , expression (10), and second to obtain the internal variables $x_{\mathcal{U}_i}$, system (11). An new ordenation of local variables is carried out in order to reduce the fill-in of Cholesky factors, METIS software [9] is used in this paper for this propose.

**Schur complement via approximate inverse**. The schur complement system can be solved using a preconditioned Krylov projection method. As the initial system is symmetric and positive definite, the Schur complement system is solved with CG accelerator. Sparse approximate-inverse preconditioner (SAI), described next, is used because its parallel benefits.

Given a systems of equations

$$Ax = b \tag{12}$$

the basic idea of approximate-inverse preconditioner is finding an approximation M to the inverse of A, by solving the optimization problem:

$$min_{M \in S} \parallel I - MA \parallel_F^2, \tag{13}$$

where S is a set of sparse matrices with a certain sparcity pattern. The optimization problem can be decoupled into $n$ minimization problems of the form

$$min_{m_j} \parallel e_j - m_j A \parallel_2^2, \ \ j = 1, 2, ..., n \tag{14}$$

where $e_j$ and $m_j$ are the jth rows of the identity matrix and $M \in S$, respectively.

The sparsity pattern of $S$ can be fixed with different strategies [7,8], in this paper a priori sparsity patterns based on the nonzero entries of $A^k$ $k = 1, 2, 3...$ are used, see reference [7]. Thershold strategy applied to $A$ for dropping small entries is carried out before evaluating $A^k$ pattern.

The entire solution stage would then consist of the following steps for processor i:

1. Use Cholesky factorization to solve $A_i v_i = b_{\mathcal{U}_i}$
2. Evaluate local contribution to new interface rhs: $c_i := E_i v_i$
3. Substract all contributions to $z$ (all-to-all communication): $z = b_{\mathcal{S}} - \sum_{i=0}^{p-1} c_i$
4. Use CG with SAI preconditioner to solve the distributed system $\mathcal{S} x_{\mathcal{S}} = z$
5. Evaluate the new internal rhs: $w_i = b_{\mathcal{U}_i} - F_i x_{\mathcal{S}}$
6. Use Cholseky factorization to solve internal system: $A_i x_{\mathcal{U}_i} = w_i$
7. $x_{\mathcal{Y}_i} = (x_{\mathcal{U}_i}, x_{\mathcal{S}_i})$

## 5. NUMERICAL EXPERIMENTS

The methods described have been implemented for the solution of Poisson equations

$$\triangle \varphi = f \tag{15}$$

derived from the pressure correction schemes of pressure based algorithms used by *TermoFluids* [2] code. Numerical experiments in rectangular boxes with Newman boundary conditions are carried out. Non-structured tetahedrical meshes are used for the discretization.

The performance of the Schur complement algorithm using *double* and *simple* interface is compared with performance of CG with inverse preconditioner. This tests have been performed on JFF cluster in CTTC center. The components of this cluster are: Intel Core2 Duo E6700 (2'67 Ghz) nodes linked with a 100 Mbit/s Ethernet. The methods compared are:

- **CGSAI**: CG accelerator with inverse preconditioner. The pattern of the inverse preconditioner is fixed as the non zero entries of $A^2$.
- **SCHURD**: Schur decompositon with *double* interface. The *inner* systems are solved with sparse Cholesky factorization and the Schur complement system via inverse preconditioner [7]. The pattern of the inverse preconditioner is the initial entries of Schur complement matrix dropping elements with absolute value smaller than $10^{-10}$.
- **SCHURS**: The same as SCHURD with *simple* interface instead of *double* interface.

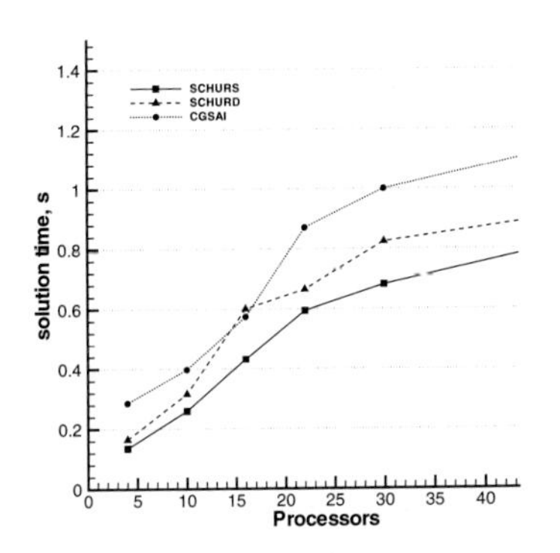


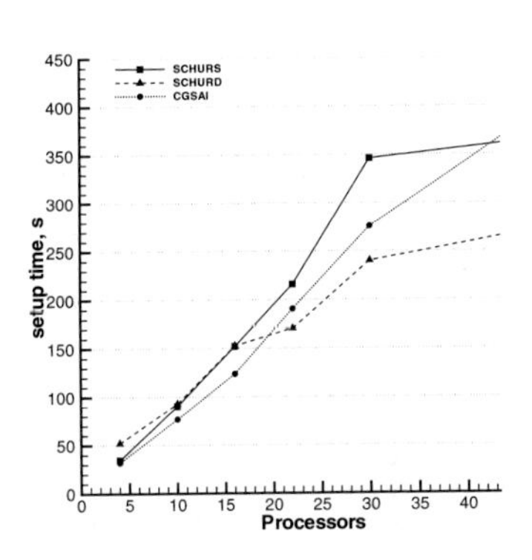


Figure 4. Solution and preprocessing times for Poisson equation with local subproblem size constant at 30.000 unknowns using three different methods CGSAI, SCHURD and SCHURS.

In figure 4 it is presented the evolution of CPU time for solution and preprocessing stages keeping the load for cpu constant at 30.000 unknowns, and increasing the number of processors from 4 to 46. In an ideal situation with perfect parallel performance, the cpu time would remain constant. The residual norm is reduced by $10^{-4}$.

In the context of Large Eddy Simulations, in a time-stepping scheme, the Poisson equation has to be solved repeatedly with different rhs on each time step while the system matrix remains constant. The preprocessing is carried out only one time while commonly the number of time steps is very large.Therefore, a large computing time can be spent in the preprocessing stage if the solution time is reduced.

It can be seen on figure 4 that the solution time grows 5.9, 5.5 and 3.9 times respectively while the size of the problem is increased 11.5 times. This growth is decelerated when the number of computers increases.

## 6. CONCLUDING REMARKS

A Schur complement method based on the direct solution of the *internal* equations and iterative solution of *interface* system has been described. Two different strategies to determine the *interface* variables are considered: *simple* and *double*. On the first case an algorithm for the load balance of interface variables is proposed, on the second case the load balance depends on the initial system distribution.

These methodologies have been implemented for the solution of Poisson equations on the context of the new CFD code *TermoFluids* [2]. Numerical experiments show that simple interface method is less expensive and the benefits of this techniques compared with other approaches such as krylov subspace methods with inverse preconditioners [7].

## REFERENCES

1. F.X.Trias, M.Soria, C.D. Perez-Segarra and A.Oliva. A Direct Schur-Fourier Decomposition for the Efficient Solution of High-Order Poisson Equations on Loosely Coupled Parallel Computers. Numerical Linear Algebra with Applications 13 (2006), 303-326.
2. O.Lehmkuhl, R.Borrell, C.D. Perez-Segarra, M.Soria and A.Oliva. *TermoFluids*: A new Parallel unstructured CFD code for the simulation of turbulent industrial problems on low cost PC Cluster. Parallel CFD Conference (2007).
3. M.Soria et al. A Direct Parallel Algorithm for the Efficient Solution of the Pressure-Correction Equation of Incompressible Flow Problems using Loosely Coupled Computers. *Numerical Heat Transfer - Part B*, vol. 41, pp. 117–138, 2002.
4. Y.Saad et al. Distributed Schur Complement Techniques for General Sparse Linear Systems. *SIAM J.SCI.COMPUT.*, vol. 21 No. 4, pp. 1337–1356, 1999.
5. Y.Saad. Iterative Methods for Sparse Linear Systems. PWS publishing, New York, 1996.
6. Tim Davis. Direct Methods for Sparse Linear Systems. SIAM Series on the Fundamentals of Algorithms, 2006.
7. Edmond Chow. Parallel Implementation and Practical Use of Sparse Approximate Inverse Preconditioners With a Priori Sparsity Patterns. *The international Journal of High Performance Computing Applications*, vol. 15, pp. 56–74, 2001.
8. T.Huckle and M.Grote. A new approach to parallel preconditioning with sparse aproximate inverses. Technical Report SCCM-94-03, Stanford University, Scientific Computing and Computacional Mathematics Program, Stanford, California, 1994.
9. G.Karypis and V.Kumar. MeTIS: A software package for partitioning unstructured graphs, partitioning meshes and computing fill-reducing ordering of sparse matrixes, version 4, 1998.

# Blood Flow Simulation in Cerebral Aneurysm: A Lattice Boltzmann Application in Medical Physics

Jörg Bernsdorf*[a] and Dinan Wang[a]

[a]NEC Laboratories Europe, NEC Europe Ltd., Rathausallee 10,
53757 St.Augustin, Germany.

From fundamental research to patient specific treatment planning computer simulations play an increasingly important role in the area of Medical Physics. In this paper we address one particular application, the simulation of blood flow in cerebral aneurysm in domains created from medical images. Our focus is on considering the correct blood flow rheology by comparing results from Newtonian and non-Newtonian flow simulations with the lattice Boltzmann method.

## 1. The Medical Background

A potential application of numerical blood flow simulation is to aid decision making processes during treatment of cardiovascular disease. One example of this is in the treatment of aneurysms. Aneurysms are extreme widenings of vessels which can be, if they rupture, life threatening. One method of treatment involves insertion of a metal frame known as a stent, to divert flow from the aneurysm. An alternative is to pack the aneurysm with wire; a procedure known as coiling. The resulting modification of the flow field triggers the process of blood clotting inside the aneurysm and in future, the flow-field following treatment can be predicted by computer simulation. This may ultimately give an insight into the success of the treatment and long-term prognosis. *In vivo* measurements of specific flow properties are possible, but usually not precise enough to predict for example, wall shear stress or pressure distribution with a sufficient spatial resolution. Since invasive treatments of the brain can be problematic, a pre-surgery risk assessment for the likelihood of rupture of the aneurysm in question is a challenging goal.

To achieve this goal, necessary steps for a precise numerical simulation of flow properties within an untreated aneurysm together with preliminary results will be presented in this paper.

The role of models such as these are currently being investigated within two European research projects; @neurIST [1] and COAST [2].

## 2. Numerical Method

### 2.1. Image Segmentation

Discretising the geometry for flow simulations from a CT or MR image is a challenging task. Depending on the applied method, the resulting geometry can vary, and advanced methods must be applied to generate suitable data for the flow simulation. This may include manipulations of the triangulated surface mesh representing the geometry. Usually, from these data, in the case of lattice Boltzmann, a voxel-mesh with adequate resolution is generated.

*j.bernsdorf@it.neclab.eu

I.H. Tuncer et al., *Parallel Computational Fluid Dynamics 2007*,
DOI: 10.1007/978-3-540-92744-0_36, © Springer-Verlag Berlin Heidelberg 2009

### 2.2. Lattice Boltzmann

The lattice Boltzmann (LB) method is based on the numerical simulation of a time, space and velocity-discrete Boltzmann-type equation. The propagation and interaction of the particles of an 'artificial computer fluid' are calculated in terms of the time evolution of a density distribution function, representing an ensemble average of the particle distribution. The flow velocity and the fluid density are derived from the moments of the (time and space-discrete) density distribution function, while the pressure is linked to the density by the speed of sound. It can be shown that these flow quantities fulfill the time dependent incompressible Navier–Stokes equations under certain conditions [3].

For simplicity, an equidistant orthogonal lattice is chosen for common LB computations. On every lattice node $\vec{r_*}$, a set of $i$ real numbers, the particle density distributions $N_i$, is stored. The updating of the lattice essentially consists of two steps: a streaming process, where the particle densities are shifted in discrete time steps $t_*$ through the lattice along the connection lines in direction $\vec{c_i}$ to their next neighbouring nodes $\vec{r_*} + \vec{c_i}$, and a relaxation step, where the new local particle distributions are computed by evaluation of an equivalent to the Boltzmann collision integrals ($\triangle_i^{Boltz}$).

For the present computations, the 3D nineteen–speed (D3Q19) lattice Boltzmann model with single time Bhatnagar–Gross–Krook (BGK) relaxation collision operator $\triangle_i^{Boltz}$ proposed by Qian et al. [4] is used.

### 2.3. Performance Optimised Implementation

The lattice Boltzmann method is said to be very efficient and easy to implement. But in most cases described in the literature, a simple full matrix implementation is used, where also the solid fraction is allocated in the computer memory. Depending on the geometry, this is a considerable waste of resources, and for vector computers, also of CPU-cycles.

### 2.4. Full Matrix vs. Sparse Implementation

In the framework of a simple full-matrix implementation, the density distribution array for the whole bounding box is allocated in memory. This results in $19 * lx * ly * lz$ REAL numbers for the D3Q19 model for a $lx * ly * lz$ lattice.

Well known methods from sparse matrix linear algebra were first applied to the lattice Boltzmann method by Schultz et al. [5], suggesting storage of the density distribution only for the fluid nodes. This requires keeping an adjacency list for the next neighbours' addresses, but (depending on the geometry) can save considerable memory. Only $N * 19$ REAL numbers for the density distribution (N=number of fluid cells) and $N * 19$ INTEGERs for the adjacency list have to be stored in case of a sparse LB implementation.

For our simulations, a pre-processor was implemented which reads the voxel geometry and generates a list of fluid nodes together with the required adjacency list. It was shown that this implementation has significant advantages with regards to memory consumption and performance, particularly for medical geometries [6].

### 2.5. Non-Newtonian Model

The literature on blood rheology gives a strong indication, that the non-Newtonian effects of blood flow must not be neglected for a variety of geometries (see, e.g. [7–9]). Particularly for estimating the rupture-risk within cerebral aneurysm, the precise knowledge of quantities as pressure distribution and wall shear stress are expected to be crucial.

For our simulations, we implemented a Carreau-Yasuda (C-Y) model:

$$\frac{\mu - \mu_\infty}{\mu_0 - \mu_\infty} = (1 + (\lambda\dot{\gamma})^a)^{(n-1)/a} , \qquad (1)$$

where $\mu_0$ and $\mu_\infty$ are the dynamic viscosities at zero and infinite shear rate respectively, $\dot{\gamma}$ is the shear rate and $\lambda$ is a characteristic viscoelastic time of the fluid. At the critical shear rate $1/\lambda$ the viscosity begins to decrease. The power law index parameters $a$ and $n$ can be determined from experimental data. In our simulations we apply the following set of parameters for blood analog fluid [7]: $\mu_0 = 0.022$ Pa s, $\mu_\infty = 0.0022$ Pa s, $a = 0.644$, $n = 0.392$, $\lambda = 0.110$ s.

## 3. LB Simulation

MRI patient data were segmented and post-processed on a triangular mesh. From these data, a lattice Boltzmann VOXEL mesh of a size $lx * ly * lz = 160 * 120 * 100$ was generated, which is of sufficient resolution to allow a mesh-converged simulation. The computational domain contains 1.58 million fluid nodes.

10,000 iterations were carried out to reach a converged result, which required 88 s CPU–time on one CPU of CCRLE's SX-8 vector–computer. This is equivalent to an update rate of 39 million lattice site updates per second.

### 3.1. Flow parameters

For the simulation presented in this study, flow at a low Reynolds number of $Re = 12$ was considered for the Newtonian case. The parameters of the C-Y model were chosen accordingly to allow for a realistic capturing of the non-Newtonian blood rheology. Since a proper definition of Reynolds number is difficult in the case of non-Newtonian flow, it was made sure that for the same flow rate the total pressure loss is equivalent to that of the Newtonian simulation. At the inlet a constant flow velocity was applied and the pressure at the outlet was kept constant. Bounce-back wall boundaries were applied, resulting in a zero flow velocity at the wall.

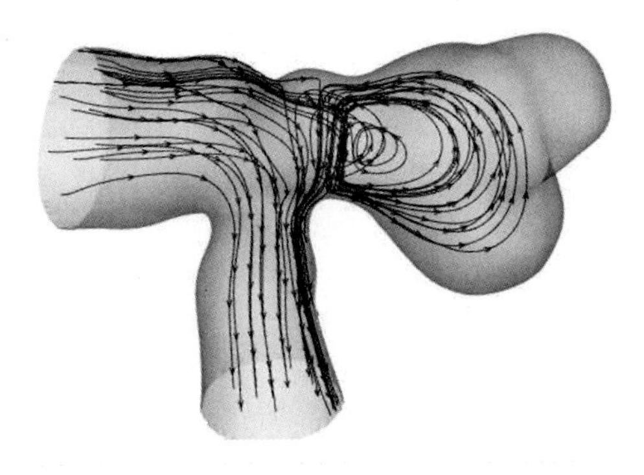

Figure 1. Streamlines indicating the complex flow inside the cerebral aneurysm.

### 3.2. Simulation results

The streamlines in Figure 1 reveal a complex flow pattern within the aneurysm. It can also be seen that only a small portion of the flow enters the aneurysm region, while the majority of trajectories directly follow the main branch. The vortex inside the aneurysm triggered by the main flow in the artery can be clearly identified.

The shear thinning behaviour of the non-Newtonian fluid, realised via the C-Y model (Eqn. 1), is expected to lead to a lower viscosity near the wall boundaries where the highest shear rates occur. This is illustrated by the simulation results in a cutting plane through the neck of the aneurysm (see Figure 2). The viscosity in this area for this particular Reynolds number lies in the range of $0.00081 Pa\,s \leq \mu \leq 0.00146 Pa\,s$. The equivalent viscosity for the Newtonian fluid (producing a similar pressure loss for the same flow rate) was set to $0,00111 Pa\,s$.

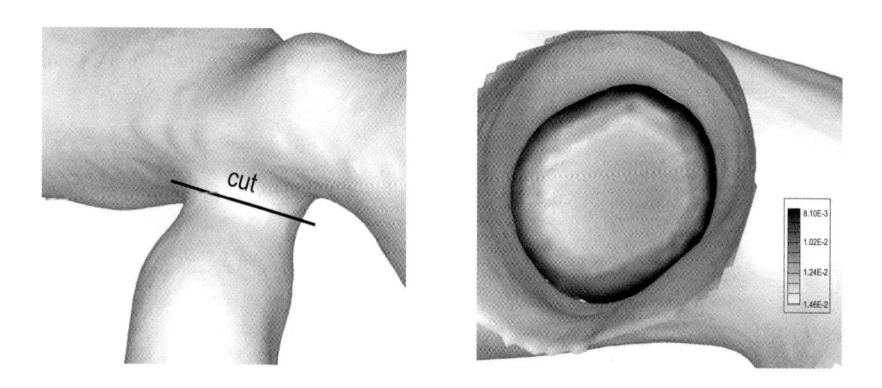


Figure 2. Cutting plane at the neck of the aneurysm (left) and viscosity distribution (right, darker colours indicate lower viscosity, quantities in Pa s, the vessel walls are also displayed).

When comparing the velocity distribution of the Newtonian and non-Newtonian flow, almost no difference can be observed (see Figure 3).

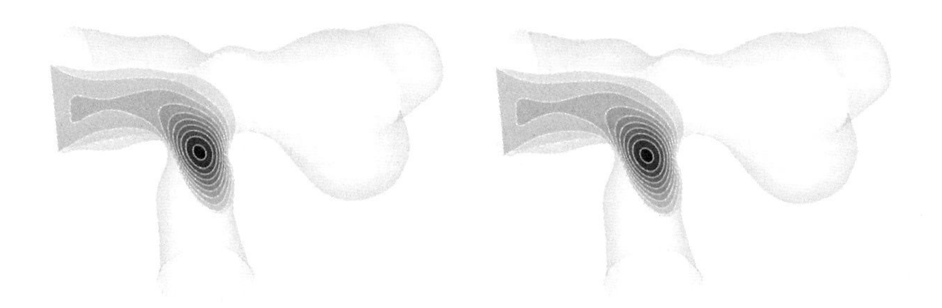

Figure 3. Velocity magnitude distribution at an x-z cutting plane inside the cerebral aneurysm. Left: Newtonian, right: non-Newtonian (darker colours indicate higher velocities).

Since the C-Y model causes the near-wall viscosity of the non-Newtonian fluid ($\mu \approx 0.00081 Pa\,s$ at the neck) to be significantly below the constant one of the Newtonian fluid ($\mu = 0,00111 Pa\,s$), it can be expected that this results in a lower non-Newtonian wall shear stress. Figure 4 compares the peak of the wall shear stress (occuring around the neck of the aneurysm) for both models. It can be observed that the maximum value is overestimated by approximately 30% for this configuration, if non-Newtonian effects are not taken into account.

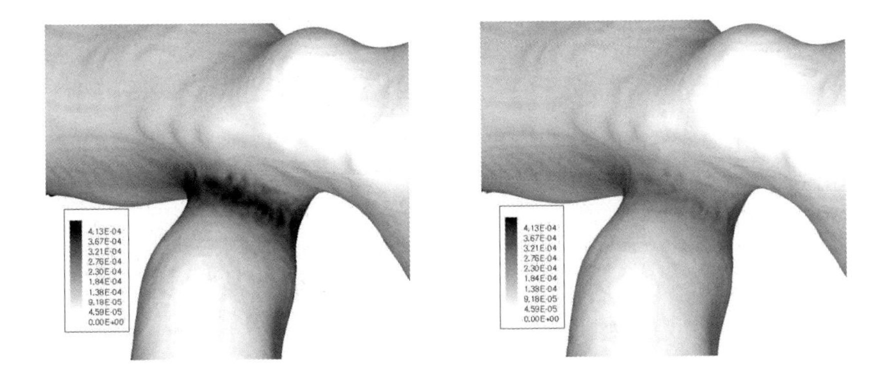


Figure 4. Wall shear stress distribution near the neck of the aneurysm. Left: Newtonian, right: non-Newtonian (darker colours indicate higher shear stress, quantities are dimensionless LB units).

## 4. Conclusion

Since the wall shear stress is an important quantity for estimating the rupture-risk of an aneurysm, the preliminary results presented here indicate that non-Newtonian effects should not be neglected within a numerical simulation.

Further work will include the investigation of the shear stress distribution for transient flow at higher Reynolds numbers for a larger variety of aneurysm geometries.

## 5. Acknowledgments

The lattice Boltzmann flow solver into which the C-Y model was integrated has been developed by the International Lattice Boltzmann Software Development Consortium. The surface mesh of the cerebral aneurysm was generated by Alberto Marzo (University of Sheffield) and Alessandro Radaelli (Universitat Pompeu Fabra, Barcelona). The voxel mesh for the lattice Boltzmann simulation was produced by Guntram Berti (NEC Europe Ltd.).

This work was generated in the framework of the @neurIST Integrated Project, which is co-financed by the European Commission through the contract no. IST-027703 and of the IST Project COAST, which is co-financed by the European Commission through the contract no. 033664.

## REFERENCES

1. http://www.aneurist.org
2. http://www.complex-automata.org
3. U.Frisch, D.D'Humières, B.Hasslacher, P.Lallemand, Y.Pomeau and J.-P.Rivet, Complex Systems **1**, 649-707 (1987)
4. Y.H.Qian, D.D'Humires, P.Lallemand, Lattice BGK models for Navier-Stokes equation, Europhysics Letters **17**, 479–484, (1992)
5. M.Schultz, M.Krafczyk, J.Tölke J. and E.Rank in: M.Breuer, F.Durst.C.Zenger (Eds.), High Performance Scientific and Engineering Computing, Lecture Notes in Computational Science and Engineering 21, Proceedings of the 3rd International Fortwihr Conference on HPSEC, Erlangen, March 12-14, 2001, 114-122, Springer (2002)
6. J.Bernsdorf, S.E.Harrison, S.M.Smith, P.V.Lawford and D.R.Hose, A Lattice Boltzmann HPC Application in Medical Physics, in: M.Resch et al. (Eds.), High Performance Computing on Vector Systems 2006, Proceedings of the High Performance Computing Center Stuttgart, Springer (2007)
7. F.J.H.Gijsen, F.N.Van de Vosse and J.D.Janssen, The influence of the non-Newtonian properties of blood on the flow in large arteries: steady flow in a carotid bifurcation model, J.Biomechanics **32**, 601–608, (1999)
8. S.Moore, T.David, J.G.Chase, J.Arnold, J.Fink, 3D models of blood flow in the cerebral vasculature, J.Biomechanics **39**, 1454–1463, (2006)
9. A.M.Artoli and A.Sequeira, Mesoscopic simulations of unsteady shear-thinning flows, *ICCS*, **Part 2**, 78–85, (2006)

# Unsteady Navier Stokes Solutions of Low Aspect Ratio Rectangular Flat Wings in Compressible Flow

**"Durmuş G. [a] , Kavsaoğlu M.Ş. [b] , Kaynak Ü.[c]"**

[a] *Anadolu University, Eskişehir,26470, Turkey*

[b] *İstanbul Technical University, İstanbul ,34469, Turkey*

[c] *University of Economics and Technology, Ankara, 06530, Turkey*

Keywords: CFD; Navier-Stokes; Multi-Block; Vortical Flows; Flow Separation.

## Introduction

Low aspect ratio rectangular flat wings are generally used in missiles. Some missiles, which are equipped with these types of wings, fly and maneuver at high angles of attacks. General description of the flow field is shown in Fig. 1. Flow separates at the sharp leading edge and forms the leading edge bubble. There are also two side edge vortices which are similar to the leading edge vortices of Delta wings. Various publications by AGARD provide a large selection of research results in this field [1-6]. Low aspect ratio rectangular flat wings were studied much less when compared to the Delta wings of the similar nature. Stahl [7] summarized some of the early research works. Winter [8] obtained pressure distributions on the suction side and force and moment measurements of various low aspect ratio rectangular flat wings at incompressible speeds. At von Karman Institute similar experiments were performed in a transonic wind tunnel for higher speeds [9, 10]. In the present study the experiments of van Westerhoven et. Al [9] and Kavsaoğlu [10] are taken as test cases for comparison of the results.

## Solution Algorithm

A parallel, multi block, time marching Navier Stokes solver is used [12]. The thin layer Navier Stokes equations are discretized by using the Beam and Warming [13] finite difference implicit algorithm. The matrix solution is carried out using a diagonally

I.H. Tuncer et al., *Parallel Computational Fluid Dynamics 2007*,
DOI: 10.1007/978-3-540-92744-0_37, © Springer-Verlag Berlin Heidelberg 2009

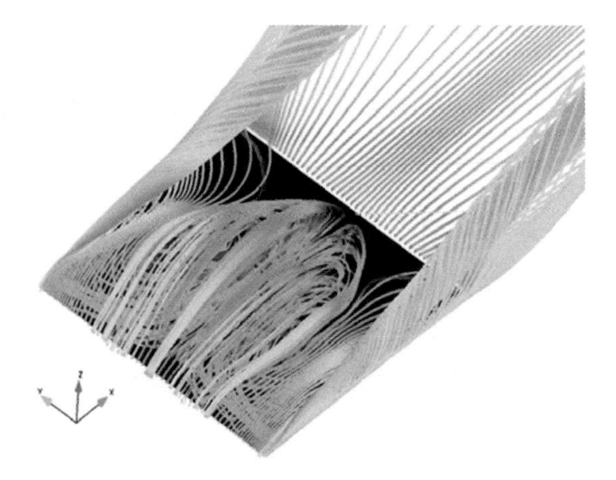

Figure 1. General description of the flow field

dominant LU-ADI factorization algorithm [14]. Each ADI operator is decomposed to the product of lower and upper bi-diagonal matrices. The method is first order accurate in time and second order accurate in space. This solution technique involves solving the time-dependent N-S equations. Multi block technique is used to reduce geometrically complex regions into several smaller, more manageable regions, called as blocks. The different blocks are solved as parallel processes. The Message Passing Interface (MPI) [16] library that is a kind of parallel processing protocol designed for distributed memory computing on clusters was used for communication between the processes. The basic concept is that a group of processors, tied together by a high-speed communications link, will solve each block of a computational domain and share the results at each time step by passing messages back and forth.In this study, above the flat plate, the turbulent eddy viscosity is calculated by using the two-layer algebraic turbulence model proposed by Baldwin and Lomax [15] with Degani Schiff modification [19]. The *f* function of the Baldwin Lomax model is calculated using the component of vorticity perpendicular to local flow direction (reduced vorticity [20]) given as $\vec{\omega}_R = \left|\vec{\omega} \times \vec{V}\right| / \left|\vec{V}\right|$. Baldwin Lomax model is also modified for the locations of side edge vortices. At a given $x$ station, first the minimum pressure point (the core of the vortex) is found. This point is set as the origin of a polar coordinate system in the search for $f_{max}$. At the wake behind the flat plate an algebraic wake type turbulence model is used [21].

## Verification Study

In order to show the validation of the code on the unsteady flow conditions, flow around the cylinder is performed as a verification study and the results are compared with the well known experimental data [22]. Geometry is a two-dimensional circular cylinder with a diameter D=1. Computational domain is composed of four blocks and totally 24000 grid points are used. The computational grid used for the cylinder problem is shown in Fig. 2. Mach number and Reynolds number was chosen as 0.4 and $1x10^4$, respectively. Time step is chosen as 0.002 in order to resolve the vortex shedding. Vortex shedding frequency based on speed of sound is calculated by using the following equation

$$1/f = t = (time\ step * ntsp * \frac{D}{a_\infty}) \quad (1)$$

The Strouhal number, non-dimensional vortex shedding frequency based on free stream velocity is defined as

$$St = f * \frac{D}{U_\infty} \quad (2)$$

In our problem, ntsp, the number of time steps required for one period, is approximately 6500 and time step is 0.002, then the Strouhal number is figured out to be 0.192, which is in agreement with the experimental value that was around 0.2.

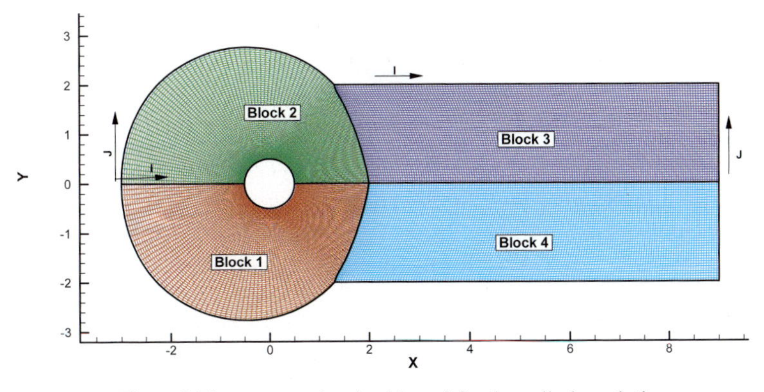


Figure 2 The computational grid used for the cylinder solution

## Computational Study

For the present computational study, an aspect ratio 1.0 flat plate with small thickness and sharp edges is considered as the test model. The flow field is assumed to be symmetric with respect to the y=0 plane. The test cases are selected from the experimental data acquired at the von Karman Institute (VKI) [9,10]. Test models are low aspect ratio, rectangular flat plates with small thickness and sharp edges. The experiments include surface oil flow measurements, force and moment measurements and surface pressure distribution measurements. Two-block, three-dimensional, structured grids shown in Fig. 3 were generated. In order to speed up the convergence of the solutions, two different grid configurations are adapted namely, the coarse and the fine grids. A three dimensional hyperbolic grid generation code [17] is employed to generate the grids.

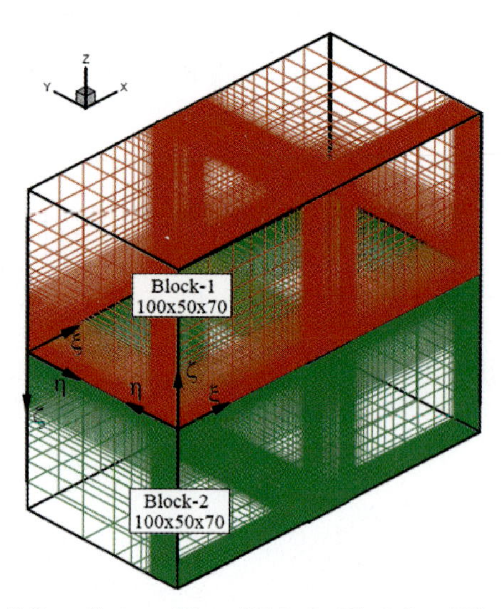


Figure 3 Overall view of the grid blocks Block 1 and Block 2

Computations are performed on Pentium IV 1500 MHz single processor workstation, which has 512 MB of memory. For the coarse grid, a two-block parallel solution with 2x46410 grid points was required 2x16 MB of RAM. Approximately 240 minutes of CPU-time was needed per 10000 iterations. On the other hand, for fine grid, a two-block parallel solution with 2x350000 grid points required 2x82 MB of RAM and approximately 1600 minutes of CPU-time was needed per 10000 iterations. The computational details are given in Table 1.

Table 1 Computational details

| | *CPU time (minute/10000 iterations)* | *Memory Requirement (Mb)* | *Number of Points* |
|---|---|---|---|
| Course Grid | 240 | 32 | 92820 |
| Fine Grid | 1600 | 164 | 700000 |

**Results and comparison**

In Fig. 4, the convergence histories of the L2 norm of the residual are shown for the case, Mach=0.55, $\alpha$ =13.5° as an example. For this case, the first 40,000 iterations were performed with the coarse grid by using local time step (steady solution). Then, flow variables were interpolated to the fine grid part of the solution and the iterations were continued by using global time step to obtain the unsteady solution. Surface pressure predictions are compared in Fig. 5 for the case Mach=0.85, $\alpha$ =13.5°. Since the flow is unsteady, the computational data presented in these figures are obtained after time averaging. At locations nearest to the leading edge, in the neighborhood of wing centerline, pressure suction levels show small deflections from the experiment.

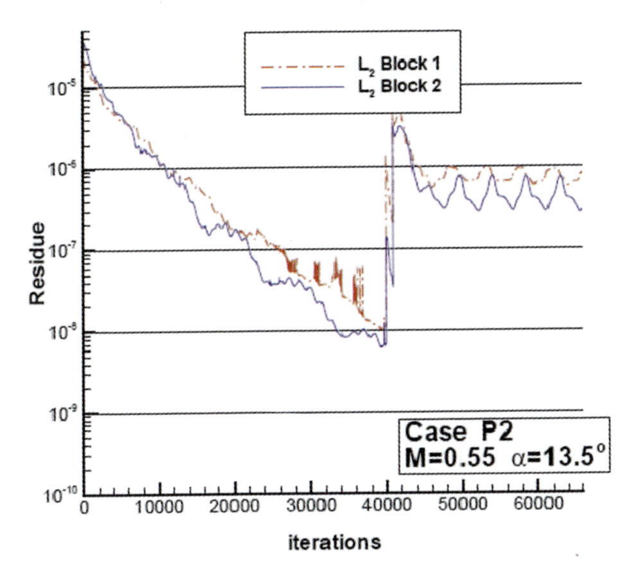


Figure 4 The convergence histories of the L2 norm of the residual, Mach=0.55, $\alpha$ =13.5°

The surface streamlines on the leeward side, obtained computationally and experimentally [10] are compared in Fig. 7 for the Case, Mach=0.42, $\alpha$ =5°. The experimentally obtained surface streamlines, were obtained by using the oil-flow method in the low speed wind tunnel [10]. The computationally obtained streamlines are drawn from the time-mean averaged data which is obtained by averaging during one period.

The predicted Normal Force's are compared with the experiment [10] at low (0.42) and high (0.85) Mach numbers. Normal Force values are obtained by the integration of the surface pressures from the top and the bottom surfaces for each case. Since the results have the oscillatory behavior, the variables were again averaged over the corresponding cycle to each case.

Figure 8 shows the averaged, the maximum and the minimum values of the Normal Force over a period of cycle at the particular angle of attack. It can be concluded that the differences between computational and experimental values are due to the differences between the pressure peak suction levels.

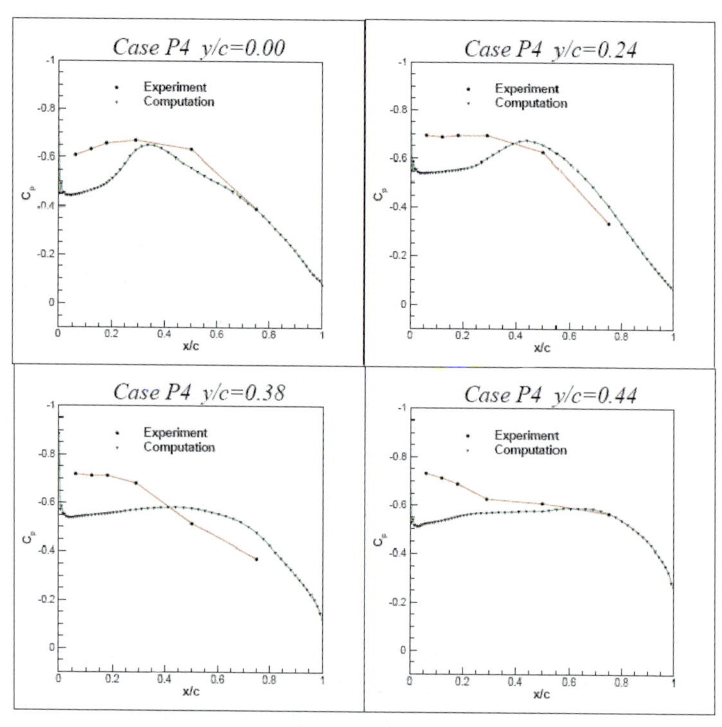


Figure 5. Comparison of computational and experimental surface pressures, Mach=0.85, $\alpha$=13.5°

Figure 6. Comparison of computational (time-mean averaged) and experimental [10] top surface streamlines, Mach = 0.42, $\alpha$=5°.

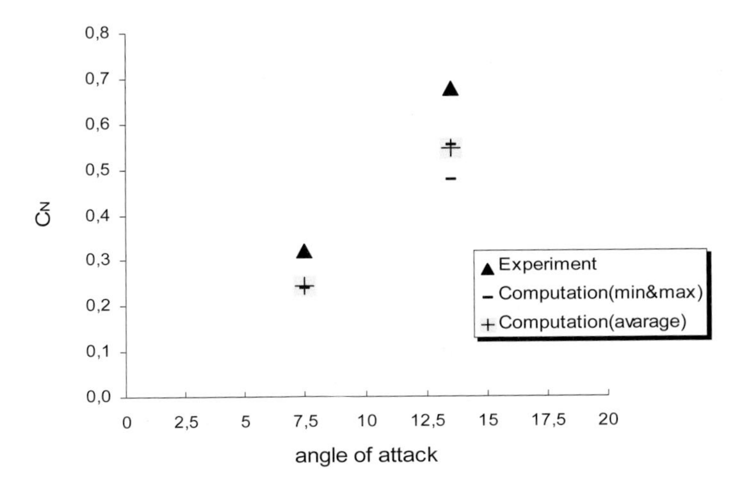


Figure 7 Comparison of computational and experimental [39] normal force coefficients, Mach = 0.85-0.87.

## Conclusion

On the lee side of a rectangular flat wing, there is a strong interference between the leading edge separation bubble and the side edge vortices making the flow field quite complicated. Moreover, it was detected that the flow is unsteady and has oscillating structure. The prediction of the surface streamlines was found to be accurate to capture the flow field structure in terms of size and formation of the primary and the secondary separation bubbles. The spanwise vortices at the side edges of the flat plate were in good agreement with the experiment, showing both the primary and the secondary vortices. It was found that the frequency of the unsteady flow, sourced by the separation bubble, increased with decreasing angle of attack and, its variation was more rapid at higher Mach numbers. It was also verified that with the increasing angle of attack, the size of the separation bubble is increased and the flow complexity is also increased showing more singular points; saddles and nodes on the surface. The periodic vortex shedding and the intermittent nature of the flow field were captured in three-dimensional time-resolved simulations. The averaged top surface pressure predictions were in good agreement except the leading edge region of the wing. This may be improved by using a more sensitive turbulence model, and by increasing the grid quality.

**References**

[1] AGARD CP-247, High Angle of Attack Aerodynamics, 1979.
[2] AGARD LS-98, Missile Aerodynamics, 1979.
[3] AGARD LS-121, High Angle of Attack Aerodynamics, 1982.
[4] AGARD CP-494, Vortex Flow Aerodynamics, 1991.

[5] AGARD CP-497, Maneuvering Aerodynamics, 1991.
[6] AGARD R-776, Special Course on Aircraft Dynamics at High Angles of Attack: experiments and Modelling, 1992.
[7] Stahl W. H., Aerodynamics of Low Aspect Ratio Wings, in AGARD LS-98, Missile Aerodynamics, 1979.
[8] Winter H., "Strömungsvorgange an Platten und profilierten Körpern bei kleinen Spannweiten". Forsch. Ing. -Wes., Vol. 6, 1935, pp. 40-50, 67-71. Also: Flow Phenomena on Plates and Airfoils of Short Span, NACA Rep. 798, 1937.
[9] van Westerhoven P., Wedemeyer E., Wendt J. F., "Low Aspect Ratio Rectangular Wings at High Incidences", Paper presented at the AGARD Symposium on Missile Aerodynamics, Trondheim, Norway, September 20-22, 1982.
[10] Kavsaoğlu M. Ş, "Flow Around Low Aspect Ratio Rectangular Flat Plates Including Compressibility", von Karman Institute For Fluid Dynamics, VKI-PR 1982-15, June 1982.
[11] Laçin F., Kavsaoğlu M., Navier Stokes Analysis of Low Aspect Ratio Rectangular Flat Wings in Compressible Flow, ICAS-94-10.3.3, 19th Congress of the International Council of the Aeronautical Sciences and AIAA Aircraft Systems Conference, September 18-23, Anaheim, CA, U.S.A.,
[12] Şen T. S., Development of a Three Dimensional Multiblock Parallel Navier Stokes Solver, PhD Thesis, Middle East Technical University, Department of Aeronautical Engineering, November 2001.
[13] Beam, R. W., and Warming, R. F., "An Implicit Finite Difference Algorithm for Hypersonic Systems in Conservation Form, Journal of Computaional Physics, Vol. 23, 1976, pp 87-110.
[14] Fujii, K., "Practical Applications of New LU-ADI Scheme for the Three Dimensional Navier-Stokes Computation of Transonic Viscous Flows," AIAA 24th Aerospace Sciences Meeting, Reno, Nevada, January, 1986.
[15] Baldwin, B. S., Lomax, H., "Thin Layer Approximation and Algebraic Model for Separated Turbulent Flows," AIAA 16th Aerospace Meeting, Huntsville, Alabama, January, 1978.
[16] MPI-2: Extensions to the Message Passing Interface, Message Passing Interface Forum, University of Tennessee, Knoxville, Tennessee, July 18, 1997.
[17] Durmuş, G., Three Dimensional Hyperbolic Grid Generation, MS Thesis, Middle East Technical University, Department of Aeronautical Engineering, September 1998.
[18] Tobak M., Peake D.J, "Topology of Three-Dimensional Separated Flows ", Annual Review of Fluid Mechanics, Vol. 14, pp. 61-85, 1982
[19] D.Degani and L.schiff "Computation of Turbulent Supersonic Flows Bodies Having Crossflow Separation", Journal of Computational Physics, Vol 66, pp. 173-196, 1986
[20] User's Guide, NPARC Flow Simulator Version 3.0, 1996, http://info.arnold.af.mil/nparc
[21] Magagnato, "Karlsruhe Parallel Program for Aerodynamics (KAPPA) Documentation", the Institute for Fluid Mechanics, University of Karlsruhe, 1996.
[22] White F. M., "Viscous Fluid Flow", McGraw-Hill, 1991.

# Case studies of solving large-scale CFD problems by means of the GasDynamicsTool software package

Alexey V. Medvedev[a]

[a]GDT Software Group, 27-79, Demonstratsii str., Tula, Russia, 300034

Numerical simulations of various CFD problems typically require extra-large computational domain usage with the number of modeling cells in the tens of thousands. The task of simulating such large models is one which many engineers and developers face, and is often a sticking point for many reasons. That's the motive for GDT Software Group to consider the development and optimization of parallel high performance modeling tools for large-scale tasks and multiple-processor systems as one of its first-rank priority goals [ 1]. The package is designed and is being continually optimized to provide the ability to simulate problems with 1-10 billions of computational cells effectively on ordinary cluster systems. The article describes attempts being made to improve software quality and performance on cluster systems with the number of CPUs up to 200 and some achievements made in production problem-solving on such systems.

GasDynamicsTool® and S-VR® are universal software packages developed by GDT Software Group for a wide range of CFD problems, including rocket systems modeling, combustion and detonation, artillery systems function, ecology problems and others. Ability of the package to work effectively on parallel computation systems is one of its notable features. Making it possible to solve as large as 10 billion-cell tasks on typical 200-CPU clusters in a most convenient and productive way is the task GDT Software Group paid special attention to last year.

## 1. Architecture peculiarities

In a new version of the GasDynamicsTool® package several changes to the parallelization scheme were made to achieve better parallel performance on cluster systems as well as on shared memory architectures. Beginning with the 5.95 version of the GasDynamicsTool package the new modular design of parallel systems support driver was introduced, which made it easy to create specially optimized versions for different kinds of interprocess communication fabrics and architectures. The goal was to achieve and prove a good parallel performance on cluster systems equipped with both low-end communication hardware (like on-board Gigabit Ethernet adapters) and middle-level fabrics (like Myrinet-2000), keeping top results on shared memory machines.

To implement the idea the schema of the application was changed to be more modular, as it is shown on Figure 1. The application itself now doesn't contain any direct calls of a specific parallel API (neither MPI, nor low level API like VAPI or MX, as well as no direct memory copy operations are present inside a solvers code itself). All the

I.H. Tuncer et al., *Parallel Computational Fluid Dynamics 2007*,
DOI: 10.1007/978-3-540-92744-0_38, © Springer-Verlag Berlin Heidelberg 2009

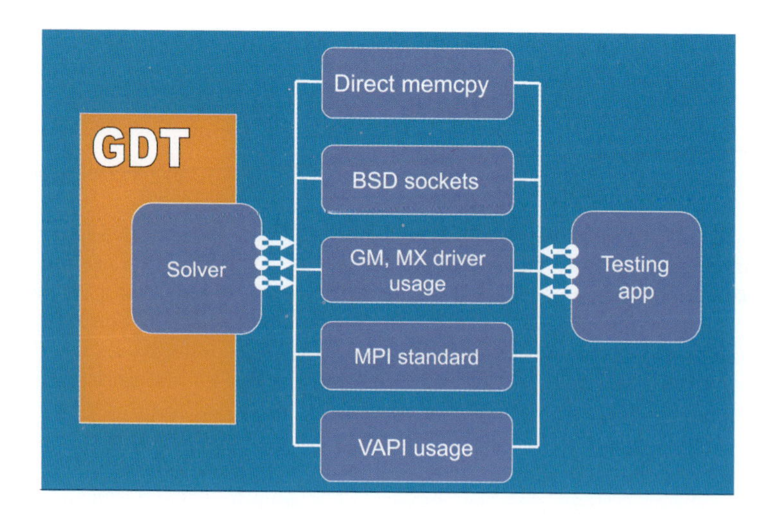


Figure 1. Software structure scheme

operations, requiring parallel processing of data are encapsulated as some abstract high-level actions, mostly of message-passing sense. Implementations of that operations are different for different kinds of parallel API to be used. Obviously, such architecture change required massive code reconsideration, but there were positive results like: code became much easier to maintain on number of different parallel architectures, change in a specific communication-level code doesn't interfere with any other code of the application, a small testing application (see Figure 1) can be implemented to make it much easier to debug the communication-level code itself.

## 2. Scalability testing

To prove the advantages of described above changes extensive scalability tests of a new application version were made on 2 MPP HPC systems. One of them was a 72-CPU Dual Xeon 2.4 GHz Beowulf-type cluster with Gigabit Ethernet communication (referred as "Ethernet Cluster" further). The other – 256-CPU Dual Xeon 3.2 GHz Beowulf-type cluster with Myrinet-2000 interconnect (referred as "Myrinet Cluster" further). This testing work was made in collaboration with Embry-Riddle Aeronautical University in Florida, US and Science and Research Center on Computer Technology, Moscow, Russia.

First, number of large and middle-scale test cases were run on both systems with the number of processors up to 72. Figure 2 shows the scalability graph for Ethernet Cluster, achieved on simple testing projects of size 110.6 million cells, 221.2 million cells, 442.4 million cells. All the projects are of the same structure and use one and the same computational model, and their mesh size is the only difference. Cartesian mesh was used in all the cases. Figure 3 shows some of scalability graphs we found on Myrinet Cluster. To summarize, Ethernet Cluster shows 70-75% or better parallel effectiveness, whereas

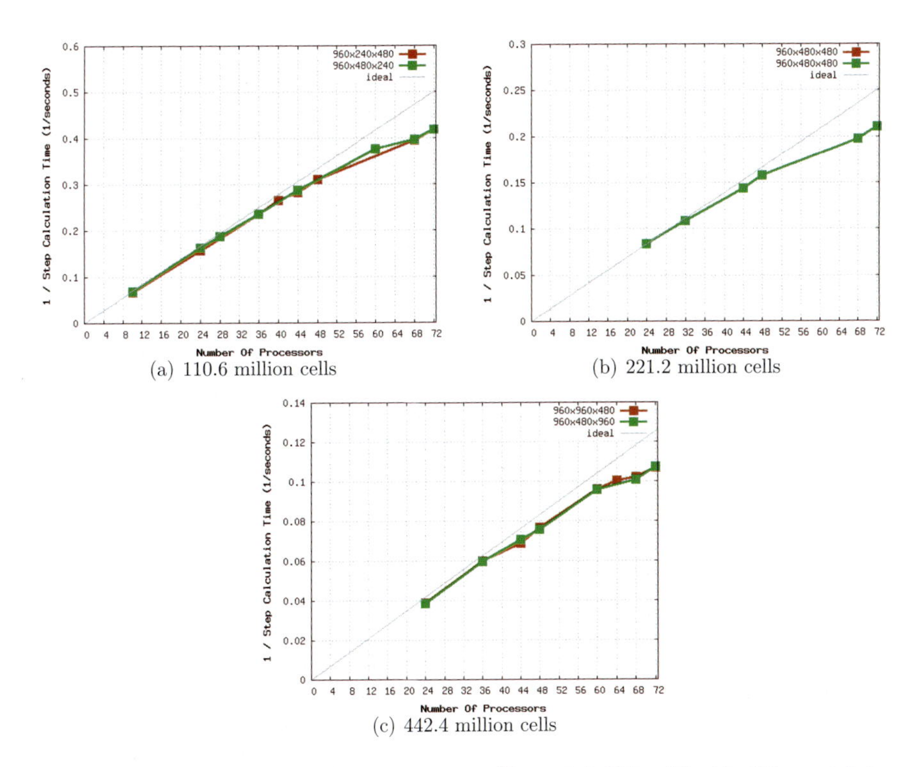

(a) 110.6 million cells
(b) 221.2 million cells
(c) 442.4 million cells

Figure 2. 72-CPU Dual Xeon cluster testcases (Xeon 2.8 GHz, Gigabit Ethernet Interconnect)

Myrinet Cluster achieves nearly 100% value of this index.

Then, number of large-scale testcases of the same sort as before were run on Myrinet Cluster to find out the code scalability on larger number of CPU. Figure 4 shows the results for 442.4; 884.7; 1769.5 million cell projects on up to 192 CPU of Myrinet Cluster. The effectiveness appears to be smoothly declining from nearly 100% to its minimal value of about 75-80%.

Figure 5 shows the collection of scalability plots were found on Myrinet Cluster during the whole testing session. Both axis are logarithm-scaled on the picture. The left plot corresponds to 110.5 million cells benchmark project, the right – 1769.5 million cells.

## 3. Large-scale problems computation

To prove the ability of the new GasDynamicaTool® package version to solve large-scale CFD problems testing calculations of 3 different practical gas dynamics problems were made. The number of cells in the 1st case was 1.92 billion cells, in 2nd case – about 1.5

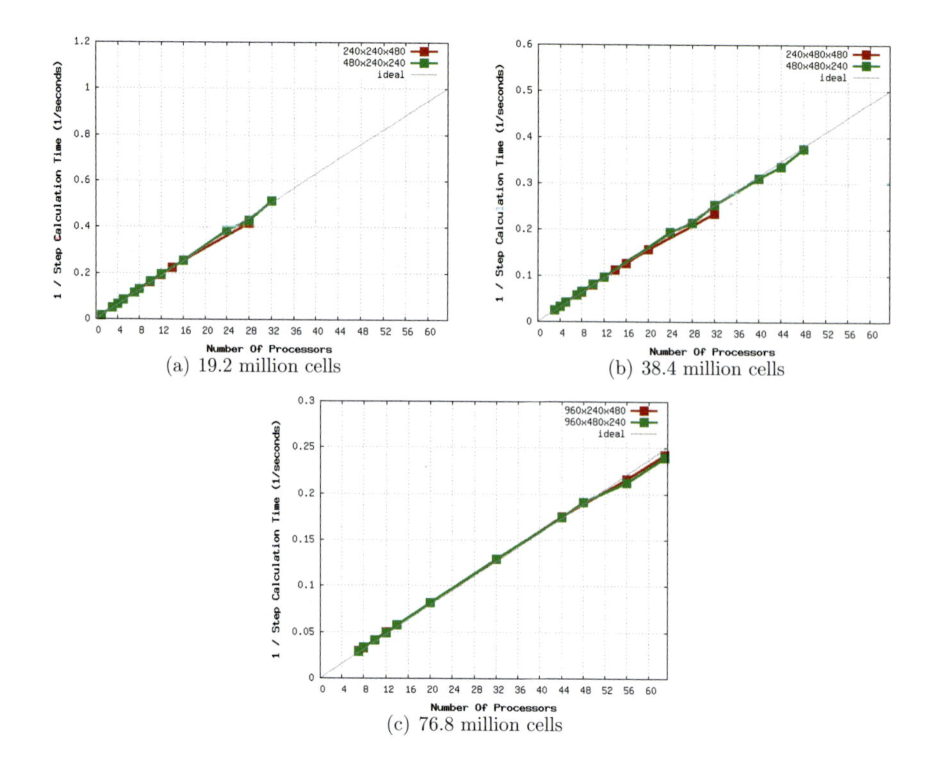


(a) 19.2 million cells

(b) 38.4 million cells

(c) 76.8 million cells

Figure 3. 256-CPU Dual Xeon cluster testcases (Xeon 3.2 GHz, Myrinet-2000 Scalable Cluster Interconnect)

billion cells, in 3rd – 3.5 billion cells. Moreover, the 100 million cells project was tested on 72-CPU cluster in the interactive "on-the-fly" [ 2] visualization mode of S-VR® package.

The results of 1.92 billion project test will be discussed here in detail. The problem of a space rocket stage separation was solved. Model definition includes a two stage rocket, stages separation is simulated in a configuration, outlined on Figure 6. As the initial conditions the rocket speed of 500 meters per second and the flight altitude of 5 kilometers was accepted. On the first stage of simulation separation engine starts (0...0.02 seconds), on the second – main engine of the 2nd rocket stage starts (0.02...0.085 seconds). The goal of the calculation is to determine flow picture after main engine start to estimate if engines work timing was correctly chosen.

The Euler equations based mathematical model was used to perform simulation to describe a non-viscous gas flow in 3D. Numerical scheme is based on Finite Volume Method with solid moving boundaries: that is an underlying concept of GasDynamicsTool's "Euler" CFD solver.

The length of a calculation domain is considerably high: no less than 30 meters is

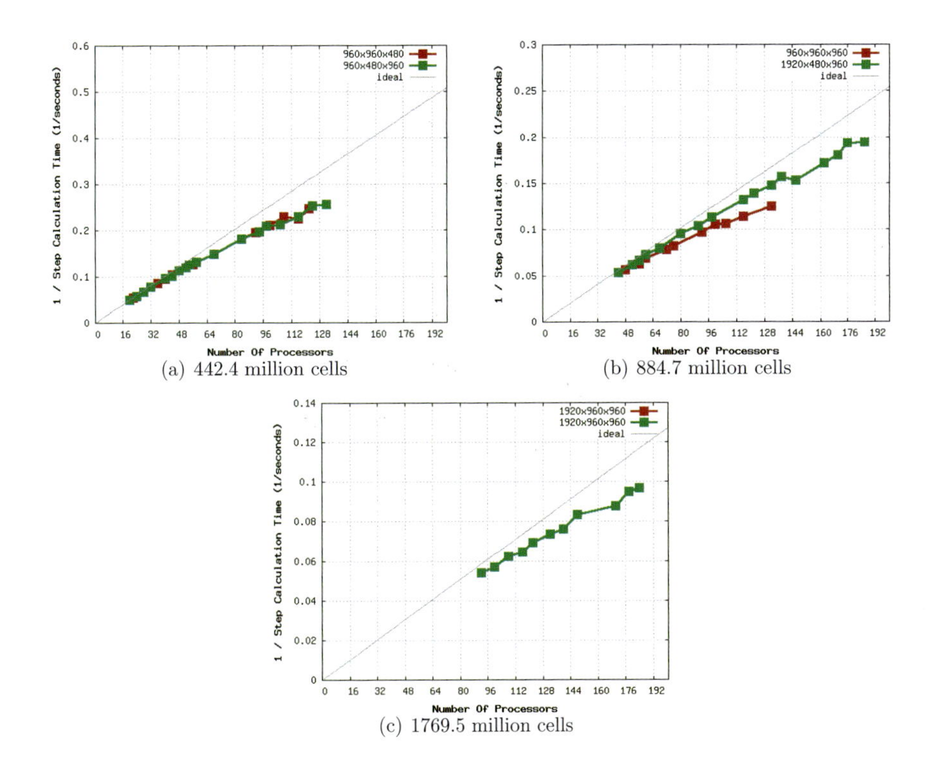


(a) 442.4 million cells

(b) 884.7 million cells

(c) 1769.5 million cells

Figure 4. 256-CPU Dual Xeon cluster testcases (Xeon 3.2 GHz, Myrinet-2000 Scalable Cluster Interconnect)

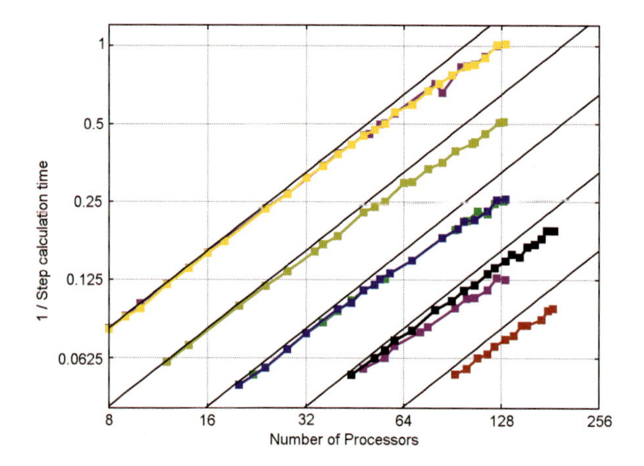


Figure 5. Aggregate scalability plot in logarithmic scale (Myrinet Cluster)

required, so the cartesian grid of 3000x800x800 has been chosen for the simulation (10 millimeter cell size). Full simulation physical time – 0.085 seconds, single scheme time step – 2 microseconds. The simulation script was divided into 3 stages: 1st is 10000 scheme steps (separation engine work), 2nd and 3rd – 8000 steps (main engine work).

Calculation was made using GasDynamicsTool 5.95 Beta on a Beowulf-type cluster with the following hardware and software configuration: 128 Dual Xeon 3.2GHz nodes, Myrinet-2000 interconnection, SAN of 4Tb total size for data storage, RedHat EL 4, Platform ROCKS cluster management system.

According to scalability and productivity benchmarking results, the prediction of full calculation time was made (see Table 1). Some of the points are not easy to predict exactly, that is why estimations of fields like "task starting time" and "initial filling time" are given roughly. Please note, that a big number of data saving stages was introduced mostly with the aim to increase the reliability of the calculation process, that means to withstand possible cluster hardware failures during the long process of calculation ("Save Data" function of GasDynamicsTool package makes it possible to restore calculation from the point corresponding to saved data). It was decided to make the number of data saving points as big as 1 Save Data operation for approximately 10 hours of calculation, because of cluster's SAN high speed.

Table 2 shows the real calculation timing. The biggest difference was found right in data saving time for the SAN productivity appeared to be several times as low as it was expected. The reason for it was related to SAN system software or hardware malfunction. That was the only thing that caused difference in values between predicted and real time, so the prediction was mostly correct.

Figure 7 shows visualization samples made by ScientificVR software package.

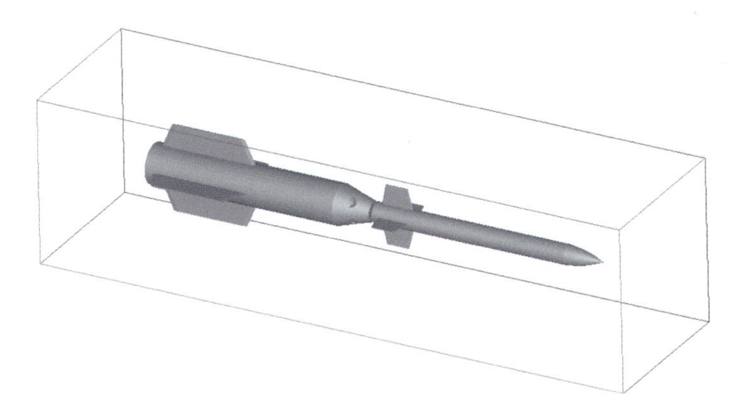

Figure 6. Two-stage rocket in calculation domain

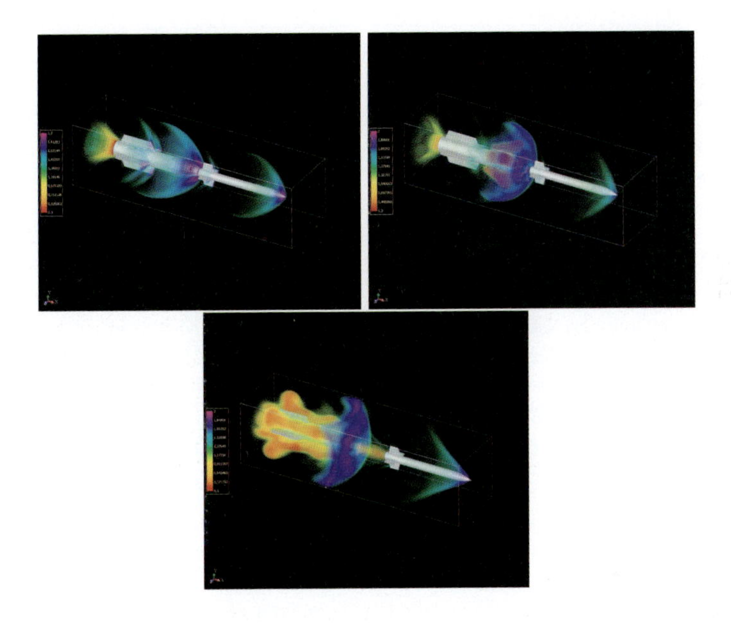

Figure 7. Calculation results, density distribution

Table 1
Calculation time prediction

| Number of CPUs to use | 170 |
|---|---|
| task starting time | 30 s |
| initial filling time (for each stage) | 3600 s |
| stages calculation times | 140000 s; 112000 s; 112000 s |
| size of 1 saved block | 57 Gb |
| data saving time (per block) | 171 s |
| FULL SIMULATION TIME | 4.5 days |

Table 2
Real calculation time statistics

| Number of CPUs to use | 170 |
|---|---|
| task starting time | 107 s |
| initial filling time (for each stage) | 2162-2283 s |
| stages calculation times | 167130 s; 135560 s; 112090 s |
| size of 1 saved block | 57 Gb |
| data saving time (per block) | **1634-2200 s (!)** |
| FULL SIMULATION TIME | 5 days |

## 4. Summary

Several testcases shown in the article prove the ability of GasDynamicsTool new version to effectively solve large-scale CFD problems on clusters with number of CPUs up to 200. The package code shows good scalability on Myrinet up to 70 CPUs, and acceptable scalability on 192 CPUs. The reason for performance decline in the later case may be caused by system software drawbacks. It is expected that implementation of a new MX-based low-level communication driver in GasDynamicsTool will correct the situation. Productivity on low-end Ethernet-based systems appeared to acceptable on number of CPUs up to 72, i.e. even simplest low cost cluster systems can be used with the package for large-scale problems solving. At once the best productivity results on hi-speed interconnect enabled systems make it possible to solve most difficult large-scale CFD problems using full computational power of modern parallel systems in the most effective way.

## REFERENCES

1. Dmitry A.Orlov, Alexey V.Zibarov, Andrey N. Karpov, Ilya Yu.Komarov, Vladimir V. Elesin, Evgeny A. Rygkov, Andrey A. Parfilov, Anna V. Antonova. CFD Problems Numerical Simulation and Visualization by means of Parallel Computation System // International conference on parallel computational fluid dynamics 2006 (Parallel CFD 2006), Busan, Korea.
2. Dmitry B. Babayev, Alexei V. Medvedev Architecture of Parallel CFD Code for Hybrid Computer Systems // Parallel CFD 2003, May 13-15, 2003, Moscow, Russia, p. 285 - 287

# Direct Monte Carlo Simulation of low-speed flows

M. Mukinovic and G. Brenner[a*]

[a]Institute of Applied Mechanics,Clausthal University,
Adolph-Roemer Str.2A, 38678 Clausthal-Zellerfeld, Germany

## 1. Introduction

The prediction of flows and other transport phenomena in and around micro scale devices challenges traditional computational techniques due to the violation of the continuum assumption. Examples are flows through Micro-Electro-Mechanical Systems (MEMS), chemical processes in micro-reactors or flows past corrugated or rough surfaces. A particular application, which motivates the present work, is the development of deposition techniques for manufacturing semi-conducting materials such as the ALD (Atomic Layer Deposition) technique. ALD is a chemical vapour deposition technique, suitable for slow and controlled growth of thin, conformal oxide or nitride films [1,2]. Gaseous precursors are admitted separately in the reactor in alternate pulses, chemisorbing individually onto the substrate, rather than reacting in the gas-phase. This technique requires low operating pressures where consequently the mean free-path in the gas is comparable high. A unique advantage of this technique is, that high conformality over toughest substrate topologies can be achieved. Thus, thin homogeneous films along highly corrugated surfaces such as micro cavities with high aspect ratio, may be produced.

Flows past and through such micro-structures constitute an emerging application field of fluid dynamics. However, despite the impressive technological and experimental progress, the theoretical understanding of microflows remains incomplete. This is partially due to the fact, that numerical simulation tools usually do not reflect the particularity of non-equilibrium flows. With the increasing performance of computers, the direct simulation Monte Carlo (DSMC) has emerged as the primary method for investigating flows at high Knudsen numbers ($K_n = \lambda/L$), where $\lambda$ is the mean free path of the gas and $L$ is some characteristic length scale of the flow. The fundamentals of the DSMC method have been documented in the well known publications of Bird [3,4]. In the meantime, the DSMC method is very popular and useful in studying hypersonic flows in aero- and astronautics, where the focus is on low density, high velocity flows. However, low speed flows present a great challenge to DSMC methods. This is mainly due to the fact, that DSMC is a statistical approach. The relevant information regarding macroscopic quantities such as flow velocity, pressure and temperature are obtained after sampling over microscopic velocities. The latter is determined by the thermal motion of the gas. In high speed, i.e. hypersonic flows, it is possible to isolate the "useful" information from the statistical "noise". In low speed flows, i.e. if the mean velocity of the gas is small compared to the thermal motion, the resulting large statistical scatter has to be reduced by greatly increasing

*The authors are grateful to the German Research Foundation (Deutsche Forschungsgemeinschaft) for financial support provided in the frame project "Anorganische Materialien durch Gasphasensynthese: Interdisziplinäre Ansätze zu Entwicklung, Verständnis und Kontrolle von CVD-Vervahren" (DFG SPP-1119).

I.H. Tuncer et al., *Parallel Computational Fluid Dynamics 2007*,
DOI: 10.1007/978-3-540-92744-0_39, © Springer-Verlag Berlin Heidelberg 2009

the sampling size. As a consequence, the application of DSMC for low speed flows is limited by CPU time constraints to few simple problems.

Several approaches have been developed in order to modify the DSMC methods for low speed flows. One possibility is the "molecular block model" DSMC [5]. An alternative approach is the "information preservation" method of Sun [6] which is utilized in the present work.

The present paper is organized as follows: After this introduction, details of the numerical algorithm are presented. Subsequently, results are presented to verify the approach and to demonstrate the increase of efficiency of the IP method compared to the standard DSMC approach. The configurations are a simple shear flow and the lid driven cavity problem. Comparisons with continuum computations are presented for low Knudsen numbers.

## 2. Computational Model

### 2.1. Molecular Dynamics model

DSMC is a molecular-based statistical simulation model for rarefied gas flows introduced by Bird [3]. The propagation and dynamics of collisions of molecules are solved numerically for a large number of computational molecules, where the latter in turn represent a large number of real atoms or molecules. For dilute gases and under the premise of molecular chaos, only binary collisions have to be considered. Intermolecular interactions and those between molecules and boundaries are modeled taking into account the conservation of momentum and energy. Using the constraint, that the computational time step is smaller that the mean collision time, the propagation and collision of molecules may be decoupled. The DSMC flowchart is presented in Figure 1.

The computational domain is divided into cells respecting the rule that each cell should be smaller than the mean free path in the gas. Additionally, each cell may contain a certain number of sub-cells which is required to realize an efficient collision procedure. Each cell is initialized with a certain number of particles (usually 20-40) with individual velocities obeying the Maxwellian distribution function.

Particles are moved in physical space according to time step and their individual velocity. The boundary conditions are enforced at this stage and macroscopic properties along solid surfaces must be sampled. Prior to the collision procedure, particles must be indexed in order of cells and sub-cell. The collision step is a probabilistic process which may be modeled using different, largely phenomenological models which are designed to reproduce real fluid behavior when examined at macroscopic level. They all have in common, that in each cell a certain number of collision pairs is selected in a random fashion using the no-time-counter (NTC) method. According to [4], the number of collision pairs $N_c$ is calculated as $N_c = \frac{1}{2\Delta V} N \overline{N} F_N \left(\sigma_T c_r\right)_{max} \Delta t$, where $N$ and $\overline{N}$ are the current and time averaged number of particles in each cell, $F_N$ is the number of real molecules, $(\sigma_T c_r)_{max}$ is the maximum product of collision cross-section and relative velocity in each cell, $\Delta t$ is time step and $\Delta V$ is cell volume. Those collision pairs (randomly selected within a common sub-cell) with a collision probability $\sigma_T c_r / (\sigma_T c_r)_{max}$ larger than a random number are selected for collision. If collisions occur, the post-collision velocities are set depending on the pre-collision velocities and depending on details of the molecular model.

Due to the fact that the initial particle distribution is far away from the steady state, a certain number of time step is required before the sampling of macroscopic properties begins. The time to reach steady state is mainly dictated by a convective time scale. The macroscopic flow properties are sampled within each cell and the corresponding macroscopic values are represented at the cell centers. For example, the macroscopic velocity ($\mathbf{c_0}$) of a single species gas is obtained from $\mathbf{c_0} = \overline{\mathbf{c}} = \sum_{i=1}^{N} \mathbf{c}_i / N$, where $\mathbf{c}_i$ is the microscopic particle velocity and $N$ is the number

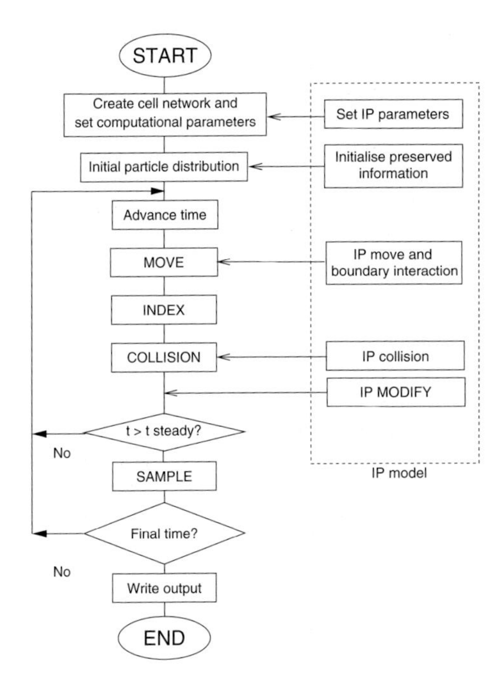


Figure 1. DSMC model flowchart and appropriate IP modification.

of sampled particles (sample size). This final sampling step is only required at the end of the simulation, i.e. after a suitable number of stream-indexing-collision steps have been finished. The number of time steps depends on the flow conditions and the statistical scatter, which is introduced due to the finite sampling size. Under the idealized assumption, that the samples are statistically independent, the thermal scatter of velocity may be estimated as $\sqrt{2RT/N}$, where $R$ is the gas constant and $T$ is the gas temperature. Taking into account that the particles in a sample are not completely statistically independent and considering the finite accuracy of the number representation in a computer, the statistical scatter is additionally increased. For air at standard conditions ($T = 290K$), a reduction of statistical scatter by $1m/s$ requires a sample size of $1.6 \cdot 10^5$. Hence, with 40 particles in one cell, this requires a sampling period of 40000 time steps. This example illustrates also the problem of simulating low-speed flows with DSMC. At high speed (e.g. supersonic) flows, the thermal or scatter velocity is small compared to the mean velocity. In low speed flows, the signal to noise ratio is of the order of unity. In order to reduce the noise, either an excessive sampling size (particle number or time steps) is required or other acceleration techniques have to be employed, such as the IP method presented in the following section.

## 2.2. Information Preservation method

To overcome the problem of statistical scatter, Fan and Shen [7] have proposed the information preservation (IP) method which is suitable for low-speed, incompressible flows. This method is implemented parallel to the DSMC procedure but does not directly interact with it. The method is based on the idea, that each simulated particle carries the so called information velocity (IP velocity). Since each simulated particle already represents an ensemble of a large number of real molecules, the IP velocity also represents an ensemble average, which is different from the mean flow velocity. The flow field is obtained by sampling the preserved velocity instead of the microscopic information, so that the statistical scatter can be greatly reduced. The preserved velocity of each simulated particle in turn is obtained from the microscopic velocity and based on conservation principles. The initial implementation of the IP method preserved only the macroscopic velocity of particles. From Sun [6], an extended IP method (DSMC-IP) was proposed, that additionally includes internal energy in preservation. Here, an equilibrium between translational temperatures and internal energy is assumed. In the IP method, microscopic movements and collisions are handled by the DSMC method but the macroscopic information of particles is updated using other algorithms. The flow field is obtained by sampling the preserved macroscopic information instead of the microscopic information, so that the statistical scatter can be greatly reduced. In the DSMC-IP method used here, each simulated particle preserves macroscopic velocity $\mathbf{V}_i$, macroscopic temperature $T_i$ and additional temperature $T_{i,a}$, while the computational cells preserve the macroscopic velocity $\mathbf{V}_c$, macroscopic temperature $T_c$ and macroscopic density $\rho_c$. The update of this information is based on physical understanding of flow and is not derived theoretically. The theory behind this method is explained in the following.

In the kinetic theory of gases, the microscopic particle velocity is defined as $\mathbf{c}_i = \mathbf{c}_0 + \mathbf{c}_i'$ and thus, the particle preserved velocity may be defined as $\mathbf{V}_i = \mathbf{c}_0 + \mathbf{c}_i''$, where $\mathbf{c}_0$ is a stream velocity and $\mathbf{c}_i'$, $\mathbf{c}_i''$ are associated scatter. Then, the particle preserved temperature is expressed as $T_i = (\mathbf{c}_i - \mathbf{V}_i)^2/3R$. The particle preserved velocity $\mathbf{V}_i$ and particle preserved temperature $T_i$ are updated in each collision step conserving momentum and energy:

$$\mathbf{V}_{1,2}'' = \frac{1 \pm C_\mu \cdot \psi}{2}\mathbf{V}_1' + \frac{1 \mp C_\mu \cdot \psi}{2}\mathbf{V}_2', \tag{1}$$

$$T_{1,2}'' = \frac{1 \pm C_\kappa \cdot \psi}{2}T_1' + \frac{1 \mp C_\kappa \cdot \psi}{2}T_2' + \frac{1 - C_\mu^2 \cdot \psi^2}{4\xi R} \cdot (V_1' - V_2')^2. \tag{2}$$

Here, superscripts $'$ and $''$ denote pre- and post-collision particle preserved information. The constants $C_\mu$ and $C_\kappa$ are species dependent and determined by numerical experiments [6], $\psi$ is the cosine of the deflection angle in the collision plane, $\xi$ is the molecule number of degrees of freedom.

The macroscopic velocity and temperature are changed not only by the momentum and energy exchange by collision but also by the pressure field. Therefore, additional updates are required. These updates are based on the Boltzmann transport equation of the form

$$\frac{\partial}{\partial t}\left(n\overline{Q}\right) + \nabla \cdot \left(n\overline{\mathbf{c}Q}\right) = \Delta\left[Q\right], \tag{3}$$

where $n$ is the number density, $Q$ is a transported quantity, and $\Delta[Q]$ is the collision integral. Applying this equation to mass ($m$), momentum ($m\mathbf{c}_i$), and energy ($1/2mc_i^2$), Sun [6] presented a model to update the preserved information which includes some approximations.

The cell preserved density satisfies the continuity equation

$$\rho_c^{t+\Delta t} = \rho_c^t - \frac{\Delta t}{\delta\Omega}\int_S \rho_c \mathbf{V_c} \cdot \mathbf{dS}, \tag{4}$$

where $\mathbf{V}_c = \overline{\mathbf{V}}_i$ is the cell preserved velocity and $\delta\Omega$ is a cell volume.

Assuming that the rate change of momentum and the viscous effects are already included in the movement and collision step, the particle preserved velocity is modified using

$$\mathbf{V}_i^{t+\Delta t} = \mathbf{V}_i^t - \frac{\Delta t}{N_P F_N m} \int_S p \mathbf{dS}, \tag{5}$$

where $N_P$ is the number of simulated particles in a cell, $F_N$ is the number of real molecules represented by the simulated particle and $p = nkT_c$ is the pressure obtained from the ideal gas law.

The modified particle preserved temperature is obtained from

$$T_i^{t+\Delta t} = T_i^t - \left( V_i^{2(t+\Delta t)} - V_i^{2(t)} + 2\frac{\Delta t}{N_P F m} \int_S p \mathbf{V}_c \cdot \mathbf{dS} \right) / \xi R. \tag{6}$$

The energy transport introduces additional approximations in the model. The DSMC-IP representation of the energy flux ($3/2kT$) is in contradiction with kinetic theory ($2kT$). Thus, each particle movement from one cell to another, preserves additional temperature $T_{i,a}$ which is borrowed from the initial cell. This borrowing is recorded as the cell additional temperature $T_{a,c}$.

The implementation of the IP method into existing DSMC codes is relatively simple. It requires allocation of additional variables for the particle and cell preserved information. The IP model, embedded in the DSMC procedure is shown in Figure 1. At the beginning of the simulation, preserved information is set according to the initial and boundary conditions. The particle preserved velocity and temperature are initialized with stream velocity and temperature respectively and the particles additional temperature is set to zero. When particle crosses a cell edge, its additional temperature is modified as $T_{i,a} = (T_i - T_{ref})/\xi$, where $T_{ref}$ is the interface temperature. When leaving the computational domain, each particle carries all preserved information along with it. The preserved information of particle reflected from the wall is also changed. For the example, a diffusely reflected particle preserve velocity and temperature of the wall. The additional temperature is updated as $T_{i,a} = (T_w - T_{ref})$, where $T_w$ is wall temperature and $T_{ref} = \sqrt{T_i \cdot T_w}$. For new particles that may enter the domain, the preserved information is set according to boundary conditions. During the collision step, the particle information is updated using equations 1 and 2. At the end of each time step, the cell additional temperature is evenly subtracted from all particles in a cell. For the next time step, this temperature is set to zero. After the new density (equation 4) and pressure are calculated, the particle preserved velocity and temperature are modified using equations 5 and 6. The cell preserved velocity and temperature are obtained as a mean value of particle preserved information ($\mathbf{V}_c = \overline{\mathbf{V}_i}$, $T_c = \overline{T_i + T_{i,a}}$). At the end of the simulation, flow field properties are obtained through time averaging of the preserved information. The flow velocity $V_f$, flow temperature $T_f$ and flow density $\rho_f$ are calculated as follows:

$$V_f = \overline{\mathbf{V}_c}, \tag{7}$$

$$T_f = \overline{T_c} + \frac{1}{\xi R} \left( \overline{\mathbf{V}_i \cdot \mathbf{V}_i} - \overline{\mathbf{V}_i} \cdot \overline{\mathbf{V}_i} \right), \tag{8}$$

$$\rho_f = \overline{\rho_c}. \tag{9}$$

This model significantly reduces statistical scatter and computational time. In [6], a systematical analysis of statistical scatter in a Couette flow is presented. For wall velocities between $0.01 - 100.0m/s$ and a Knudsen number of $Kn = 0.01$, with a sample size of 1000 particles per cell, the scatter is reduced to about 0.5 % of wall velocity while DSMC scatter is always about $11m/s$ for the same sample size.

## 3. Results

The results presented in this paper are obtained with the DSMC-IP code under development at our institute. This is a general purpose two dimensional code written in Fortran 90 programming language compiled with Intel Fortran compiler 9.1. The DSMC core is based on Birds example code [4] extended with the IP model from Sun [6]. The grid is rectilinear and block structured. The results presented in the following two subsections are intended to verify the present numerical approach and to demonstrate the gain of performance when using the IP method. Finally, results for the lid driven cavity are shown. This test case already represents a simplified model of a surface corrugation.

### 3.1. Plane Couette flow with energy transport

The two-dimensional flow between two parallel plates in shearing motion is considered. In the continuum limit, a linear velocity distribution due to the no-slip at the walls and a constant shear stress is expected. For low Knudsen numbers, the DSMC method should recover this solution. For larger Knudsen numbers, wall slip is expected which reduces the shear stress in the flow. Figure 2a presents the velocity distribution depending on the Knudsen number obtained with the DSMC-IP code. The wall velocity is $u_{wall} = 1m/s$ and the wall distance is $\delta = 1m$. In order to show the influence of the viscous dissipation for $Kn = 0.01$, the wall velocity was increased up to $u_{wall} = 300m/s$. Figure 2b presents the temperature distribution. Here, the DMSC solution is compared with the DSMC-IP solution to show the reduction of statistical scatter.

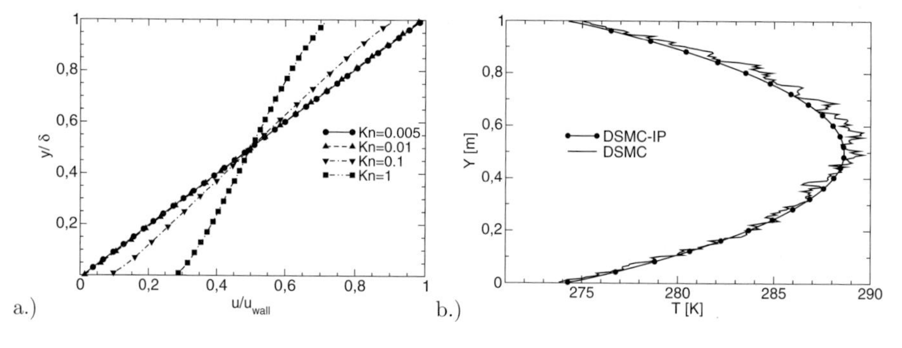


Figure 2. a.) Velocity distribution in a plane Couette flow as function of Knudsen number obtained from the DSMC-IP method. ($u_{wall} = 1m/s$). b.) Temperature distribution as function of wall velocity for $Kn = 0.01$ and $u_{wall} = 300m/s$.

### 3.2. Two dimensional channel flow with pressure gradient

If the flow is driven by a pressure gradient, the plane Poiseuille flow with a parabolic velocity profile is obtained, see Figure 3a. In the continuum limit a constant friction factor times Reynolds number $f \cdot Re = 24$ is expected. The friction factor is the nondimensional pressure gradient defined as $f = dp/dx \cdot 2\delta/\rho\bar{u}^2$. The Reynolds number is $Re = \bar{u}\delta/\nu$, where $\bar{u}$ is the mean velocity and $\nu$ is the kinematic viscosity. Depending on the Knudsen number, a decrease of the pressure gradient is expected due to increasing wall slip. This is confirmed in Figure 3b, where $f \cdot Re$ is shown versus the Knudsen number.

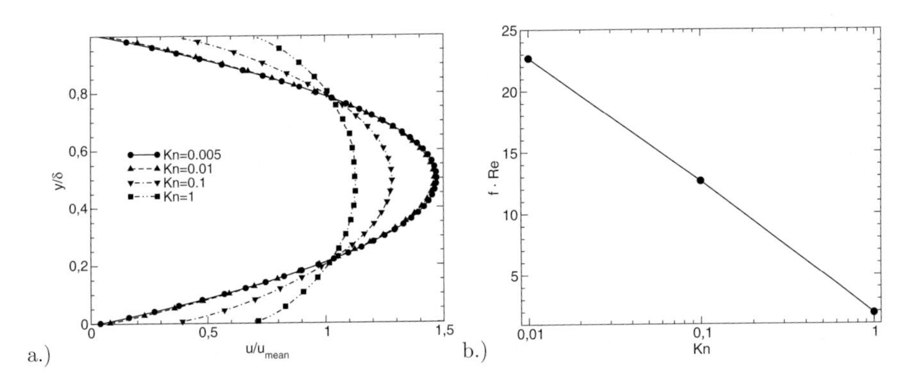


Figure 3. a.) Velocity profile depending on Knudsen number in a plane Poiseuille flow, b.) Variation of friction coefficient flow as function of the Knudsen number.

### 3.3. Flow in a micro lid-driven cavity

The nitrogen flow in a lid-driven cavity for $Kn = 0.01$ and $Re = 50$ is considered. The computational mesh consists of $200 \times 200$ cells. Each cell contains about 30 particles. Thus, in total, 1200000 particles are simulated. In Figure 4 isolines of velocity magnitude $U$ are presented. For comparison, the DSMC-IP and Navier-Stokes (NS) solution are shown in Figure 4a. At the same time as the DSMC-IP solution is obtained, the respective DSMC solution is presented in Figure 4b. The agreement of DSMC-IP and NS solution is quite well taking into account that NS solver uses no-slip boundary conditions for walls. The DSMC solution shows the great influence of statistical scatter for the insufficiently large sample size (600000 particles per cell). For this

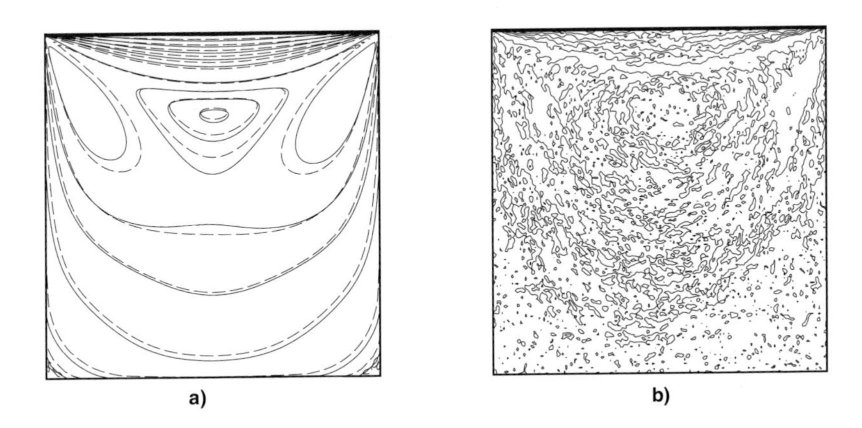


Figure 4. The isolines of the velocity magnitude. $Re = 50$, $Kn = 0.01$, $u_{lid} = 11.1m/s$. a) DSMC-IP (solid lines) and Finite volume method (dashed lines) solution. b) DSMC solution.

case, computational times of the particular parts of the code (see Figure 1) are presented in Table 1. Separate runs of the DSMC and the DSMC-IP version of the code for 1000 time steps are performed on a Intel Xeon processor at 3.73 GHz. The DSMC-IP version results in about 30% larger computational time costs per time step but the DSMC alone requires much larger sample size and thus computational time.

Table 1
Computational time for the particular parts of the DSMC-IP code.

| | MOVE | INDEX | COLLISION | IP MODIFY | SAMPLE | TOTAL TIME |
|---|---|---|---|---|---|---|
| DSMC | 368.28 | 168.06 | 125.82 | 0.00 | 127.05 | 789.21 |
| DSMC-IP | 404.32 | 161.87 | 160.53 | 180.35 | 126.13 | 1033.20 |

## 4. Conclusion

The goal of the present paper was to present modifications to a DSMC approach that allows to predict low velocity, rarefied gas flows including energy transport. This modification is based on the IP method. The background of the work is the simulation of rarefied flows in chemical reactors used for atomic layer deposition. The results show the validity of the approach compared to classical continuum methods in the low Knudsen number limit. It is shown, that the IP concept greatly improves the statistical scatter in low speed flows. The implementation of this model for multicomponent gases in the state as it is, leads to numerical instabilities and unphysical results, especially for the mixtures with a large differences in molecular mass. Thus, some effort should be put in the analysis and the modification of the IP model.

## REFERENCES

1. M. Leskela, M. Ritala, Atomic layer deposition (ALD): from precursors to thin solid films, This Solid Films, 409, pp. 138–146, 2002.
2. S. D. Elliot, Predictive process design: a theoretical model of atomic layer deposition, Comp.Mat. Sciences, 33, pp. 20-25, 2005.
3. G. A. Bird, Approach to Translational Equilibrium in a Rigid Sphere Gas, Physics of Fluids, 6 (10), pp. 1518-1519, 1963.
4. G. A. Bird, Molecular Gas Dynamics and the Direct Simulation of Gaws Flows, Oxford Science Publishing, 1994.
5. L. S. Pan, T. Y. Ng, D. Xu and K. Y. Lam, Molecular block model direct simulation Monte Carlo method for low velocity microgas flows, J. Micromech, Microeng., 11, pp- 181-188, 2001.
6. Q. Sun, Information Preservation Methods for Modeling Micro-Scale Gas Flow, PhD thesis, The University of Michigan, 2003.
7. J. Fan, C. Shen, Statistical Simulation of Low-Speed flows in Transition Regime. In *Proceedings of the 21th International Symposium on Rarefied Gas Dynamics*, edited by R. Brun, et al., Marseille, France, pp. 245-252, 1999.

# Parallel computing of 3D separated stratified fluid flows around a sphere

**Paul V. Matyushin, Valentin A. Gushchin**

*Institute for Computer Aided Design of the Russian Academy of Sciences 19/18, 2nd Brestskaya str., Moscow 123056, Russia*

Keywords: stratified fluid; sphere wake; direct numerical simulation; vortex structures; visualization; flow regimes; internal waves; Reynolds number; internal Froude number

## 1. INTRODUCTION

In the natural environments (atmosphere and hydrosphere) and in the process installations the density of the fluid is variable. Even a small density variation ($\Delta\rho/\rho \sim 10^{-4} - 10^{-8}$) causes considerable alterations in the fluid motion which are absent in the homogeneous fluid (see experimental results [1-2]). The numerical studies of the non-homogeneous (density stratified) fluids are very rare [3]. Therefore at the present paper the stratified viscous incompressible flows around a sphere are investigated by means of the direct numerical simulation (DNS) on the massively parallel computers with a distributed memory at the following ranges of the internal Froude $Fr$ and Reynolds $Re$ numbers: $0.004 \leq Fr \leq 10$, $10 \leq Re \leq 1000$ ($Fr = U/(N \cdot d)$, $Re = U \cdot d/\nu$, where $U$ is the scalar of the free stream velocity, $N$ is a buoyancy frequency, $d$ is a sphere diameter and $\nu$ is the kinematical viscosity).

In the case of the homogeneous fluid ($Fr = \infty$) there is only one non-dimensional parameter ($Re$). With increasing of $Re$ the number of the degrees of freedom is increased too. At $Re = 200$ the axisymmetrical wake is transforming in the steady double thread wake. At $270 < Re < 400$ the separation of one edge of a vortex sheet (surrounding the recirculation zone) is observed. At $Re > 400$ the alternative separation of the opposite edges of the vortex sheet and the irregular rotation of this sheet are observed [4-5]. Thus the continuous changing of the wake vortex structure is observed with increasing of $Re$. The detailed *formation mechanisms of vortices* (FMV) in the sphere wake at $200 \leq Re \leq 1000$ have been described in [5] (*for the first time*). In particular it was shown that the detailed FMV for $270 < Re \leq 290$, $290 < Re \leq 320$ and $320 < Re \leq 400$ are different. The main differences have been observed in the recirculation zone. At $290 < Re \leq 320$ a new flow regime has been discovered [5].

I.H. Tuncer et al., *Parallel Computational Fluid Dynamics 2007*,
DOI: 10.1007/978-3-540-92744-0_40, 

In the case of the stratified fluid the full 3D vortex structure of the flow around a sphere is not clear for many flow regimes [1-2]. In addition the clear understanding of the continuous changing of the full 3D sphere wake vortex structure with decreasing of *Fr* is also absent. At the present paper this continuous changing of the wake structure is investigated at $Re = 100$ by using DNS and the β-visualization technique.

## 2. NUMERICAL METHOD

The density of the linearly stratified fluid $\rho(x, y, z)$ is equal to $\rho_0(1 - x/(2C) + S)$ where $x, y, z$ are the Cartesian coordinates; $z, x, y$ are the streamwise, lift and lateral directions ($x, y, z$ have been non-dimensionalized by $d/2$); $C = \Lambda/d$ is the scale ratio, $\Lambda$ is the buoyancy scale, which is related to the buoyancy frequency $N$ and period $T_b$ ($N = 2\pi/T_b$, $N^2 = g/\Lambda$); $g$ is the scalar of the gravitational acceleration; $S$ is a perturbation of salinity.

The density stratified incompressible viscous fluid flows are simulated on the basis of the Navier-Stokes equations in Boussinesq approximation (1) – (2) and the diffusion equation for the stratified component (salt) (3) with four dimensionless parameters: $Fr$, $Re$, $C \gg 1$, $Sc = \nu/\kappa = 709.22$ where $\kappa$ is the salt diffusion coefficient.

$$\frac{\partial \mathbf{v}}{\partial t} + (\mathbf{v} \cdot \nabla)\mathbf{v} = -\nabla p + \frac{2}{Re}\Delta \mathbf{v} + \frac{C\,S}{2Fr^2}\frac{\mathbf{g}}{g} \quad (1)$$

$$\nabla \cdot \mathbf{v} = 0 \quad (2)$$

$$\frac{\partial S}{\partial t} + (\mathbf{v} \cdot \nabla)S = \frac{2}{Sc \cdot Re}\Delta S + \frac{v_x}{2C} \quad (3)$$

In (1) – (3) $\mathbf{v} = (v_x, v_y, v_z)$ is the velocity vector (non-dimensionalized by $U$), $p$ is a perturbation of the pressure (non-dimensionalized by $\rho_0 U^2$).

The spherical coordinate system $R, \theta, \varphi$ ($x = R\sin\theta\cos\varphi$, $y = R\sin\theta\sin\varphi$, $z = R\cos\theta$, $\mathbf{v} = (v_R, v_\theta, v_\varphi)$) and the O-type grid are used. On the sphere surface the following boundary conditions are used:

$$v_R = v_\theta = v_\varphi = 0,$$

$$\left.\frac{\partial \rho}{\partial R}\right|_{R=d/2} = \left.\left(\frac{\partial S}{\partial R} - \frac{1}{2C}\frac{\partial x}{\partial R}\right)\right|_{R=d/2} = 0.$$

On the external boundary the following boundary conditions are used: 1) for $z < 0$: $v_R = \cos\theta$, $v_\theta = -\sin\theta$, $S = 0$, $v_\varphi = 0$; 2) for $z \geq 0$: $v_R = \cos\theta$, $v_\theta = -\sin\theta$, $S = 0$, $\partial v_\varphi/\partial R = 0$.

For solving of the equations (1) – (3) the Splitting on physical factors Method for Incompressible Fluid flows (SMIF) with the hybrid explicit finite difference scheme (second-order accuracy in space, minimum scheme viscosity and dispersion, monotonous, capable for work in the wide range of the Reynolds and Froude numbers) based on the Modified Central Difference Scheme (MCDS) and the Modified Upwind Difference Scheme (MUDS) with the special switch condition depending on the velocity sign and the sign of first and second differences of the transferred functions has been used [6]. The Poisson equation for the pressure has been solved by the Preconditioned Conjugate Gradients Method.

## 3. CODE PARALLELIZATION

The parallelization of the algorithm has been made and successfully applied on the following massively parallel computers with a distributed memory with the switch Myrinet-2000: MBC-1000 and MBC-15000BM (based on the Intel Xeon (2.4 GHz) and on the PowerPC 970 (2.2 GHz) processors correspondingly). The code has been parallelized by using the domain decomposition in the radial direction. The computational domain has been divided into the spherical subdomains corresponding to the parallel processor units. These units have been divided into the odd and even ones in order to exchange the boundary data between the odd and even processors. The Speed up of the code for MBC-1000 is demonstrated in Fig. 1.

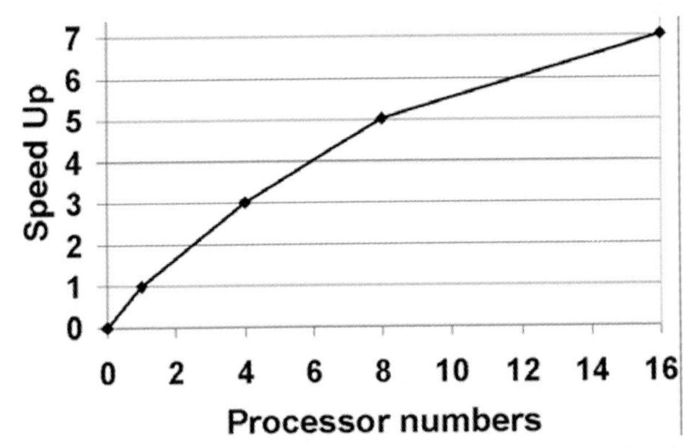


Fig 1. Speed up for the switch Myrinet-2000 (MBC-1000) for the computational grid 80x50x100.

## 4. VISUALIZATION OF VORTEX STRUCTURES

For the visualization of the 3D vortex structures in the sphere wake the isosurfaces of $\beta$ have been drawn, where $\beta$ is *the imaginary part of the complex-conjugate eigen-values of the velocity gradient tensor* **G** [7]. $\beta$ has a real physical meaning. Let us consider a local stream lines pattern around any point in a flow (where $\beta > 0$) in a reference frame **x** moving with the velocity of this point ($\mathbf{v} = d\mathbf{x}/dt \approx \mathbf{G}\,\mathbf{x}$, where **v** is a velocity of a fluid particle in the considered reference frame **x** and $t$ is time). It's easy to demonstrate (see the theory of the ordinary differential equations) that the local stream lines pattern in the considered reference frame **x** is closed or spiral, and $\beta$ *is the angular velocity of this spiral motion*. The good efficiency of this $\beta$-visualization technique has been demonstrated in [5].

## 5. RESULTS

### 5.1. The diffusion-induced flow around a resting sphere.

In the beginning the code for DNS of the 3D separated stratified viscous incompressible flows around a sphere has been tested in the case of a resting sphere in a continuously stratified fluid. Due to axial symmetry of the problem around the vertical axis and the axial symmetry of the chosen spherical coordinate system around the axis $z$, the axis $z$ was vertical for the considered axisymmetrical problem (Fig. 2). It means that for example $v_x$ in the last term of the equation (3) was replaced by $v_z$.

As a result of this test it was shown (*for the first time*) that the interruption of the molecular flow (by the resting sphere) not only generates the axisymmetrical flow on the sphere surface (from the equator to the poles) but also creates the short unsteady internal waves [8-9]. At first a number of these waves is equal to $t/T_b$. For example at $t = 2 \cdot T_b$ four convective cells with the opposite directions of the vorticity (two waves) are presented in the stream lines distribution (Fig. 2, right panel). A base cell, whose size is determined by the radius of the sphere, is located near the sphere surface. At time more than $37 \cdot T_b$ the sizes and arrangement of cells are stabilized, and only the base cell and two thin adjacent cells with a thickness 2.2 mm are observed both in the salinity perturbation field $S$ and in the stream lines pattern. In other words *the high gradient sheets of density* with a thickness 2.2 mm are observed near the poles of the resting sphere. The similar high gradient sheets of density have been observed before a moving sphere (near the poles) at $Fr \leq 0.02$.

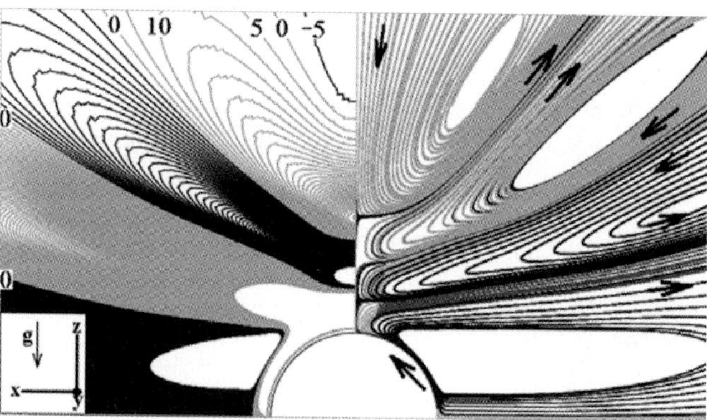


Fig. 2. The diffusion-induced flow around a resting sphere ($d = 2$ cm, $T_b = 6.34$ s) at $t = 2 \cdot T_b$. Isolines of the perturbation of salinity $S \cdot 10^{12}$ (left panel) and the instantaneous stream lines (right panel). The darker isolines correspond to negative $S$.

### 5.2. 3D vortex structures of the separated stratified fluid flows around a sphere.

Let us consider the continuous changing of the complex 3D vortex structure of the flow around a sphere with decreasing of the internal Froude number $Fr$ (from $\infty$ to 0.005) at $Re = 100$ (Fig. 3). At $Fr \geq 10$ the axisymmetrical one-thread wake [4-5] is observed. The vortex structure of this flow can be divided into two parts: the vortex ring in the recirculation zone and the shear layer (a vortex sheet surrounding the recirculation zone). With decreasing of *Fr* (from 10 to 2) the vortex ring is deformed in an oval (Fig. 4a); the four vortex threads are formed in the wake (Fig. 3a). In the vertical plane the part of fluid is supplied in the recirculation zone. Then this fluid goes through the core of the vortex oval and is emitted downstream in the horizontal plane. The 3D instantaneous stream lines which are going near the sphere surface go around this vortex oval and form the four vortex threads (Fig. 3a).

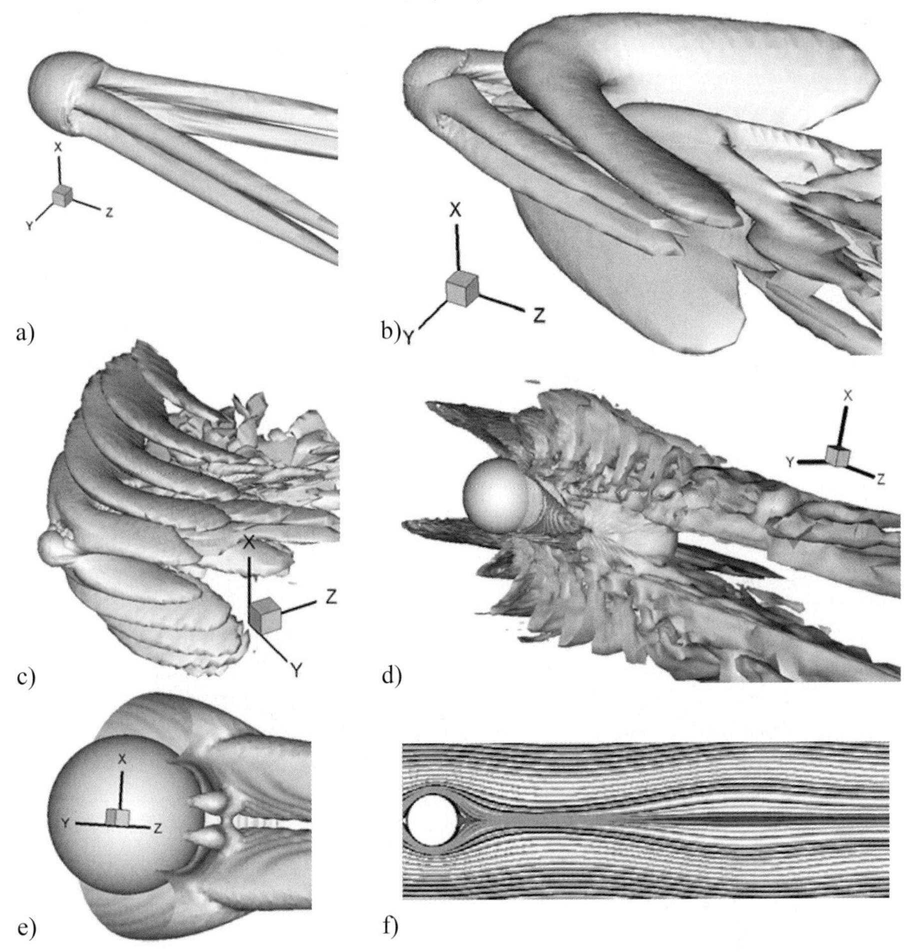


Fig. 3. The vortex structure of the sphere wakes at $Re = 100$: a) $Fr = 2$ ($\beta = 0.055$), b) $Fr = 1$ ($\beta = 0.02$), c) $Fr = 0.5$ ($\beta = 0.02$), d) $Fr = 0.08$ ($\beta = 0.005$), e) $Fr = 1$ ($\beta = 0.1$) and the instantaneous stream lines in the vertical plane: f) $Fr = 1$.

With decreasing of *Fr* from 2 to 0.4 the angle between the vortex threads in the horizontal plane is increased and the internal waves with length $\lambda/d \approx 2\pi{\cdot}Fr$ (in the vertical plane) appear above and bellow these threads (Fig. 3b-c, f). With decreasing of $Fr$ from 2 to 0.9 *the oval vortex* in the recirculation zone (Fig. 4a) is transformed into the quasi-rectangle with the long horizontal and short vertical vortex tubes (Fig. 3e). At $Fr \approx 0.9$ the horizontal vortex tubes are "cut" in the vertical plane $x$-$z$ and the cut ends of these tubes are redirected downstream (along the $z$ axis). Thus *the vortex* in the recirculation zone is *transformed into two symmetric vortex loops* (Fig. 4c). With decreasing of $Fr$ from 2 to 0.9 four additional (secondary) vortex threads induced by the four primary vortex threads are formed near the $z$ axis (between the primary threads, Fig. 3b). At $Fr \approx 0.6$ the four legs of the two symmetric vortex loops are connected with the corresponding induced vortex threads; the flow inside these legs is redirected downstream and the vortex loops are shifted from the sphere surface; the primary separation line on the sphere surface vanish (Fig. 5a). Thus at $0.4 < Fr < 0.6$ the primary and secondary vortex threads and the many arc-shaped horizontal vortices associated with the internal waves are observed in the fluid (Fig. 3c). The recirculation zone is absent.

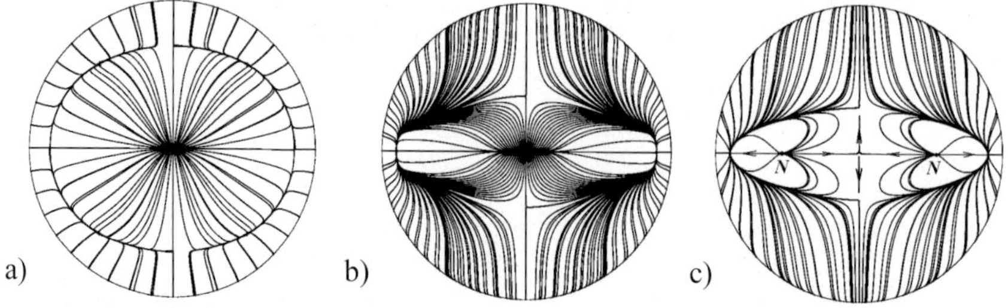


Fig. 4. The skin friction patterns on the sphere lee side at $Re = 100$: *a-c* - $Fr = 2, 1, 0.8$.

Beginning at $Fr \approx 0.4$ a new recirculation zone is formed from the "wave crest" which is situated very close to the sphere (Fig. 5b). The wake vortex structure looks like the structure at Fig. 3c. Beginning at $Fr \approx 0.25$ a quasi-two-dimensional structure *bounded by lee waves* is formed near the horizontal plane $y$-$z$ (Fig. 3d, 5c). In addition the four long large quasi-horizontal vortex threads are formed in the wake (Fig. 3d). The previous primary and secondary vortex threads observed at $Fr > 0.4$ were shifted upstream and *transformed into* the high gradient sheets of density. Thus at $Fr < 0.25$ *two steady vertical vortices* are attached to the sphere. At $0.03 \leq Fr < 0.25$ the ring-like primary separation line is observed (Fig. 5c). At $Fr < 0.03$ the cusp-like primary separation line is observed. In both cases four singular points (two nodes (in the vertical $x$-$z$ plane) and two saddles (in the horizontal $y$-$z$ plane)) belong to the primary separation line (Fig. 5c).

At $Re > 120$ and $0.25 < Fr < 0.4$ the wake is unsteady in the vicinity of the horizontal plane $y$-$z$. At $Re > 120$ and $Fr < 0.25$ the edges of two vertical vortex sheets (on each side of the quasi-2D recirculation zone) are detached alternatively. The Strouhal numbers $St = fd/U = 0.19, 0.197, 0.24, 0.2, 0.22$ (where $f$ is the frequency of shedding) for $(Fr, Re) = (0.03, 750), (0.04, 200), (0.05, 500), (0.11, 350), (0.125, 200)$ correspondingly are in a good agreement with [1] where for $0.03 < Fr < 0.28$ and $120 < Re < 500$ $St$ has been equal to 0.2.

## 6. CONCLUSIONS

The continuous changing of the complex 3D sphere wake vortex structure of the stratified viscous fluid with decreasing of $Fr$ from $\infty$ to 0.005 has been investigated at $Re = 100$ (Fig. 3) (*for the first time*) owing to the mathematical modelling on the massively parallel computers with a distributed memory. At $0.6 < Fr < 0.9$ the gradual disappearance of the recirculation zone is observed. It is the most interesting and complex transformation of the wake vortex structure. At $Fr < 0.4$ a new recirculation zone is formed from the "wave crest" which is situated very close to the sphere.

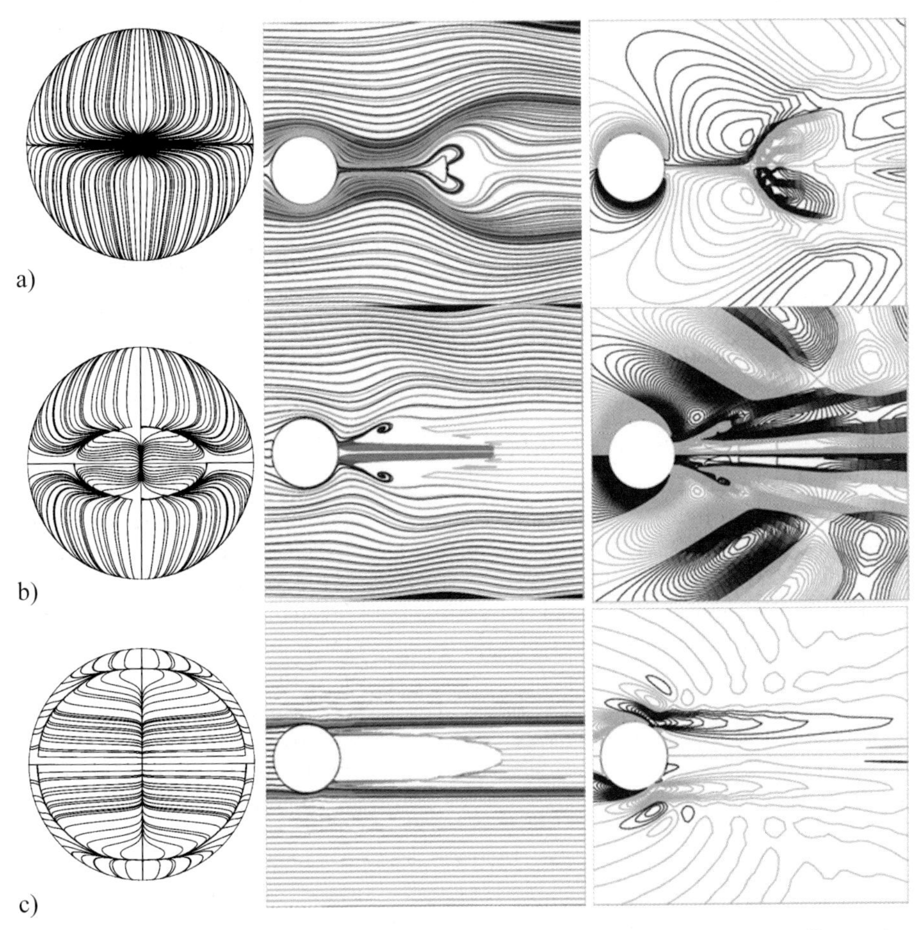


Fig. 5. The skin friction patterns on the sphere lee side (left), the instantaneous stream lines and isolines of the perturbation of salinity $S$·(right) in the vertical plane at $Re = 100$: a) $Fr = 0.5$ ($dS = 5 \cdot 10^{-5}$), b) $Fr = 0.3$ ($dS = 10^{-5}$), c) $Fr = 0.08$ ($dS = 2 \cdot 10^{-7}$ is the distance between isolines). The darker isolines correspond to negative $S$.

With increasing of *Re* (from 100 to 1000) the number of the degrees of freedom of the flow is increased too. At the same time the stratification stabilizes the flow in the sphere wake. As a result of DNS at $0.004 \le Fr \le 10$ and $10 \le Re \le 1000$ the following nine flow regimes were simulated:

1) $10 < Fr < \infty$, $Re < 200$ ("*One-thread wake*");
2) $2 < Fr < 10$, $Re < 300$ ("*Four-thread wake*");
3) $0.9 < Fr < 2$, $Re < 400$ ("*Non-axisymmetric attached vortex* in the recirculation zone");
4) $0.6 < Fr < 0.9$, $Re < 500$ ("*Two symmetric vortex loops* in the recirculation zone");
5) $0.4 < Fr < 0.6$ ("*Lack of any recirculation zone*");
6) $0.25 < Fr < 0.4$, $Re < 120$ ("*New recirculation zone formation. Steady vertical and horizontal vortices*");
7) $0.25 < Fr < 0.4$, $Re > 120$ ("*Unsteady vertical and steady horizontal vortices*");
8) $Fr < 0.25$, $Re < 120$ ("*Two steady attached vertical vortices* bounded by lee waves");
9) $Fr < 0.25$, $Re > 120$ ("*Vertical vortices shedding*").

The obtained characteristics of the flows (such as the drag coefficient, horizontal and vertical separation angles) are in a good agreement with [1-2] and some other experimental and numerical results.

This work has been supported by Russian Foundation for Basic Research (grants № 05-01-00496, 08-01-00662), by the program № 14 of the Presidium of RAS and by the program № 3 of the Department of Mathematical Sciences of RAS.

**REFERENCES**

1. Q. Lin, W.R. Lindberg, D.L. Boyer and H.J.S. Fernando, "Stratified flow past a sphere", *J. Fluid Mech.*, **240**, 315-354 (1992).
2. J.M. Chomaz, P. Bonneton, E.J. Hopfinger, "The structure of the near wake of a sphere moving horizontally in a stratified fluid", *J. Fluid Mechanics,* **254**, 1-21 (1993).
3. H. Hanazaki, "A numerical study of three-dimensional stratified flow past a sphere", *J. Fluid Mech.*, **192**, 393-419 (1988).
4. V.A. Gushchin, A.V. Kostomarov and P.V. Matyushin, "3D Visualization of the Separated Fluid Flows", *Journal of Visualization*, **7** (2), 143-150 (2004).
5. V.A. Gushchin and P.V. Matyushin, "Vortex formation mechanisms in the wake behind a sphere for $200 < Re < 380$", *Fluid Dynamics*, **41** (5), 795-809 (2006).
6. V.A. Gushchin and V.N. Konshin, "Computational aspects of the splitting method for incompressible flow with a free surface", *Journal of Computers and Fluids*, **21** (3), 345-353 (1992).
7. M.S. Chong, A.E. Perry and B.J. Cantwell, "A general classification of three-dimensional flow field", *Phys. Fluids*, **A 2** (5), 765-777 (1990).
8. V.G. Baydulov, P.V. Matyushin and Yu.D. Chashechkin, "Structure of a diffusion-induced flow near a sphere in a continuously stratified fluid", *Doklady Physics*, **50** (4), 195-199 (2005).
9. V.G. Baydulov, P.V. Matyushin and Yu.D. Chashechkin, "Evolution of the diffusion-induced flow over a sphere submerged in a continuously stratified fluid", *Fluid Dynamics*, **42** (2), 255-267 (2007).

# $C(p,q,j)$ Scheme with Adaptive time step and Asynchronous Communications

T. PHAM [a] and F. OUDIN-DARDUN[a] *

[a]CDCSP/ICJ UMR5208-CNRS, Université Lyon 1, F-69622 Villeurbanne Cedex

We address the challenge of speeding-up the solving of ODE/DAE/EDP nonlinear systems in term of elapsed time with using parallel computer. As modern time integrators use adaptive time steps to pass the stiffness of the solution, we propose to extend the $C(p,q,j)$ introduced in Garbey and all. (J. Comput. Phys. Vol. 161 No. 2 401-427 ) in order to use adaptive time step between the coupled subsytems of equations and to perform asynchronous communications to send the data needed for extrapolation. The details of implementation will be given as well as numerical and efficiency performances.

## 1. Introduction and motivation

The idea of the $C(p,q,j)$ scheme is based on extrapolation technique on previous time steps with a delay $p$ and used this extrapolation during $q$ time steps. This allows to overlap communication by computation. In [2] a system of Navier-Stokes equations under the Boussinesq approximation is coupled with reaction-convection-diffusion equations. The times steps of the two schemes were constant and verify CFL conditions. When one has to deal with stiff system, time integrator has to use adaptive time step [6]. In case of two subsystems with different time scales the use of the same time step for the two systems becomes prohibitive. In this work, we investigate the extension of $C(p,q,j)$ scheme in order to use adaptive time steps and the consequence on the communication schedule between the two subsystems.

## 2. Systems of ODEs and $C(p,q,j)$ scheme

### 2.1. $C(p,q,j)$ scheme with constant time steps

We consider a system of two coupled systems of differential equations;

$$u'(t) = f(t,u,v) \tag{1}$$

$$v'(t) = g(t,u,v) \tag{2}$$

A $k$-order BDF(backward differentiation formula) scheme with variable time steps for each subsystem takes the form :

$$\sum_{i=0}^{k} \alpha_{n,i} u_{n+1-i} + h_n f(t_{n+1}, u_{n+1}, v_{n+1}) = 0 \tag{3}$$

The similar formula holds for other subsystem. $\alpha_j$ are computed as function of the current and past step sizes $h_j$. The system is completely coupled due to the presence of $v_{n+1}$ in $f$ and $u_{n+1}$

*This work was funded by the ANR Technologie Logicielle 2006 PARADE


DOI: 10.1007/978-3-540-92744-0_41, 

in $g$.

Then we replace $v$ from $f$ and $u$ from $g$ by their polynomial approximations. e.g.:

$$u_{n+1} - u_n = hf(t_{n+1}, u_{n+1}, v^*_{n+1}) \quad (4)$$

$$v_{n+1} - v_n = hg(t_{n+1}, u^*_{n+1}, v_{n+1}) \quad (5)$$

We can solve independently each subsystem. A simple example of a second order extrapolation scheme: $u^*_{n+1} = (p+1)u_{n-p+1} - pu_{n-p}$ where $u^*_{n+1}$ computed from two previous solutions. So far we defined the constant time steps $C(p,q,j)$ scheme:

- $p$ : number of steps delayed from the predictions
- $q$ : number of successive steps predicted from the same data points
- $j$ : number of points used

Full detail of the numerical scheme is described in [2]

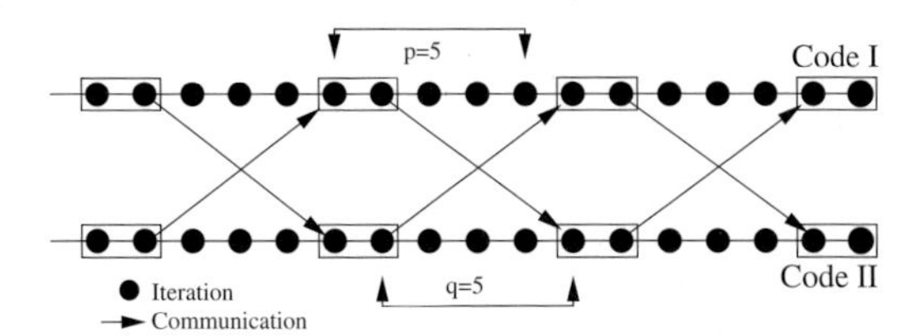


Figure 1. Communication Schedule for the C(5, 5, 2) scheme

## 2.2. New stategy for $C(p,q,j)$ scheme

We propose to extend the scheme to adaptative time steps. We use the $j$-th order extrapolation scheme using Newton Divided Difference(NDD) formula:

$$v^*(t) = \sum_{k=0}^{j} v[t_0 \dots t_k](t - t_0)\dots(t - t_k)$$

by constructing the NDD array:

$$\begin{array}{lllll} t_0 & v[t_0] & & & \\ t_1 & v[t_1] & v[t_0 t_1] & & \\ t_2 & v[t_2] & v[t_1 t_2] & v[t_0 t_1 t_2] & \\ \vdots & \dots & & \dots & \\ t_j & v[t_j] & & \dots & v[t_0, t_1, \dots, t_j] \end{array}$$

where $v[t_i] = v(t_i)$ and

$$v[t_i, \dots, t_{i+k}] = \frac{v[t_i, \dots, t_{i+k-1}] - v[t_{i+1} \dots, t_{i+k}]}{t_i - t_{i+k}}$$

This extrapolation technique does not require constant time steps and allows the current solver to take internal time steps. When received a new value in history, we shift the NDD array one rank and only the last line needs to be evaluated. When a subsystem performs its time step at the time $t_{m+1}$- based on NDD array, we evaluate according to NDD formula at the new time node $t_{m+1}$.
This choice of j-th extrapolation by NDD array, each extrapolation has the error formula (in the case of equidistant points):

$$e_j(t_{n+1}) = \frac{f^{(j+1)}(\xi)}{(j+1)!} \prod_{i=0}^{j} (t_{n+1} - t_{n-i}) \tag{6}$$

The main advantages of this technique are:

- Different time steps between solvers: adaptative time steps can be applied for each subsystem
- Dealing with stiffness for difference components
- Once received the next time step, we shift the NDD array up one rank and recompute only the last line

The gain in computation:

- Expecting computation gain in Jacobian saving:
  - Cost for whole matrix-vector multiplication : $n^3$
  - Cost for local evaluation : $n_u^3 + n_v^3$ (where $n = n_u + n_v$)
- Expecting computation gain in non-linear solver:
  - Implicit scheme results in a non-linear system of the form $u_{n+1} = \mathcal{F}(u_{n+1})$, requires Newton iterations. The local Jacobian matrix has better condition number and requires less iterations than the global system.
- Extrapolation requires only $n(j-1)$ *flops* at each update

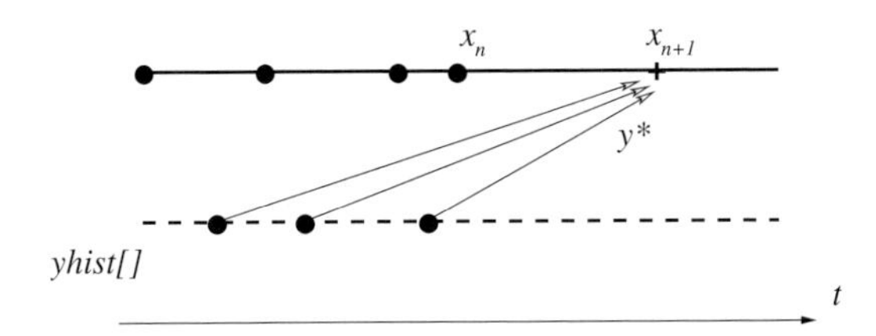


Figure 2. A simple scheme using 3-point extrapolation in adaptative time steps: here we show the history array *yhist* for extrapolating extra components $y*$ to solve for $x_{n+1}$

**2.3. Extrapolation quality control**

The first natural idea when predicting coupled terms for a current subsystem is how to control error made from polynomial extrapolation instead of having true terms. An rigourous error formula are under investigation. Alternatively to this a priori error estimate, we can make a posteriori checking for extrapolation. The idea consists to compare the solution predicted by extrapolation with the one truly obtained by solving the corresponding subsystem.
We take for instance the first subsystem $u$, since we have different time nodes for $u$ and $v$, we denote the index for $u$ by $i$, for $v$ by $k$. We have two sequences of time steps: $t_i, i = 1, \dots$ and $t_k, k = 1, \dots$. We are solving for the next time step $t_{i+1}$, and the other solver does $v_{k+1}$. The current solver for $u$ received some previous components $v_{k-p-j}, \dots, v_{k-p}$ and store them in the divided difference array. The coupled term $v^*$ needed for solving $u_{i+1}$ is predicted by:

$$v^*(t_{i+1}) = \sum_{l=0}^{j} v[t_{k-p-j}, \dots, t_{k-p-j+l}](t - t_{k-p-j}) \dots (t - t_{k-p-j+l})$$

in another words,$v^*(t_{i+1})$ is the solution predicted by the first subsystem. In fact, the solution $v$ is also solved in the second subsystem at the time $v(t_{k+1})$. We wish to control the error as $\|\tilde{v}(t_{k+1}) - v^*(t_{k+1})\|$:

- $v$ is not evaluated at $t_{i+1}$ in general since different time nodes between subsystems, we may have to solve $v$ in a domain covering the node $t_{n+1}$ and get $\tilde{v}$ by interpolation of appropriated order to the numerical integrator. This can be easily done in many ODE solvers.
- Once received the solution from the other time integrator to compare with predictions, we may have passed few steps away. What if we have gone wrong with the prediction- we store up to some reasonable state variables to redo the wrong step. This technique makes sure that we control the error of the extrapolated terms in current subsystem

We give this pseudo algorithm:

1: Solving for $u_{n+1}$, history stored from $v_{k-p-j}$ to $v_{k-p}$
2: **if** Received new value $v_{k-p+1}$ **then**
3: Evaluate $v^*(t_{k-p+1})$ based on $v_{k-p-j}, \dots, v_{k-p}$
4: **if** $\|v_{k-p+1} - v^*_{k-p+1}\| > \delta$ ($\delta$ "not too large" fixed at priori) **then**
5: Restart the solver back to the step $t_{k-p}$
6: **end if**
7: **end if**

This posteriori error control makes sure that no important error due to predicted terms may results in important pertubation for a subsystem. Remark also this restart technique is a compromise for the method performance. A restart in solver at the time $t_i$ will cancel all computation done up to $t_i$ for all solvers- therefore very computationally expensive. So we need to have history data up-to-date to avoid this restarting scheme.

**2.4. New parallel implementation for $C(p, q, j)$ scheme**

When performing communication between solver, we use asynchronous communication operation(ISend/IRecv). When advance solver in a time step, we allow delay to communication to be completed. Thus communication ovelaps computation. But what if we made an extrapolation far away from the history. Poor extrapolation leads to wrong solution and requires to be restarted. We then fixed a history limit $h_{lim}$ for history. When trying to make a time step

$t_{next} > t_{hist} + h_{lim}$, we finalize the latest pending receive(call $Wait$ for the last $IRecv$)
The overall algorithm of exchanging messages is given below:

1: **for** steps **do**
2: **while** *(Found a new message)* **do**
3: ready to receive (by Irecv) and update if completed
4: **end while**
5: Call SOLVER to get a new step
6: Send the last solution: make an Isend
7: **if** (History is out of date) **then**
8: Wait for newer message
9: **end if**
10: **end for**

The SOLVER here is CVODE- a part in SUNDIALs: Suite of Nonlinear and Differential/Algebraic Equation Solver. Developed by *Center for Applied Scientific Computing- LLNL*. CVODE is an extention rewritten in C of VODE (Fortran). CVODE uses:

- Newton inexact methods for solving non-linear system
- Implicit ADAMS scheme for non-stiff ODEs and BDF for stiff system
- Adaptive orders and adaptive time steps

More details on CVODE can be found in [1]. The exchanging message -restarting algorithm is implemented within the code CVODE.

## 3. Numerical and performance results

### 3.1. HIRES test case

Our preliminary results use the test case High Irradiance Response (HIRES problem) originates from plant physiology. The problem consists of the following eight equations:

$$\begin{aligned}
u_1' &= -1.71u_1 + 0.43u_2 + 8.32u_3 + 0.0007 \\
u_2' &= 1.71u_1 - 8.75u_2 \\
u_3' &= -10.03u_3 + 0.43u_4 + 0.035u_5 \\
u_4' &= 8.32u_2 + 1.71u_3 - 1.12u_4 \\
u_5' &= -1.745u_5 + 0.43u_6 + 0.43u_7 \\
u_6' &= -280.0u_6u_8 + 0.69u_4 + 1.71u_5 - 0.43u_6 + 0.69u_7 \\
u_7' &= 280.0u_6u_8 - 1.81u_7 \\
u_8' &= -280.0u_6u_8 + 1.81u_7
\end{aligned}$$

We test with 2 solvers: the first one solves equations $\{u_1, u_2, u_4, u_6\}$, the second does remaining components $\{u_3, u_5, u_7, u_8\}$.

In the figure 4 : we compare the reference solution to the solution given by our scheme using 5-th order extrapolation for extra-components. Individual relative errors compared to reference solution are in the scale of $1e-5$. The relative tolerance is fixed at $1.e-8$.

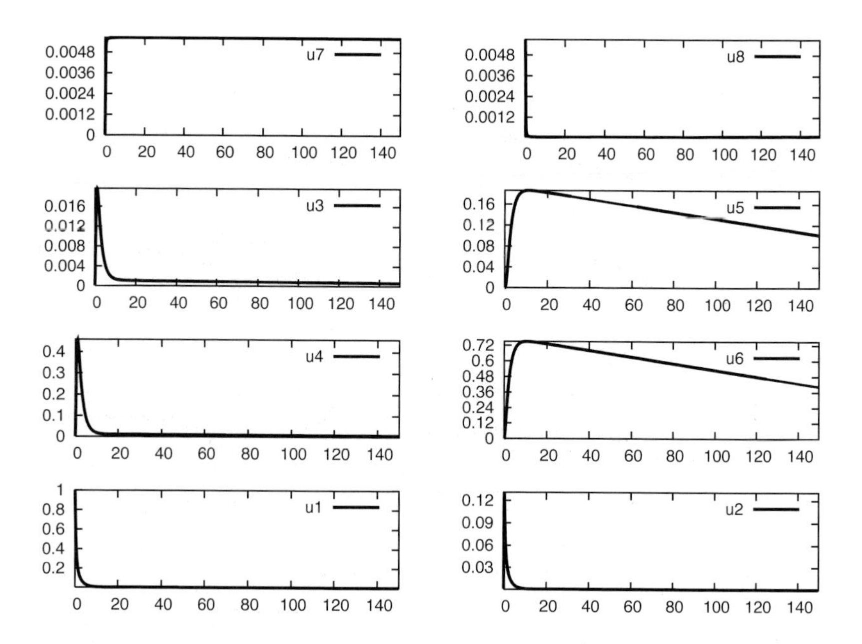


Figure 3. HIRES problem: reference solution

**3.2. EMEP test case**

EMEP problem originally from the the chemistry part of the EMEP MSC-W ozone chemistry model - Norwegian Meteorological Institute-Oslo-Norway. This model involving 140 reactions with a total of 66 species. The resulting equations are stiff with different scales reactions. The relative tolerance is fixed at $1.e-9$.
Performance comparison

- On the time interval $[14400, 72000]$, using the same discretization
  - Single solver for the global system: 3.71s
  - 2 solvers: 3.05s

We pick the error behave for 6 interested components following different orders of extrapolation, see figure 5. Our numerical results get better when we increase the extrapolation order $j = 1$ to 3. But when using $j = 4$, we encountered wrong results. This can be explained by instability when increasing in order of extrapolation. We can see that in this specific case, the case of $j = 3$ gives the best result.

**3.3. Example provided from PDEs**

The following example is a pair of kinetics-advection-diffusion PDE, which represent a simplified model for the transport, production, and loss of ozone and the oxygene singlet in the upper

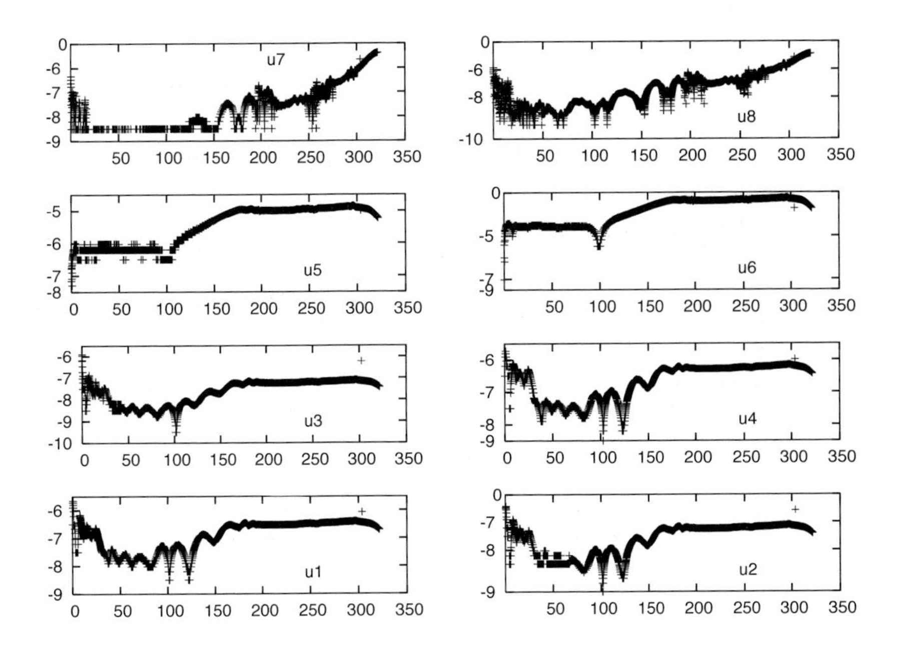


Figure 4. HIRES problem: relative error to the reference solution using 5-th order extrapolation-solver's relative tolerance fixed at $1.e-8$

atmosphere:

$$\frac{\partial c^i}{\partial t} = K_h \frac{\partial^2 c^i}{\partial x^2} + V \frac{\partial c^i}{\partial x} + \frac{\partial}{\partial y} K_v(y) \frac{\partial c^i}{\partial y} + R^i(c^1, c^2, t)$$

where the superscipts $i$ are used to distinguish the two chemical species, and where the reaction terms are given by:

$$\begin{aligned} R^1(c^1, c^2, t) &= -q_1 c^1 c^3 - q_2 c^1 c^2 + 2q_3(t) c^3 + q_4(t) c^2 \\ R^2(c^1, c^2, t) &= q_1 c^1 c^3 - q_2 c^1 c^2 - q_4(t) c^2 \end{aligned}$$

The spacial domain is $0 \leq x \leq 20$, $30 \leq y \leq 50$ (in $km$). The various constants are:

$$\begin{aligned} K_h &= 4.0 \cdot 10^{-6} \\ V &= 10^{-3} \\ K_v(y) &= 10^{-8} exp(\frac{y}{5}) \\ q_1 &= 1.63 \cdot 10^{-16} \\ q_2 &= 4.66 \cdot 10^{-16} \\ c^3 &= 3.7 \cdot 10^{16} \end{aligned}$$

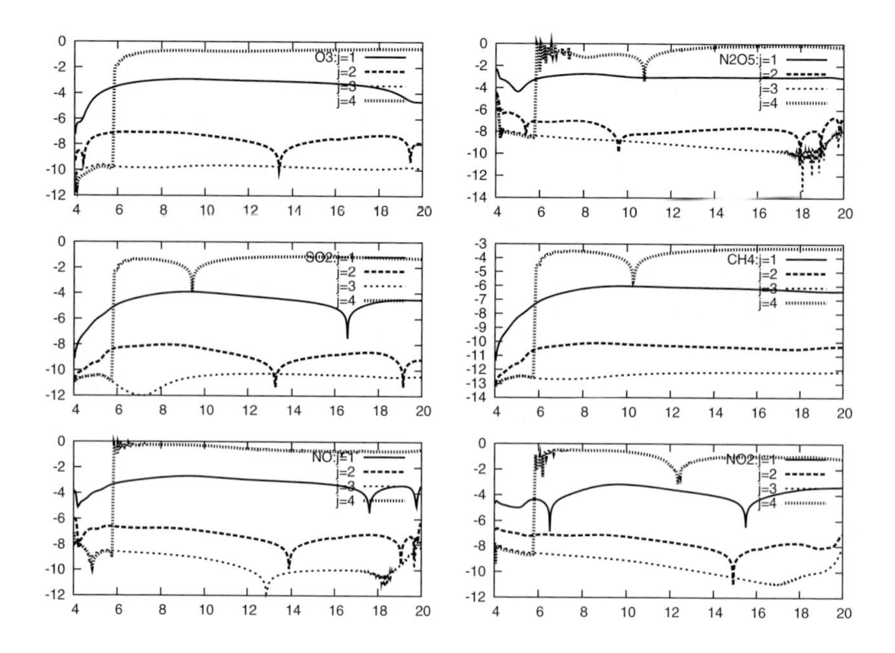


Figure 5. EMEPproblem: relative error(log scale) to the reference solution using different order extrapolations $j$, solver's relative tolerance fixed at $1.e-9$

Using standard central finite difference to uniformly discreatize $100 \times 100$ mesh, by 2 species, giving an ODE system of 20000 components. The partitioning for parallel solvers is done by geometric partitioning. The original domain is equally divided into 2, 4, 8 subdomains.

Table 1
Performance on parallel computers for the PDEs test case: using SGI Altix 350 system: 16 Intel Intanium 2 processors

| No. of Processors | 1 | 2 | 4 | 8 |
|---|---|---|---|---|
| Time elapsed(in *sec*) | 13.30 | 8.28 | 4.31 | 2.98 |

The ODE system size 20000

## 4. Concluding remarks

Our adaptative $C(p, q, j)$ scheme well performs stiff and non still problems thanks to adaptative time steps. The performance of this method is also validated for several test cases. The mathematical properties can be found in [3] and [4]. The authors thank to *Agence National de la Recherche*-France who has funded for this work.

## REFERENCES

1. G.D. Byrne and A.C. Hindmarsh, A Polyalgorithms for the numerical Solution of Ordinary Differential Equations, ACM Transactions on Mathematical software, V. 1 No. 1 71-96 (1975).
2. M. Garbey and D. Tromeur-Dervout, A parallel adaptive coupling algorithm for systems of differential equations, J. Comput. Phys., V. 161 No. 2 401 427 (2000)
3. S. Skelboe, Accuracy of Decoupled Implicit Integration Formulae, SIAM J. Sci. Comput., V. 21 2206 2224 (2000)
4. S. Skelboe and J. Sand, Stability of Backward Euler Multirate Methods and Convergence of Waveform Relaxation, BIT, V. 32 350 366 (1992)
5. S. Skelboe, Methods for Parallel Integration of Stiff Systems of ODEs, BIT, V.32 689 701 (1992)
6. E. Hairer and G. Wanner, Solving Ordinary Differential equations II - Stiff and DAE problems, Springer No. 14 (2002)

# Parallel Coupling of Heterogeneous Domains with KOP3D using PACX-MPI

Harald Klimach[a], Sabine P. Roller[a], Jens Utzmann[b] and Claus-Dieter Munz[b]

[a]High Performance Computing Center Stuttgart (HLRS), University of Stuttgart, 70550 Stuttgart, Germany

[b]Institute for Aerodynamics and Gasdynamics (IAG), University of Stuttgart, Pfaffenwaldring 21, 70569 Stuttgart, Germany

This paper outlines how coupled heterogeneous domains can be distributed in a supercomputing environment using PACX-MPI. KOP3D is a coupling program in the field of aero acoustics, a typical multi-scale application, since on the one hand it has to account for the small vortical structures as the source of the noise and on the other for the long wave length of the acoustic waves. The amount of energy, and with that the pressure, in the sound generating flow is by orders of magnitude higher than the amount of energy carried by the acoustic waves. The acoustical wavelength is much larger than the diameter of the noise-generating vortices, and even the computational domains may vary in a wide range from the geometry which is in the order of meters for an airfoil to the distance between a starting or landing plan and the observer on the ground which is in the order of about 1 km. Through the use of heterogeneous domain decomposition it is possible to reduce the computation time needed to simulate large flow fields by adaptation of the equations, the discretization, the mesh and the time step to the local requirements of the flow simulation within each sub-domain. These locally adapted computations may result in different requirements on the computer architecture. On the other side, in a supercomputing network there are generally different computer architectures available. By matching the sub-domains to the best suited available architectures an even shorter time to solution may be gained. The parallel version of the KOP3D coupling scheme is shown and the benefits of running the simulation distributed on the vector machine NEC-SX8 and on the scalar Itanium II (IA64) machine are demonstrated. A 2D and a 3D testcase are presented.

## 1. KOP3D

The coupling scheme of KOP3D was mainly developed, to enable the direct simulation of the generation of sound waves and their propagation in a large domain at the same time. A detailed description of the scheme is given in [1]. The scheme is used to couple solvers with explicit time marching discretizations. To reduce the number of needed elements, high order schemes are used in the solvers on each domain. The coupling is able to preserve this high order across the domain interfaces.

In KOP3D there is one part for domains with structured meshes, and one for those with unstructured meshes. Implemented discretization methods are Finite Differences (FD), Finite Volume (FV) and Discontinuous Galerkin (DG) schemes. Generally DG methods are used for

I.H. Tuncer et al., *Parallel Computational Fluid Dynamics 2007*,
DOI: 10.1007/978-3-540-92744-0_42, © Springer-Verlag Berlin Heidelberg 2009

domains with unstructured meshes and FV are used for those with structured meshes. Finite Differences can be used on structured meshes of domains with linearized Euler equations.

The coupling between these sub-domains is realized by ghost cells which are computed by interpolation in the corresponding neighbor domains. This is done in a very generalized manner by interpolating the values on the points needed for the Gaussian integration in the ghost cell. The Gaussian integration is done after all the values at the needed points are interpolated. By splitting the calculation of the mean-value in the ghost cells into those two steps, it doesn't matter, how the ghost cell is positioned on the neighbors. Each point can be calculated independently in the domain where it is found in.

For parallel computation of the sub-domains, a list of all points for Gaussian integration, which can be found in a neighbor, has to be created and these lists exchanged between all neighbor domains.

The coupling scheme in KOP3D is not only capable of spatial decomposition, it also accounts for high efficiency in the temporal coupling. Typically, each domain uses its own time step given by the local stability condition. Then a sub-cycling has to be done, where the domains with finer discretization run for several time steps, while the coarser domains perform only one. Data is exchanged only when two neighboring domains reach the same time level. Between those interaction times the ghost cells are updated by an extrapolation in time, using the Cauchy-Kowalevskaja procedure, also known as Lax-Wendroff procedure [2] to construct time derivatives from spatial derivatives.

In the view of parallel computations, this results in less communication, as the domains don't need to exchange data at every single time step. Hence, the coupling between different domains is quite loose and can be feasible even with quite low network requirements. This makes it well suited for inter-cluster communication.

## 2. PACX-MPI

The PACX-MPI library was developed to enable the usage of heterogeneous clusters of supercomputers within a single run of an application. It is described in more detail in [3]. From the application point of view there is no logical difference between a run in an ordinary MPI Environment and a run on PACX-MPI. All processors are made available within the global communicator, regardless of the physical position of the processor. Internally the library uses on each machine the "native" MPI library which is optimized for the given architecture. The inter-cluster communication is carried out by daemons. Any communication between processors on different clusters is tunneled through those two processes. The resulting process layout is depicted in figure 1. The advantage is that PACX-MPI can account for different network protocols within each machine and between the clusters. This is important, if the one architecture provides a fast infiniband interconnect while the other is running e.g. gigabit ethernet.

The two processes for inter-cluster communication do the local distribution of messages, apply data conversions if necessary and compress the data, that is to be send over the slower network connection. This setup is especially well suited for loosely coupled applications but it even serves strongly coupled applications quite well [5].

## 3. RUNNING KOP3D ON PACX-MPI

As presented above, KOP3D offers an appropriate method for coupling heterogeneous domains of flow simulations with relatively little communication, where each domain may have different numerical requirements. PACX-MPI on the other hand offers an elegant way to run an application distributed across different super-computing clusters with different architectures.

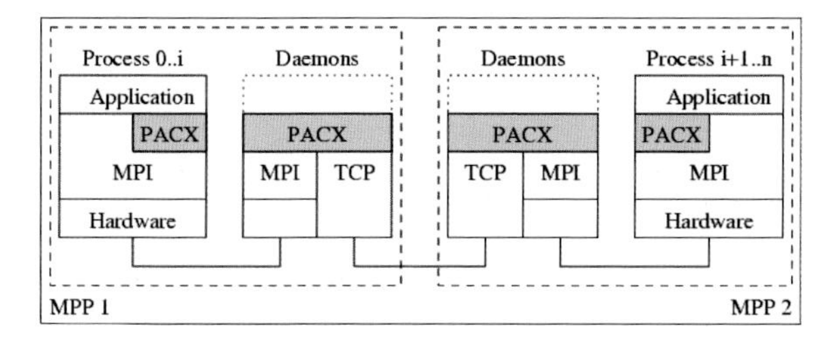


Figure 1. Process Layout in PACX-MPI, taken from [4] with kind permission of Rainer Keller.

Our goal is it to take advantage of the heterogeneity of both the simulated flow and the supercomputing environment by fitting the requirements of the simulation to the properties of the used hardware.In order to do this we added a property to the running processes by which we can decide on which architecture it is actually running. This is defined upon startup of the process and can then be used during the distribution of the domains onto the available processes to map each domain on the best suited architecture. Especially we can run the structured domains on a vector machine and the unstructured ones on a scalar machine. Up to now, this division is done at startup time, therefore the proper ratio between the calculation loads in each domain type has to be known a priori to get a balanced run.

### 3.1. Communication Layout

KOP3D actually consists of several parts. There is the coupling itself and the solvers within each domain. Each of those coupled domains may be calculated across several processors themselves, which leads to an inherent hierarchy of communication levels corresponding to the domains and there subdivision (what we call sections). Each section is uniquely defined by the value pair of the global rank of the processor it is calculated on and the domain it resides in.

All communication between neighboring sections is done by MPI point to point messages. Each domain uses its own communicator for internal communication between sections inside it. The exchange of data for the Gaussian integration points between sections in different domains at a meeting sub-cycle is done in a global communicator.

This layout of KOP3D maps quite well with the existing network setup. Generally there is only a weak interconnection between different clusters and strong connections of processors within the same cluster, thus a good communication performance can be gained, when the domain boundaries coincide with the cluster boundaries. A single domain should not be distributed between different clusters.

### 3.2. Results of 2D Simulations on PACX-MPI

A first example to demonstrate the ability to run in the heterogeneous environment was a plane sinusoidal acoustic pressure fluctuation, which is scattered at a cylinder with perfectly reflecting walls. The calculation area is divided into one unstructured domain around the cylinder surrounded by a structured domain.

During the run with PACX-MPI the structured domain was calculated on the NEC SX-8 at the HLRS and the unstructured domain on the Itanium II based Linux cluster, which usually serves as front end to the vector system.

However, for settings like this in 2D, no real advantage in sense of calculation time can be achieved by the heterogeneous calculation, since the problem size is to small for effective computation on the vector system NEC SX-8 and small enough to almost fit into the cache of the scalar Itanium II machine. Therefore, this first test case served as a proof of concept only and showed already, that PACX-MPI may handle the communication between the different domains without big performance impacts.

For problems with 2D nonlinear equations the benefit is much higher: here, the unstructured domains perform pretty bad on vector architectures, while the structured domains are quite costly on the scalar machines in comparison to the vector architecture. Therefore the aim is to use the best suited architecture for both types of domains.

As an example that uses 2D nonlinear Euler equations we calculated the single airfoil gust response by Scott [6] from the *4th Computational Aeroacoustics (CAA) Workshop on Benchmark Problems* (Problem 1 from Category 3). Here, the computational domain is divided into a small unstructured domain around the airfoil within a structured domain. On both sub-domains, the nonlinear Euler equations are solved. Finally those two domains are embedded into a large far field domain in which the linearized Euler equations are solved on a structured mesh. The configuration of the meshes is shown in figure 2.

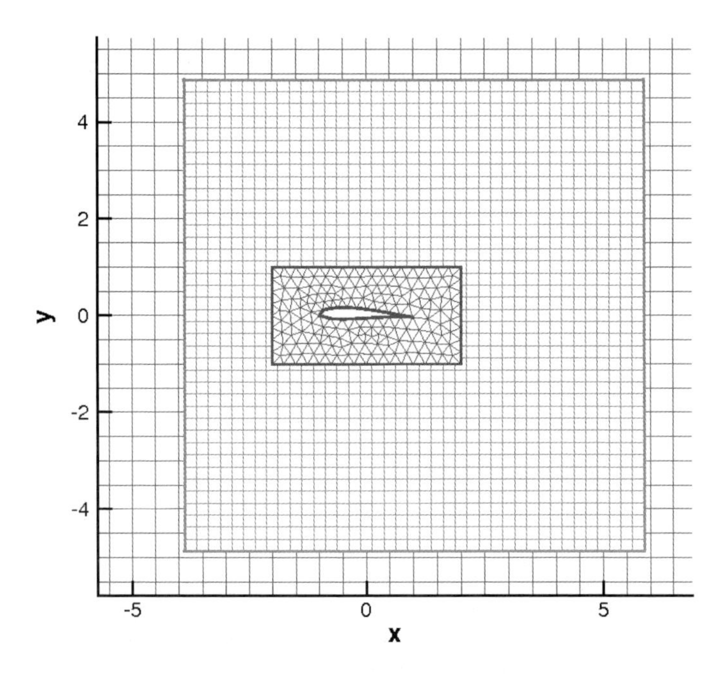


Figure 2. Domain configuration for the airfoil gust response problem.

We compare the calculation time needed for simulations with a) all domains on NEC SX-8, b) with all domains on the Linux cluster and c) the distributed run, where the unstructured domain is calculated on the Linux cluster and the structured domains are calculated together on the vector NEC SX-8 machine. The comparison of the simulations on each platform respectively, and the distributed PACX-MPI run is presented in table 1.

Table 1
Comparison of calculation times for the single airfoil gust response problem.

| | Linux Cluster | NEC SX-8 | PACX |
|---|---|---|---|
| Unstructured | 324 s | 840 s | 326 s |
| Structured | 196 s | 7 s | 7 s |
| Total | 520 s | 847 s | 333 s |

Hence, by using PACX-MPI we are able to run each part of the overall simulation on the architecture best fitted for the specific task and thereby reduce the needed total runtime. It can be observed that the structured part is performing extremely well on the NEC SX-8. It just needs roughly 4% of the runtime on the Itanium. The unstructured part on the other hand shows only poor performance with about 260% of the runtime on the Itanium when it is run on the NEC SX-8.

Due to the fast interconnect between the two machines, the increased communication time compared to the intra-cluster connections is negligible. Especially, as the layout of the KOP calculations reduces the necessary usage of this weak interconnection.

### 3.3. Results of the 3D Simulation

As an 3D example we use a similar testcase to the cylinder scattering in 2D: The noise emitted from a point is scattered at a perfectly reflecting sphere. In 3D the number of cells in the outer structured mesh is drastically increased, and the needed computational power and memory is correspondingly much higher when compared to the 2D setup.

The computational volume is devided into two domains, an unstructured mesh around the sphere embedded into a structured mesh simulating the far field. The sphere has a radius of one meter and the whole simulated volume has an extent of $51 \times 41 \times 41$ meters. The structured mesh has about 10 million cells and the unstructured mesh is built with 9874 cells. Both domains are computed using a spatial discretization scheme of order 8. In the domain with the unstructured mesh a Discontinuous Galerkin scheme is used. A Finite Difference scheme is used on the structured mesh.

This is a well suited problem for our heterogeneous setup, as the structured 3D mesh can be computed with high performance on the vector machine. The main time consuming routine in the structured part runs on the NEC SX-8 with a performance of nearly 14 GFlop/s which equals to 87.5% of the peak performance of 16 GFlop/s. Whereas the biggest part of the unstructured mesh is running in routines with less then 100 MFlop/s on the NEC-SX8. This is caused by the bad vectorization in this part. It may be overcome, but that would require a major rewrite of the existing code. For the time being, this part is much slower on the vector machine when compared to the execution on a scalar processor.

The simulation of 2 seconds on a single Itanium II (IA64) processor takes in total 9452 seconds of real time for computing. On a single NEC SX8 processor the same simulation takes 9242 seconds. So it would not make much sense to run the whole simulation on the more

Table 2
Comparison of calculation times for the sphere scattering test case.

| | Linux Cluster | NEC SX-8 |
|---|---|---|
| Unstructured | 3041 s | 7519 s |
| Structured | 5891 s | 792 s |
| Coupling | 512 s | 926 s |

expensive vector machine. However as shown in table 2 it would be a big advantage to run the structured part on the vector machine, and thereby exploiting the high performance in this part on this architecture. Table 2 shows in the third line the time necessary to compute the coupling information. As can be seen, this is not good vectorized and runs only with poor performance on the vector machine but for larger simulated volumes the proportion of this bad performing part will ever get smaller. By comparing the computational needs of each domain on its appropiate architecture it turns out 1.8 IA64 processors would be needed for each SX8 processor to achieve a balanced run.

Table 3
Runtimes for coupled PACX-MPI computations.

| | 1 SX8 / 1 IA64 | 1 SX8 / 2 IA64 |
|---|---|---|
| **Elapsed time** | **3064 s** | **1765 s** |
| Unstructured | 2966 s | 3224 s |
| Structured | 790 s | 790 s |
| Coupling | 2373 s | 1276 s |
| Wait in Coupling | 1404 s | 236 s |

In table 3 the runtimes of the simulation in a coupled PACX-MPI run are shown. By using one NEC SX8 processor for the structured domain and one Intel Itanium II processor the elapsing time is reduced to 3064 seconds. Which is the time needed to compute the solution of the unstructured mesh on the Intel Itanium II processor. The computation of the structured mesh can be done completely within this time by a single NEC SX8 processor. However the vector machine has a long idle time while waiting on informations from the unstructured neighbor due to the big imbalance. This results in the much increased coupling time of 2373 seconds in total. 1404 seconds of this cpu time is spent waiting on the neighbor. Despite this, the total elapsed time is still better then the optimal time of 4726 seconds, that could be achieved with an ideal load balancing on two scalar processors.

With two Intel Itanium II processors and one NEC SX8 processor, which better meets the optimal factor of 1.8, the idling time spent waiting on the neighbor is drastically reduced and shifted to the less expensive IA64 architecture. The needed elapsing time in this setup is reduced to 1765 seconds which is around 56% of the time that would be needed when using three processors on either architecture.

This 3D example nicely shows how the simulation can profit from the heterogeneous network of supercomputing clusters. For simulations with an even larger far field area, like the ones that we like to simulate with hundreds of meters, the proportion of the unstructured mesh may become quite small. But still running in it on the relatively expensive NEC-SX8 processor with less then one percent of the peak performance leads to inadequate long runtimes, which can be avoided by just putting this part on a cheaper and for this part even faster PC-Cluster. So to get an acceptable time to solution for those huge problems without wasting ressources the option to use different architectures is a feasible way to go.

### 3.4. Outlook

As the data presented in the above section shows, the problem obviously is the load balancing between the different architectures. The load distribution has to be known before the simulation is started, and it is not easily possible to change it, as no processors can be shifted from one machine to the other. Despite this problem of perhaps not ideally balanced load distributions a big advantage can be gained by using the appropiate architecture for each kind of problem.

In order to really distribute a simulation across different clusters a coallocation facility is needed to get concurrent timeslots in the queueing systems of the involved clusters. There is currently work going on in the *Distributed European Infrastructure for Supercomputing Applications* (DEISA) to provide this feature in combination with the fast interconnect between the different european supercomputing sites. We hope to exploit these facilities with KOP3D in combination with PACX-MPI as soon as they become fully available.

We want to do bigger 3D simulations with different physical phenomena involved. Especially a simulation of the Von Karman vortex street behind a sphere is planned. Including the whole range of effects from the boundary layer to the aero acoustic wave propagation.

Another topic that will be investigated further is the overhead generated by PACX-MPI. More detailed measurements of the communication time are already planned. With the simulations presented in this paper no real difference between intra-cluster and inter-cluster communication had been observed. But the communication over this weak link should be as low as possible anyway.

## REFERENCES

1. J. Utzmann, T. Schwartzkopf, M. Dumbser, and C.-D. Munz. Heterogeneous domain decomposition for computational aeroacoustics. *AIAA Journal*, 44(10):2231–2250, October 2006.
2. P.D. Lax and B. Wendroff. Systems of conservation laws. *Communications in Pure and Applied Mathematics*, 13:217–237, 1960.
3. Thomas Beisel, Edgar Gabriel, and Michael Resch. An extension to MPI for distributed computing on MPPs. pages 75–82, 1997.
4. Rainer Keller and Matthias Mueller. PACX-MPI project homepage. `http://www.hlrs.de/organization/amt/projects/pacx-mpi`.
5. Thomas Boenisch and Roland Ruehle. Adapting a CFD code for metacomputing. 1998.
6. J.R. Scott. Single airfoil gust response. In *Proceedings of 4th Computational Aeroacoustics (CAA) Workshop on Benchmark Problems*, September 2004. NASA/CP-2004-212954.

# Numerical simulation of 3D turbulent flows around bodies subjected to vortex-induced and forced vibration

**Dmitri K. Zaitsev,[a] Nikolai A. Schur,[a] Evgueni M. Smirnov[a]**

*[a]Dept. Aerodynamics, St.-Petersburg State Polytechnic University, Polytechnicheskaya 29, St.-Petersburg, 195251, Russia*

A parallel CFD technique is applied to fluid structure interaction problems. Namely, the vortex-induced vibration of an elastically mounted circular cylinder is investigated, and the flow generated by oscillations of a thin flexible blade (modeling a piezo-fan) is considered. The flow around vibrating bodies is computed with the deforming mesh approach based on the ALE formulation, and the hydrodynamic force computed is used to predict the body motion/deformation. The turbulence is simulated via a RANS/LES vortex-resolving approach. In the cylinder case characterized by two-dimensional geometry, both 2D and 3D formulations are used, with the spanwise periodicity conditions imposed for 3D simulation. A comparison with experimental data has proven that the 2D simulation is inadequate.

Numerical simulation; turbulent flow; fluid-structure interaction; vortex-induced and forced vibration; RANS-LES approach; deforming grid; parallel computations

## 1. INTRODUCTION

Numerical simulation of turbulent flows around vibrating bodies is a challenging CFD problem because of huge computer resources needed. The problem is even more complex if the flow feedback on the body motion (deformation) has to be taken into account. For tasks of such a kind, the simulation time with a present-day PC can exceed several months. Parallel computing on a relatively cheap cluster system makes it possible to reduce the computational expenses down to an acceptable level.

In the present work, the flow around vibrating bodies is simulated using the deforming mesh technique based on the arbitrary Lagrangian-Eulerian (ALE) formulation. The hydrodynamic force components computed at each physical time step are used to predict the body motion (deformation), i.e. the fully coupled fluid structure interaction problem is considered. The turbulence is modeled via a RANS/LES vortex-

I.H. Tuncer et al., *Parallel Computational Fluid Dynamics 2007*,
DOI: 10.1007/978-3-540-92744-0_43, 

resolving approach. All the computations are performed using a well-validated in-house code.

The vortex-induced vibration (VIV) of a circular cylinder is extensively studied during several decades. A comprehensive review of experimental results as well as phenomenon discussions can be found in [1,2]. Recent numerical simulations are reported in [3-7]. Oscillations of a freely vibrating cylinder in the free-stream flow originate from the hydrodynamic force periodic change due to the vortex shedding. When the shedding frequency is close to the natural vibrating frequency of the elastically mounted cylinder, the oscillations rise up. In turn, these resonant oscillations affect considerably the flow so that the synchronization or lock-in phenomenon occurs. In this regime, the vortex shedding follows the cylinder vibration rather than the Strouhal relationship known for the von Karman vortex street. Moreover, hysteresis behavior is observed sometimes due to the non-linear interaction between the fluid and the cylinder motion.

The flow induced by oscillations of a thin flexible blade (e.g. with piezoelectric excitation) can be used for cooling some components of electronic devices. Piezoelectric fans that have recently emerged as an alternative to traditional fans are characterized by very low power and low noise. Results of experiments on visualization of the flow generated by a piezoelectric fan are reported in [8,9]. An attempt to simulate the phenomenon in the framework of 2D formulation was undertaken in [10]. Presented in this paper, the results of 3D simulation of the flow from a piezoelectric fan seem to be the first ones.

## 2. NUMERICAL ASPECTS

### 2.1. Deforming grid approach

In the present study, the flow around vibrating bodies is computed using the deforming mesh approach (e.g. [11]). In this approach, the original grid is modified at every instant following a current position (shape) of the body, without any change of the grid structure. To work with a deforming grid, the arbitrary Lagrangian-Eulerian formulation is used, i.e. the flow governing equations are written for an arbitrary moving/deforming volume $\Omega$ (the grid cell) as follows:

$$\frac{d}{dt}\int_{\Omega}\rho d\Omega+\int_{S}\rho(\vec{v}-\vec{v}_b)\cdot\vec{n}dS=0 \quad (1)$$

$$\frac{d}{dt}\int_{\Omega}\rho\vec{v}d\Omega+\int_{S}\rho\vec{v}(\vec{v}-\vec{v}_b)\cdot\vec{n}dS=\int_{S}\left(\underline{\underline{\boldsymbol{\tau}}}\cdot\vec{n}-p\vec{n}\right)dS \quad (2)$$

$$\frac{d}{dt}\int_{\Omega}\rho\phi d\Omega+\int_{S}\rho\phi(\vec{v}-\vec{v}_b)\cdot\vec{n}dS=\int_{\Omega}Q_{\phi}d\Omega-\int_{S}\vec{G}_{\phi}\cdot\vec{n}dS \quad (3)$$

Here the mass- and momentum balance in the flow is represented by equation (1) and (2) respectively, whereas (3) is a general transport equation suitable for any scalar $\phi$ (e.g. turbulence characteristics).

It should be emphasized that the manner of discretization of the time derivatives and the surface integrals in the above equations has to ensure strict satisfaction of the kinematic relation

$$\frac{d}{dt}\int_{\Omega} d\Omega - \int_{S} \vec{v}_b \cdot \vec{n} dS = 0 \qquad (4)$$

referred to as the space conservation law. Otherwise, artificial sources or sinks of mass, momentum etc. caused by the grid movement would appear in equations (1-3) and destroy the solution. In fact, the space conservation law prescribes the way of computing the cell face velocity, $v_b$, that has to be linked with the volumes swept by the cell faces instead of simple differentiation of grid co-ordinates, see [11] for details.

The deforming mesh approach is a flexible and efficient technique that is relatively easy to implement in any CFD code supporting time-dependent computations. In fact, the code modification affects only the convective terms and the time derivative approximation. An algorithm of the grid deformation has to be implemented in the code as well. For the present study, an explicit algorithm of block-structured grid deformation in accordance with movement of grid-block boundaries has been elaborated.

**2.2. Turbulence modeling**

In both the problems considered, the flow is dominated by large-scale vortices shed from the vibrating body. These vortices interact with each other, break apart, and provide intensive mixing that is not reproduced by any turbulence models assigned to close the Reynolds-averaged Navier-Stokes equations (so called RANS models, like $k$-$\varepsilon$, $k$-$\omega$, etc.). In particular, it is well known that the RANS models fail to predict the flow past a circular cylinder with a proper accuracy. So one has to apply some of the vortex-resolving turbulence models, like Direct Numerical Simulation (DNS), Large Eddy Simulation (LES), or RANS/LES hybridization.

In the present work the RANS/LES approach known as the Detached Eddy Simulation (DES) is used, with the formulation given in [12] that is based on the Menter SST (MSST) turbulence model. According to this approach, the original MSST model [13] is used for unsteady RANS computations close to the wall (in the boundary layer) whereas farther from the wall the model shifts smoothly (depending on the cell size the local turbulence characteristics) to its sub-grid scale mode. Note that selection of the MSST turbulence model is not a principal point here, and other RANS turbulence models can be used for DES predictions [12].

**2.3. Code SINF**

All the computations presented in this paper were performed using the 3D steady/unsteady Navier-Stokes code SINF (Supersonic to INcompressible Flows). This in-house code is based on the second-order finite-volume spatial discretization using the cell-centered variable arrangement and body-fitted block-structured grids. Both matching and non-matching grid interfaces are available. The artificial compressibility technique and/or the SIMPLEC method are used for the pressure-velocity coupling in the case of incompressible fluid flows [14]. Three-layer second-order scheme is implemented for physical time stepping.

Parallelization of the code is based on the MPI standard and the domain decomposition strategy according to the grid block structure. So, the inter-process communication concerns mainly connection of the adjacent grid blocks. To achieve a high efficiency of parallelization, connection of two blocks, say *A* and *B*, treated by different processes is organized as follows. First, an image, *A'*, of near-interface layers of block *A* is created in the process hosting block *B* by means of MPI copying. Then the connection of blocks *A'* and *B* is performed in the serial mode. After that all the necessary data are copied from *A'* to *A* via MPI. For a number of test computations (using grids up to 20 million cells and clusters up to 24 processors) the parallelization efficiency is higher than 0.8.

A more detailed description of code SINF and some examples of its application can be found in [14-16].

## 3. SIMULATION RESULTS

### 3.1. Vortex-induced vibration of a cylinder

In this section, an elastically mounted circular cylinder subjected to transverse vortex-induced vibrations is considered, see Fig.1. The cylinder motion is computed from the dynamic equation (5). With the velocity and length scales, *U* and *D*, the dimensionless equation can be written as (5a). So, in addition to the Reynolds number, $Re=UD/\nu$, the problem is determined by the relative mass of the cylinder, $m^*$, the damping factor, $\zeta$, and the normalized velocity of the free stream, $U^*$.

$$m\ddot{y} + C\dot{y} + Ky = F_y \tag{5}$$

$$\ddot{\bar{y}} + \frac{2\zeta\dot{\bar{y}}}{U^*} + \frac{\bar{y}}{(U^*)^2} = \frac{2\pi C_y}{m^*}, \quad C_y = \frac{2F_y}{\rho U^2 DL} \tag{5a}$$

$$m^* = \frac{4m}{\pi\rho D^2 L}, \quad \zeta = \frac{C}{2\sqrt{mK}}, \quad U^* = \frac{U}{D}\sqrt{\frac{m}{K}} \tag{6}$$

In accordance with experimental data [17], the cylinder mass is $m^*=3.3$ and the damping factor is $\zeta=4\cdot10^{-3}$. The normalized velocity, $U^*$, is varied from 2 to 11 so that the vortex-shedding frequency is around the cylinder natural frequency, $f_n$. Along with the velocity range, the Reynolds number varies from 1200 to 7000 that corresponds to regimes with laminar boundary layer separation and a turbulent wake.

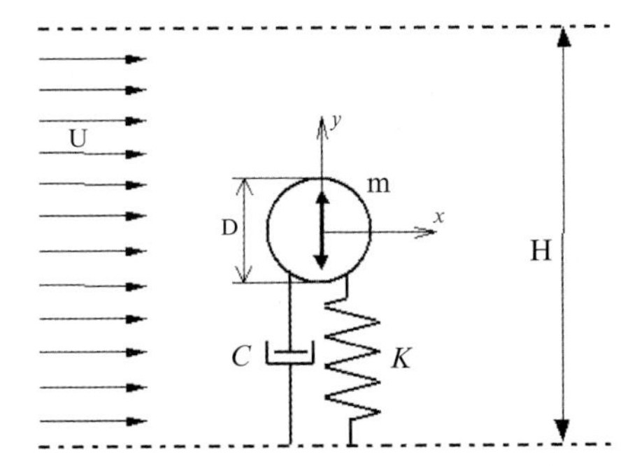


Fig.1. Scheme for simulation of vortex-induced vibration of an elastically mounted cylinder

Despite the problem geometry is two-dimensional, a 3D formulation is used since the flow in the wake is known to be essentially three-dimensional. The computational domain is of $2D$ to $6D$ along the cylinder axis (with periodicity conditions at the boundaries) and of $34D$ in the streamwise direction (the preceding part is of $7D$). The channel blockage, $D/H$, is estimated as 10%. The near-wall grid spacing is $0.002D$ ($y^+$ is below unity). At least 4 large-scale vortices are well resolved in the wake. The overall grid size is up to 0.6 million cells.

Typical simulation results are illustrated in Fig.2. For the case presented ($U^*=4$) an average magnitude of vibration is about $0.6D$ (random peaks exceed $0.7D$). As well, one can notice that the drag coefficient oscillates around the value of 2.8, and the lift force behaves extremely chaotic, with peaks ranging from 0.2 to 3. Note that for the fixed cylinder case the drag force coefficient is known to be about 1.0 and the actual lift coefficient doesn't exceed 0.5. It points to a very considerable effect of the cylinder motion on the flow. Because of the chaotic character of the oscillations one has to compute a very long time interval (up to several hundred cycles) for getting statistically independent data.

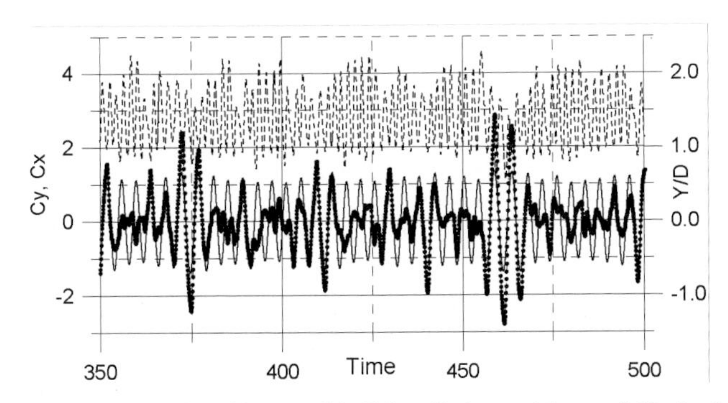


Fig.2. Sample time history of (solid) cylinder position, and (dashed) drag and (symbols) lift coefficients

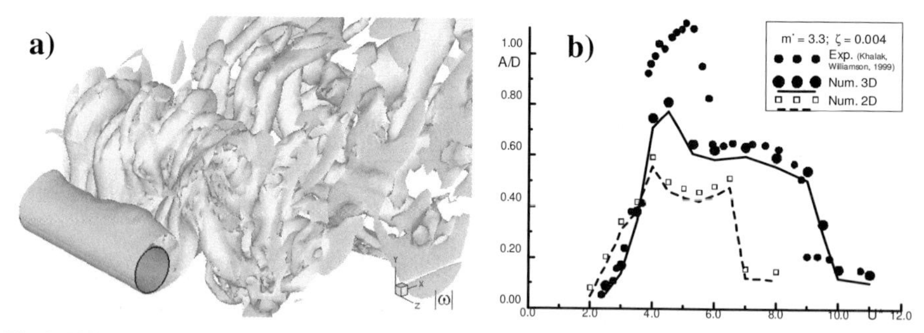


Fig.3. VIV of a cylinder: (a) 3D structure of the wake, and (b) the peak amplitude response versus the normalized velocity

Shown in Fig.3a, an iso-surface of the vorticity illustrates the 3D structure of the flow past the oscillating cylinder. Note that a typical spanwise size of large-scale vortices is about $2D$ . This gives in particular an estimation of the least spanwise size of the computational domain.

Main results of the computations are summarized in Fig.3b. The amplitude response of the cylinder is plotted against the normalized velocity of the free stream. Both 3D and 2D simulation results are presented, and the experimental data [15] are given as well. One can observe that both the 2D and 3D simulation data exhibit the resonant amplification of the vibration magnitude in the lock-in regime. However in the 2D computations the lock-in range is too narrow, and the resonant amplitude is underestimated. The 3D simulation provides a much better agreement with the experimental data though there is still some discrepancy in the region of the “upper branch” of the response curve.

To check if the above mentioned discrepancy is due to an insufficient spanwise size of the computational domain, additional test computations were performed at $U^*$=4.5 for an increased spanwise size (from $2D$ to $4D$, and then to $6D$). This resulted in some increase of the lift force, and the cylinder motion was in somewhat less chaotic, but the peak amplitude remained nearly the same. Refinement of the grid and reduction of the time step also brought a negligible effect. So the reasons behind the discrepancy under consideration are still not clear.

### 3.2. Flow induced by a piezoelectric fan

In this Section, the flow induced by oscillations of a flexible cantilevered blade is considered, see Fig.4. The oscillations are forced by prescribed harmonic movement of a blade point near its fixed end. The excitation frequency, ω, equals to the first resonant frequency of the structure; the oscillation amplitude at the bade tip is adjusted to be 10% of the blade length, *L*. The blade is treated infinitely thin, and the blade width is a quarter of its length. The problem is treated as fully three-dimensional (without applying the mirror symmetry condition at the middle plane *y*=0). The computational domain was a cube with the side size of 4*L* (i.e. forty times the tip vibration magnitude). The computational grid consisted of 1.6 million cells.

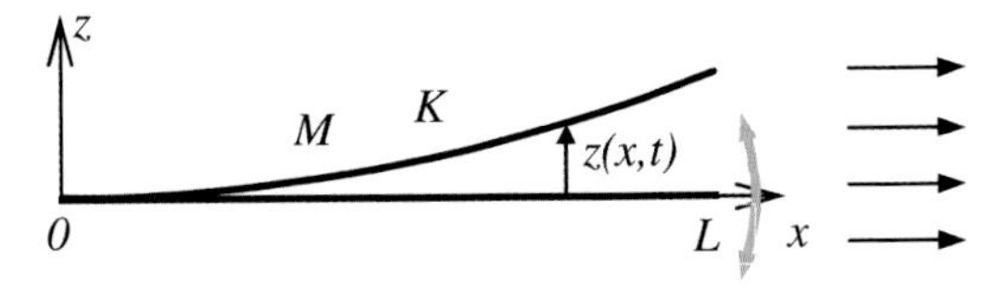


Fig.4. Scheme for simulation of flow induced by a piezoelectric fan

The blade motion/deformation is computed from the dynamic equation (7). With $L$ and $\omega$ as scales it can be written in the dimensionless form (7a) involving the blade characteristic parameters: relative mass, $M^*$, and stiffness, $K^*$. The first mode of oscillations is associated with the value of $K^*$=0.0809. A reasonable estimation of the blade mass is $M^*$=10. The Reynolds number is evaluated as $Re=\omega L^2/\nu=2\cdot 10^5$.

$$M\ddot{z} + Kz_x^{(IV)} = q(x,t) \qquad (7)$$

$$\ddot{\bar{z}} + K^* \cdot \bar{z}_x^{(IV)} = \frac{\bar{q}(x,t)}{M^*}, \quad \bar{q}(x,t) = \frac{2q}{\rho\omega^2 L^4} \qquad (7a)$$

$$M^* = \frac{M}{\rho L^2}, \quad K^* = \frac{K}{M\omega^2 L^4} \qquad (8)$$

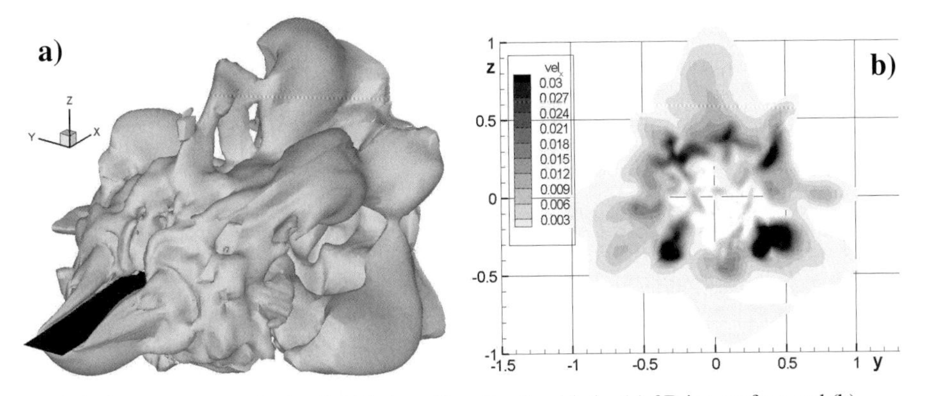


Fig.5. An instantaneous velocity field induced by vibrating blade: (a) 3D iso-surface and (b) distribution over a cross-section of $x$=const

An instantaneous distribution of the $x$-component of the computed flow velocity is illustrated in Fig.5. Shown in Fig.5a, an iso-surface of the velocity highlights a rather chaotic 3D flow dominated by large-scale vortices that provide intensive mixing and, consequently, a rapid divergence of the jet (the half-angle is about 45°). This spatial configuration of the flow is consistent with experimental observations [9].

The x-velocity distribution over the jet cross-section positioned at the distance of 0.3$L$ from the blade tip is shown in Fig.5b. One can observe a stagnation zone at the jet axis and four local maxima where the $x$-velocity exceeds 30% of the tip oscillation velocity. Mutual positions of the four maxima as vertices of a nearly regular square correlate well with the area swept by the blade tip. This suggests that these maxima are due to complex 3D interaction of the vortices shed from the tip and two side edges of the blade. So, the 2D formulation (adopted, in particular, in [10]) is unable to provide an adequate simulation of the jet formation. In the above mentioned stagnation zone, temporary local reverse flows exist. Further downstream the reverse flows disappear completely because of turbulent mixing.

It should be emphasized also that for the problem considered in this Section, the flow formation time is extremely long because of sluggish migration of large vortices at the jet periphery. To speed up the transient process, a weak co-flow (about 5% of typical jet velocity) was imposed at the inlet boundary of the domain. Nevertheless, we had to compute almost 200 cycles to achieve a developed flow regime that suitable for further averaging and statistical analysis. The computations took about one month CPU time using 11 processors of an Opteron-based cluster.

**Acknowledgements.**

This work was partially supported by the Russian Federation Program for Support of Leading Scientific Schools (grant NSh-376.2006.8).

**References**

1. A. Khalak, C.H.K. Williamson. J. Fluids and Structures. 13 (1999), 813-851
2. R.D. Gabbai, H. Benaroya. J. Sound and Vibration, 282 (2005), 575–616
3. H.M. Blackburn, R.N. Govardhan, C.H.K. Williamson. J. Fluids and Structures 15 (2000) 481-488
4. J.R. Meneghini, F. Saltara, at all. Europ. J. Mechanics. B/Fluids. 23 (2004) 51–63
5. H. Al-Jamal, C. Dalton, J. Fluids and Structures 19 (1) (2004) 73–92
6. C. Evangelinos, D. Lucor, G.E. Karniadakis, J. Fluids and Structures 14 (3) (2000) 429–440
7. E. Guilmineau, P. Queutey. Proc. 7th Int. Conf. on Flow-Induced Vibration, Lucerne, Switzerland (2000) 257-264
8. P. Burmann, A. Raman, S.V. Garimella. IEEE trans. on components and packaging technologies, 25 (4) 2003, 592-600
9. M. Wait, S. Basak, S.V. Garimella. TCPT-2005-086, Computer and Information Technology, Purdue Univ., 2005, 19p
10. T. Acikalin. ME 608 Final Project, School of Mechanical Engineering, Purdue University, 2002, 8p
11. J.H. Ferziger, M. Peric. Computational methods for fluid dynamics. Berlin: Springer, 1999, 389p
12. A.Travin, M.Shur, M.Strelets, P.R.Spalart. In: Advances in LES of Complex Flows (Proc. EUROMECH Colloquium 412), Kluwer Academic Publishers, 2002. Vol.65. P.239.
13. F.R. Menter. AIAA Journal, 32 (1994), 1598-1605
14. E.Smirnov, D.Zaitsev. In: ECCOMAS 2004. (CD-ROM proceedings), 13p.
15. A.I. Kirillov, V.V. Ris, E.M. Smirnov, D.K. Zaitsev. In: Heat Transfer in Gas Turbine Systems (Annals of the N.Y.Acad.Sci., Vol.934). - N.Y.Acad.Sci., N.Y., 2001, 456-463
16. E.M. Smirnov, N.G. Ivanov, A.G. Abramov, et al. In: Parallel CFD. Advanced Numerical Methods Software and Application (Proc. ParCFD-03). Elsevier. 2004. 219-226
17. A. Khalak, C. H. K.Williamson, J. Fluids and Structures. 10 (1996) 455-472

# Parallel simulation of type IIa supernovae explosions using a simplified physical model

J. M. McDonough[a] and J. Endean[b]

[a]Departments of Mechanical Engineering and Mathematics
University of Kentucky, Lexington, KY 40506-0503 USA

[b]Department of Mathematics
Rice University, Houston, TX 77005 USA

A simplified model for simulation of type IIa supernovae explosions is presented; preliminary results are given, and parallel speedups obtained from OpenMP parallelization on a symmetric multi-processor and on a blade cluster are reported.

## 1. INTRODUCTION

Supernovae are among the most intriguing and important objects in the known universe. They result during collapse of massive stars (greater than roughly eight solar masses), and occur in several identifiable types with the type depending on the initial stellar mass, composition and other physical properties. It is well known that a particular type of supernova (type Ia) serves as a "standard candle" for measuring distances in the universe; but it is far more important that destruction of stars in supernovae explosions spreads matter (especially the heavy elements) throughout the universe, as discussed by Woosley and Janka [1]. It has been estimated that matter in our solar system represents the outcome of as many as four generations of supernovae explosions; clearly, without such explosions our galaxy and solar system could not have come into being in their present form, and life would probably not exist on planet Earth—if there were such a planet.

In this study we employ a simplified physical model to simulate the explosion of a type IIa supernova. These are particularly rich in the heavier elements, e.g., iron and silicon, but also oxygen, and it is believed that their destruction is non-nuclear in the sense that it is not the result of either a fission or fusion reaction. High-energy physics aspects are involved, of course, but it is believed that they mainly supply heat to drive hydrodynamic instabilities that ultimately tear away the supernova's outer atmosphere in a process analogous to an extremely intense (because of the very high energies involved) solar wind-like phenomenon. Hence, type IIa supernovae provide interesting problems to which computational fluid dynamics (CFD) can be directly applied.

On the other hand, the incredible size of supernovae (often 15 times more massive than our sun, and with correspondingly large radii) poses severe difficulties for simulations that must involve hydrodynamic instabilities because evolution of these requires relatively fine

I.H. Tuncer et al., *Parallel Computational Fluid Dynamics 2007*,
DOI: 10.1007/978-3-540-92744-0_44, © Springer-Verlag Berlin Heidelberg 2009

grid resolution leading to numerical analytic problems that are beyond the capability (both storage and speed) of present high-end supercomputers. In recent years this problem has been addressed in at least three ways: *i*) often, only 2-D simulations are performed; *ii*) the inviscid Euler equations, rather than the viscous Navier–Stokes (N.–S.) equations, have been employed, and *iii*) parallelization has been relied upon.

It is important to point out that if hydrodynamic instabilities are truly responsible for supernovae explosions (and there are other candidates, e.g., electromagnetic effects), then neither of the first two "remedies" for the computational difficulties can be viewed as acceptable. In particular, it is well known that there are fundamental differences between 2-D and 3-D solutions of the N.–S. equations—as fundamental as lack of (long-time) existence of strong solutions, and use of the Euler equations in their place raises further questions concerning validity of the nature of predicted hydrodynamic instabilities. Different instabilities can occur in the two different cases (although there are common ones as well), but in addition to this is the issue of turbulence. It is well known, in terms of both physics and mathematics, that viscosity (dissipation) is required in any correct physical model of turbulence. Hence, in light of these considerations, we have chosen to employ the 3-D, viscous, compressible N.–S. equations for the present study. This is despite the fact that we recognize viscous effects cannot be resolved on current computing hardware for problems posed on extremely large spatial domains such as this, but we believe it is time to begin using fully-correct treatments and developing corresponding software so as to be ready to use new, more powerful computers as they become available.

## 2. PHYSICAL PROBLEM

Details of the initial stages of type IIa supernovae are well known [1]. In particular, as fusion reactions subside in the core of a massive star, gravitational forces acting on the stellar atmosphere can no longer be supported by interior pressure, and gravitational collapse ensues. For stars of greater than approximately eight solar masses, the consequence can be a proto-neutron star (PNS) of radius $R_{ns}$ (Fig. 1) still possessing an atmosphere of relatively heavy elements including iron, silicon. oxygen and others. As the iron core is compressed, the inner most part becomes the PNS and is incompressible and essentially impenetrable. As a consequence, a shock wave forms during this process as supersonic incoming matter strikes the PNS outer surface. This shock wave propagates outward through the atmosphere of the developing supernova but stalls at a particular radius where the speed of incoming matter matches that of the outgoing shock (see Burrows [2]). This is depicted in the sketch of Fig. 1. If this physical configuration could be maintained, there probably would be no supernova explosion—at least not of the nature envisioned on the basis of physical observations and numerical simulations.

But further energy is input to the so-called gain region shown in Fig. 1 via neutrino heating from neutrinos emitted from the neutrinosphere shown immediately outside the PNS, with outer radius $R_\nu$, in the figure. This energy input results in nonuniform, unstably-stratified heating of the gain region, leading, in turn, to hydrodynamic instabilities which re-energize the stalled shock (Burrows et al. [3]). The shock then propagates outward, blowing off the outer stellar atmosphere in a manner quite similar to a very strong solar wind—but leaving behind a neutron star.

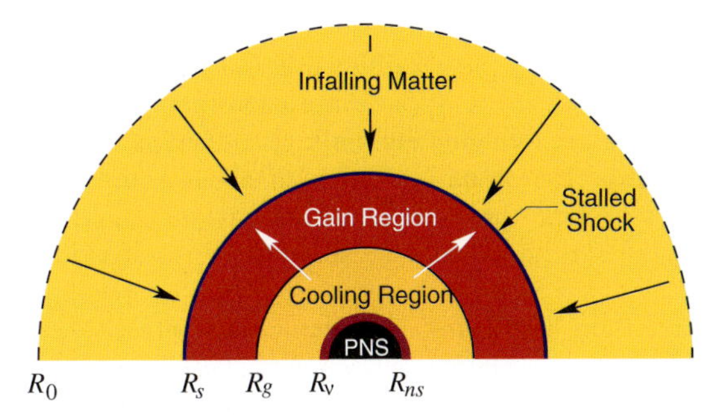


Figure 1. Type IIa supernova structure during period of stalled shock.

To produce a more tractable problem we have chosen to simulate a physical domain extending only from the gain radius $R_g$ out to a radius $R_0 = 4 \times 10^5$ km (significantly less than that of the sun) as shown in Fig. 1. But for relative simplicity with respect to unknown boundary conditions at $R_g$, we actually compute all the way to the center of the supernova—where the equations of motion are not valid; at the end of each time step we replace computed results interior to $R_g$ with values published previously and believed to be reliable (in particular, those computed by Janka [4]).

## 3. GOVERNING EQUATIONS

As already noted, the equations employed in this study are the 3-D, viscous, compressible N.–S. equations, but now also including a gravitational source term in both the momentum and energy equations, and a radiation source term in the energy equation. The equations then take the generic form

$$\boldsymbol{\mathcal{Q}}_t + \nabla \cdot \boldsymbol{\mathcal{F}}(\boldsymbol{\mathcal{Q}}) = \boldsymbol{\mathcal{S}}\,, \tag{1}$$

where $\boldsymbol{\mathcal{Q}} \equiv (\rho, \rho u, \rho v, \rho w, \rho\mathcal{E})^T$ is the vector of conserved (in the inviscid case) variables, and $\boldsymbol{\mathcal{F}}$ represents the matrix of flux components containing both inviscid and viscous contributions which we denote as, e.g., $\boldsymbol{\mathcal{F}}_2 \equiv \boldsymbol{\mathcal{F}}_{2,I} - \boldsymbol{\mathcal{F}}_{2,V}$, for the $x$-momentum equation, with

$$\boldsymbol{\mathcal{F}}_{2,I} = (\rho u^2 + p, \rho uv, \rho uw)\,, \qquad \text{and} \qquad \boldsymbol{\mathcal{F}}_{2,V} = (\tau_{xx}, \tau_{xy}, \tau_{xz})\,.$$

Here the $\tau$s are the viscous stresses, defined, e.g., as

$$\tau_{xx} \equiv 2\mu(u_x - \frac{1}{3}\nabla \cdot \boldsymbol{U})\,, \tag{2a}$$

$$\tau_{xy} \equiv \mu(u_y + v_x)\,, \tag{2b}$$

with $u$ and $v$ being components of the velocity vector $\boldsymbol{U} = (u, v, w)^T$, $\mu$ the dynamic viscosity, $\rho$ the density, $p$ the pressure and subscripts on the right-hand sides of Eqs.

(2) denoting partial differentiation with respect to spatial coordinates. Similarly, the subscript $t$ in Eq. (1) represents partial differentiation with respect to time, and $\nabla\cdot$ is the usual divergence operator in an appropriate coordinate system. We note that the Stokes hypothesis relating the dynamic and bulk viscosities as $\lambda = -2/3\mu$ has been used to obtain the form of Eq. (2a). Finally, the gravitational source terms, included in $\boldsymbol{\mathcal{S}}$, are the usual Newtonian ones; these are known to provide qualitatively-correct results in the present case, and can be rescaled to account for relativistic effects as shown by Mezzacappa [5].

The radiation source term used herein is a simple parametrization intended to account for neutrino heating of the gain region shown in Fig. 1. The precise nature of this term is still a subject of much investigation, and it involves considerable high-energy particle physics. But our goal in the present study is to attempt to understand hydrodynamic instabilities caused by energy input to the gain region, so high-energy physics details are not so important. Thus, we use the following form for the contribution of radiation to the source term of the energy equation:

$$\mathcal{S}_{5,R} = \dot{Q}_\nu = \frac{\varphi(1-\varphi)^{(n-1)}\dot{E}_\nu}{4\pi(r_n + \Delta r/2)^2}\Delta x \Delta y\,. \tag{3}$$

Here, $\dot{Q}_\nu$ is the fraction of total neutrino energy $\dot{E}_\nu$ input rate deposited per unit time in each grid cell of size $\Delta x\Delta y\Delta z$ contained in a spherical shell of radius $r_n$ and thickness $\Delta r = \sqrt{\Delta x^2 + \Delta y^2 + \Delta z^2}$. The factor $\varphi$ represents the fraction of (remaining) neutrinos deposited in any given shell, and is an assigned (and thus parametrized) input, which depends on details of neutrino physics and probably is not constant across the entire gain region. In this preliminary study we neglect such details and set $\varphi$ to a constant (usually, $\varphi = 0.75$). Similarly, $\dot{E}_\nu$ is rate of neutrino energy crossing the gain radius $R_g$ from all types of neutrinos. This quantity, itself, must generally be computed from first-principles high-energy physics analyses. At present, there is no completely accepted approach for this, so we employ what are considered to be reasonable values (see [4] for more information on neutrino reactions). The computations reported herein were performed with $\dot{E}_\nu = 10^{20}$ Joules/sec.

## 4. NUMERICAL PROCEDURES

The numerical procedures employed for the simulations reported here are quite standard. Backward-Euler temporal integration is used in conjunction with centered spatial differencing, providing formal first-order accuracy in time and second order in space in smooth regions of flow. MacCormack–Baldwin artificial dissipation is employed in a flux-modification form (see Hirsch [6]) to handle shock capturing in a very simple way. This is a quite robust method introduced a considerable time ago for aerodynamics calculations, and it works for a wide range of values of dissipation coefficients. Although, in general, one expects these to change with changes in spatial grid size, we have used the same values on all grids in the current study: 0.5, 0.2, 0.2, 0.2, and 0.8, respectively, for the continuity equation, the three momentum equations and the energy equation.

Nonlinearities are treated via quasilinearization, and the discrete equations are efficiently solved sequentially via Douglas and Gunn [7] time splitting along Cartesian spatial directions, which also leads to straightforward parallelization. We note, however, that sequential solution of the individual equations in the system (1) is not standard for compressible N.–S. formulations. It was done in the present case with a goal of further parallelization as can be accomplished for the incompressible equations. But there is some evidence, as has been often observed in earlier work of other investigators, that the stability limitations imposed by this approach can be too severe to permit its use in fully-operational general codes.

## 5. RESULTS and DISCUSSION

In this section we provide results of three distinct types, viz., grid-function convergence for the numerical procedure, short-time flow field evolution of the physical problem, and results of parallelization attempts on two different computers.

**Grid-Function Convergence.** We noted in the preceding section that the numerical methods we have employed should produce results that are second-order accurate in space and first order in time in regions of flow where solutions to Eqs. (1) are sufficiently smooth. To test this we have employed computations on grids consisting of $26^3$, $51^3$ and $101^3$ executed with (physical) time step sizes fixed at $5\times10^{-7}$, $1.25\times10^{-7}$, and $3.125\times10^{-8}$ sec., respectively, on the three grids.

To simplify the overall analysis we have chosen to compare only a single computed quantity, the temperature, $T$, from the three calculations; and we have done this only at the final time, 5 $\mu$sec, of this set of calculations. This choice of physical variable is a natural one for two reasons. First, it is the temperature distribution that drives hydrodynamic instabilities of the type expected here, so it is important to assess how well this is being resolved. Second, the temperature is not computed directly—it is not an inviscidly conserved quantity—and, instead, must be computed from a formula containing <u>all</u> of the conserved-variable solution components. Hence, it carries information on errors from all parts of the complete vector solution of Eqs. (1). To carry this out, we have searched computed output for each grid to find the maximum value of $T$, which turned out to be a monotonically decreasing function of grid spacing. From these max-norm data we computed the error ratio from

$$r_{err} = \frac{T^h_{max} - T^{h/2}_{max}}{T^{h/2}_{max} - T^{h/4}_{max}},$$

where $h$ is grid spacing. The value corresponding to the $26^3$ grid is $h \simeq 30.77$ km, a quite large value. Nevertheless, the temperature data lead to $r_{err} \simeq 3.23$, which is indicative of reasonable agreement with the formal accuracy of the methods being employed since these have a theoretical value $r_{err} = 4.0$ for the combination of spatial and temporal discretizations employed here.

**Early-Time Supernova Evolution.** In this section we present results computed on the $101^3$ uniform grid with time step $\Delta t = 1\times10^{-7}$ sec. to display the nature of the supernova evolution during early times after initial stalling of the shock. It is clear that the grid resolution is not sufficient to provide a good assessment of whether hydrodynamic instabilities might ultimately result in an "explosion" that sends the outer atmosphere of the star into interstellar space. On the other hand, we will be able to see that the behavior of our model for neutrino heating of the gain region is reasonable, and at the same time the amount of shock smearing due to use of artificial dissipation (as opposed to employing some more sophisticated high-resolution method) will itself limit the possibility of hydrodynamic instabilities, at least very near the shock.

Figure 2 displays a sequence of solution time slices showing dimensionless temperature distribution taking on values from 0.0 (darkest) to 1.0 (white) in a plane through the center of the (nearly) spherical star. As noted earlier, relevant computations were performed only starting at the inner radius of the gain region; thus, the high-temperature PNS and the immediately adjacent cooling region show fictitious (not computed) but reasonable shading. The first time slice, part (a) of the figure, corresponds to a time very close to

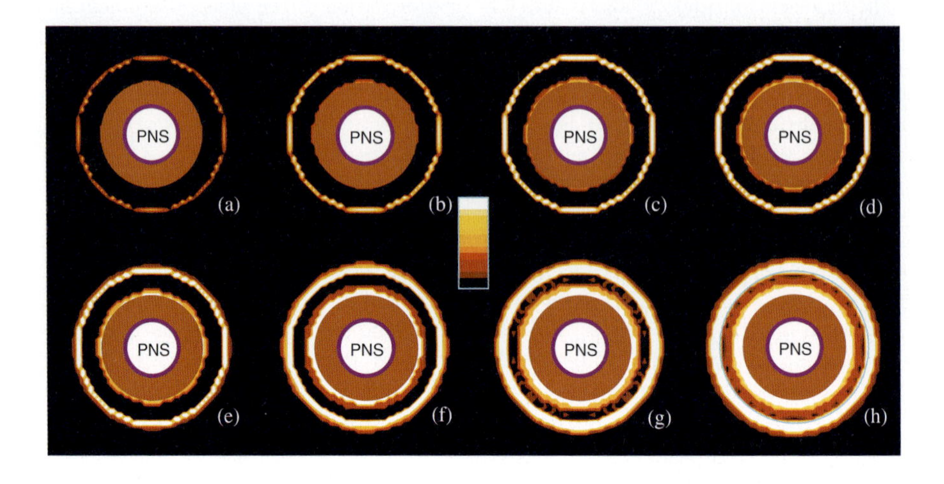


Figure 2. Early-time evolution of type IIa supernova.

the beginning of the calculation ($2 \times 10^{-6}$ sec. after stalling of the shock). The shock appears as the outer ring in the figure, and it is to be noted that calculations are carried out considerably beyond this location, in fact, extending all the way to the edge of the figure. Figure 2(b) shows the analogous central slice for a time of $6 \times 10^{-6}$ sec. It is clear that the shock has begun to strengthen, and careful examination of the figure shows early indications of heating of the gain region near its inner radius. Continued shock strengthening (as well as—unfortunately—smearing) is indicated in parts (c) through (e) of the figure corresponding to times of $1 \times 10^{-5}$, $1.5 \times 10^{-5}$ and $2 \times 10^{-5}$ sec. Heating of the interior parts of the gain region is also increasing in these successive figures.

In Fig. 2(f) can be seen the first clear evidence of re-initiation of shock movement; this corresponds to a time of $4 \times 10^{-5}$ sec, which is somewhat earlier than seen in previous simulations. We believe this is being caused at least in part by under resolution of our spatial discretization. In particular, we have found during the grid-function convergence tests reported above that shock propagation speeds were generally higher on coarse grids, and propagation also began earlier in time on these grids. Furthermore, from the perspective of physics, the re-initiation has occurred without any evidence of hydrodynamic instabilities. As noted above, it seems unlikely that these could be resolved—at least their early stages—on the coarse grids employed here because it is evident in this part of the figure that further heating of the gain region is occurring in such a manner as to promote thermally-driven instabilities only deep in this region, and not near the shock.

Parts (g) and (h) of the figure, corresponding to times of $7 \times 10^{-5}$ and $1 \times 10^{-4}$ sec., respectively, show continued shock propagation and now quite significant smearing, as well as further heating of the interior of the gain region. Unfortunately, it is clear that shock smearing and shock temperature overshoot is such as to confine the region of potential thermal instabilities to fairly deep in the gain region during this period. Despite this, the shock has been restarted and has moved a noticeable distance, as can be seen by comparing the original shock location shown in Fig. 2(h) as a gray ring just outside the extent of gain region heating with the outer reaches of shock heating at the current time. For the early times involved (total time scale for a type IIa supernova explosion is believed to be between one and two seconds), this shock location is not unreasonable (see, e.g., 2-D results of Miller et al. [8]), and it raises the question of whether some initial movement can occur simply due to energizing the gain region via neutrino heating, as we have done here, without producing thermal convection needed to generate hydrodynamic instabilities. Further calculations employing much finer gridding and better shock-capturing methods will be needed to resolve this issue.

**Parallelization Results.** Figure 3 provides speedups obtained using OpenMP parallelization focusing only on Do loops of the Fortran 90 code running on a Hewlett Packard SuperDome SMP and on an IBM HS21 Blade Cluster. We have chosen to initially employ OpenMP because portions of the code are still under development, and use of MPI is sufficiently difficult that we prefer to delay this until the code is in a more mature state. Results are reasonably good through eight processors on the HP SuperDome which, based on past experience with several other codes (see, e.g., Xu et al. [9]), is not unexpected. It was found that using 16 processors resulted in considerably less than doubling of the speedup achieved with eight processors (although there was some performance improvement), so we chose to not make runs with more than eight processors.

Results for the IBM Blade Cluster are very disappointing and difficult to explain. The same code and OpenMP parallelization was used for the IBM speedup results shown in the figure as on the HP SuperDome, and the IBM machine is considerably newer; however, other users of this hardware have experienced similar problems. We must conclude that this machine probably should not be used for OpenMP parallelization, with the implication that considerable human effort is required to prepare parallel code that is suitable for execution on this hardware—effort that might be better spent on other aspects of problem solving in general.

## 6. SUMMARY AND CONCLUSIONS

In this study we have implemented a fully-viscous 3-D compressible Navier–Stokes code to simulate onset of type IIa supernovae explosions using a parametric gain-region neu-

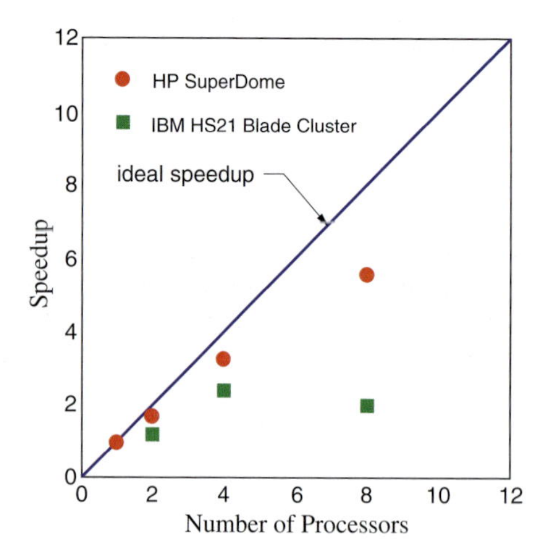


Figure 3. Parallel speedup using OpenMP.

trino heating model. We conducted grid-function convergence tests to validate the code, and we have demonstrated re-initiation of shock movement at times not inconsistent with those reported in the extant literature. On the other hand, we have observed that this is not the result of hydrodynamic instabilities because the current calculations are not sufficiently resolved to accurately represent such detail. Nevertheless, we now have a code that will permit valid representations of instabilities of viscous origin, including turbulence, in supernovae explosions as computing power improves to the point of permitting such calculations. We demonstrated satisfactory parallel performance of the code using OpenMP on a HP SuperDome through eight processors, but scaling beyond this number was not good. In addition, OpenMP parallelization performance on an IBM Blade Cluster was very poor, suggesting that only MPI should be employed for this type of architecture.

## REFERENCES

1. S. E. Woosley and H.-Th. Janka (2005). *Nature Phys.* **1**, 147.
2. A. Burrows (1996). *Nucl. Phys.* **602**, 151.
3. A. Burrows, J. Hayes and B. A. Fryxell (1995). *Astrophys. J.* **450**, 830.
4. H.-Th. Janka (2001). *Astron. & Astrophys.* **368**, 527.
5. A. Mezzacappa (2005). *Annu. Rev. Nucl. Part. Sci.* **55**, 467.
6. C. Hirsch (1990). *Numerical Computation of Internal and External Flows, Vol. 2*, 279.
7. J. Douglas and J. E. Gunn (1964). *Numer. Math.* **6**, 428.
8. D. S. Miller, J. R. Wilson and R. W. Mayle (1993). *Astrophys J.* **415**, 278.
9. Ying Xu, J. M. McDonough and K. A. Tagavi (2006). In *Parallel Computational Fluid Dynamics*, Deane et al. (Eds.), North-Holland Elsvier, Amsterdam, 235.

# A Fast Parallel Blood Flow Simulator

B. Hadri [a], M. Garbey [a]

[a]Department of Computer Science,
University of Houston, Houston, TX 77204, USA

**This paper presents a fast parallel blood flow simulator with a long term goal that is to provide a close to real time simulator that can be used automatically with MRI data in clinical conditions. To achieve this performance, our method relies on three techniques; a level set method to obtain the geometry of the artery, the $L_2$ penalty method to deal with complex geometry, and a fast domain decomposition solver designed for parallel processing.**

## 1. Introduction

Most Western countries are facing increasing rates of cardiovascular diseases (CVD) and it has been the leading cause of death and disability over the past 90 years. [16].

Hemodynamic simulation is becoming an important tool on the understanding of cardiovascular diseases[15]. Because hemodynamic has a very complex interplay with biochemistry and gene expressions, a first step might be to provide extensive data set of blood flow simulation in real clinical conditions that can be post-processed systematically with statistical and data mining methods. Our long term goal is to provide an hemodynamic simulator that can be routinely used to accumulate a large data base of medical data and eventually assist endovascular surgeons in their decision process.

Because of the extreme variety of blood flow conditions for the same patient, we are only interested in providing instant first order approximation of some main quantities of interest in cardiovascular disease such as the shear stress and the pressure on the wall. Statistic and sensitivity analysis will post-process those many runs.

The goal is to provide a parallel, fast and robust numerical solver of blood flow that can easily interface with medical imaging. Our solver uses an immersed boundary technique that relies on the $L_2$ penalty approach pioneered by Caltagirone [2,3]. Furthermore, it combines nicely with a level set method based on the Mumford-Shah energy model [5] to acquire the geometry of a large vessel from medical images. Finally, we use a domain decomposition algorithm called Aitken Schwarz [4,7] that has high numerical efficiency and scales well on low cost Beowulf Clusters.

The plan of this paper is as follows. We describe in section 2 the methodology and formulation of our Navier-Stokes solver. Then in section 3, we analyze the performances of the parallel hemodynamic simulator .

I.H. Tuncer et al., *Parallel Computational Fluid Dynamics 2007*,
DOI: 10.1007/978-3-540-92744-0_45,  

## 2. Fast prototyping of the Navier-Stokes problem

The flow solver is an immersed boundary like method [12]. The wall boundary condition is immersed in the Cartesian mesh thanks to a penalty term added to the momentum equation. Thanks to this method, there is no issue on mesh generation that imposes automatically the no slip boundary condition on the wall. The drawback is that one gets a lower order approximation of the solution [2]. However, due to the uncertainty on the medical imaging data as well as a number of biological unknowns such as the tissue constitutive laws, especially in the presence of a cardiovascular disease, we expect that the limit on the accuracy of our NS solver will not be the restricting factor in real clinical conditions.

In this section, we will focus on the Navier-Stokes(NS) equations combined with the L2 penalty method, followed by the image segmentation in order to extract the geometry of the artery and finally we show how we solve the equations with an efficient domain decomposition technique.

### 2.1. NS formulation

For fast prototyping of incompressible NS flow, the penalty method, introduced by Caltagirone and co-workers [3], is used for since it is easy to implement and applies naturally to flow in a pipe with moving walls [13]. Furthermore, the Eulerian approach is considered and one can combine fast solvers for regular Cartesian grid solution with some form of fictitious domain decomposition.

The flow of incompressible fluid in a parallelepiped domain $\Omega = (0, L_x)\times(0, L_y)\times(0, L_z)$ with prescribed values of the velocity on $\partial\Omega$ obeys the NS equations:

$$\partial_t U + (U.\nabla)U + \nabla p - \nu\nabla.(\nabla U) = f, \text{ in } \Omega, \text{ and } div(U) \;=\; 0, \text{ in } \Omega$$

We denote by $U(x, y, z, t)$ the velocity with the components $(u_1, u_2, u_3)$ , $p(x, y, z, t)$ as the normalized pressure of the fluid and $\nu$ as the kinematic viscosity.

With an immersed boundary approach the domain $\Omega$ is decomposed into a fluid subdomain $\Omega_f$ and a wall subdomain $\Omega_w$. In the $L_2$ penalty method the right hand side $f$ is a forcing term that contains a mask function $\Lambda_{\Omega_w}$ such that

$$\Lambda_{\Omega_w}(x, y, z) = 1, \text{ if } (x, y, z) \in \Omega_w, \text{ 0 elsewhere},$$

and is defined as follows

$$f = -\frac{1}{\eta}\Lambda_{\Omega_w}\,\{U - U_w(t)\}.$$

$U_w$ is the velocity of the moving wall and $\eta$ is a small positive parameter that tends to zero.

A formal asymptotic analysis helps us to understand how the penalty method matches the no slip boundary conditions on the interface $S_w^f = \bar{\Omega}_f \bigcap \bar{\Omega}_w$ as $\eta \to 0$. Let us define the following expansion:

$$U = U_0 \;+\; \eta\, U_1,\; p \;= p_0 \;+\; \eta\, p_1.$$

Formally we obtain at leading order, $\frac{1}{\eta}\Lambda_{\Omega_w}\,\{U_0 - U_w(t)\} = 0$, that is $U_0 = U_w$, in $\Omega_w$. The leading order terms $U_0$ and $p_0$ in the fluid domain $\Omega_f$ satisfy the standard set of NS equations:

$$\partial_t U_0 + (U_0.\nabla)U_0 + \nabla p_0 - \nu\nabla.(\nabla U_0) = 0, \text{ in } \Omega_f \text{ and } div(U_0) = 0, \text{ in } \Omega.$$

At next order we have in $\Omega_w$, $\nabla p_0 + U_1 + Q_w = 0$, where

$$Q_w = \partial_t U_w + (U_w.\nabla)U_w - \nu\nabla.(\nabla U_w).$$

Further the wall motion $U_w$ must be divergence free, and at next order we have in $\Omega_f$,

$$\partial_t U_1 + (U_0.\nabla)U_1 + (U_1.\nabla)U_0 + \nabla p_1 - \nu\nabla.(\nabla U_1) = 0, \text{ with } div(U_1) = 0.$$

To acquire the geometry of a large vessel from medical images, the level set method based on the Mumford-Shah energy model [5], has been implemented which allows us to deal with complex geometries such as carotid bifurcation or the circle of Willis.

### 2.2. Image segmentation

Various imaging modalities, such as ultrasound, computer tomography (CT) or magnetic resonance imaging (MRI) are used to acquire images of blood vessels. CT or MR angiography images are then post-processed to extract the exact geometry of the arteries that are specific to each individual. This process is crucial for the accuracy in Computational Fluid Dynamic (CFD) computations.

The mask function $\Lambda_{\Omega_w}$ of the penalty method, is obtained directly with an image segmentation technique based on the level set method. Since the contours of the image are not necessarily sharp, it is preferable to use the level set method presented in [5] and based on the Mumford-Shah Model. For completeness we refer to the review papers [11,14] for a more comprehensive description of the level set method in the framework of image analysis.

As a result of the image segmentation, we can simulate the blood flow through a patient-specific artery. With the appropriate boundary conditions, we have all the information to solve the NS equations. The main goal of this project is to get the fastest results. however, a short computation time to obtain a solution of the NS equation requires an efficient solver .

### 2.3. Resolution of the NS equations

To solve the incompressible NS equations, we use a projection method [6] for the time step as follows:

- Step 1: prediction of the velocity $\hat{u}^{k+1}$ by solving

$$\frac{\hat{u}^{k+1} - u^{k,*}}{\Delta t} - \nu\Delta u^k = f^{k+1} - \nabla p^k$$

in $(0, L_x) \times (0, L_y)$ with the boundary condition $\hat{u}^{k+1} = g$ on $\partial\Omega$. The ratio $\frac{\hat{u}^{k+1} - u^{k,*}}{\Delta t}$ is a first order approximation of the total derivative in time and $u^{k,*}$ comes from the methods of characteristics. Eventually the diffusion term in this equation can be made implicit depending on the local ratio between $\nu$ and the space mesh size.

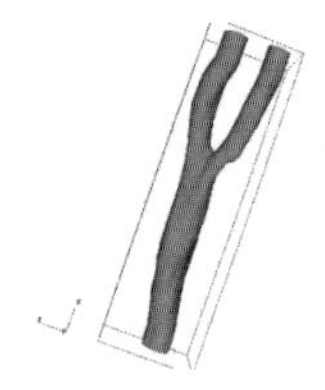

Figure 1. Geometry of the carotid with a stenos

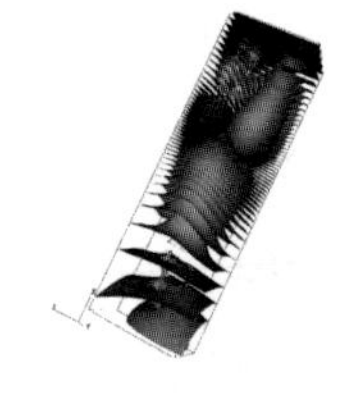

Figure 2. Pressure inside the carotid

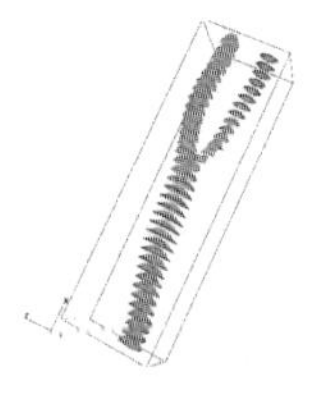

Figure 3. Velocity inside the carotid

- Step 2 : projection of the predicted velocity to the space of divergence free functions.

$$-div\nabla\delta p = -\frac{1}{\Delta t}div(\hat{u}^{k+1})$$

$$u^{k+1} = \hat{u}^{k+1} - \Delta t\delta p; \quad p^{k+1} = p^k + \delta p$$

The NS calculation decomposes into three steps that are the prediction of the flow speed components, the solution of a Poisson problem for the pressure, and eventually the computation of the shear stress along the wall. The Poisson problem requires most of the floating points operations. Indeed, in an incompressible NS code, the resolution of the pressure correction, that corresponds to a Poisson problem, is the most time consuming routine. Depending on the size of the problem, between 50% up to 97% of the elapsed time of the whole code is dedicated to the resolution of this linear system [10]. This is exemplified for 16 millions unknowns, the resolution of the pressure solver for our incompressible NS 3D simulation represents almost 90% of the elapsed time of the sequential processing. It is then critical to have a fast parallel solver for the pressure correction.

In order to use parallel computers, we use a domain decomposition algorithm called Aitken Schwarz that has high numerical efficiency and scales well on low cost Beowulf Clusters[4,9]. The next section presents sequential and parallel performances.

## 3. Performance analysis

A large number of geometries corresponding to real medical test cases as well as artificial benchmark problems have been executed for code verification purpose.

For example, Figure 1 shows a carotid bifurcation in the neck that has a severe stenos. Because we use an immersed boundary technique, the pressure field is computed in the all domain of computation, i.e., inside and outside the lumen. Figure 2 shows an example of the pressure field we computed with the steady flow solution shown in Figure 3. This solution has unrealistic free outlet boundary conditions. In this particular simulation, the Reynolds number is moderate and is about 400 near the inlet.

Let us first analyze the sequential version of the NS code. In Table 1, we report the acceleration of the method depending on the number of subdomains and the grid size. For some configurations, we get an acceleration of the method close to 12 times. We can

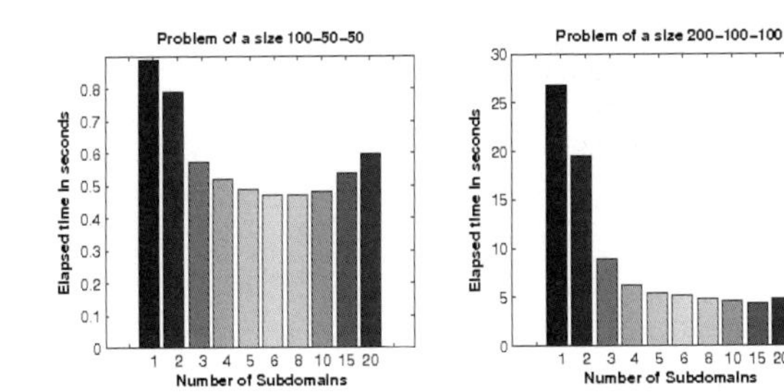


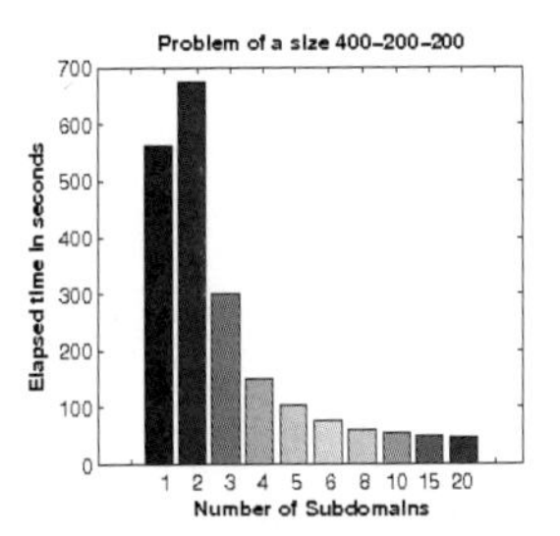


Figure 4. Performance analysis on the sequential code

Table 1
Acceleration of the method depending on the number of subdomains and the grid size

| | Number of Subdomains | | | | | | | | | |
|---|---|---|---|---|---|---|---|---|---|---|
| Grid Size | 1 | 2 | 3 | 4 | 5 | 6 | 8 | 10 | 15 | 20 |
| 100-50-50 | 1.00 | 1.12 | 1.55 | 1.71 | 1.82 | 1.88 | 1.90 | 1.85 | 1.65 | 1.49 |
| 200-100-100 | 1.00 | 1.37 | 3.02 | 4.30 | 5.01 | 5.21 | 5.61 | 5.80 | 6.10 | 5.64 |
| 400-200-200 | 1.00 | 0.83 | 1.87 | 3.75 | 5.43 | 7.44 | 9.38 | 10.34 | 11.67 | 11.79 |

notice that if we multiply by 8 the grid size of the problem, the elapsed time is multiplied by 10. Indeed, for $2.5^5$ unknowns, we spent 0.5 seconds per time step, for $2^6$ unknowns around 5 seconds per time step and $16^6$ unknowns close to 50 seconds per time step. We have a quasi linear behavior of the elapsed time when we increased the size of the problem.

Figure 4 represents the elapsed time in second for one time step depending on the number of subdomains and the grid size for a sequential code. We observe that we reach the best performance with an optimal number of subdomains that growths with the size of the discretization mesh. In general, the parallel algorithm run on a single processor is (much) faster than the sequential algorithm, i.e., with no domain decomposition. It appears that the Aitken-Schwarz algorithm is Cache friendly[7,8] since the elapsed time generally decreases when the number of subdomain grows until the best performance with an optimal number of subdomain is reached.

Regarding the parallel version, the experiments have been performed on a SUN cluster. The systems gathers 24 $X2100$ nodes, 2.2 GHz dual core AMD Opteron processor, 2 GB main memory each, with an Infiniband Interconnect. In Table 2, we report the elapsed time in seconds for the resolution of the pressure depending on the number of processors. Through these results, we can show that we are able to simulate in parallel in few minutes a complete cardiac cycle on a large artery. Indeed, the values that we have reported represent the elapsed time for only one step time, and knowing that a cardiac cycle can be represented approximately by 100 step time, we can conclude from

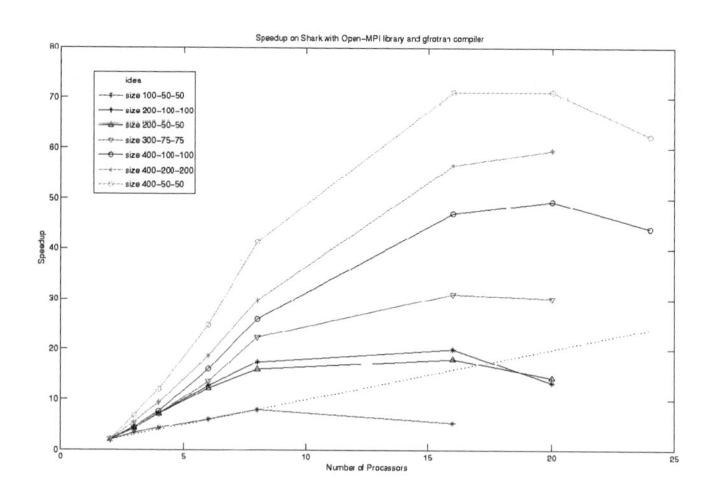


Figure 5. Speedup based on the 2 subdomains performance

Table 2
Elapsed time for the resolution of the pressure

| | Number of Processors | | | | | |
|---|---|---|---|---|---|---|
| **Grid Size** | 2 | 3 | 4 | 6 | 8 | 16 |
| 100-50-50 | 0.52 | 0.1 | 0.24 | 0.17 | 0.13 | 0.2 |
| 200-50-50 | 2.08 | 0.98 | 0.58 | 0.34 | 0.26 | 0.23 |
| 400-50-50 | 12.80 | 3.72 | 2.13 | 1.03 | 0.62 | 0.36 |
| 200-100-100 | 8.91 | 4.21 | 2.46 | 1.4 | 1.02 | 0.89 |
| 400-100-100 | 34.8 | 15.5 | 9.08 | 4.34 | 2.68 | 1.48 |
| 400-200-200 | 188 | 67.7 | 40.2 | 20.16 | 12.65 | 6.65 |

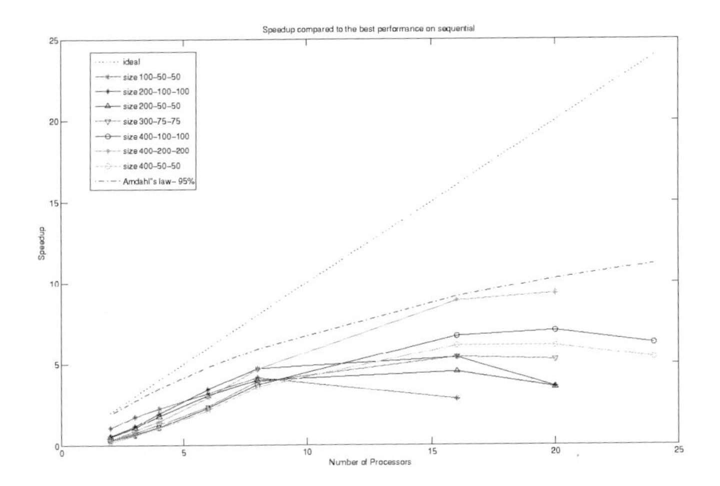


Figure 6. Speedup based on the best performance on sequential

the Table 2 that for a 4 million unknowns, the simulation with 20 processors will take less than 3 minutes. Figure 5 shows the speedup based on the 2 subdomains performance. We can notice that we have a super-linear speedup. This can be explained by the fact that the computation with 2 subdomains on 2 processors was not fast compare to the other configuration. The more subdomain we have, the smaller is the grid size of the problem, which results in a better use of the cache. Nevertheless, Figure 6 shows the speedup based on the sequential code, running the best parallel algorithm; on a single processor, with the optimum number of subdomain, as described Table 1. Then, we do have not anymore a super-linear speedup. In the pressure solver, the acceleration of Aitken-Schwarz step is done by only one processor [9]. The Amdahl's law states[1] that the maximum speedup of the parallelized version is $\frac{1}{(1-f)+\frac{1-f}{p}}$ times faster than the non-parallelized implementation, with f the fraction of the parallelized code and p the number of processor. From Figure 6, we conclude that our code is 95% parallelized.

## 4. Conclusion

The methodology to combine a $L_2$ penalty method, a level set method, and a fast domain decomposition solver allows us to get fast and scalable simulation. Based on the performance of our algorithm design, we are able to simulate one complete cardiac cycle in only few minutes on a Beowulf cluster interconnected by a gigabit ethernet switch.

**Acknowledgments**

Research reported here was partially supported by Awards 0305405 and 0521527 from the National Science Foundation.

## REFERENCES

1. Gene Amdahl, "Validity of the Single Processor Approach to Achieving Large-Scale Computing Capabilities", AFIPS Conference Proceedings, (30), pp. 483-485, 1967.
2. P.Angot, C.H.Bruneau and P.Fabrie,A Penalisation Method to Take into Account Obstacles in Viscous Flows,Numerische Mathematik,1999,81,pp 497-520.
3. E. Arquis and J.P. Caltagirone,Sur Les Conditions Hydrodynamiques au Voisinage d'une Interface Milieu Fluide-Milieux Poreux: Application a la Convection Naturelle,CRAS, Paris II,1984,299,pp1-4.
4. N. Barberou and M. Garbey and M. Hess and M. Resch and T. Rossi and J.Toivanen and D.Tromeur Dervout, Efficient metacomputing of elliptic linear and non-linear problems,Journal of Parallel and Distributed Computing, 63, p564-577, 2003.
5. T.F. Chan and L.A.Vese,Active Contours without Edges,IEE Transaction on Image Processing,2001,10 (2),pp 266-277.
6. A.J. Chorin, The numerical solution of the Navier-Stokes equations for an incompressible fluid, Bull. Amer. Math. Soc., Vol 73, pp 928, 1967.
7. M.Garbey and D.Tromeur Dervout, On some Aitken like acceleration of the Schwarz Method, Int. J. for Numerical Methods in Fluids, 40, p1493-1513, 2002.
8. M.Garbey, Acceleration of the Schwarz Method for Elliptic Problems, SIAM, J. SCI COMPUT., 26, 6, p 1871-1893, 2005.
9. M.Garbey, B.Hadri and W.Shyy, Fast Elliptic Solver for Incompressible Navier Stokes Flow and Heat Transfer Problems on the Grid, 43rd Aerospace Sciences Meeting and Exhibit Conference,Reno January 2005, AIAA-2005-1386, 2005.
10. David Keyes, Technologies and Tools for High-Performance Distributed Computing-Seventh Workshop on the DOE, Advanced CompuTational Software (ACTS) Collection,August 2006
11. S.Osher and N. Paragios,Geometric Level Set Methods in Imaging, Vision and Graphics, Springer Verlag, ISBN 0387954880, 2003.
12. C.S.Peskin,The Immersed Boundary Method, Acta Numerica pp1-39, 2002
13. K. Schneider and M.Farge, Numerical Simuation of the Transient Flow Behavior in Tube Bundles Using a Volume Penalization Method. Journal of Fluids and Structures, 20, pp 555-566, 2005.
14. J.A.Sethian, Level Set Methods and Fast Marching Methods: evolving interfaces in computational geometry, fluid mechanics, computer vision, and materials science, Cambridge Monographs on Computational Mathematics, P.G.Ciarlet, A.Iserles, R.V.Kohn and M.H.Wright editors, 1999.
15. Tambasco M, Steinman DA. Path-dependent hemodynamics of the stenosed carotid bifurcation. Ann Biomed Eng. 2003 Oct;31(9):1054-65.
16. T. Thom, N. Haase, W. Rosamond, V.J. Howard, J. Rumsfeld, T. Manolio et al. Heart Disease and Stroke Statistics - 2006 Update. Circulation, 113(6):e85-e151, 2006.

# Parallel simulation of flows in porous media using adaptive locally-refined meshes*

Boris Chetverushkin[a], Natalia Churbanova[a], Anton Malinovskij[b], Anton Sukhinov[c] and Marina Trapeznikova[a]

[a] Institute for Mathematical Modeling RAS, Miusskaya Square 4, Moscow 125047, Russia

[b] Moscow State University of Technology "Stankin", Vadkovskij per. 1, Moscow 127994, Russia

[c] Moscow Institute of Physics and Technology, Institutskij per. 9, Dolgoprudny, Moscow Region, 141700, Russia

The work is aimed at numerical investigation of multi-phase contaminant transport in heterogeneous porous media at simulation of oil recovery processes. An original algorithm for construction of hierarchical locally-refined Cartesian meshes with adaptation to the solution in its high-gradient regions is developed and implemented on distributed memory multiprocessors using dynamic load balancing.

**Keywords:** Porous media flows; Multiphase contaminant transport; Adaptive Cartesian mesh refinement; Distributed memory multiprocessors; Dynamic load balancing

## 1. INTRODUCTION

The work is aimed at development of accurate and efficient parallel algorithms for simulation of multi-phase fluid flows in heterogeneous porous media when one of fluids contains a contaminant. Simulation of such flows is of great practical importance because it is necessary to predict them, for example, while developing oil recovery technologies or investigating ecology problems.

* This work is supported by INTAS (Ref. Nr. 03-50-4395) and by the Russian Foundation for Basic Research (Grants 05-01-00510, 05-01-00750, 06-01-00187).


DOI: 10.1007/978-3-540-92744-0_46, 

The classical Buckley-Leverett model [1] extended by the equation for the contaminant concentration [2] governs the process under consideration. Numerical implementations are based on finite difference approximations over rectangular staggered grids. The well known difficulty is smearing sharp fronts of solutions. A widely used tool for accuracy increase is employment of adaptive mesh refinement [3-5]. Authors of the present paper propose their own new original procedure for construction of Cartesian hierarchical locally-refined meshes with dynamic adaptation to solutions in the high-gradient regions. An attractive advantage of the proposed algorithm is possibility to keep the total number of mesh cells nearly constant during the mesh evolution. Starting steps towards the procedure development are reflected in [6]. There a convection problem with the known exact solution is described as a test for the mesh refinement algorithm validation. The current research is devoted to improvement of the algorithm and its implementation on distributed memory multiprocessor systems using dynamic load balancing. The problem of passive contaminant transport in an oil-bearing stratum at water flooding is studied numerically using the proposed approach.

## 2. CONTAMINANT TRANSPORT IN OIL FIELDS: PROBLEM STATEMENT AND NUMERICAL IMPLEMENTATION

The process of oil recovery by means of non-piston water displacement is simulated. It is supposed that the oil field is covered by a regular network of vertical water injection and oil production wells and a contaminant arrives with water through a number of wells (for example, salty water or water containing indicators or reagents). The oil-bearing stratum is thin so that the 2D (plane) problem is considered. Flow of two phases – water (w) and oil (o) in a porous medium is governed by the classical Buckley-Leverett model [1]. Fluids are immiscible and incompressible, the medium is undeformable, phase flows comply with the Darcy law, capillary and gravity forces are neglected. To account for contaminant transport the model is extended by the equation for the contaminant concentration [2]. As phases are immiscible the passive contaminant has a nonzero concentration only in water.

For computations the complete statement is reduced to the next system of equations:

$$m\frac{\partial s_w}{\partial t} + \mathrm{div}\left(F(s_w)K(s_w)\mathbf{grad}P\right) = \begin{cases} q \times F(\bar{s}) - \text{at sources;} \\ q \times F(s_w) - \text{in the whole domain;} \end{cases} \tag{1}$$

$$\mathrm{div}\left(K(s_w)\mathbf{grad}P\right) = q\,; \tag{2}$$

$$\frac{\partial\left(m s_w c + a(c)\right)}{\partial t} + \mathrm{div}\left(\mathbf{W}_w c + \mathbf{S}(\mathbf{W}_w)\right) = Q_c\,; \tag{3}$$

The Darcy velocity of water $\mathbf{W}_w$ complies with the generalized Darcy law:

$$\mathbf{W}_w = -k\frac{k_w(s_w)}{\mu_w}\mathbf{grad}P \tag{4}$$

The "diffusion flow" $\mathbf{S}(\mathbf{W}_w)$ caused by the convective diffusion is written like this:

$$S_i = -D_{ij}\frac{\partial c}{\partial x_j}, \quad i, j \in \{x, y\}. \tag{5}$$

For the effective tensor of the convective diffusion $D$ the next phenomenological formula after V.N. Nikolaevskij [7] is used:

$$D_{ij} = ((\lambda_1 - \lambda_2)\delta_{ij} + \lambda_2 n_i n_j)W_w, \tag{6}$$

where $W_w = |\mathbf{W}_w|$, $\mathbf{n} = \mathbf{W}_w / W_w$, $\lambda_1$ and $\lambda_2$ are some positive factors (measured in meters) to determine the diffusion asymmetry – their order of magnitude agrees with the reference size of the medium micro heterogeneity. One can see that even in an isotropic medium there is a preferential direction caused by the Darcy velocity vector.

The water saturation $s_w$, the pressure $P$ and the contaminant concentration $c$ are sought in a subdomain of symmetry cut from an unbounded uniform stratum. The next notations are used: $m$ is the porosity, $q$ describes debits of wells, $Q_c$ describes sources of the contaminant, $F(s_w)$ is the Buckley-Leverett function, $K(s_w)$ is a nonlinear factor including the absolute permeability $k$, relative phase permeabilities $k_w(s_w)$, $k_o(s_w)$ and dynamic viscosities $\mu_w$, $\mu_o$. Adsorption $a(c)$ is not taken into account in the present computation yet. The more detailed statement including formulas for relative permeabilities is reflected in [6]. Calculations have been performed for different sets of parameters (see some values below in Sect. 5).

For numerical implementation conservative finite difference schemes over rectangular staggered grids are used. Equations (1)-(2) are solved by the IMplicit Pressure – Explicit Saturation (IMPES) method. Solution of transport equation (1) faces significant difficulties due to existence of a discontinuity in the saturation function. In the current research equation (1) is approximated by the explicit upwind scheme. The iterative local relaxation (ad hoc SOR) method with red-black data ordering is applied to solve the elliptic pressure equation (2) [8]. To provide the second order of approximation on time as well as on space the two-level scheme with the weight $1/2$ is proposed for approximation of equation (3) with the consequent solution by ad hoc SOR with special multicolor data ordering on the nine-point stencil [6]. The nine-point stencil is used for approximation as equation (3) contains mixed derivatives in the term "$\text{div}(\mathbf{S}(\mathbf{W}_w))$" (with the account of (5) and (6)).

In previous works the above problem was solved as in homogeneous as in heterogeneous permeability fields for different well disposition schemes using uniform rectangular computational meshes [8, 6]. In the present research the numerical approach is generalized for the case of Cartesian locally-refined adaptive meshes.

## 3. ADAPTIVE HIERARCHICAL CARTESIAN MESH REFINEMENT

A special algorithm has been developed by the authors to construct Cartesian hierarchical locally-refined meshes with dynamic adaptation to solutions in their high-gradient regions.

The mesh consists of square cells. Each cell can be divided into four cells of the identical size, four adjacent cells of the identical size can be combined in one, only those cells can be merged which once formed a single cell. Each cell $C_i$ has size $h_i$ and stores value $V_i$ describing the mean value of a calculated function within the cell. Sizes of cells having common points should not differ more than twice. The mesh is stored as the quaternary tree (see Fig. 1) where nodes are cells which have been split, and leaves are cells on which calculations are performed at the current time level. It is significant that the maximum quantity of mesh cells can be fixed by the user.

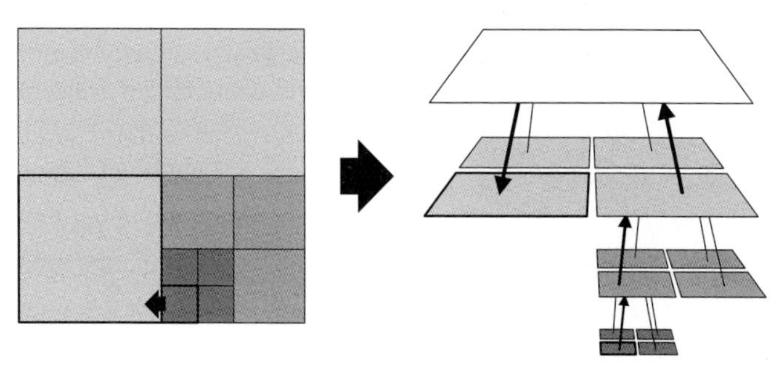

Figure 1. Structure of data as the quaternary tree

The continuously differentiable interpolation function $f(x,y)$ should be built on the mesh satisfying the following restriction:

$$\frac{1}{h_i^2}\iint_{C_i} f(x,y)dxdy = V_i, \quad 0 \le i \le N-1 . \tag{7}$$

For each cell the nine-point stencil is constructed – see Fig.2. In each point $p_i$ of this stencil approximations of the function $f$ and its partial derivatives are defined by analytical solution of systems of linear equations. Then one can use the obtained values for interpolation, for approximation of differential equations and for determination which cells have to be combined or split. Interpolation points contain sufficient information, thus, computations are performed independently of the adjacent mesh structure. Detailed formulas are given in [9].

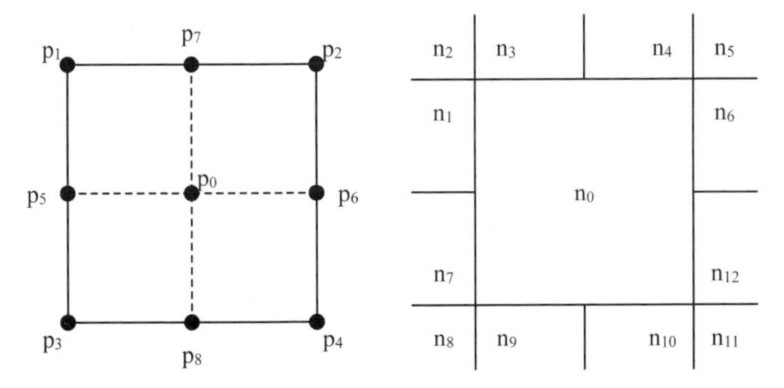


Figure 2. Interpolation points and neighbors

To get high accuracy the mesh should be adapted to the solution at each time step. The algorithm of mesh refinement involves the notion of the data variation as the difference between the maximum and the minimum values of the mesh function within the cell. The variation is not known exactly but it can be evaluated using interpolation. The idea of mesh restructuring consists in minimizing the maximal variation over the mesh when the maximum number of cells is fixed:

$$\max_{i}\left(\max_{0\le j\le 8}\left(f_j^i\right)-\min_{0\le j\le 8}\left(f_j^i\right)\right)\to \min . \tag{8}$$

## 4. PARALLEL IMPLEMENTATION OF MESH REFINEMENT

The parallel algorithm of mesh refinement is implemented using the domain decomposition technique. The 2D computational domain is divided to square subdomains $d_{kl}$ called "clusters". Clusters are dynamically distributed over processors as unbreakable units to obtain load balancing and data exchange minimization and to achieve high efficiency of parallelization. The number of clusters can differ from the number of processors $NP$. Each cluster is covered by an adaptive mesh. The number of cells $C(d_{kl})$ and computational complexity of each cluster change in time due to adaptation of meshes. Each cluster is assigned to a processor according to the following criterion:

$$\max_{i=1,\ldots,NP} N_i + \alpha_1 \cdot \max_{i=1,\ldots,NP} L_i + \alpha_2 \cdot \max_{i=1,\ldots,NP} G_i \to \min . \tag{9}$$

Here:

$N_i$ – the number of cells on the $i$-th processor;

$L_i$ – the total length of the clusters interface between the $i$-th processor and other processors (measured in cells);
$G_i$ – the amount of cells to be transferred to/from the $i$-th processor to achieve the necessary clusters distribution (load balancing overhead expenses);
$\alpha_1$ and $\alpha_2$ – some weights to be defined empirically.

## 5. NUMERICAL RESULTS

First of all the developed tools for adaptive mesh refinement were approved by solving a convection problem with the known exact solution [6]. The computational domain is the unit square, at the initial moment the required function equals 0.0 everywhere except the circle of diameter 0.25 where it equals 1.0, the circle center has coordinates (0.25,0.25), the velocity is vector (1,1) in the whole domain. At time 0.7 the circle should reach coordinate (0.75,0.75). Comparison of results obtained on uniform and on adaptive meshes demonstrated the next time advantage of adaptive meshes: the run time on the adaptive mesh was 1.5 times greater than on the uniform mesh with the same amount of cells but to achieve the same precision on the uniform mesh 16 times more cells were needed. Thus, having the same precision, solution of the given test problem could be obtained about 10 times faster using the adaptive mesh.

The above applied problem of contaminant transport in an oil-bearing stratum has been solved on a multiprocessor system using three separate adaptive meshes to store fields of the water saturation $s_w$, the pressure $P$ and the contaminant concentration $c$.

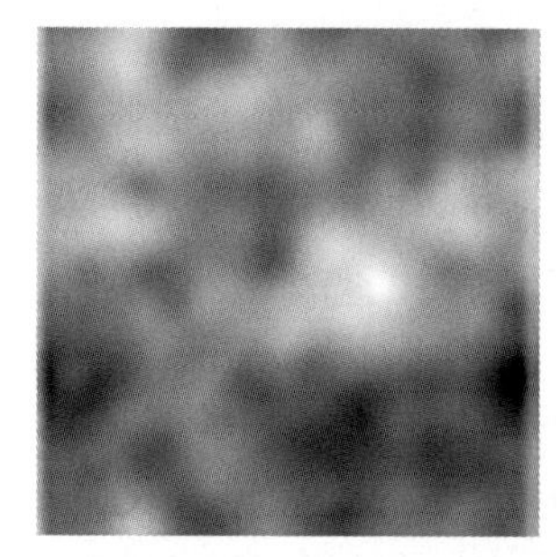

Figure 3. The absolute permeability pattern (dark areas – fine sand, light areas – coarse sand)

The typical five-spot well configuration has been predicted. Fig. 3 shows the medium pattern: the absolute permeability $k$ is a function (not a tensor as the medium anisotropy is not taken into account yet) varying from $10^{-8}$ m$^2$ (coarse sand) to $10^{-12}$ m$^2$ (fine sand), the porosity $m = 0.2$. This pattern is considered as a computational domain with the boundary conditions of symmetry with the water injection well in the left bottom corner and the oil production well in the right top corner, the well debit $q = 300$ m$^3$/day.

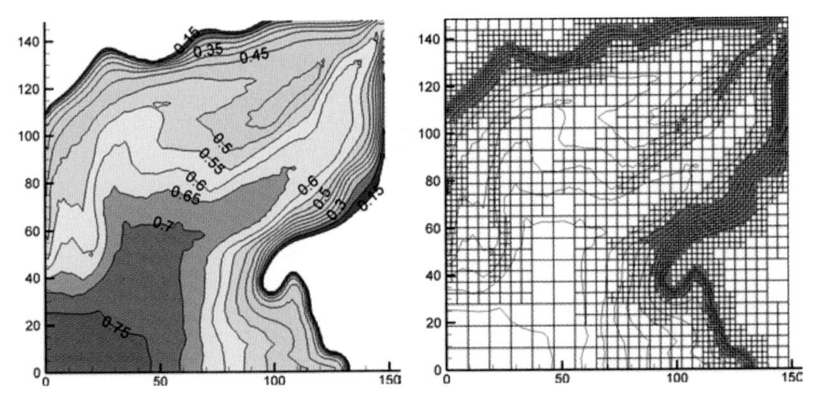


Figure 4. Water saturation field and mesh (4096 cells)

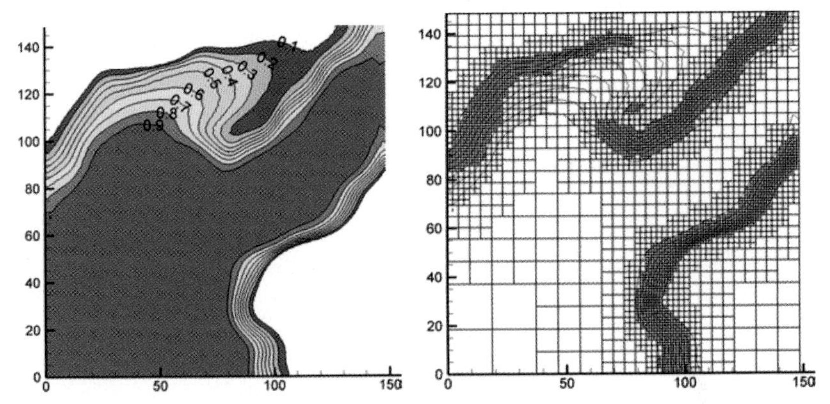


Figure 5. Contaminant concentration field and mesh (4096 cells)

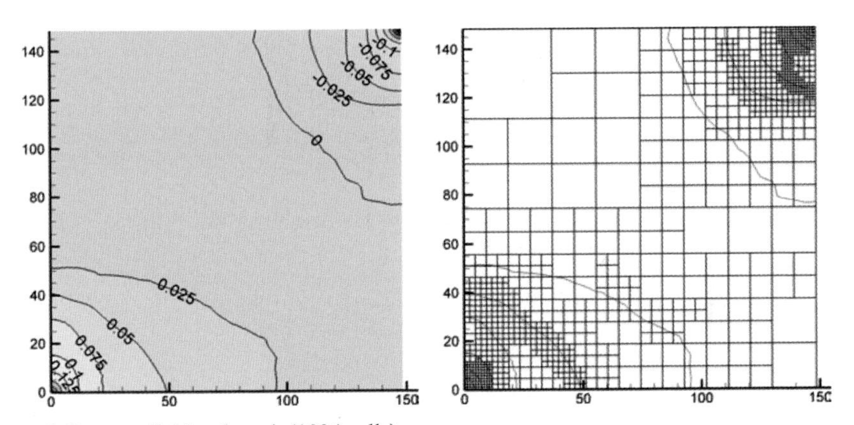


Figure 6. Pressure field and mesh (1024 cells)

Results for this medium and corresponding meshes obtained after 100 days of oil recovery are presented in Fig. 4 – 6. Maxima of the saturation, the concentration and the pressure correspond to the water injection well which supplies the stratum with the contaminant. Moving fronts have complex shapes and one can see the so-called "fingering". Meshes are refined at each time level and adapted to the fronts.

Calculations have been performed on the self-made cluster consisting of 14 two-processor two-kernel modules with the peak performance of 300 GFlops. The developed parallel mesh refinement tools allow determining the optimal processor number for the given quantity of mesh cells from the standpoint of efficient load balancing. For example, for a coarse enough mesh consisting of 4096 cells, which is used to obtain the water saturation, the optimal number of processors is 3.

## 6. CONCLUSION

The problem of passive contaminant transport in an oil-bearing stratum at water flooding has been studied numerically using a new parallel algorithm of 2D adaptive mesh refinement.

In the nearest future the developed mesh refinement tools will be implemented in the form of a package in order to use it widely for solving problems of practical interest.

## REFERENCES

1. R.E. Ewing (Ed.), The mathematics of reservoir simulations, SIAM, Philadelphia, 1983.
2. V.M. Entov and A.F. Zazovskij, Hydrodynamics of processes for oil production acceleration, Nedra, Moscow, 1989. - In Russian.
3. M.J. Berger and M.J. Aftosmus, "Aspects (and Aspect Ratios) of Cartesian Mesh Methods", In: *Proc. of the 16$^{th}$ Int. Conf. on Numerical Methods in Fluid Dynamics,* Springer-Verlag, Heidelberg, Germany (1998).
4. J.A. Trangenstein, "Multi-scale iterative techniques and adaptive mesh refinement for flow in porous media", *Advances in Water Resources*, **25** (2002) 1175-1213.
5. L.H. Howell and J.B. Bell, "An adaptive mesh projection method for viscous incompressible flow", *SIAM J. Sci. Comput.* **18**, 4 (1997) 996-1013.
6. B.N. Chetverushkin, N.G. Churbanova, A.A. Soukhinov and M.A. Trapeznikova, "Simulation of Contaminant Transport in an Oil-Bearing Stratum at Water Flooding", In: *CD-Rom Proc. of ECCOMAS CFD 06*, P. Wesseling, E. Onate and J. Periaux (Eds.), TU Delft, The Netherlands (2006).
7. V.N. Nikolaevskij, Mechanics of porous and fissured media, Nedra, Moscow, 1984. - In Russian.
8. M.A. Trapeznikova, N.G. Churbanova and B.N. Chetverushkin, "Parallel elliptic solvers and their application to oil recovery simulation", In: *HPC'2000, Grand Challenges in Computer Simulation, Proc. of Simulation Multi Conf.*, SCS, San Diego, CA (2000) 213-218.
9. B.N. Chetverushkin, N.G. Churbanova, M.A. Trapeznikova, A.A. Soukhinov and A.A. Malinovskij, "Adaptive Cartesian Mesh Refinement for Simulation of Multiphase Flows in Porous Media", *Computational Methods in Applied Mathematic*, 2 (2008) – to appear.

# Performance Evaluation of Two Parallel, Direct Sparse Solvers for an Aeroacoustic Propagation Model

Y. Özyörük[a] and E. Dizemen[a]

[a]Department of Aerospace Engineering
Middle East Technical University
06531 Ankara, Turkey

This paper deals with parallel performance assessment of two state-of-the-art and publicly available direct sparse matrix solvers, namely SuperLU and MUMPS, for solution of acoustic propagation and radiation through turbomachinery exhaust ducts on structured meshes. The aeroacoustic model is based on the linearized Euler equations written in frequency domain. Perfectly matched layer equations are used at the duct inlet and far-field boundaries. The governing equations and the boundary conditions are discretized using finite differences. The direct solution of the resulting linear system of equations has a very high demand particularly on computer memory. Therefore, a distributed memory computing approach is employed. It is demonstrated that the MUMPS solver is more suitable for the present type problems.

## 1. INTRODUCTION

Although significant advances have been achieved in aircraft engine technologies, increasing air traffic and community sensitivity keep placing an important pressure on aircraft engine manufacturers to lower noise radiating from engines further. Exhaust noise is one of the dominant components contributing to the overall noise emission from aircraft, especially at take-off. Therefore, significant research is conducted to understand noise generation and propagation mechanisms in air-breathing engines [1–4].

In general, prediction of propagation of any generated noise component through engine exhaust ducts is not straightforward due to many physical factors involved. First of all, highly non-uniform, turbulent flows exist in the exhaust ducts, both core and bypass. Also acoustic treatment panels, namely liners, are employed at the duct walls for noise attenuation that in turn require proper wall conditions. In addition, the propagated noise radiates to far-field through highly sheared, turbulent layers of the exhaust jets.

Among the noise components propagating through the ducts and radiating to far-field that contribute to the overall noise signature of an aircraft engine is tonal noise generated by rotor-stator interactions. Also any additional linear acoustic field component in the engine ducts can be decomposed into its Fourier modes. By assuming that amplitudes of pressure and induced velocity perturbations in the ducts are in general significantly smaller than their mean counterparts, and the time scales associated with them are disparate from those of the basic flows both in the ducts and in the jet shear layers, we can predict propagation and radiation of the acoustic modes by solving the linearized Euler equations. Two solution approaches are possible. We can

I.H. Tuncer et al., *Parallel Computational Fluid Dynamics 2007*,
DOI: 10.1007/978-3-540-92744-0_47, 

compute the acoustic field directly in time domain or frequency domain. Time-domain solutions are, however, hindered by capture of the inherent shear layer instabilities. Therefore, this paper solves the linearized Euler equations directly in the frequency domain for the frequencies of interest, as suggested by Ref. [5]. Two state-of-the-art sparse solvers which are in public domain, namely MUMPS and SuperLU are used for the direct, parallel solution of the resulting linear system of equations. The MUMPS sparse solver, written in Fortran, is based on a multifrontal LU decomposition [6,7]. The SuperLU code is written in C and performs an LU decomposition with partial pivoting and triangular system solves through forward and back substitution [8].

The paper assesses these codes in direct solution of the aforementioned aeroacoustic model. Memory and CPU usage requirements are compared and discussed. In the following section, the aeroacoustic propagation model is briefly discussed. Then, results for example calculations are shown, and a performance assessment study is conducted.

## 2. AEROACOUSTIC MODEL

The aeroacoustic model is based on the frequency-domain linearized Euler equations. Acoustic modes are successfully introduced into the computational domain (ducts) through the perfectly matched layer (PML) equations of Hu [9]. The same type of boundary conditions are used at the far-field boundaries to let the outgoing waves without reflection. Any flow quantity is assumed to have the form $q = \bar{q} + q'$, where $q$ is the primitive flow quantity (density, three velocity components and pressure), $\bar{q}$ is the mean flow quantity, and $q'$ is the perturbed flow quantity due to acoustic propagation. Cylindrical-polar coordinates $(x, r, \theta)$ are employed. The waves are assumed periodic both in time and azimuthal direction. That is, any perturbed flow quantity $q'$ is assumed to have the form $q'(t, x, r, \theta) = Re[\hat{q}(\omega, x, r)e^{i(\omega t - m\theta)}]$, where $\hat{q}$ is the complex amplitude of the perturbed quantity, $i = \sqrt{-1}$, $\omega$ is the circular frequency, and $m$ is the azimuthal mode order. Then, the three-dimensional equations for a single spinning mode order $m$ at circular frequency $\omega$ are effectively reduced to two dimensions, axial and radial coordinates $(x, r)$, for an axisymmetric geometry. Body-conforming structured meshes are used. The equations in the mapped domain $(\xi, \eta)$ are written in the form

$$i\omega\hat{\mathbf{q}} + [A]\frac{\partial\hat{\mathbf{q}}}{\partial\xi} + [B]\frac{\partial\hat{\mathbf{q}}}{\partial\eta} + [C]\hat{\mathbf{q}} = \hat{\mathbf{r}} \tag{1}$$

where $\hat{\mathbf{q}}$ is the vector of the complex, dependent, flow perturbations (density, axial, radial and azimuthal velocity, and pressure), $[A]$, $[B]$, and $[C]$ are $5 \times 5$-matrices which are function of the background flow information, i.e. $\bar{\mathbf{q}}$ and the transformation metrics, and $\hat{\mathbf{r}}$ is the right-hand side vector resulting from driving the acoustic field with the PML equations. The details of the aeroacoustic code, named FLESTURN, is given in Ref. [10].

Direct discretization of these equations using finite differences results in a linear system of equations that may be written

$$\mathbf{Ax} = \mathbf{b} \tag{2}$$

where $\mathbf{A}$ is the sparse coefficient matrix, $\mathbf{x}$ is the solution vector, and $\mathbf{b}$ is the right-hand side vector.

### 2.1. Discretization Algorithms

There are three discretization algorithms available in the developed aeroacoustic propagation code. These are the standard, 4th-order finite difference, the dispersion-relation-preserving (DRP) algorithm, and the linear B-spline, Galerkin discretization algorithm. The stencils used by these algorithms for a totally interior grid point are shown in Fig. 2.1. It is quite clear that the structure of the coefficient matrix **A** will depend on the stencil used. On a structured mesh, the DRP [11] algorithm results in the widest bandwidth, and therefore, is expected to require the highest amount of memory for a given mesh. The B-spline Galerkin discretization algorithm uses 9 points but has a relatively compact stencil. Therefore, the coefficient matrix is expected to have higher number of non-zero entries than the standard 4th-order algorithm, but effectively the resulting bandwidth for the coefficient matrix is expected to be smaller than the other two algorithms. Therefore, factorization of the coefficient matrix resulting from the B-spline Galerkin discretization is not expected to be much more memory demanding than that the other two discretization algorithms.

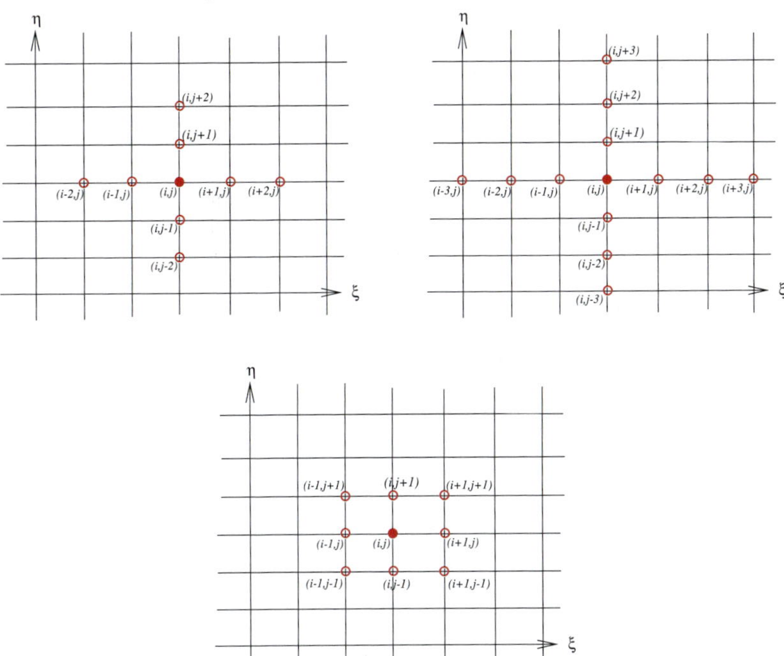


Fig 2.1: Stencils: Top left - 4th-order difference; Top right - DRP; Bottom - Linear B-spline Galerkin.

## 3. RESULTS and DISCUSSION

The assessment of the two direct sparse solvers SuperLU and MUMPS is made on an example problem involving acoustic propagation and radiation through a semi-infinite annular duct with infinite centerbody, as shown in Fig. 3.1. The 4th-order finite-difference scheme and 3 different size meshes are used for the assessment and comparisons of the two codes. The mesh sizes, the number of unknowns, and the resulting number of non-zero entries in the sparse coefficient matrix are shown in Table 1. The first mesh (case) has $201 \times 101$, the second mesh has $401 \times 201$, and the third mesh has $801 \times 401$ grid points. In other words, the number of unknowns (degree of freedom in a finite element sense) is increased by a factor of 4 from one mesh to the next. It is clear from the table that there is almost a linear correlation between the number of unknowns and the non-zero entries of the coefficient matrix resulting from the discretization of the aeroacoustic model using the 4th-order finite difference scheme. Such a discretization on a uniform mesh gives the pattern for the non-zero element locations shown in Fig. 3.2. Because the computational mesh is structured, the entries of the coefficient matrix look well organized.

Computations are done on a parallel platform that contains 12 dual core 64 bit AMD Opteron processors, each having 4 GB of RAM. The machines are connected to each other with a one gigabit switch.

Table 2 shows the influence of the process decomposition on the performance of SuperLU on mesh 1. While a good load balancing is achieved by SuperLU for arbitrary partitioning, there appears a significant effect on the CPU time by the decomposition structure. SuperLU performs better for row-based partitioning of the coefficient matrix.

Table 3 compares the performances of SuperLU and MUMPS. This comparison is made for all the three mesh cases using 8 dedicated processors, with a forced, $8 \times 1$ partitioning for SuperLU and internally controlled partitioning for MUMPS. The peak memory usage and the wall clock times were recorded for each run. It is clear that the memory need by SuperLU gets quickly out of hand with the increasing problem size, while the rate at which this occurs is rather smaller for MUMPS. Also a significant difference is observed in the CPU time usage. MUMPS outperforms SuperLU in terms of both memory and CPU time usage. However, it has been observed that there may occur significant differences in the peak memory need between the MUMPS processes, while SuperLU distributes the load more evenly in terms of memory requirement.

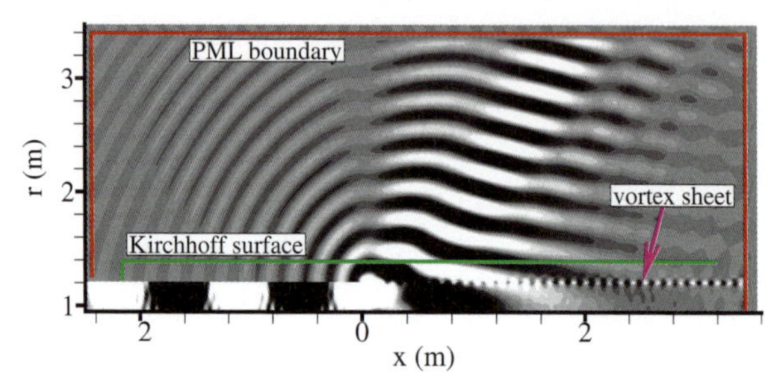


Fig 3.1: Example acoustic field.

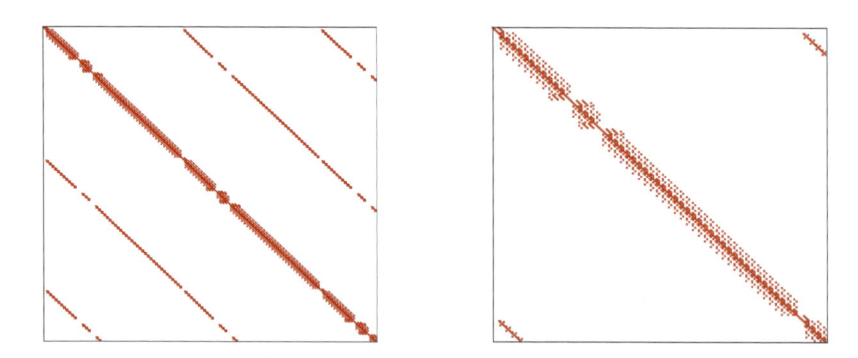

Fig 3.2: Sparse matrix structure. Left: full, Right: close-up view.

Table 1
Test grids and corresponding unknowns and sizes of coefficient matrix.

| Case | Mesh size | Number of unknowns | Number of non-zero entries |
|---|---|---|---|
| 1 | $201 \times 101$ | 101,505 | 724,963 |
| 2 | $401 \times 201$ | 403,005 | 2,943,359 |
| 3 | $801 \times 401$ | 1,606,005 | 11,808,512 |

Table 2
SuperLU performance test on mesh 1.

| Process partitioning (rows×columns) | Wall clock time | Peak memory usage per process |
|---|---|---|
| $8 \times 1$ | 2 min, 31 s | 85 Mb |
| $4 \times 2$ | 3 min, 28 s | 84 Mb |
| $2 \times 4$ | 5 min, 24 s | 78 Mb |

## 4. CONCLUSIONS

A parallel performance study for two state-of-the-art, parallel, direct sparse solvers, namely SuperLU and MUMPS that are in public domain has been performed for prediction of noise propagation and radiation from an aircraft engine like geometry solving the linearized Euler equations in the frequency domain. It has been shown that for this kind of problems the MUMPS sparse solver is superior over the SuperLU solver in terms of both memory usage and computational speed. However, when the MUMPS solver is let decompose the problem by itself, the peak memory usage may differ significantly from one process to another, limiting the size of the solvable problem, whereas SuperLU does a better job in distributing the task. Nevertheless, the former code requires less memory for the same size problem.

## 5. ACKNOWLEDGMENTS

This work has been sponsored by the European Commission through the FP6 STREP project TURNEX, coordinated by B. J. Tester and monitored by D. Chiron.

## REFERENCES

1. X. Zhang, X. X. Chen, C. L. Morfey, and P. A. Nelson, Computation of spinning modal radiation from unflanged duct. *AIAA Journal*, **42(9)**, pp.1795-1801, 2004.
2. G. Gabard and R. J. Astley. Theoretical models for sound radiation from annular jet pipers: far- and near-field solutions. *Journal of Fluid Mechanics*, **549**, pp.315-342, 2006.
3. A. Demir and S. Rienstra, Sound radiation from an annular duct with jet flow and a lined centerbody. AIAA Paper 2006-2718, 12th AIAA Aeroacoustics Conference, Cambridge, MA, May 2006.
4. J. Kok, Computation of sound radiation from cylindrical ducts with jets using a high-order finite-volume method. AIAA Paper 2007-3489, 13th AIAA/CEAS Aeroacoustics Conference, Rome, Italy, May 2007.
5. A. Agarwal, P. J. Morris, and R. Mani, The calculation of sound propagation in nonuniform

Table 3
SuperLU versus MUMPS on 8 processors. SuperLU with $8 \times 1$ partitioning; MUMPS with internal default partitioning.

| Case | SuperLU Peak memory per process (Mb) | SuperLU Wall clock time (seconds) | MUMPS Peak memory per process (Mb) | MUMPS Wall clock time (seconds) |
|---|---|---|---|---|
| 1 | 85 | 152 | 126 | 25 |
| 2 | 442 | 666 | 659 | 76 |
| 3 | 3994 | 6371 | 2533 | 451 |

flows: Suppression of instability waves. AIAA paper 2003-0878, 41st Aerospace Sciences Meeting, Reno, Nevada, Jan. 2003.
6. http://graal.ens-lyon.fr/MUMPS/
7. P. R. Amestoy, I. S. Duff, J.-Y. L. L'Excellent, and J. Koster, A fully asynchronous multifrontal solver using distributed dynamic scheduling. *Journal of Matrix Analysis and Applications*, **23**, pp.15-41, 2001.
8. http://crd.lbl.gov/ xiaoye/superlu/
9. Q. F. Hu, A perfectly matched layer absorbing boundary condition for linearized Euler equations with non-uniform mean flow. *Journal of Computational Physics*, **208**, pp. 469-492, May 2005.
10. Y. Özyörük, E. Dizemen, S. Kaya, A. Aktürk, A frequency domain linearized Euler solver for turbomachinery noise propagation and radiation. AIAA/CEAS paper 2007-3521, 13th AIAA/CEAS Aeroacoustics Conference, Rome, May 2007.
11. C. K. W. Tam and J. C. Webb, Dispersion-Relation-Preserving finite difference schemes for computational acoustics. *Journal of Computational Acoustics*, **107**, pp.261-281, 1993.

# Three Dimensional Smoke Simulation on Programmable Graphics Hardware

Gökçe Yıldırım [a,b], H. Yalım Keleş [a,c] and Veysi İşler [a]

[a]Department of Computer Engineering, METU, Ankara, Turkey

[b]Simsoft, Ankara, Turkey

[c]Tübitak Bilten, METU, Ankara, Turkey

The complexity of the calculations underlying Fluid Dynamics makes it difficult to achieve high-speed simulations. In this paper, a gaseous fluid, smoke, is simulated in a three dimensional environment with complex obstacles by solving Navier-Stokes equations using a semi-Lagrangian unconditionally stable method. The simulation is performed both on CPU and GPU. Owing to the advantage of programmability and parallelism of GPU, smoke simulation on graphics hardware is shown to run significantly faster than the corresponding CPU implementation.

## 1. INTRODUCTION

There are two main approaches used in fluid simulations: Lagrangian particle-based and Eulerian grid-based approaches. Many different methods have been developed on both approaches [1–5]. In particle-based methods, fluid is composed of particles which dynamically change over time as a result of external forces. Simplicity of particles makes these methods advantageous [6]. Moreover, particle-based methods guarantee mass conservation. On the other hand, an important drawback of particle-based simulations is the difficulty of representing smooth surfaces of fluids using particles [7]. The second kind of approaches used in fluid simulations are Eulerian grid-based methods. In these methods, the spatial domain is discretized into small cells to form a volume grid. It is easy to analyze the fluid flow with Eulerian methods. Euler equations neglect the effects of the viscosity of the fluid which are included in Navier-Stokes equations (NSEs).

In this paper, three dimensional simulation of smoke is performed on both CPU and GPU. Data in this simulation is represented on a three dimensional grid of cells. For the physical simulation of smoke behavior, NSEs are solved using a semi-Lagrangian unconditionally stable method [3] on both CPU and GPU, and the performances are compared. Owing to the parallelism in graphics hardware, smoke simulation performed on GPU is shown to run significantly faster than the corresponding CPU implementation. One of the reasons for GPU's higher performance is the use of flat 3D textures that reduces the number of rendering passes, since the whole three dimensional data is processed in one render. Moreover, due to programmability at the fragment level, packing attributes into a single RGBA texel at the fragment level decreases the number of rendering passes by

I.H. Tuncer et al., *Parallel Computational Fluid Dynamics 2007*,
DOI: 10.1007/978-3-540-92744-0_48, © Springer-Verlag Berlin Heidelberg 2009

reducing the number of textures to be processed. Furthermore, double-buffered offscreen floating point rendering targets are utilized, which decreases context switches resulting in higher performance.

Arbitrary boundary conditions are also handled in this paper. This is accomplished by grouping the grid nodes into different types and generating the texture offsets for the velocity and scalar variables of these nodes. The variables found in the calculated texture offsets are then used to modify the variables on boundaries according to the specified boundary conditions [8].

Vertex and fragment programs are written in "C For Graphics (Cg)" in this paper since it is a platform-independent and architecture neutral shading language. Fragment programs written in Cg process in parallel each fragment data. This is the main reason of the significant performance of GPU over CPU as CPU works by iterating over each grid cell.

The closest works to ours are [9,8]. These works are explained in Section 2. Our simulation is different from these works in the dimensionality of the environment [9] and the programming language used for programming at the fragment and vertex level [8].

The outline of this paper is as follows: In Section 2, existing works about fluid simulations are presented. Then, in Section 3, the equations of fluid dynamics used in this smoke simulation as well as the details of GPU implementation are explained. Moreover, in this section, boundary processing is given. In Section 4, CPU and GPU results are compared and discussed. Finally, in Section 5, the paper is concluded with a list of future works.

## 2. RECENT WORK

In the processing of fluid flow on GPU, this paper is mainly influenced by [9,8]. In [9], the whole computational processing was accelerated by packing the vector and scalar variables into four channels together to reduce the number of rendering passes. As for the boundary conditions, a more general method that can handle arbitrary obstacles in the fluid domain was provided. This method was extended to three dimensions in [8]. Our work is different from these in that our simulation is three dimensional (unlike [9]) and we used Cg as the programming language (unlike [8]) for fragment and vertex level programming.

## 3. IMPLEMENTATION

Our smoke simulation is based on Stam's semi-Lagrangian method [3]. Stability for any time step is guaranteed by means of this method. However, since numerical dissipation is inherent in the semi-Lagrangian method, vorticity confinement force is added to reduce dissipation like in [10]. Furthermore, forces due to thermal buoyancy are also calculated to simulate motion of smoke physically [11].

### 3.1. Fluid Flow Equations

In this paper, smoke is assumed to be an incompressible, constant density fluid, and uniform cartesian grids of limited size are employed. Cell-centered grid discretization is used for the description of attributes such as velocity, density, temperature and pressure.

In other words, these attributes are defined at cell centers. The cell-centered approach is simple and decreases the number of computations [3].

NSEs describe fluid flow fully using mass and momentum conservation [1]. To simulate the behavior of smoke, the evolution of velocity of incompressible smoke over time, denoted by $\mathbf{u} = (u,\ v,\ w)$ is given by incompressible NSEs:

$$\nabla \cdot \mathbf{u} = 0, \quad (1)$$

$$\frac{\partial \mathbf{u}}{\partial t} = -(\mathbf{u} \cdot \nabla)\mathbf{u}) - \frac{1}{\rho}\nabla p + \nu \nabla^2 \mathbf{u} + \mathbf{F}, \quad (2)$$

where $\mathbf{u}$ is velocity, $t$ is time, $\rho$ is density, $p$ is pressure, $\nu$ is kinematic viscosity, and $\mathbf{F}$ is the external force2) [3]. For the evolution of density $\rho$ and temperature $T$ moving through the velocity field, the following equations are used respectively as in [10]:

$$\frac{\partial \rho}{\partial t} = -(\mathbf{u} \cdot \nabla)\rho) - k_\rho \nabla^2 \rho + S_\rho, \quad (3)$$

$$\frac{\partial T}{\partial t} = -(\mathbf{u} \cdot \nabla)T) - k_T \nabla^2 T + S_T. \quad (4)$$

Advection terms in equations 2, 3 and 4 are solved using Stam's semi-Lagrangian unconditionally stable method [3]. In the calculation of diffusion terms, Jacobi iteration method is used [12]. As a result of the calculations of velocity, temperature and density evolution in each time step, density quantities for each grid cell are used as transparency factors, and changing them according to the computed motion of the smoke leads to the simulation of the flow of the smoke.

Temperature is an important factor that governs smoke motion. As the gas rises when heated, it causes internal drag, and a turbulent rotation is produced. This effect is known as thermal buoyancy [11]. Force due to thermal buoyancy affects the smoke motion. Calculation of buoyant force is as follows:

$$\mathbf{F}_{buoy} = -\alpha \rho \mathbf{z} + \beta (T - T_{amb})\mathbf{z}. \quad (5)$$

Physically, the mixture of the smoke and air contains velocity fields with large spatial deviations accompanied by a significant amount of rotational and turbulent structure on a variety of scales. Nonphysical numerical dissipation decreases these interesting flow features. Therefore, vorticity confinement force [10] is added as an external force to add these flow effects back:

$$\mathbf{F}_{conf} = \varepsilon h(\mathbf{N} \times \omega), \quad (6)$$

where $\omega$ and $\mathbf{N}$ denote vorticity and normalized vorticity vectors, respectively, and are defined as follows:

$$\omega = \nabla \times \mathbf{u}, \quad (7)$$

$$\mathbf{N} = \frac{\nabla \mid \omega \mid}{\mid \nabla \mid \omega \mid\mid}. \quad (8)$$

The buoyant force and the vorticity confinement force together with the user force comprise $\mathbf{F}$ in equation 2.

### 3.2. Hardware Implementation

The smoke simulation in this paper is three dimensional, i.e., data is represented on a three dimensional grid. The natural representation for this grid on CPU is an array whereas the analog of an array on GPU is a texture. Textures on GPU are not as flexible as arrays on CPU. However, their flexibility is improving with the evolution in graphics hardware. Textures on current GPUs support all the basic operations necessary to implement a three dimensional smoke simulation. Since textures usually have four color channels, they provide a natural data structure for vector data types with components up to four [13].

In this simulation, vector values such as velocity and scalar values such as temperature and density are stored in textures. For velocity values, three channels of textures are occupied by the three components of RGBA four channels of a single texel. In the case of scalar values such as temperature and density, we can exploit the fact that the same operations are performed to compute them. Owing to this exploitation, to reduce the number of passes on GPU at the fragment level, we pack scalar values into RGBA four channels of a single texel as in [8].

Different from two dimensional fluid simulations, one of the difficulties in three dimensional fluid simulations is the representation of three dimensional grid using textures. In this paper, flat 3D textures are used for the representation of three dimensional volume as a texture [13].

As well as data representation, the read operation differs on GPU. On CPU, the array is read using an offset or index. However, on GPU, textures are read using a texture lookup.

In CPU implementation, all steps in the algorithm are performed using loops. To iterate over each grid cell, three nested loops are used. At each cell, the same computation is performed. However, the processing of the algorithm is different on GPU. The fragment pipeline is designed to perform identical computations at each fragment so that it appears as if there is a processor for each fragment and all fragments are updated simultaneously [14]. This is the natural consequence of parallelism of GPU.

The main algorithm flow for velocity values on GPU is shown in figure 1(a). It is very similar to the flow of velocity update on CPU except that here, instead of an array, a texture is updated. In figure 1(b), the main algorithm flow for scalar values such as density and temperature is displayed. In the main algorithm flow for scalar values, one texture is used to store both values.

#### 3.2.1. Implementation of boundary condition

Implementation of Boundary Condition (IBC) is an important issue in smoke simulation due to the existence of natural boundaries such as walls and obstacles in nature. Different methods for IBC exist [15,13,8]. In our simulation, we process boundary conditions as in [8]. In this method, the voxels are grouped into different types according to their positions relative to the boundary cells and then the offsets and modification factors for vector and scalar attributes are calculated. Using the calculated offsets and factors, attributes of boundary voxels are modified. In this way, arbitrary interior boundaries and obstacles are handled. Processing of boundary conditions is explained below (see [8] for more details). The equation which expresses Dirichlet, Neumann and mixed boundary

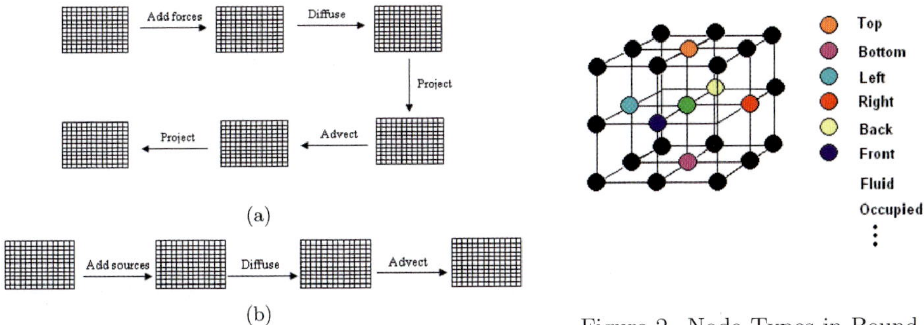


Figure 1. **(a)** Algorithm Flow for Velocity Texture on GPU. **(b)** Algorithm Flow for Texture of Scalar Values on GPU.

Figure 2. Node Types in Boundary Processing. The figure illustrates the node types relative to the center voxel (green). The node types denote the relative orientation of the boundary according to the center voxel.

conditions is as follows:

$$\alpha\phi + b\frac{\partial\phi}{\partial\mathbf{n}} = c, \tag{9}$$

where $\phi$ denotes any attribute value such as density and velocity. After the classification of voxels into two types (0 for obstacle, 1 otherwise) according to the occupation of obstacles, voxel values are modified using the value of a surrounding voxel that is found using equation 9. In [8], for the discretization of equation 9, the normal direction is simplified into 28 directions as shown in figure 2. By the discretization of equation 9, the formula, $\phi_{boundary} = d\phi_{offset} + e$, serves in the calculation of modification factors that are used in the update of boundary attributes. The node type of a voxel is calculated as follows:

$$\begin{aligned} NodeType = {} & Obstacle(i,j,k) * 64 + Obstacle(i+1,j,k) * 1 + Obstacle(i\text{-}1,j,k) * 2 + Obstacle(i,j\text{-}1,k) * 4 + \\ & Obstacle(i,j+1,k) * 8 + Obstacle(i,j,k+1) * 16 + Obstacle(i,j,k\text{-}1) * 32, \end{aligned} \tag{10}$$

where *Obstacle(i,j,k)* is equal to 0 if the voxel is occupied by obstacle or 1 otherwise. Using this coding scheme, 7 nodes are sufficient to determine the 28 directions. In this coding scheme, 6 nodes denote neighbors of node at i, j, k. Seventh is the node itself. Three different values (-1, 0, 1) can be assigned for each of i, j, k, which results in 27 different directions and the 28th meaning the node occupied with fluid. The result of the coding scheme can be found as a table in [8]. The 64 results of the coding scheme are arranged into 28 directions and stored as 3D position offsets, which are then used to generate the actual texture offsets stored in flat 3D textures. These actual texture offsets give the information for the relevant surrounding voxel. The velocity factors are calculated in parallel according to the boundary conditions. After the actual texture offsets and velocity factors are calculated, we process the boundary conditions for velocity attributes and scalar values such as density and temperature.

Figure 3 shows the Cg code of one of the fragment programs, which handles boundary conditions in our simulation. In this program, the boundary cells are set using offset texture. In lines 2-6, parameters are given: *gridCoord* denotes grid coordinates, *texCoord* denotes texture coordinates, *texAttr* denotes attribute (velocity or scalar) texture, *texOffsets* denotes texture offsets and *texeAttrFactor* denotes attribute (velocity or scalar) factor texture. In line 7, texture offset coordinates are found. In the following lines, the texture attribute in the calculated texture offset coordinate is multiplied by the attribute factor value and the result is assigned to the current cell.

```
1.  float4 setBoundaryWithOffsets (
2.      in half2 gridCoord :WPOS,
3.      in float4 texCoord                 :TEXCOORD0
4.      uniform samplerRECT texAttr        :TEXUNIT0,
5.      uniform samplerRECT texOffsets     :TEXUNIT1,
6.      uniform samplerRECT texAttrFactor  :TEXUNIT2)
    {
7.      float2 texOffsetCoords = f2texRECT(texOffsets, texCoord.xy);
8.      return (f4texRECT(textAttrFactor, texCoord.xy) *
9.          f4texRECT(textAttr, gridCoord.xy + texOffsetCoords));
    }
```

Figure 3. Cg Code of a Fragment Program that Handles Boundary Processing.

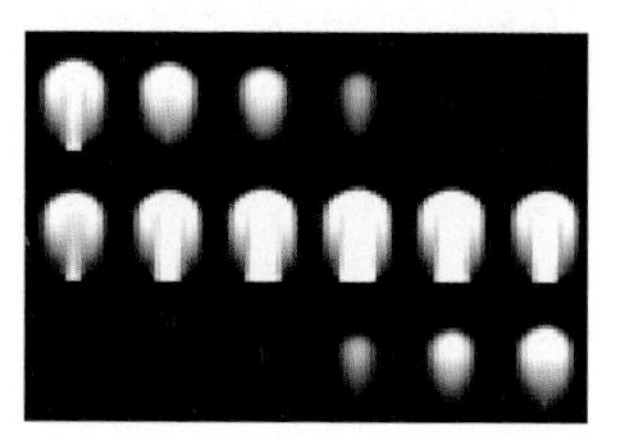

Figure 4. Flat 3D Texture of Smoke Flowing in a Grid of 16×32×16 Voxels.

## 4. RESULTS AND DISCUSSION

All experiments have been performed on a PC with a Pentium 4 3.0 GHz CPU and 1 GB main memory. The graphics chip is GeForce FX 5700 with 256 MB video memory. The operating system is Windows XP. The simulations are run on both CPU and GPU to compare the computation times of fluid dynamics equations. All the computations in this paper are based on 32-bit float precision to make it suitable for real-world problems.

Flat 3D textures are used to represent data on GPU. Figure 4 shows the flat 3D texture of a flowing smoke in upward direction in a grid of 16×32×16 voxels.

Figure 5(a-d) shows four sequential frames of smoke flowing in a grid of 16×32×16 voxels. In the scene, there is a density source at the bottom of the grid and the velocity is mainly in upward direction, which simulates an air sprayer at the bottom. In figure 5(a), smoke is released from a source of virtual air sprayer at the bottom and in Figure 5(b-d), smoke continues to flow up with the fluid flow effects and interacting with inner obstacles and outer walls. In the sequence of figures, the processing of boundary conditions can be seen easily. In figure 5(e-g), three sequential cross sections of a house, in which smoke starts to flow from the first floor and continues up to the other floors, are shown. Smoke interacts with inner walls and enters through holes of the house in this sequence. As it can be seen in these figures, the smoke simulation is physically realistic. During rendering process, we do not use shading effects. With shading effects, the whole simulation would look much more realistic.

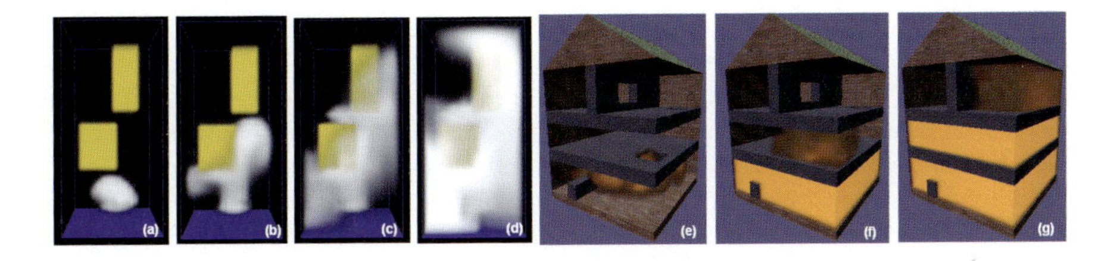


Figure 5. **(a-d)** Smoke Flowing in a Grid of 16×32×16 Voxels with Inner Obstacles in Upward Direction in Movement Sequence of (a), (b), (c) and (d). The boundary is determined by the rectangle prism whose edges are drawn as blue lines and the yellow obstacles. **(e-g)** Cross Section of a House in Which Smoke Flows in Movement Sequence of (e), (f) and (g). The boundary is determined by the walls and the floors of the house.

Table 1
Comparison of Performance on CPU and GPU.

| Grid Dimensions | Average CPU Time (ms) | Average GPU Time (ms) | SpeedUp |
|---|---|---|---|
| 16×16×16 | 40 | 50 | 0.8 |
| 32×32×32 | 1012 | 275 | 3.7 |
| 64×64×64 | 29424 | 2028 | 14.5 |
| 128×128×128 | 316020 | 17302 | 18.2 |

The experiments have been done on grids of different sizes. Table 1 lists the comparison of the performances of the same algorithm on GPU and CPU on the same platform. To achieve reliable test results, on both CPU and GPU implementation, all steps are included and the same number of iterations is executed to solve diffusion terms. Table 1 shows that with the grid of dimensions 16×16×16, the performance of CPU is better than GPU in terms of fluid flow computations. This is because of the high proportion of messages passed between CPU and GPU in fragment programs. However, with the increase in grid size, the proportion of messages passed between CPU and GPU in fragment programs decreases. Supporting this idea, with the increase in grid size, performance on GPU exceeds performance on CPU about 18 times in table 1. Since the texture size is limited with GPU video memory, the implementation could not be tested on grids with larger sizes. Consequently, the experiments on CPU and GPU prove higher performance of GPU on physically based calculations such as smoke simulations.

## 5. CONCLUSION AND FUTURE WORKS

In this paper, simulation of smoke in three dimensional environments with complex obstacles is performed on both CPU and GPU. For the physical simulation of smoke behavior, NSEs are solved using a semi-Lagrangian unconditionally stable method. For the implementation at the vertex and fragment level on GPU, we prefer Cg, which is a platform-independent and architecture neutral shading language.

The performance of smoke simulation on CPU and GPU is compared. We have shown that although CPU is better than GPU in small grid resolutions (16×16×16) due to high proportion of message passing between CPU and GPU, GPU simulates smoke faster for bigger grids; for a 128×128×128 grid, up to 18 times speed-up is achieved on GPU.

We are currently working on accelerating smoke flow further on GPU. This may be achieved by decreasing the number of passes further and optimizing the GPU instructions. We would like to further extend our work to simulate other fluids such as liquids and fire. Moreover, in the near future we would like to implement the effects of dynamic obstacles to the fluid flow on GPU.

## REFERENCES

1. N. FOSTER, D. METAXAS, Realistic animation of liquids., Graphical Models and Image Processing 58 (5) (1996) 471–483.
2. M. MULLER, D. CHARYPAR, M. GROSS, Particle-based fluid simulation for interactive applications., Proc. SIGGRAPH symposium on Computer animation (2003) 154–159.
3. J. STAM, Stable fluids., Proc. SIGGRAPH '99 (1999) 121–128.
4. J. STAM, E. FIUME, Turbulent wind fields for gaseous phenomena., Proc. SIGGRAPH '93 (1993) 369–376.
5. J. STAM, E. FIUME, Depiction of fire and other gaseous phenomena using diffusion processes., Proc. SIGGRAPH '95 (1995) 129–136.
6. T. REEVES, Particle systems - a technique for modeling a class of fuzzy objects., Proc. SIGGRAPH 83.
7. S. CLAVET, P. BEAUDOIN, P. POULIN, Particle-based viscoelastic fluid simulation., Symposium on Computer Animation (2005) 219–228.
8. Y. LIU, X. LIU, E. WU, Real-time 3d fluid simulation on gpu with complex obstacles., Proc. Pacific Graphics (2004) 247–256.
9. E. WU, Y. LIU, X. LIU, An improved study of real-time fluid simulation on gpu., Journal of Visualization and Computer Animation (2004) 139–146.
10. R. FEDKIW, H. V. JENSEN, J. STAM, Visual simulation of smoke., Proc. SIGGRAPH '01 (2001) 15–21.
11. N. FOSTER, D. METAXAS, Modeling the motion of a hot turbulent gas., Proc. SIGGRAPH '97 (1997) 181–188.
12. J. BOLZ, I. FARMER, E. GRINSPUN, P. SCHRDER, Sparse matrix solvers on the gpu: Conjugate gradients and multigrid., ACM Transactions on Graphics 22 (3) (2003) 917–924.
13. M. J. HARRIS, W. BAXTER, T. SCHEUERMANN, A. LASTRA, Simulation of cloud dynamics on graphics hardware., Proc. ACM SIGGRAPH/Eurographics Workshop on Graphics Hardware (2003) 12–20.
14. NVIDIA, Gpu programming. `http://developer.nvidia.com` (2005).
15. W. LI, Z. FAN, X. WEI, A. KAUFMAN, Gpu-based flow simulation with complex boundaries., TR-031105, Computer Science Department, SUNY at Stony Brook.

# An Approach for Parallel CFD Solutions of Store Separation Problems

**E. Oktay[a], O. Merttopcuoglu[b], and H.U. Akay[c]**

[a]EDA- Engineering Design and Analysis Ltd. Co.
Ankara, Turkey

[b]ROKETSAN -Missile Industries Inc.
Ankara, Turkey

[c]Department of Mechanical Engineering
Indiana University-Purdue University Indianapolis
Indianapolis, Indiana 46202 USA

**ABSTRACT**

A new fast and accurate parallel algorithm is developed for solution of moving body problems, with specific reference to store separation problems. The algorithm starts with the development of separate meshes for the moving body (store) and the aircraft wing, which are then connected by using mesh blanking and mesh filling algorithms automatically. Following the partitioning of the connected meshes for parallel computing and obtaining a steady state flow solution, the separation starts by using a dynamically deforming mesh algorithm coupled with the six-degree of freedom rigid body dynamics equations for the store. The solutions continue until severe mesh distortions are reached after which automatic remeshing and partitioning are done on a new mesh obtained by blanking and filling operations to continue with the solutions. As the store reaches far enough distances from the aircraft, the algorithm switches to a relative coordinates eliminating any need for mesh deformations and remeshing. The developed algorithms and the results are discussed with a sample problem, including the parallel efficiency on distributed computers.

*Keywords: Store Separation, Moving Boundary, Parallel CFD, Moving Grid.*

**INTRODUCTION**

Moving body flow problems such as the simulation of the separation of an external store from underneath of an aircraft wing to compute its influence on the performance

I.H. Tuncer et al., *Parallel Computational Fluid Dynamics 2007*,
DOI: 10.1007/978-3-540-92744-0_49, 

of the aircraft are inherently difficult to solve due to excessive computations and memory needed as well as excessive deformations of the meshes. The problem is further complicated for parallel computing when the meshes have to be moved during the computations spoiling the connectivity and load balancing of the solution blocks. Yet, these calculations are important since aerodynamicists need to predict the movements of stores before flight tests, to avoid any disasters emanating from faulty designs. Store separation problem is essentially a simulation problem of multi bodies in relative motion and the determination of forces and moments acting upon the moving body because of the airflow and its consequent displacements. The modeling of this problem requires the interactive solution of unsteady flow and rigid-body equations together.

In this paper, a new fast and accurate method to solve an external store separation problem in parallel computing clusters is presented. First, the moving-boundary problem is solved using a deforming mesh algorithm and ALE formulation [1-2] on pre-partitioned blocks starting from steady state solution, whose input comes from the dynamics of the rigid-body, and the resultant unsteady fluid flow around the body is determined. Then, using the resultant force and moment terms in the rigid-body equations, the motion and the trajectory of the body are calculated. The solutions are carried out in this manner, as long as the solution mesh is not distorted. Whenever the mesh distortion reaches a specified value which the solver cannot handle, the solution is proceeds by generating a new local mesh automatically and the solutions on the new mesh are calculated. When the moving the body reaches a position sufficiently far from the effect of the fixed body, our parallel solver (FAPeda) [3] automatically switches itself to the relative coordinate solution [4-5]. In this way, there is no need for either deforming mesh or blanking and remeshing. The solution strategy and the results are discussed with a sample problem, illustrating the accuracy of the algorithms as well as the parallel efficiency on distributed computers.

## GENERAL APPROACH

For the method developed, meshes for the moving and fixed bodies are generated separately, after which they are overlapped to form an overlapped mesh as in the overset or Chimera mesh method [6-7]. However, instead of interpolating between the two meshes at the overlapped regions, our method further blanks the overlaps and connects the two meshes. One of the major contributions of this study has been to develop a very fast automatic mesh generation in small empty regions occurring after blanking of a part of the mesh occupied by the moving body and transferring the previous solution space on to the newly created mesh accurately using a dual time approach.

The method has no interpolation requirement, which is usually a source of inaccuracies in both space and time integration of the conventional Chimera mesh method. It also has no difficulties in the case of multi body overlappings as in the Chimera method. Blanking process requirement is less than that of the Chimera method because of the deforming mesh algorithm. Since the deforming mesh algorithm handles small displacements, it does not require blanking and consequently mesh generation. With these features, this algorithm is very appropriate for moving boundary solutions.

Furthermore, the code can automatically switch itself to a relative coordinate solution methodology when the moving body is sufficiently far from the effects of the fixed body. In this case, neither mesh deformations nor blanking and meshing are needed. Therefore, the solution becomes very fast. This part of the solver is useful for tracking the store rather than the separation problem.

Simulation of the flow is realized in terms of the unsteady solution of Euler equations, while a domain partitioning parallel solution strategy is used. The Euler solver is based on a cell-centered finite volume scheme [8] and uses backward-Euler implicit integration method [1-3 and 9]. The solver uses an unstructured mesh, which is continuously deforming according to a spring-analogy scheme as the distance from the wing to the load increases.

## GOVERNING EQUATIONS

Euler equations for an unsteady compressible fluid flow in an Arbitrary Lagrangian Eulerian (ALE) system can be expressed in integral form as follows [1]:

$$\frac{\partial}{\partial t}\iiint_{\Omega} \boldsymbol{Q}\,dV + \iint_{\partial\Omega} \boldsymbol{F}\cdot\boldsymbol{n}\,dS = 0 \qquad (1)$$

where, the velocity of the fluid is expressed as relative to the local velocity of the mesh and

$$\boldsymbol{Q} = [\rho, \rho u, \rho v, \rho w, e]^{\mathrm{T}}, \qquad \boldsymbol{F}\cdot\boldsymbol{n} = [(\boldsymbol{V}-\boldsymbol{W})\cdot\boldsymbol{n}]\begin{bmatrix}\rho\\ \rho u\\ \rho v\\ \rho w\\ e+p\end{bmatrix} + p\begin{bmatrix}0\\ n_x\\ n_y\\ n_z\\ W_n\end{bmatrix} \qquad (2)$$

Also, in the above $\mathbf{n} = n_x\boldsymbol{i} + n_y\boldsymbol{j} + n_z\boldsymbol{k}$ is the unit vector normal to the surface, $V = u\boldsymbol{j} + v\boldsymbol{j} + w\boldsymbol{k}$ fluid velocity vector, $\boldsymbol{W} = \boldsymbol{i}\,\partial x/\partial t + \boldsymbol{j}\,\partial y/\partial t + \boldsymbol{k}\,\partial z/\partial t$ is the grid velocity vector, and $W_n = W\cdot n = n_x\,\partial x/\partial t + n_y\partial y/\partial t + n_z\,\partial z/\partial t$ is the component of grid velocity normal to the surface, $p$ is pressure, and $e$ is the total energy. From the gas law, the pressure is related to the total energy as follows:

$$p = (\gamma - 1)\left[ e - \frac{1}{2}\rho(u^2 + v^2 + w^2) \right] \tag{3}$$

For the movement of the moving body (store), the dynamic equations of the rigid-body are express as:

$$m(\frac{d\boldsymbol{V}_c}{dt} + \boldsymbol{\omega}_c \times \boldsymbol{V}_c) = m\boldsymbol{g} + \boldsymbol{F}_c \tag{4}$$

$$I\frac{d\boldsymbol{V}_c}{dt} + \boldsymbol{\omega}_c \times (I\boldsymbol{V}_c) = \boldsymbol{M}_c \tag{5}$$

Here, $\boldsymbol{V}_c$ and $\boldsymbol{\omega}_c$ are velocity and angular velocity vectors of the moving body, respectively. $\boldsymbol{F}_c$ and $\boldsymbol{M}_c$ are force and moment vectors affecting on the body and calculated by following surface integrals, respectively:

$$\boldsymbol{F}_c = \iint_S p\boldsymbol{n}dS \tag{6}$$

$$\boldsymbol{M}_c = \iint_S p\boldsymbol{r} \times \boldsymbol{n}dS \tag{7}$$

where $p$ is pressure distribution on the moving body surface $S$.

## DEFORMING MESH ALGORITHM

The deforming mesh algorithm models the mesh movements of interior points to conform to the known movements of the boundaries. The algorithm used in this study has been previously developed by Batina [10]. In this algorithm, the computational mesh is moved to conform to the final position of the boundary. The algorithm treats the computational mesh as a system of interconnected springs at every mesh point constructed by representing each edge of every cell by a linear spring. The spring stiffness for a given edge is taken as inversely proportional to the length of the edge. The mesh points on the outer boundary of the domain are held fixed while the final location of the points on the boundary layer edge (i.e., moved Euler grid points on the body) is given by the body motion. The static equilibrium equations in the $x$, $y$, and $z$ directions that result from a summation of forces are solved iteratively using Jacobi iterations at each interior node $i$ of the mesh for the displacements $\Delta x_i$, $\Delta y_i$ and $\Delta z_i$, respectively [1].

## MESH BLANKING AND MESH FILLING OPERATIONS

The minor unstructured grid covering the moving body first overlaps with the major unstructured mesh including the other fixed body. To create new mesh after the distortion caused by excessive mesh deformations, the overlapped portion of the major mesh with the minor mesh is blanked. It is obvious that, there is an unavoidable requirement of an efficient and reliable search algorithm if the significant number of elements of such a three-dimensional mesh is considered for a time dependent solution. For this purpose, the neighbor-to-neighbor jump search algorithm [7] is efficiently utilized in the present method. The procedure is schematically shown in Figure 1 for a triangular grid. The only insufficiency of this efficient search method is that it may fail for convex domains depending on the starting point. This problem has been overcome by adding subsidiary mesh points into the moving body and outside of the computational domain [7].

The small empty region occurring after blanking of a part of the mesh occupied by the moving and fixed bodies is filled by tetrahedral cells based on Delaunay triangulation [11]. In this method, the local element lengths are automatically set using boundary triangles and dimension of the hole. The mesh blanking and filling algorithms are depicted in the schematic shown in Figure 2. The solution domain on the blanked region is first predicted from the previous solution using the quadratic Shepard method for trivariate interpolations [12] and then corrected with respect to external flux distribution using the Lagrange multiplier method with the dual time stepping approach. This way, the flux conserving in the remeshed region is computed without any interpolations.

## PARALLELIZATION OF THE CODE

For parallelization of the code, a domain-decomposition approach is used, where the flow domain is subdivided into a number of subdomains equal or more than the number of available processors. The interfaces serve to exchange the data between the blocks. The governing equations for flow or mesh movements are solved in a block solver, which updates its interface solver with newly calculated nodal variables at each time step. The interface solver in each block, in turn, communicates with the interface solver of the neighboring block for the same interface. Each interface solver also updates its block after receiving information from its neighbor. In each block, the nodes on the interfaces are flagged as either receiving or sending nodes. The ones on the interior side of the blocks are sending nodes; the ones on the exterior side are receiving nodes. The sending nodes send their data to the corresponding nodes in the neighboring blocks; the receiving nodes receive the data from the corresponding nodes in the neighboring block. The communication between the blocks is achieved by means of the message-passing library MPI. The computational domain that is discretized with an unstructured mesh is partitioned into subdomains or blocks using a program named General Divider, which is a general unstructured mesh-partitioning code [13]. It is capable of partitioning both structured (hexahedral) and unstructured (tetrahedral) meshes in three-dimensional geometries. The interfaces between partitioned blocks are of matching and overlapping type. The interfaces serve to exchange data among the blocks. The partitioning is done following the merging of moving body and fixed-body meshes by blanking and remeshing operations yielding a nearly balanced solution blocks. The same block structure is used during mesh deformations via ALE until severe distortions are reached at which stage a new merging operation is performed by repeating blanking and remeshing operations followed by partitioning.

## TEST CASE

For the validation purpose of the present method, the simulation results are compared with the wind-tunnel test data of AFRL-WPS model (Air Force Research Laboratory - Generic Wing, Pylon, and Moving Finned Store) which are given in Reference [14]. Figure 2 shows the mesh coupling stages around the WPS configuration and Figure 3 shows various store positions during separation and pressure distribution on the surfaces. The comparisons of the computed trajectory with the experimental data for the store's center of gravity are shown in Figure 4, in terms of the linear displacements and Euler angles, showing good agreements between the solution results and the test data. Parallel efficiency of the code using up to 6 Intel P4 dual core processors, with 3.2 GHz clock speed and 2 GB RAM memory are shown in Figure 5, showing a reasonably good speedup for even a mesh of 400,000 finite volume cells. While the problem with a single processor took 1731 seconds of elapsed time for 100 number of time steps the 12 processor solution took only 219 seconds, which translates into 65% efficiency. The relatively small size of the original mesh and the unbalanced solution block sizes reached due to the changes in the number of cells in the blocks that undergo mesh blankings and fillings, contribute to the less than ideal parallel efficiencies in this case. It is expected that the efficiency of the scheme will be best realized for larger size problems and viscous turbulent flows.

## CONCLUSIONS

The method developed here for parallel computing of store separation problems yields accurate and fast solutions. The combined blanking, filling, and partitioning algorithm provides significant advantages over the classical overset algorithm, since no interpolations are needed between moving and fixed meshes in this case. Reasonable parallel efficiencies are obtained even for relatively small size meshes. The work is in progress in extending the method to viscous flow calculations. The advantages experienced here for Euler equations are expected to extend to viscous calculations too.

## 9. REFERENCES

1. Oktay, E., and Akay, H.U. and Uzun, A., "A Parallelized 3D Unstructured Euler Solver for Unsteady Aerodynamics," *Journal of Aircraft*, Vol. 40, No. 2, pp. 348-354, 2003.
2. Akay, H.U., Oktay, E., Li, Z. and He, X., "Parallel Computing for Aeroelasticity Problems," AIAA Paper: 2003-3511, 33rd *AIAA Fluid Dynamics Conference*, June 20-23, 2003, Orlando, FL.
3. Akay, H.U., Baddi A., Oktay, E., "Large-Scale Parallel Computations of Solid-Fluid Interaction Problems for Aeroelastic Flutter Predictions," AIAC-2005-

002, *Ankara International Aerospace Conference*, August 22-25, 2005, Ankara, Turkey.

4. Kandil, O.A. and Chuang, H.A., "Computation of Steady and Unsteady Vortex-Dominated Flows with Shock Waves," *AIAA Journal*, Vol. 26, pp. 524-531, 1998.
5. Hi, X., "Parallel Computations of Solid-Fluid Interactions Using Arbitrary Lagrangian-Eulerian and Relative Coordinate Formulations," *Master's Thesis*, Purdue University, May 2004.
6. Benek, J.A, Buning, P.G., and Steger, J.L., "A 3-D Chimera Grid Embedding Technique," *AIAA Paper* 85-1523, June 1985.
7. Nakahashi, K. and Togashi, F., "Intergrid-Boundary Definition Method for Overset Unstructured Grid Approach," *AIAA Journal*, Vol. 38, pp. 2077-2084, 2000.
8. Frink, N.T., Parikh, P., and Pirzadeh, S., "A Fast Upwind Solver of the Euler Equations on Three-Dimensional Unstructured Meshes," *AIAA Paper*, 91-0102, 1991.
9. Anderson, W.K., "Grid Generation and Flow Solution Method for Euler Equations on Unstructured Grids," Journal of Computational Physics, Vol. 110, pp. 23-38, 1994.
10. Batina, J.T., "Unsteady Euler Algorithm with Unstructured Dynamic Mesh for Complex Aircraft Aerodynamic Analysis," *AIAA Journal*, Vol. 29, No.3, pp. 327-333, 1991.
11. Karamete, K., "A General Unstructured Mesh Generation Algorithm with its Use in CFD Applications," *Ph.D. Thesis*, METU, Dec. 1996.
12. Renka, R.J., "Quadratic Shepard Method for Trivariate Interpolation of Scattered Data," *Transactions on Mathematical Software*, Vol. 14, No. 2, p. 151, 1988.
13. Bronnenberg, C.E., "GD: A General Divider User's Manual - An Unstructured Grid Partitioning Program," *CFD Laboratory Report*, IUPUI, 1999.
14. Fox, J.H., "Chapter 23: Generic Wing, Pylon, and Moving Finned Store," In "Verification and Validation Data for Computational Unsteady Aerodynamics," *RTO Technical Report*, RTO-TR-26, 2000.

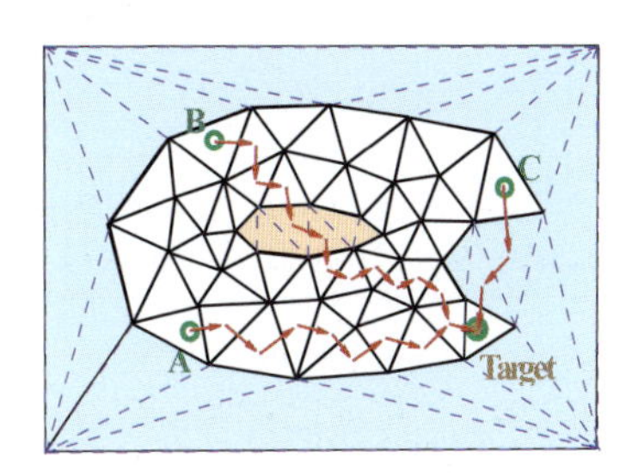


Figure 1: Neighbor-to-neighbor search

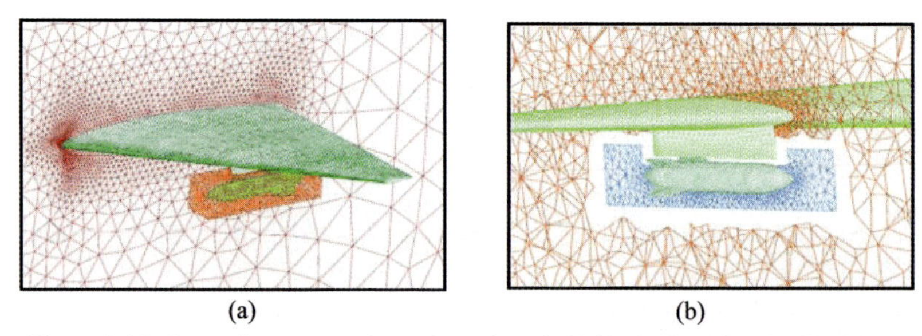


Figure 2: Mesh coupling stages: a) overlapped mesh, b) blanked mesh to be filled later

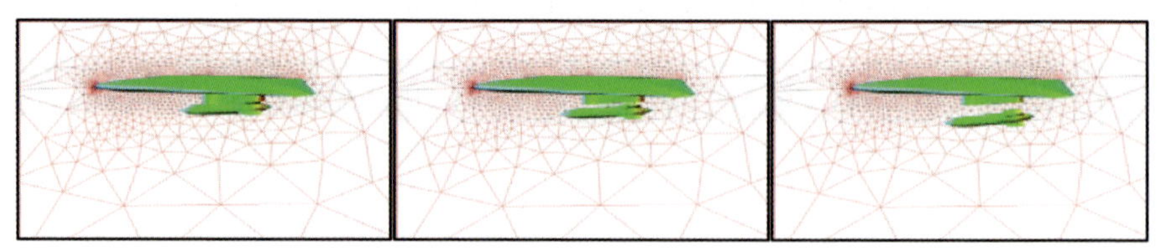

Figure 3: Computed store positions and pressure distribution

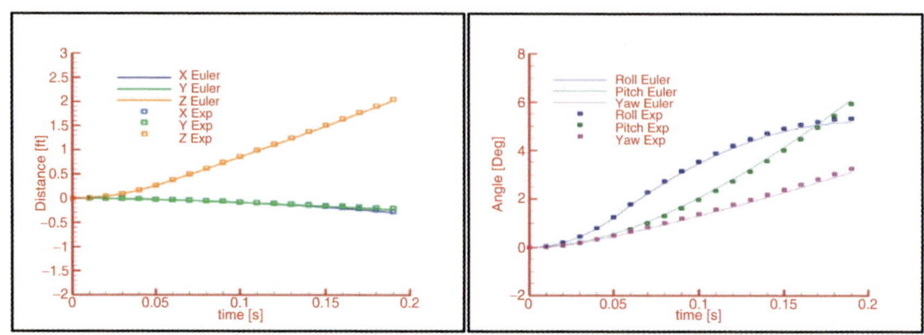


Figure 4: Computed and measured trajectory of the store

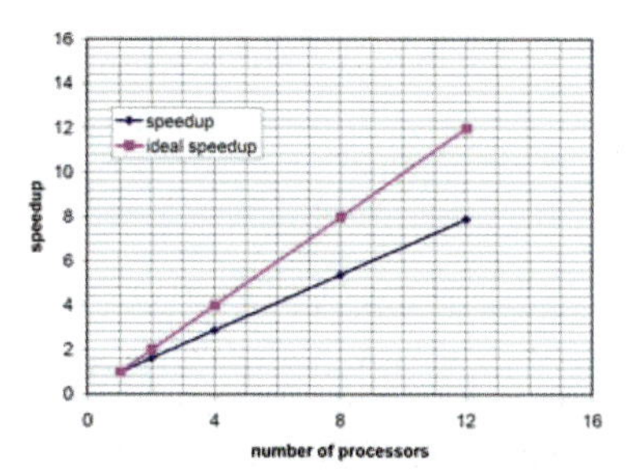


Figure 5: Parallel efficiency of the code

# Hybrid Parallelism for CFD Simulations: Combining MPI with OpenMP

**E. Yilmaz, R.U. Payli, H.U. Akay, and A. Ecer**

Computational Fluid Dynamics Laboratory
Department of Mechanical Engineering
Indiana University-Purdue University Indianapolis (IUPUI)
Indianapolis, Indiana 46202 USA
*http://www.engr.iupui.edu/cfdlab*

**ABSTRACT**

In this paper, performance of *hybrid* programming approach using MPI and OpenMP for a parallel CFD solver was studied in a single cluster of multi-core parallel system. Timing cost for computation and communication was compared for different scenarios. Tuning up the MPI based parallelizable sections of the solver with OpenMP functions and libraries were done. BigRed parallel system of Indiana University was used for parallel runs for 8, 16, 32, and 64 compute nodes with 4 processors (cores) per node. Four threads were used within the node, one for each core. It was observed that MPI performed better than the *hybrid* with OpenMP in overall elapsed time. However, the *hybrid* approach showed improved communication time for some cases. In terms of parallel speedup and efficiency, *hybrid* results were close to MPI, though they were higher for processor numbers less than 32. In general, MPI outperforms *hybrid* for our applications on this particular computing platform.

Keywords: Parallel CFD, Hybrid Parallelism, OpenMP, MPI.

## 1. INTRODUCTION

Parallel computing on shared memory systems using OpenMP (www.openmp.org) and distributed memory systems using MPI (www.lam-mpi.org, www-unix.mcs.anl.gov/mpi) are not new. However, introduction of multi-core systems and Grid computing paradigm opened a new direction to combine shared memory parallel systems and distributed memory systems for the high performance computing applications. This approach, often called *Hybrid Parallelism*, exploits parallelism beyond the single level by using MPI in the coarse-grain parallelism level and OpenMP in the fine-grain level. While MPI performs parallelism between the compute nodes (or

I.H. Tuncer et al., *Parallel Computational Fluid Dynamics 2007*,
DOI: 10.1007/978-3-540-92744-0_50, 

clusters in the Grid computing paradigm), OpenMP takes the job for the loop level parallelism. Jost, et. al. [1] have shown that 'the benefit of the *hybrid* implementations is visible on a slow network' since 'slow networks lead to a decrease in the performance of the pure MPI implementation'. While some application have very well structured loop systems suitable for data decomposition among cores of the computing nodes, other applications may use complex computational loops which increase dependence, hence significantly reduce applicability of OpenMP implementation. Eventually, the type of the applications dictates kind of *hybrid* parallelization approach. Usually, MPI processes are multi-threaded using OpenMP work-sharing directives within each MPI process. However, it is not always straightforward to get maximum performance out of the mixed environments as in pure MPI or OpenMP. Unfortunately, there are several 'mismatch problems between the *hybrid* programming schemes and the *hybrid* hardware architectures' as Rabenseifner [2] has demonstrated that 'different programming schemes on clusters of Symmetric Multiprocessors (SMP) show different performance benefits or penalties'. Chow and Hysom [3] demonstrated that 'the hybrid programs can perform better for small size problems' and related this fact with 'the lower latency when network interface is not shared' for IBM SP clusters. Hence, the problem size can play an important role in overall efficiency as well. Main motivation for the *hybrid* parallelism lies in the fact that the intra-node communication is faster than the inter-node communication; on the other hand, intra-node can have lower bandwidth than inter-node.

## 2. MULTICORE ARCHITECTURE

To better comprehend performance and efficiency issues in the multi-core computing, we have to go beyond the programming aspect and explore the multi-core hardware architecture first. Processors of the parallel systems have been identified by two types in terms of access to memory and cache: 1) Non-Uniform Memory Access (NUMA) is a configuration where a processor can have direct access to the memory, 2) in SMPs, several processors are allowed to access to a shared memory. While the NUMA architecture was designed to surpass the scalability limits of SMP architecture, SMP works fine for a relatively small number of CPUs. High number of CPUs compete for access to the shared memory bus, causing bottlenecks. Figure 1 shows basics of how memory and cache are related with each other under those two configurations. At least this aspect of the hardware would have significance to the application programmers. In NUMA, transfer message size to the neighbour processors should be watched, because access to the neighbour memory will be through the bus. The processors in one node in SMP have direct access to the local memory while SMP needs additional synchronisation to avoid limitations on bandwidth as all processors access to shared memory. To get the benefit of both, hardware vendors offer mixed configurations as well.

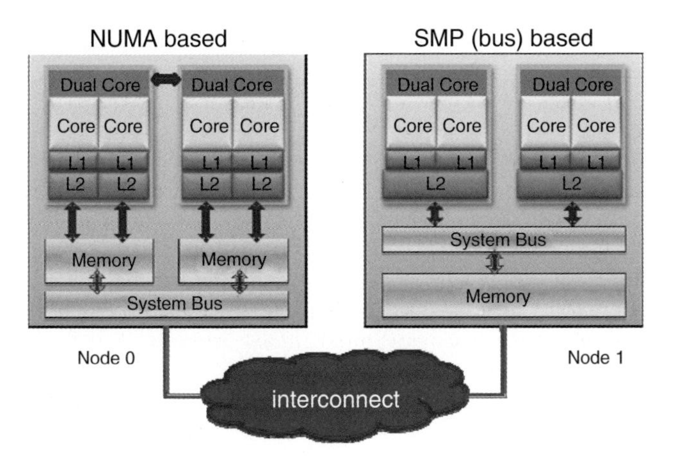


FIGURE 1: CHIPSET CONFIGUARTION FOR PARALLEL SYSTEMS

## 3. HYBRID PROGRAMMING

The most practical way of implementing *Hybrid* programming is to start from an existing MPI implementation of the CFD application. Basically, each partitions of mesh and data will be multi-threaded by using OpenMP compiler directives. Unlike the distributed data implementation in MPI, *in OpenMP,* computation loops are partitioned to create threads. Then, these threads run eventually on the processor cores of each compute node. Unfortunately, designing an efficient OpenMP coding is not an easy task. Loops and segments of the code should be carefully analyzed for the dependency of the shared vectors between the cores. Sometimes restructuring of the code and computation loops might be necessary to get the maximum benefit. Loop partitionings are done using different scheduling; static, dynamic, runtime, or guided. Simple partitioning is to divide equally between the cores or processors (static). At the end of the loop or program segment, synchronization might be necessary to advance the solution and pass the data to the other thread or processors. In that case, *Barriers* are set to ensure that the data are not lost or overwritten. To avoid duplicate computations, only the parallel section of the code should be used by OpenMP directives. Therefore, parallel section is initialized before it is used. In some cases, excessive parallel section initialization will increase the overhead. A representative *hybrid* coding for our application is given below:

```
Program Pacer3D
        call mpi_init ( ierr )                                  Initialize MPI
        call mpi_comm_size ( mcworld , ntasks   , ierr )
        call mpi_comm_rank ( mcworld , myid   , ierr )
.......................
        call SET_OMP_NUM_THREADS(4)                             Set number of threads to 4
!$OMP PARALLEL                                                  OpenMP parallel initialized
        nthreads = OMP_GET_NUM_THREADS()                        Get thread information
        idthreads = OMP_GET_THREAD_NUM()
!$OMP END PARALLEL
.......................                                         MPI block-to-block
        call mpi_sendrecv (...............)                     communication
................
!$OMP PARALLEL                                                  OpenMP parallel initialized.
!$OMP& PRIVATE (ip,fact) SHARED                                 OpenMP Loop
(npoin,help1,help2)                                             parallelization:
! $OMP DO SCHEDULE(DYNAMIC)                                     This section of the program
        do ip = 1 , npoin                                       is divided into 4 threads by
                fact        = 1.0 / help2(ip,3)                 schedule (dynamic) option
                help1 (ip,1) = 36.0*abs(help1(ip,1)/ &          and will be run on 4 cores.
                                    help2(ip,2))                Local variables are defined
                help1 (ip,2) = fact * help1 (ip,2)              by private, and global
        enddo!                                                  vectors/variables are defined
!$OMP ENDDO                                                     by shared. Barrier is used to
!$OMP END PARALLEL                                              synchronize with other
        frequency1 = help (1,1)                                 threads at the end of the
        frequency2 = help (1,2)                                 parallel section.
!$OMP BARRIER
.......................
        call mpi_finalize ( ierr )                              Finalize MPI
        END
```

## 4. RESULTS

### 4.1. CFD Application

Our CFD code [4] in this study is based on the finite volume discretization of the three dimensional Euler flow equations. It uses explicit time integration and unstructured tetrahedral grids. Parallelization is based on message passing interface. The test case is an external transonic flow over Onera M6 wing with 3.7M tetrahedral cell elements.

### 4.2. Computing Platform

Computations have been performed on BigRed computing platform which has the following features:

**Compute Node**
768 JS21 Bladeserver nodes
PowerPC 970MP processors
2 x 2.5GHz dual-core
8GB 533MHz DDR2 SDRAM
1 x Myricom M3S-PCIXD-2-I (Lanai XP)
(plus 4 **User Nodes** with same configuration)

**Storage Nodes**
Power5+ processor
2 x 1.65GHz dual-core
8 GB 533MHz DDR2 SDRAM
2 x Emulex LP10000 PCI-X/133MHz FC adapters

### 4.3. Evaluating OpenMP Options

We have tested our code for different options of the *Schedule* directive for the loops as well as locations to initialize the parallel section, see Table 1. When used for an individual loop default value of *Schedule,* which is *static,* is overwritten. Time measurements show that the use of *Barrier* has almost no influence in total elapsed time unlike our initial expectations. However, the use of *Barrier* at certain section of the CFD code is necessary to ensure that different parallel units of the code advance with the same values, hence avoiding overwrite of the data. Also, the use of combined parallel section initialization introduces only negligible increase in timing. Scheduling options might have significant effect on the timing. We observed that using *dynamics* scheduling gives worst timing. In scheduling with *dynamics* option, loops are divided into chucks where the size is specified by the users. Choosing smaller size would end up with more number of threads than cores. This will add more overhead to the timing.

Table 1: OpenMP performance for various options

| Operations | Time (sec) / step |
|---|---|
| Combined parallel sections | 0.58 |
| Individual parallel sections | 0.57 |
| Use of Barrier between parallel sections | 0.57 |
| No Barrier | 0.57 |
| Use of *Schedule* for Do loops | |
| Schedule, dynamics | 1.40 |
| Schedule, static (loop size / # of threads for each core) | 0.57 |
| Schedule, guided (loop size / # of threads for each core) | 0.59 |
| Schedule, runtime | 0.57 |

### 4.4. Parallel Performance

To test parallel performance, we run our application code up to 256 blocks. Maximum 256 processors were used in this test. In the *hybrid* case, four threads are generated for each node. For MPI case, we run it over 256 compute nodes not to account for intra-node communications into effect. Figure 2 shows parallel efficiency for both cases. The *Hybrid* has demonstrated better efficiency for less number of nodes, however after

32 nodes, the *MPI* performs better. The same trend is visible in speedup comparison as shown in Figure 3. Both cases have demonstrated super linear speedup, which is typical of this particular application (Payli, et al. [5]). Timing measurements were done using *cpu_time* function of the Intel Fortran for CPU time of the blocks and *mpi_wtime* was used for communication time measurements for the communication part only and overall total elapsed time measurements. Table 2 summarizes our timing measurements for both cases for 32 to 256 processors. Except for the 128 processor case, the *hybrid* measures less communication time compared to MPI. Note that the number of blocks for the pure MPI case is identical with the number of processors while in the *Hybrid* case, it is the number of cores times the number of nodes.

We also observed that in the *hybrid* implementation with OpenMP, multi-threading smoothens out the waiting time fluctuations in MPI block-to-block communications by looking at standard deviations of the communication time of the blocks. This is due to the less number of the block-to-block communications between the compute nodes.

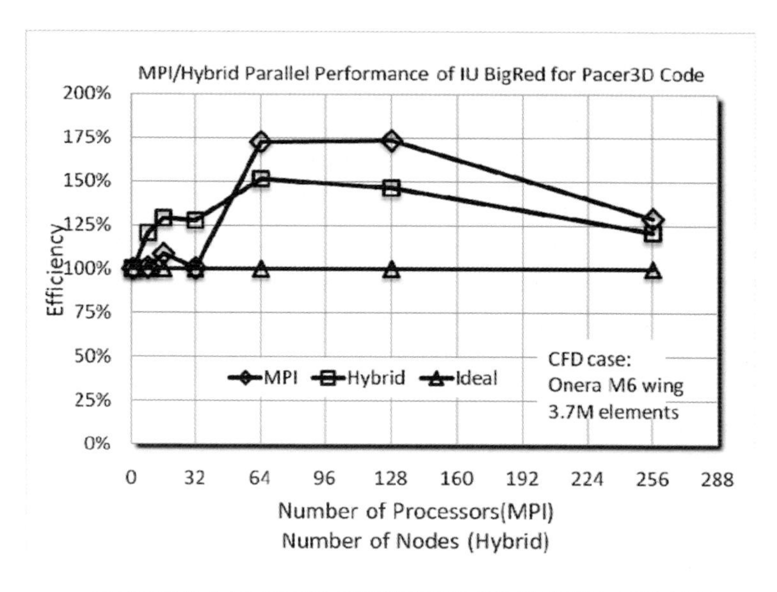


FIGURE 2: PARALLEL EFFCIENCY OF HYBRID AND MPI

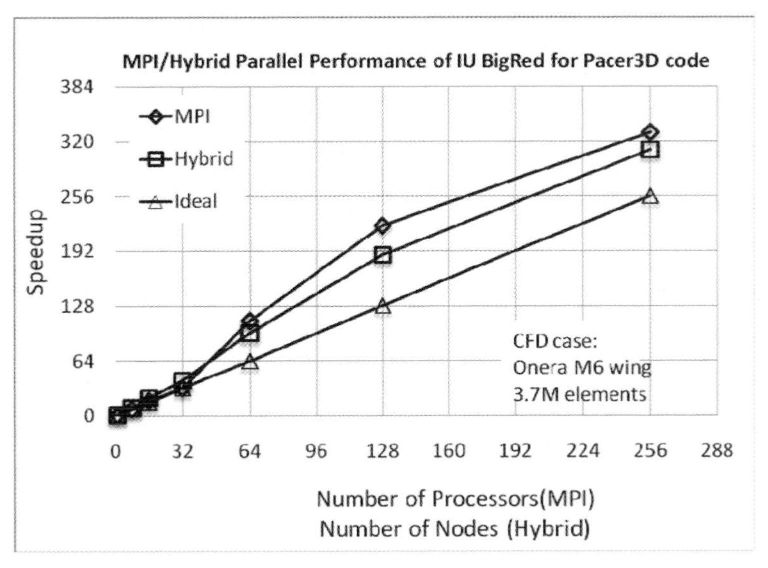


FIGURE 3: SPEEDUP FOR HYBRID AND MPI

Table 2: Timing comparison for Hybrid versus MPI in seconds

| MPI | | | | Hybrid | | | |
|---|---|---|---|---|---|---|---|
| # of procs | Elapsed Time | CPU Time | Comm Time | # of node | Elapsed Time | CPU Time | Comm Time |
| 32 | 1.292 | 1.291 | 0.376 | 8 | 1.555 | 1.041 | 0.130 |
| 64 | 0.377 | 0.376 | 0.073 | 16 | 0.724 | 0.497 | 0.062 |
| 128 | 0.188 | 0.186 | 0.043 | 32 | 0.367 | 0.237 | 0.081 |
| 256 | 0.126 | 0.124 | 0.051 | 64 | 0.154 | 0.111 | 0.030 |

## 5. CONCLUSIONS

Our study using one type of multi-core hardware and mixing OpenMP with MPI for a particular application has shown that pure MPI performs better than the hybrid implementation considering overall elapsed timing. However, we have also observed that the OpenMP improves the communication time compared to MPI for some cases. In addition, parallel performance for hybrid is higher for less number partitions (8, 16, 32), however the pure MPI takes over the advantage after 32 blocks. Finally, the hybrid implementation shows less fluctuations in waiting time compared to MPI block-to-block communications, which is another indication that why we have better performance for the less number of partitions. More work will identify possible improvement areas as

Multicore dominates the parallel computing in the near future. Next research related with the hybrid parallel computing would involve using different computing platforms as well as MPI/OpenMP tune-up for communication-computation overlap strategies.

## 6. ACKNOWLEDGMENTS

This research was supported in part by the National Science Foundation under Grants No. ACI-0338618l, OCI-0451237, OCI-0535258, and OCI-0504075 and was supported in part by the Indiana METACyt Initiative. The Indiana METACyt Initiative of Indiana University is supported in part by Lilly Endowment, Inc. and by the Shared University Research grants from IBM, Inc. to Indiana University.

## 7. REFERENCES

1. Jost, G., Jin, H. Mey, D., and Hatay, F.F, "Comparing the OpenMP, MPI, and Hybrid Programming Paradigms on an SMP Cluster," NAS Technical Report NAS-03-019, and the Fifth European Workshop on OpenMP (EWOMP03), Aachen, Germany, 2003.
2. Rabenseifner, R., "Hybrid Parallel Programming: Performance Problems and Chances," 45th CUG Conference (www.cug.org) 2003, Columbus, Ohio, USA, 2003.
3. E. Chow and D. Hysom, "Assessing Performance of Hybrid. mpi/openmp Programs on SMP Clusters," Technical Report, UCRL-JC-143957, Lawrence Livermore National laboratory, Livermore, CA, 2001.
4. Yilmaz, E., Kavsaoglu, M.S., Akay, H.U., and Akmandor, I.S., "Cell-vertex Based Parallel and Adaptive Explicit 3D Flow Solution on Unstructured Grids," International Journal of Computational Fluid Dynamics, Vol. 14, pp. 271-286, 2001.
5. Payli, R.U., Akay, H.U., Baddi, A.S., Yilmaz, E., Ecer, A., and Oktay, E., "Computational Fluid Dynamics Applications on TeraGrid," Parallel Computational Fluid Dynamics," Edited by A. Deane, et al., Elsevier Science B.V., pp. 141-148, 2006.

# Impact of the TeraGrid on Large-Scale Simulations and Visualizations

**Resat U. Payli, Erdal Yilmaz, Hasan U. Akay, and Akin Ecer**

Computational Fluid Dynamics Laboratory
Department of Mechanical Engineering
Indiana University-Purdue University Indianapolis (IUPUI)
Indianapolis, Indiana 46202 USA
*http://www.engr.iupui.edu/cfdlab*

**ABSTRACT**

TeraGrid is a National Science Foundation supported computing grid available for scientists and engineers in U.S. universities and government research laboratories. It is the world's largest and most comprehensive distributed cyberinfrastructure for open scientific research. With its high-performance data management, computational, and visualization resources, TeraGrid has opened up new opportunities for scientists and engineers for solving large-scale problems with relative ease. To assess the performance of different resources available on the TeraGrid, parallel performance of a flow solver was tested on Indiana University's IBM e1350 and San Diego Supercomputing Center's IBM BlueGene/L systems, which are two of the TeraGrid computational resources with differing architectures. The results of a large-scale problem are visualized on a parallel visualization toolkit, ParaView, available on the TeraGrid to test its usability for large scale-problems.

*Keywords: TeraGrid; Grid Computing; Parallel CFD; Parallel and Distributed Visualization.*

## 1. INTRODUCTION

Performing large-scale simulations with Computational Fluid Dynamics (CFD) codes and visualizing the obtained results are challenging tasks. These applications require a great deal of processing power, fast communication between the processors, and large storage systems. The high-performance data management, computational, and

I.H. Tuncer et al., *Parallel Computational Fluid Dynamics 2007*,
DOI: 10.1007/978-3-540-92744-0_51, 

visualization resources of TeraGrid [1] have made this kind of simulations and visualizations possible.

TeraGrid is an open scientific discovery infrastructure combining leadership class resources at nine partner sites to create an integrated, persistent computational resource. Using high-performance network connections, the TeraGrid integrates high-performance computers, data resources and tools, and high-end experimental facilities around the country. These integrated resources include more than 102 teraflops of computing capability and more than 15 petabytes (quadrillions of bytes) of online and archival data storage with rapid access and retrieval over high-performance networks. Through the TeraGrid, researchers can access over 100 discipline-specific databases. With this combination of the resources, the TeraGrid is the world's largest, most comprehensive distributed cyberinfrastructure for open scientific research.

TeraGrid is coordinated through the Grid Infrastructure Group (GIG) at the University of Chicago, working in partnership with the Resource Provider sites: Indiana University, Oak Ridge National Laboratory, National Center for Supercomputing Applications, Pittsburgh Supercomputing Center, Purdue University, San Diego Supercomputer Center, Texas Advanced Computing Center, University of Chicago/Argonne National Laboratory, and the National Center for Atmospheric Research.

We have demonstrated CFD applications on the TeraGrid previously [3-5]. In this paper, we will present performance of our parallel flow solver [2] on Indiana University's IBM e1350 and San Diego Supercomputing Center's IBM BlueGene/L systems which are two of the TeraGrid computational resources. We will also show how to visualize the results of these large-scale simulations with open source visualization toolkit ParaView's [6] parallel and distributed client/server mode by using some capabilities of the TeraGrid environment, such as Global Parallel File System Wide Area Network (GPFS-WAN).

## 2. SIMULATION RESOURCES

### 2.1. INDIANA UNIVERSITY'S IBM e1350 SYSTEM

Indiana University's e1350 system has 768 IBM JS21 compute nodes, each with two dual-core 2.5 GHz PowerPC 970 MP processors, 8 GB memory, and the nodes are interconnected with PCI-X Myrinet-2000. Also each processor core has 64 KB L1 Instruction cache, 32 KB Data cache, and 1 MB L2 cache.

### 2.2. SDSC'S IBM BlueGene/L SYSTEM

San Diego Supercomputing Center's IBM BlueGene/L system has 3072 compute nodes and 384 I/O nodes. Each node consist of two PowerPC processors that run at 700 MHz and share 512 MB of Memory, 3-D torus for point-to point message passing and a global tree for collective message passing. It also has 32KB, 32-byte line, 64-way L1 cache 16 128-byte lines L2 cache, which act as prefetch buffer, and 4 MB 35 cycles shared L3 cache.

Two compute modes are available on the BlueGene/L system:

1. In coprocessor (CO) mode only one processor per node performs computation while the other processor performs communication and I/O.
2. In virtual mode (VN) both processors in a compute node perform computation as well as communication and I/O.

The VN mode is usually faster than the CO mode. However in the VN mode the memory per processor is half that of the CO mode. Codes which require more memory runs only in the CO mode.

## 3. VISUALIZATION RESOURCE

UC/ANL's IA-32 TeraGrid Linux Visualization Cluster consists of 96 nodes with dual Intel Xeon processors, with 4 GB of memory and nVidia GeFORCE 6600GT AGP graphics card per node. The cluster is running SuSE Linux and is using Myricom's Myrinet cluster interconnect network.

## 4. TERAGRID GPFS-WAN STORAGE SYSTEM

TeraGrid Global Parallel File System-Wide Area Network (GPFS-WAN) is a 700 TB storage system which is physically located at SDSC, but is accessible from the TeraGrid platforms on which it is mounted such as SDSC's BlueGene/L and UC/ANL's Visualization Cluster.

## 5. SIMULATIONS

PACER3D [2] is a distributed memory parallel finite volume Euler flow solver on unstructured tetrahedral meshes. The flow solver uses cell-vertex based numerical discretization and has capability for adaptive remeshing. It employs artificial dissipation terms in flux calculations. Local time stepping is used for steady state flow solutions. To accelerate steady-state solutions, local time stepping, residual averaging technique, and enthalpy damping are used.

PACER3D was ported to IU's IBM e1350 and SDSC's IBM BlueGene/L systems and optimized for both systems without any special modifications. The virtual mode (VN) is used while simulations are carried on the BlueGene/L system in order to better utilize the hardware.

To show the scalability of the parallel solver on IU's IBM e1350 and SDSC's IBM BlueGene/L systems, flow around an aircraft configuration is considered. This geometry contains 18 million tetrahedral elements and more than 3 million grid points. The geometry is partitioned into 64, 128, 256, 512, 1024, and 2048 solution blocks.

Speedup charts based on wall clock time up to 2048 processors for IBM e1350 and IBM BlueGene/L are given in Figure 1. Also Table 1 shows the timing for each time steps of the PACER3D on both systems.

IBM e1350 parallel speedup results show that the speedup is identical with the ideal speedup up to 128 processors. After that, the speedup is super linear up to 1024 processors. For 2048 processors it is under the ideal speedup. The reason we are getting the super linear speed up is the small memory requirement of the solver. This small memory requirement increases the cache utilization efficiency which is likely getting higher cache hit rate up to 1024 processors. As seen in Table 1, for 2048 processors, the CPU and communication times are very close, that is why the speedup is below the ideal speed up. This indicates that the problem size in this case is not big enough to run on more than 1000 processors when IU's IBM e1350 system is used.

IBM BlueGene/L parallel speedup results show that the speedup is identical with ideal speedup up to 256 processors after that, it is slightly higher than the ideal speedup. Because of the internetwork configuration of the BlueGene/L and small memory requirement of the PACER3D we did get good scalability.

## 6. VISUALIZATION

ParaView is an open source visualization software which offers an advantage of running visualization processes in distributed and parallel fashion for processing large data sets. It runs parallel on distributed and shared memory systems using Message Passing Interface (MPI). ParaView uses the data parallel model. In this model, data are broken into the pieces to be processed by different processors which allow visualization of large data sets. This capability makes ParaView a suitable a visualization application for the TeraGrid environment.

The visualization data sets, which are the results of running the simulations on SDSC's BlueGene/L system, are stored on the GPFS-WAN storage system. The GPFS-WAN system can be accessible from the UC/ANL's Visualization Cluster without moving the data to the UC/ANL's Visualization Cluster local storage.

Figure 2 shows a ParaView visualization session. While ParaView servers (data and render) run on the UC/ANL cluster in parallel, the client side of the ParaView runs as a single process on the local desktop. Each data server reads piece of the data set from GPFS-WAN storage system and generates the geometric model. Each render server takes the geometric model and renders it. Once all the results are rendered, these images composited to form a final image and this image sent to the client. The final image is displayed on the local display.

Figure 3 shows Process ID scalars which indicate that how the data distributed across the processors. A unique scalar value with different color assigned to each pieces of data according to which processor it resides on. The total number of visualization processors used in this case is 64.

## 7. CONCLUSIONS

Broad range of compute, visualization and data storage capabilities of the TeraGrid gives the researcher flexible giant virtual computing environment. In this environment several large-scale simulations can be carried out on variety of platforms. Visualization capabilities of the TeraGrid made possible to visualize large data sets. Storing results of the simulations to GPFS-WAN storage system prevents moving the large data sets from simulation platform to visualization platform.

We have demonstrated possibility of the scalability of a parallel CFD application on TeraGrid resources. While IBM BlueGene/L scales very well for the range we studied, IBM e1350's performs lower only for 2048 number of the processors. This can be attributed to the differences in machine architecture such as communication network speed between the nodes and intra-nodes and actual clock speed as well, since the test case is identical for both computer systems. However, the overall elapsed time is much lower for the e1350 system than the IBM BlueGene/L due to higher processor speed – less than one-third at 1024 processors.

## 8. ACKNOWLEDGMENTS

This research was supported in part by the National Science Foundation under Grant Nos. ACI-0338618l, OCI-0451237, OCI-0535258, and OCI-0504075 and by the Indiana METACyt Initiative. The Indiana METACyt Initiative of Indiana University is supported in part by the Lilly Endowment, Inc. and in part by the Shared University Research grants from IBM, Inc. to Indiana University. This research was also supported in part by the National Science Foundation through TeraGrid resources provided by SDSC and ANL. TeraGrid systems are hosted by NCSA, SDSC, ANL, PSC, TACC, Indiana University, Purdue University, NCAR and ORNL.

## 9. REFERENCES

1. www.teragrid.org.
2. E. Yilmaz, M.S. Kavsaoglu, H.U. Akay, and I.S. Akmandor, "Cell-vertex Based Parallel and Adaptive Explicit 3D Flow Solution on Unstructured Grids,"

*International Journal of Computational Fluid Dynamics*, Vol. 14, pp. 271-286, 2001.

3. E. Yilmaz, R.U. Payli, H.U. Akay, and A. Ecer, "Efficient Distribution of a Parallel Job across Different Grid Sites," *Parallel Computational Fluid Dynamics*, Edited by K.J. Hyuk, et al., Elsevier Science B.V., 2007 (in print).
4. R.U. Payli, H.U. Akay, A.S. Baddi, E. Yilmaz, A. Ecer, and E. Oktay, "Computational Fluid Dynamics Applications on TeraGrid," *Parallel Computational Fluid Dynamics*, Edited by A. Deane, et al., Elsevier Science B.V., pp. 141-148, 2006.
5. R.U. Payli, E. Yilmaz, H.U. Akay, and A. Ecer, "From Visualization to Simulation: Large-scale Parallel CFD Applications on TeraGrid," *TeraGrid Conference: Advancing Scientific Discovery* (poster presentation and demonstration), Indianapolis, IN, June 12-15, 2006.
6. www.paraview.org.

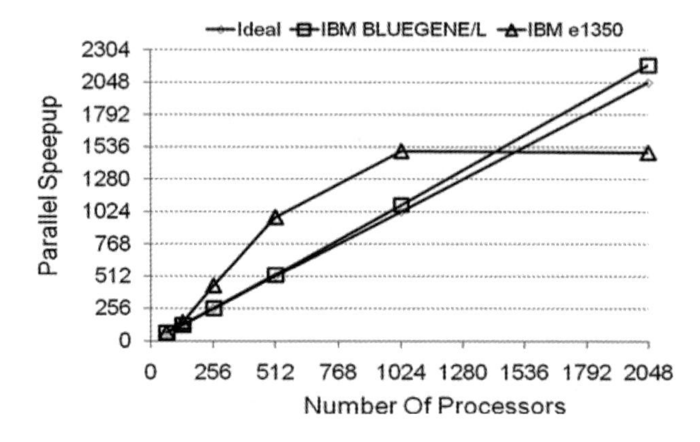


Figure 1: Speedup for IBM e1350 (IU) and IBM BlueGene/L (SDSC)

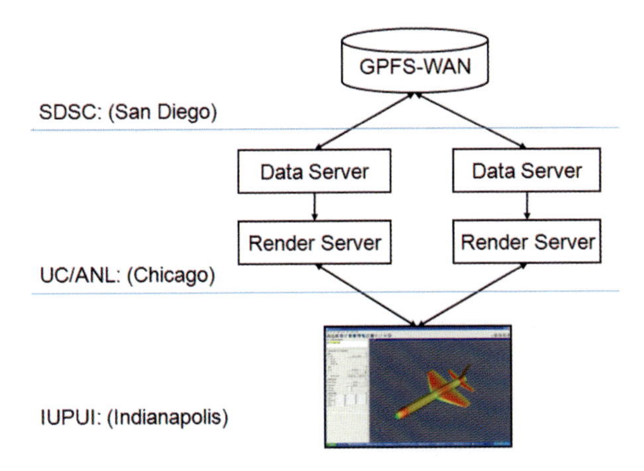


Figure 2: Execution flow of ParaView on UC/ANL visualization resource

Figure 3: ParaView Process ID colors for 64 processors

Table 1: Timing on IBM BlueGene/L and IBM e1350 for each time step of Pacer3D in seconds

| | IBM BlueGene/L (SDSC) | | | IBM e1350 (IU) | | |
|---|---|---|---|---|---|---|
| | **Elapsed** | **CPU** | **Communication** | **Elapsed** | **CPU** | **Communication** |
| **64** | 5.243 | 5.243 | 0.928 | 2.238 | 2.238 | 0.543 |
| **128** | 2.635 | 2.635 | 0.583 | 0.967 | 0.967 | 0.433 |
| **256** | 1.307 | 1.307 | 0.280 | 0.329 | 0.328 | 0.197 |
| **512** | 0.644 | 0.644 | 0.168 | 0.147 | 0.146 | 0.099 |
| **1024** | 0.311 | 0.311 | 0.097 | 0.095 | 0.095 | 0.073 |
| **2048** | 0.154 | 0.154 | 0.058 | 0.096 | 0.096 | 0.086 |

# Parallel CFD Simulations of Unsteady Control Maneuver Aerodynamics

**Jubaraj Sahu**[a]

*[a]U.S. Army Research Laboratory,Aberdeen Proving Ground, Maryland 21005-5066, USA*

Unsteady flow; free flight aerodynamics; virtual fly-out; control maneuver

## 1. ABSTRACT

This paper describes a multidisciplinary computational study undertaken to model the flight trajectories and the free-flight aerodynamics of finned projectiles both with and without control maneuvers. Advanced capabilities both in computational fluid dynamics (CFD) and rigid body dynamics (RBD) have been successfully fully coupled on high performance computing (HPC) platforms for "Virtual Fly-Outs" of munitions. Time-accurate Navier-Stokes computations have been performed to compute the unsteady aerodynamics associated with the free-flight of a finned projectile at a supersonic speed using a scalable unstructured flow solver on a highly parallel Linux Cluster. Some results relating to the portability and the performance of the flow solver on the Linux clusters are also addressed. Computed positions and orientations of the projectile along the flight trajectory have been compared with actual data measured from free flight tests and are found to be generally in good agreement. Computed results obtained for another complex finned configuration with canard-control pitch-up maneuver in a virtual fly-out show the potential of these techniques for providing the actual time-dependent response of the flight vehicle and the resulting unsteady aerodynamics for maneuvering projectiles.

## 2. INTRODUCTION

As part of a Department of Defense High Performance Computing (HPC) grand challenge project, the U.S. Army Research Laboratory (ARL) has recently focused on the development and application of state-of-the art numerical algorithms for large-scale simulations [1-3] to determine both steady and unsteady aerodynamics of projectiles with and without flow control. Our objective is to exploit computational fluid dynamics (CFD) techniques on HPC platforms for design and analysis of guided projectiles

I.H. Tuncer et al., *Parallel Computational Fluid Dynamics 2007*,
DOI: 10.1007/978-3-540-92744-0_52, 

whether the control forces are provided through canard control or jet control such as the micro adaptive flow control systems for steering spinning projectiles [3]. The idea is to determine if these control devices can provide the desired control authority for course correction for munitions. Knowledge of the detailed aerodynamics of maneuvering guided smart weapons is rather limited. Multidisciplinary computations can provide detailed fluid dynamic understanding of the unsteady aerodynamics processes involving the maneuvering flight of modern guided weapon systems [4]. The computational technology involving computational fluid dynamics (CFD) and rigid body dynamics (RBD) is now mature and can be used to determine the unsteady aerodynamics associated with control maneuvers.

Our goal is to perform real-time multidisciplinary coupled CFD/RBD computations for the flight trajectory of a complex guided projectile system and fly it through a "virtual numerical tunnel" similar to what happens with the actual free flight of the projectile inside an aerodynamics experimental facility. Recent advances in computational aerodynamics and HPC make it possible to solve increasingly larger problems including these multidisciplinary computations at least for short time-of-flights. A few highly optimized, parallelized codes [5,6] are available for such computations. Various techniques, including structured, unstructured, and overset methods can be employed to determine the unsteady aerodynamics of advanced guided projectiles and missiles. In the present study, a real time-accurate approach using unstructured methodology has been utilized to perform such complex computations for both a spinning projectile and a finned projectile. The maneuver of a projectile through flow control adds another complexity to the trajectory and the unsteady aerodynamics computations. Trajectory computations for the finned projectile have been accomplished both without and with aerodynamic flow control through the use of canard control surfaces. The emphasis in the present research is to provide insight into the time-dependent interaction of these canard maneuvers and the resulting unsteady flow fields and control forces that affect the flight dynamics response of the vehicle. In addition, the objective was to accomplish this through coupled CFD and RBD methods i.e. through virtual fly-outs.

Calculations for the finned configurations were performed using a scalable parallel Navier-Stokes flow solver, CFD++ [7,8]. This flow solver is a multipurpose, unstructured, 3-D, implicit, Navier-Stokes solver. The method used is scalable on SGI, IBM SP4, and newer computers such as the Linux Cluster. The method incorporates programming enhancements such as dynamic memory allocation and highly optimized cache management. It has been used extensively in the parallel high performance computing numerical simulations of projectile and missile programs of interest to the U.S. Army. The advanced CFD capability used in the present study solves the Navier Stokes equations [9] and incorporates unsteady boundary conditions for simulation of the synthetic microjets [10,11] as well as a higher order hybrid Reynolds-Averaged Navier-Stokes (RANS)/Large Eddy Simulation (LES) turbulence model [12-15] for accurate numerical prediction of unsteady turbulent flows. Sahu [3] used these advanced techniques and performed numerical flow field computations for both steady and unsteady jets for a spinning projectile configuration at a low subsonic speed. The present numerical study is a big step forward which now includes numerical simulation of the actual fight paths of the finned projectiles both with and without flow control using coupled CFD/RBD techniques. The following sections describe the coupled numerical procedure and the computed results obtained for the finned bodies at supersonic speeds both with and without control maneuvers.

## 3. COMPUTATIONAL METHODOLOGY

A complete set of three-dimensional (3-D) time-dependent Reynolds-averaged Navier-Stokes equations are solved using the finite volume method [7,8]:

$$\frac{\partial}{\partial t}\int_V \mathbf{W}dV + \oint[\mathbf{F}-\mathbf{G}]\cdot dA = \int_V \mathbf{H}dV \qquad (1)$$

where **W** is the vector of conservative variables, **F** and **G** are the inviscid and viscous flux vectors, respectively, **H** is the vector of source terms, V is the cell volume, and A is the surface area of the cell face.

Second-order discretization was used for the flow variables and the turbulent viscosity equations. Two-equation [15] turbulence models were used for the computation of turbulent flows. Dual time-stepping was used to achieve the desired time-accuracy. Grid was actually moved to take into account the spinning motion of the projectile and grid velocity is assigned to each mesh point. To account for the spin, the grid speeds are assigned as if the grid is attached to the projectile and spinning with it. Similarly, to account for rigid body dynamics, the grid point velocities can be set as if the grid is attached to the rigid body with six degrees of freedom (6 DOF). For the rigid body dynamics, the coupling refers to the interaction between the aerodynamic forces/moments and the dynamic response of the projectile/body to these forces and moments. The forces and moments are computed every CFD time step and transferred to a 6DOF module which computes the body's response to the forces and moments. The grid point locations and grid point velocities are set from the dynamic response.

## 4. PARALLEL COMPUTATIONAL ISSUES

The CFD++ CFD flow solver includes three unification themes: unified physics, unified grid, and unified computing. The "unified computing" capability includes the ability to perform scalar and parallel simulations with great ease. The parallel processing capability allows us to run on a wide variety of hardware platforms and communications libraries including MPI, PVM etc. The code is compatible with and provides good performance on standard Ethernet (e.g. 100Mbit, 1Gbit, 10Gbit) as well as high performance communications channels of Myrinet and Infiniband etc.

It is very easy to use CFD++ on any number of CPUs in parallel. For parallel multi CPU runs, the extra file required is a domain-decomposition file. It uses the METIS tool developed at the University of Minnesota. The code runs in parallel on many parallel computers including those from Silicon Graphics, IBM, Compaq (DEC and HP), as well as on PC workstation clusters. Excellent performance (see Figure 1 for the timings on a 4-million mesh) has been observed up to 64 processors on Silicon Graphics O3K (400 MHz), IBM SP P3 (375 MHz), IBM SP P4(1.7GHz), and Linux PC cluster (3.06 GHz). Computed results on the new Linux PC cluster seem to show 2 to 4-fold reduction in CPU time for number of processors larger than 16. Similar good performance is also achieved on the Linux PC cluster for a larger 12-million mesh (Figure 2) up to 128 processors.

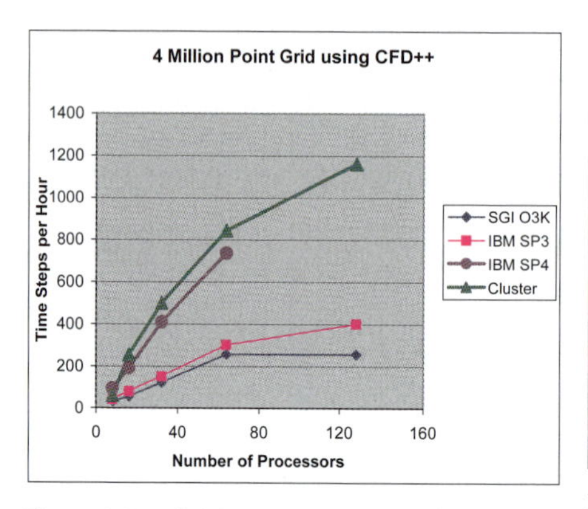


Figure 1. Parallel Speedups (4-million grid).

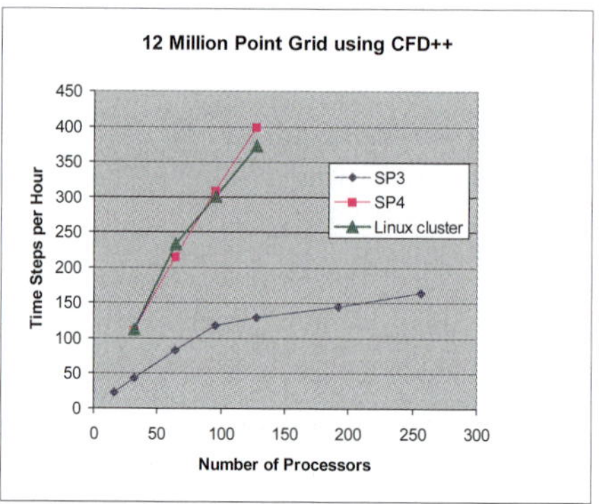


Figure 2. Parallel Speedups (12-million grid).

The computational algorithms implemented in CFD++ synergistically help achieve all three unification goals. In particular, the parallel processing capability is fully compatible with all types of meshes including structured and unstructured, steady and moving and deforming, single and multi block, patched and overset types. Inter-CPU communications are included at the fine grid level as well as all the multigrid levels to help ensure high degree of robustness consistently observed in using CFD++, independent of the number of CPUs being employed.

## 5. RESULTS

### 5.1. Finned Projectile CFD with 6-DOF

Time-accurate numerical computations were also performed using Navier-Stokes and coupled 6-DOF methods to predict the flow field and aerodynamic coefficients, and the flight paths of a finned projectile at supersonic speeds.

The supersonic projectile modeled in this study is an ogive-cylinder-finned configuration (see Figure 3). The length of the projectile is 121 mm and the diameter is 13mm. Four fins are located on the back end of the projectile. Each fin is 22.3 mm long and 1.02 mm thick. The computational mesh for the 25-mm projectile model is a C-grid (see Figure 4) consisting of seven zones. The first zone encompasses the entire projectile body, from the tip of the nose to the end of the fins. In general, most of the grid points are clustered in the afterbody fin region. Figure 4 shows a 3-D view of the full projectile mesh. The total number of grid points is 1.5 million for the full grid.

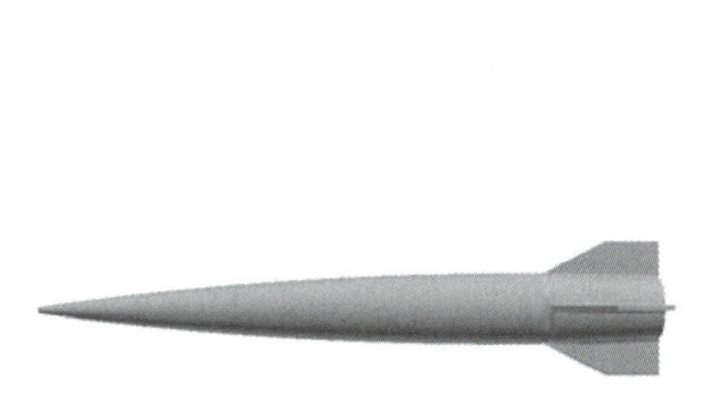

Figure 3. Computational Finned Configuration.

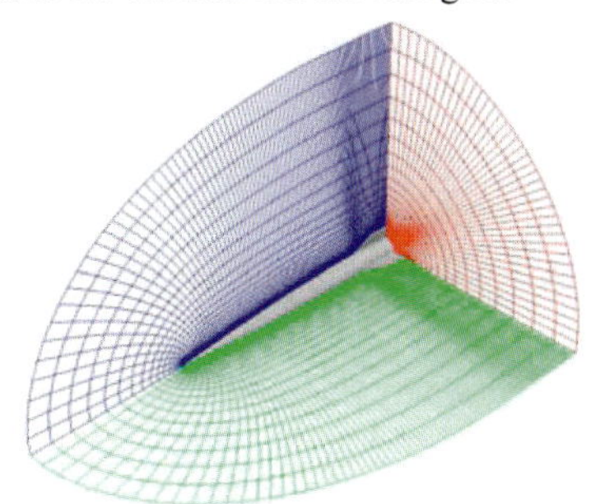

Figure 4. Computational Grid.

The first step here was to obtain the steady state results for this projectile at a given initial supersonic velocity. Also imposed were the angular orientations at this stage. Corresponding converged steady state solution was then used as the starting condition along with the other initial conditions for the computation of coupled CFD/RBD runs. Numerical computations have been made for these cases at an initial velocities 1037 and 1034 m/s depending on whether the simulations were started from the muzzle or a small distance away from it. The corresponding initial angles of attack were, $\alpha = 0.5^{\circ}$ or $4.9^{\circ}$ and initial spin rates were 2800 or 2500 rad/s, respectively.

Figure 5 shows the computed z-distance as a function of x (the range). The computed results are shown in solid lines and are compared with the data measured from actual flight tests. For the computed results the aerodynamic forces and moments were completely obtained through CFD. One simulation started from the gun muzzle and the other from the first station away from the muzzle where the actual data was measured. The first station was located about 4.9 m from the muzzle. Both sets of results are generally found to be in good agreement with the measured data, although there is a small discrepancy between the two sets of computed results. The z-distance is found to increase with increasing x-distance. Figure 6 shows the variation of the Euler pitch angle with distance traveled. Both the amplitude and frequency in the Euler pitch angle variation are predicted very well by the computed results and match extremely well with the data from the flight tests. Both sets of computations, whether it started from the muzzle or the first station away from the muzzle, yield essentially the same result. One can also clearly see the amplitude damped out as the projectile flies down range i.e. with increasing x-distance.

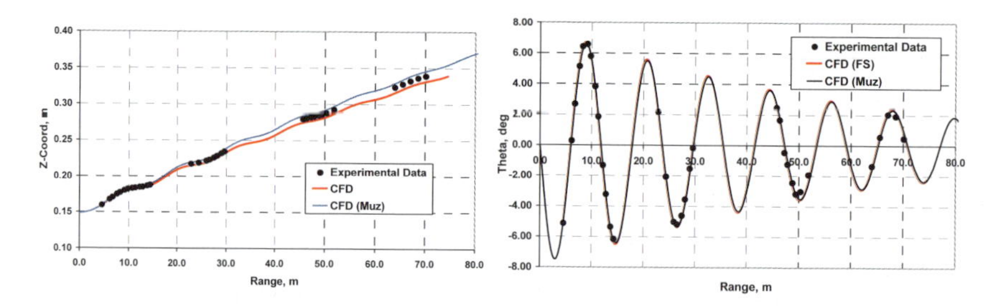


Figure 5. Computed z-distance vs. range.

Figure 6. Euler pitch angle vs. range.

### 5.2. Complex Projectile in Supersonic Flight with Canard maneuver

Another case considered in the study is a complex canard-controlled finned projectile. Here, the control maneuver is achieved by the two horizontal canards in the nose section (Figs. 7-9). Unstructured Chimera overlapping grids were used (see Fig. 7) and solutions have been obtained for several canard deflection cases. Figure 8 shows the computed pressure contours at M = 3.0 and $\alpha = 0°$ for a canard deflection of 20 deg. Although not shown here, this produces lift that can be used to obtain increased range. A typical result is shown in Figure 9 for the

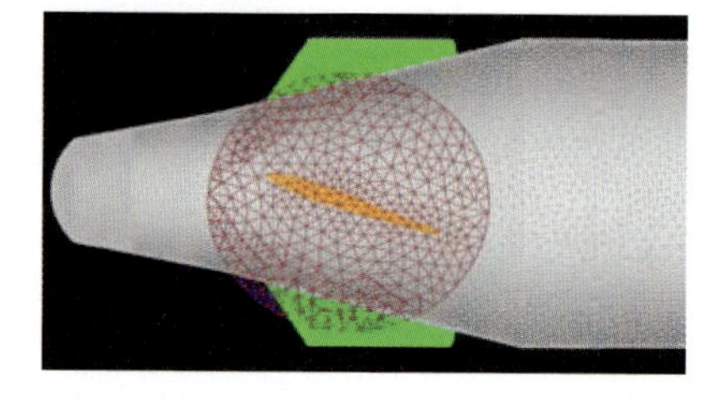

Figure 7. Unstructured Chimera mesh in the nose region (side view).

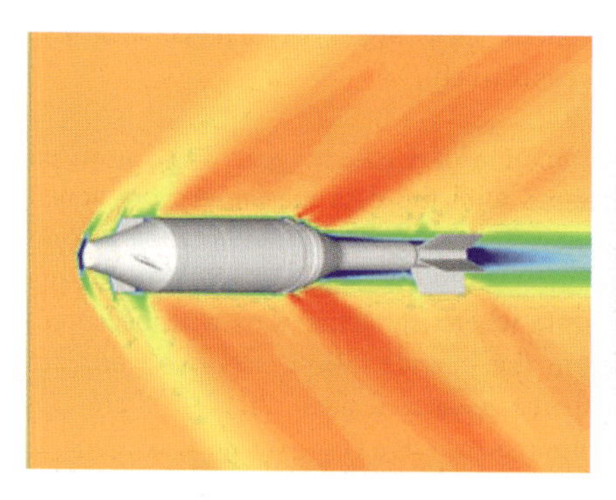

Figure 8. Computed pressure contours,

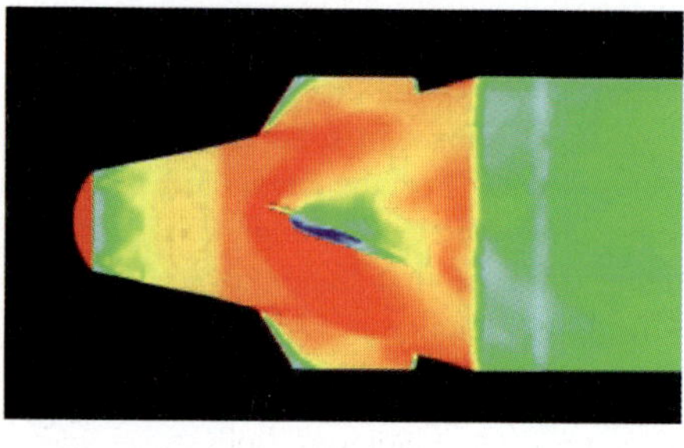

Figure 9. Computed surface pressure contours in the nose section, M = 3.0, $\alpha = 0°$.

canard deflection of 20° (high pressure region shown in red and low pressure region in blue).

Some results for a "pitch maneuver" are shown in Figures 10 and 11. In this case, the two horizontal canards are rotated down 10° in 0.01 sec, held there for the next 0.01 sec, and then deflected back to their horizontal positions in the next 0.01 sec. This maneuver generates a lot of lift force (see Figure 10) until the end of this virtual fly-out simulation (time = 0.145 sec). This results in the nose of the projectile pitching up and the z-distance of the center of gravity of the projectile increasing from 0 to 0.5 meter (see Figure 11). Also, shown in this figure is the Euler pitch angle which goes from 0 to a peak value of about -16° (nose-up corresponds to negative Euler pitch angle). These results clearly show a large effect on the time-dependent response of the projectile subject to a canard maneuver.

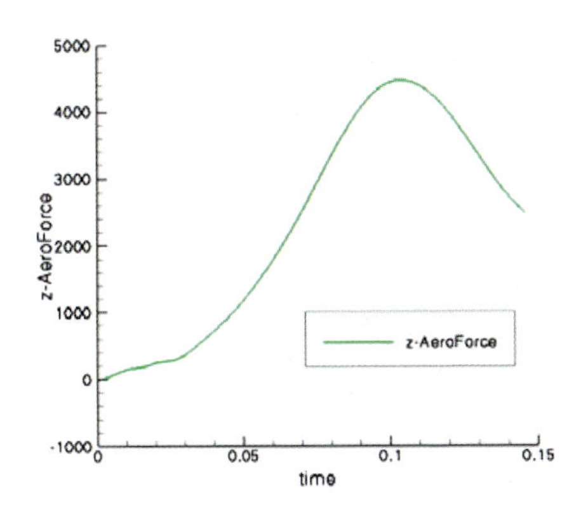


Figure 10. Time history of lift force.

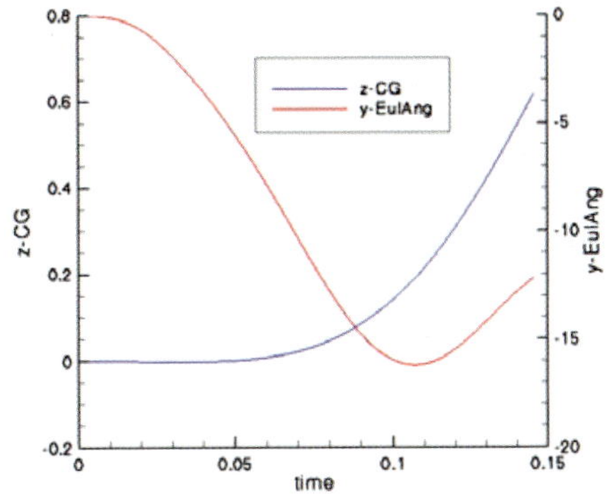

Figure 11. Time history of z-distance (center of gravity) and Euler pitch angle.

## 6. CONCLUDING REMARKS

This paper describes a new coupled CFD/RBD computational study undertaken to determine the flight aerodynamics of finned projectiles using a scalable unstructured flow solver on various parallel computers such as the IBM, and Linux Cluster. Advanced scalable Navier-Stokes computational techniques were employed to compute the time-accurate aerodynamics associated with the free flight of the finned projectiles at supersonic velocities both with and without control maneuvers. High parallel efficiency is achieved for the real time-accurate unsteady computations. For the finned configuration without control maneuver, coupled CFD/RBD calculations show the flight trajectory and the unsteady aerodynamics associated with its flight. Computed positions and orientations of the projectile have been compared with actual data measured from free flight tests and are found to be generally in very good agreement. Computed results obtained for another complex finned configuration with canard-control pitch-up maneuver in a virtual fly-out show the potential of these techniques for providing the actual time-dependent response of the flight vehicle and the resulting unsteady aerodynamics for maneuvering projectiles.

**REFERENCES**


1. J. Sahu, Pressel, D., Heavey, K.R., and Nietubicz, C.J., "Parallel Application of a Navier-Stokes Solver for Projectile Aerodynamics", Proceedings of Parallel CFD'97 Meeting, Manchester, England, May 1997.
2. J. Sahu, Edge, H.L., Dinavahi, S., and Soni, B., "Progress on Unsteady Aerodynamics of Maneuvering Munitions" Users Group Meeting, Albuquerque, NM, June 2000.
3. J. Sahu, "Unsteady Numerical Simulations of Subsonic Flow over a Projectile with Jet Interaction", AIAA paper no. 2003-1352, Reno, NV, 6-9 Jan 2003.
4. J. Sahu, H.L. Edge, J. DeSpirito, K.R. Heavey, S.V. Ramakrishnan, S.P.G. Dinavahi, "Applications of Computational Fluid Dynamics to Advanced Guided Munitions", AIAA Paper No. 2001-0799, Reno, Nevada, 8-12 January 2001.
5. Peroomian, O. and Chakravarthy S., "A `Grid-Transparent' Methodology for CFD," AIAA Paper 97-0724, Jan. 1997.
6. Meakin, R. And Gomez, R., "On Adaptive Refinement and Overset Structured Grids", AIAA Paper No. 97-1858-CP, 1997.
7. P. Batten, U. Goldberg and S. Chakravarthy, "Sub-grid Turbulence Modeling for Unsteady Flow with Acoustic Resonance", AIAA Paper 00-0473, Reno, NV, January 2000.
8. O. Peroomian, S. Chakravarthy, S. Palaniswamy, and U. Goldberg, "Convergence Acceleration for Unified-Grid Formulation Using Preconditioned Implicit Relaxation." AIAA Paper 98-0116, 1998.
9. T. H. Pulliam and J. L. Steger, "On Implicit Finite-Difference Simulations of Three-Dimensional Flow" *AIAA Journal*, vol. 18, no. 2, pp. 159–167, February 1982.
10. B. L. Smith and A. Glezer, "The Formation and Evolution of Synthetic Jets." *Journal of Physics of Fluids*, vol. 10, No. 9, September 1998.
11. M. Amitay, V. Kibens, D. Parekh, and A. Glezer, "The Dynamics of Flow Reattachment over a Thick Airfoil Controlled by Synthetic Jet Actuators", AIAA Paper No. 99-1001, January 1999.
12. S. Arunajatesan and N. Sinha, "Towards Hybrid LES-RANS Computations of Cavity Flowfields", AIAA Paper No. 2000-0401, January 2000
13. R.D. Sandberg and H. F. Fasel, "Application of a New Flow Simulation Methodology for Supersonic Axisymmmetric Wakes", AIAA Paper No. 2004-0067.
14. S. Kawai and K. Fujii, "Computational Study of Supersonic Base Flow using LES/RANS Hybrid Methodology", AIAA Paper No. 2004-68.
15. U. Goldberg, O. Peroomian, and S. Chakravarthy, "A Wall-Distance-Free K-E Model With Enhanced Near-Wall Treatment" *ASME Journal of Fluids Engineering*, Vol. 120, 1998.
16. Sahu, J., "Unsteady CFD Modeling of Aerodynamic Flow Control over a Spinning Body with Synthetic Jet." AIAA Paper 2004-0747, Reno, NV, 5-8 January 2004.

# Parallel solution of flows with high velocity and/or enthalpy gradients

**Ulgen Gulcat and Ali Dinler**

*Istanbul Technical University,Faculty of Aeronautics & Astronautics, Maslak 34469, Istanbul/TURKEY*

In this study, near wall solution of a hypersonic flow with high enthalpy gradient is obtained numerically. Since there are high temperature gradients, the governing equations of motion are modified with additional terms containing the derivatives of velocity and enthalpy. The finite difference is used in discretization of the equations. The resulting algebraic equations are solved with block elimination or with the Block Krylov technique. The domain decomposition method is implemented for the parallel solution. At the interfaces, the matching and overlapping nodes are used. In order to reduce the number of iterations Aitken's acceleration is implemented at the interfaces. With the aid of this acceleration, almost 90% parallel efficiency is achieved with 64 processors running on MPI.

Keywords: high gradient flows; Aitken's acceleration

## 1. INTRODUCTION

In recent years, constitutive equations of fluid mechanics are under revision because of flows with very high velocity and/or high enthalpy gradients. Examples of such cases are, i) high velocity narrow jet injected into a very cold or very hot gas, ii) very high velocity external flow near a hot wall. Presence of very high gradients in these flow fields introduce extra mass diffusion terms in continuity, extra stresses in momentum and extra stress work in energy equations, [1], which is similar to the approach of [2].
In this work, the two and three dimensional boundary layer of a flat plate immersed in a hypersonic flow is numerically solved with the classical boundary layer equations and with the equations containing the extra terms involving high gradients of enthalpy. The free stream Mach number is 8, and the surface to free stream temperature ratio is 4. The finite difference technique is utilized for the discretization of the computational domain. Since the governing equations are parabolic in character, the marching procedure is

I.H. Tuncer et al., *Parallel Computational Fluid Dynamics 2007*,

DOI: 10.1007/978-3-540-92744-0_53, 

applied in the main flow direction. The domain decomposition technique is implemented for the parallel solution of the algebraic equations. At the domain interfaces, matching and overlapping grids are utilized. MPI is used as the means of parallel processing tool for 226 CPU HP distributed with each 2GB memory.

## 2. FORMULATION

The formulations of compressible three dimensional boundary layer equations with extra terms will be given and the numerical formulation will also be provided in this section.

### 2.1 Governing equations.

Continuity:
$$\frac{\partial}{\partial x}(\rho u)+\frac{\partial}{\partial y}(\rho v)+\frac{\partial}{\partial z}(\rho w)=\mathcal{A} \tag{1}$$

x-momentum:
$$\rho\left(u\frac{\partial u}{\partial x}+v\frac{\partial u}{\partial y}+w\frac{\partial u}{\partial z}\right)=\frac{\partial}{\partial z}\left(\mu\frac{\partial u}{\partial z}\right)+\mathcal{B} \tag{2}$$

y-momentum:
$$\rho\left(u\frac{\partial v}{\partial x}+v\frac{\partial v}{\partial y}+w\frac{\partial v}{\partial z}\right)=\frac{\partial}{\partial z}\left(\mu\frac{\partial v}{\partial z}\right)+\mathcal{C} \tag{3}$$

Energy:
$$\rho\left(u\frac{\partial h}{\partial x}+v\frac{\partial h}{\partial y}+w\frac{\partial h}{\partial z}\right)=\frac{\partial}{\partial z}\left(\frac{\mu}{\Pr}\frac{\partial h}{\partial z}\right)+\mu\left[\left(\frac{\partial u}{\partial z}\right)^2+\left(\frac{\partial v}{\partial z}\right)^2\right]+\mathcal{D} \tag{4}$$

$\mathcal{A}$, $\mathcal{B}$, $\mathcal{C}$, and $\mathcal{D}$ are the extra terms containing the enthalpy gradients, and are as follows

$$\mathcal{A}=-\frac{\partial}{\partial y}\left(\frac{\mu}{2h}\frac{\partial h}{\partial y}\right)-\frac{\partial}{\partial z}\left(\frac{\mu}{2h}\frac{\partial h}{\partial z}\right),\quad \mathcal{B}=\frac{\partial}{\partial y}\left(\frac{\mu v}{2h}\frac{\partial h}{\partial y}\right)+\frac{\partial}{\partial z}\left(\frac{\mu u}{2h}\frac{\partial h}{\partial z}\right)$$

$$\mathcal{C}=\frac{\partial}{\partial y}\left(\frac{\mu v}{2h}\frac{\partial h}{\partial y}\right),\quad \mathcal{D}=\frac{\mu v}{2h}\frac{\partial h}{\partial y}\frac{\partial v}{\partial y}+\frac{\mu u}{2h}\frac{\partial h}{\partial z}\frac{\partial u}{\partial z}$$

Here, $u$ is the velocity component in the main flow, $v$ is the velocity component in span wise directions and $w$ is the velocity component in direction normal to the flat plate, $\rho$ is density, $\mu$ is viscosity, $h$ is enthalpy and Pr is the Prandtl number.

### 2.2 Numerical formulation

The finite difference discretization of Equations (2-4) results in three different penta-diagonal matrix equations to be solved in parallel. The continuity equation on the other hand can be solved for w velocity explicitly. The penta-diagonal equation looks like

$$a_{j,k}X_{i,j-1,k}+b_{j,k}X_{i,j,k-1}+c_{j,k}X_{i,j,k}+d_{j,k}X_{i,j,k+1}+e_{j,k}X_{i,j+1,k}=r_{j,k}\,, \tag{5}$$

wherein the coefficients of the unknown $X$ are in terms of the mesh sizes, the velocity components, the diffusion coefficients, and i, j, k are indices in x, y and z directions respectively.
The $w$ component of the velocity vector is obtained explicitly from the continuity equation which is discretized as given in [3,4].

## 3. DOMAIN DECOMPOSITION

During the parallel computation, the domain decomposition technique is used. At the interfaces, matching and overlapping grids are utilized as shown in Figure 1-2.

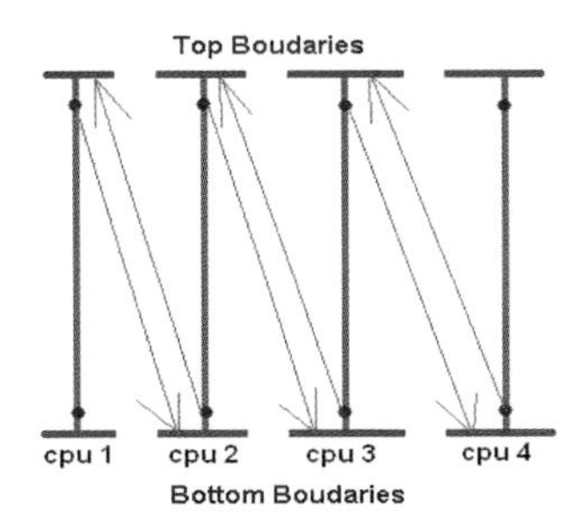


Figure 1. 1-D Overlap

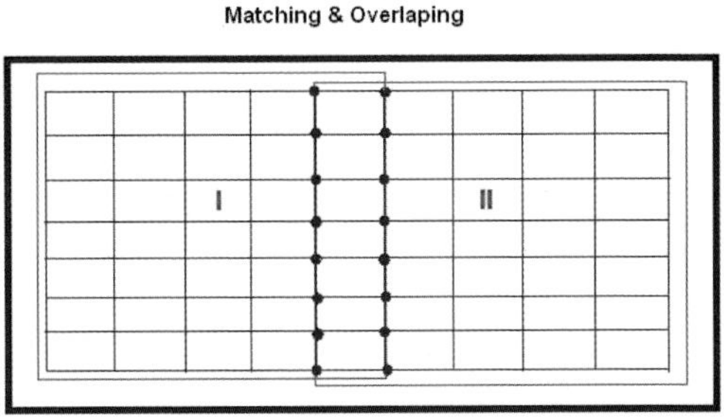

Matching & Overlaping

Figure 2. 2-D Overlap

In order to reduce the number of iterations, the interface values are modified with Aitken's acceleration technique, [5]. The final value to be used in this acceleration technique is given by

$$x_{new} = x_n - \frac{(x_n - x_{n-1})^2}{x_n - 2x_{n-1} + x_{n-2}},$$

where $x_n$, $x_{n-1}$, $x_{n-2}$ are last three values.

## 4. SOLUTION METHOD

Equation (5) constructs a penta-diagonal matrix $M$ that has two noteworthy features. First, it is a nonsymmetrical block tri-diagonal matrix and second, off-diagonals are diagonal matrices. Shown in Figure 3 is the form of the resulting penta-diagonal matrix wherein $A_i$ and $C_i$ are the sub matrices containing only diagonal elements whereas $B_i$ is a sub matrix of tri-diagonal character. Using this property of the matrix equation, direct inversion is possible.

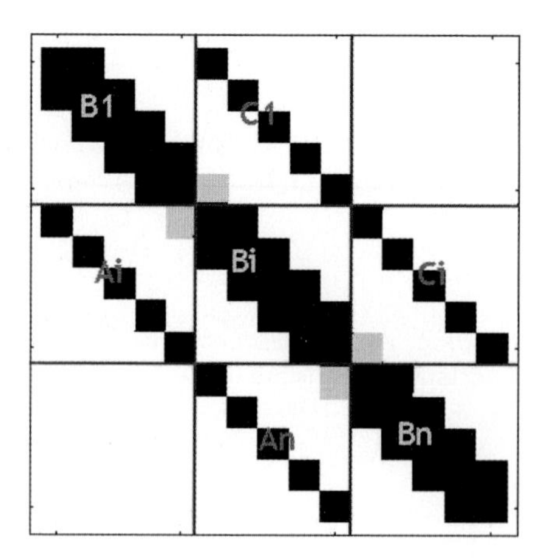


Figure 3. Penta-diagonal coefficient matrix $M$

Here, $A_i = \begin{pmatrix} \gamma_1^{\ i} & & & \\ & \gamma_2^{\ i} & & \\ & & \cdots & \\ & & & \gamma_p^{\ i} \end{pmatrix}$, $C_i = \begin{pmatrix} \alpha_1^{\ i} & & & \\ & \alpha_2^{\ i} & & \\ & & \cdots & \\ & & & \alpha_p^{\ i} \end{pmatrix}$ are $pxp$ diagonal matrices.

We need to solve $Mx = f$ linear system where $f$ is the single right hand side. So, we developed and implemented a penta-diagonal solver in order to obtain direct inversion of matirix $M$ as follows;

**Penta-diagonal solver:** If $M$ is an nxn square matrix, $p$ is the block size, $m$ is the number of blocks and $I_p$ is $pxp$ identity matrix then we can decompose matrix $M$ as $LU=M$ where

$$L = \begin{pmatrix} I_p & & & & \\ \beta_2 & I_p & & & \\ & \cdots & \cdots & & \\ & & \beta_{n-1} & I_p & \\ & & & \beta_n & I_p \end{pmatrix}, \quad U = \begin{pmatrix} D_1 & C_1 & & & \\ & D_2 & C_2 & & \\ & & \cdots & \cdots & \\ & & & D_{n-1} & C_{n-1} \\ & & & & D_n \end{pmatrix}$$

Since $C_i$ matrices are given we can find $D_i$ and $\beta_i$ matrices as follows:

$D_1 = B_1$, $\beta_i = A_i D^{-1}{}_{i-1}$, $i=2,\ldots n$. $\overline{d}_{ij}$ 's are elements of $D_i^{-1}$. $f_i$ is $mx1$ vector. In open form

$$\beta_i = \begin{pmatrix} \gamma_1^i & & & \\ & \gamma_2^i & & \\ & & \cdots & \\ & & & \gamma_p^i \end{pmatrix} \begin{pmatrix} \overline{d}_{11}^{\,i-1} & \overline{d}_{12}^{\,i-1} & \cdots & \overline{d}_{1p}^{\,i-1} \\ \overline{d}_{21}^{\,i-1} & \overline{d}_{22}^{\,i-1} & \cdots & \overline{d}_{2p}^{\,i-1} \\ \cdots & \cdots & \cdots & \cdots \\ \overline{d}_{p1}^{\,i-1} & \overline{d}_{p2}^{\,i-1} & \cdots & \overline{d}_{pp}^{\,i-1} \end{pmatrix} = \begin{pmatrix} \gamma_1^i \overline{d}_{11}^{\,i-1} & \gamma_1^i \overline{d}_{12}^{\,i-1} & \cdots & \gamma_1^i \overline{d}_{1p}^{\,i-1} \\ \gamma_2^i \overline{d}_{21}^{\,i-1} & \gamma_2^i \overline{d}_{22}^{\,i-1} & \cdots & \gamma_2^i \overline{d}_{2p}^{\,i-1} \\ \cdots & \cdots & \cdots & \cdots \\ \gamma_p^i \overline{d}_{p1}^{\,i-1} & \gamma_p^i \overline{d}_{p2}^{\,i-1} & \cdots & \gamma_p^i \overline{d}_{pp}^{\,i-1} \end{pmatrix}$$

$D_i = B_i - \beta_i C_{i-1}$, Similarly, using the entries of the diagonal matrix $C_{i-1}$

$$\beta_i C_{i-1} = \begin{pmatrix} \gamma_1^i \overline{d}_{11}^{\,i-1} \alpha_1^{i-1} & \gamma_1^i \overline{d}_{12}^{\,i-1} \alpha_2^{i-1} & \cdots & \gamma_1^i \overline{d}_{1p}^{\,i-1} \alpha_p^{i-1} \\ \gamma_2^i \overline{d}_{21}^{\,i-1} \alpha_1^{i-1} & \gamma_2^i \overline{d}_{22}^{\,i-1} \alpha_2^{i-1} & \cdots & \gamma_2^i \overline{d}_{2p}^{\,i-1} \\ \cdots & \cdots & \cdots & \cdots \\ \gamma_p^i \overline{d}_{p1}^{\,i-1} \alpha_1^{i-1} & \gamma_p^i \overline{d}_{p2}^{\,i-1} \alpha_2^{i-1} & \cdots & \gamma_p^i \overline{d}_{pp}^{\,i-1} \alpha_p^{i-1} \end{pmatrix}$$ and we obtain $D_i$.

**Penta-diagonal solver algorithm:**

LU decomposition
1. $D_1 = B_1$
2. For $i$=2,...,$n$ Do:
3. $\beta_i = A_i D^{-1}{}_{i-1}$
4. $D_i = B_i - \beta_i C_{i-1}$
5. End Do
Forward Substitution
6. $y_1 = f_1$
7. For $i$=2,...,$m$ Do:
8. $y_i = f_i - \beta_i y_{i-1}$
9. End Do
Backward Substitution
10. $x_m = D^{-1}{}_m y_m$
11. For $i$=$m$-1,...,1 Do:
12. $x_i = D^{-1}{}_i (y_i - C_i x_{i+1})$
13. End Do

## 4. RESULTS and DISCUSSION

The procedure given in detail above is used to solve the hypersonic flow past a finite flat plate shown in Figure 4. The free stream Mach number of the flow is 8 and the wall temperature to free stream temperature ratio is 4. The boundary conditions are prescribed on the figure.

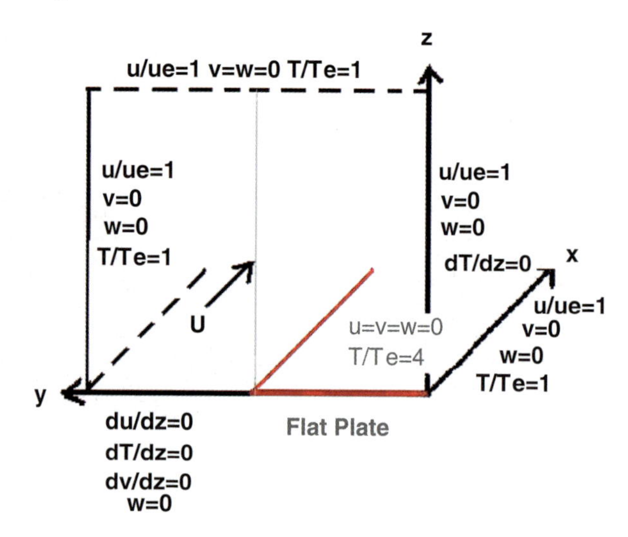


Figure 4. 3-D Flow Domain

The flow field is discretized with a grid size of 60x128x40. Marching is used in x direction. The velocity and the temperature fields computed without the extra terms are shown in Figures 5-6 at x=0.3. Figure 6 shows the chord wise and the span wise temperature gradients.

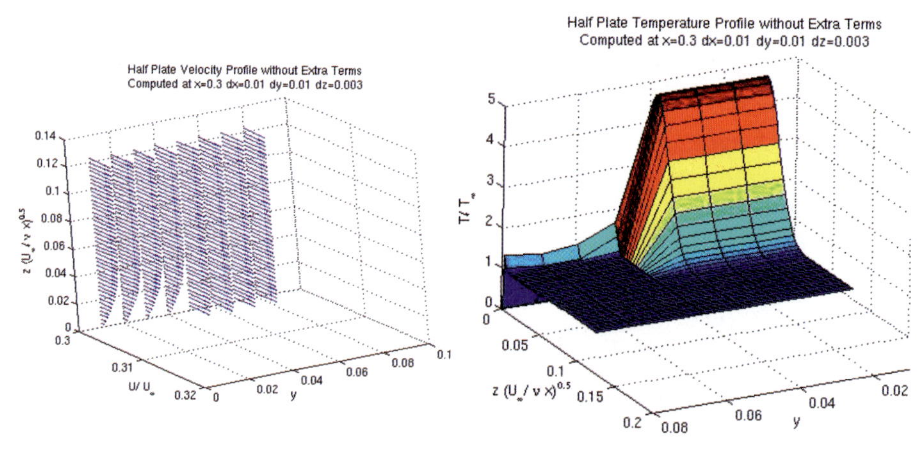


Figure 5. Velocity Profiles at x=0.3

Figure 6. Temperature at x=0.3

Shown on Figures 7 and 8 are the velocity and the temperature fields with extra terms with which the thickness of the boundary layer, i.e. $z(U_\infty/\nu\gamma)^{0.5} = 0.08$, is almost twice the thickness shown in Figure 5-6, i.e. $z(U_\infty/\nu\gamma)^{0.5} = 0.04$, for which the extra terms are not present.

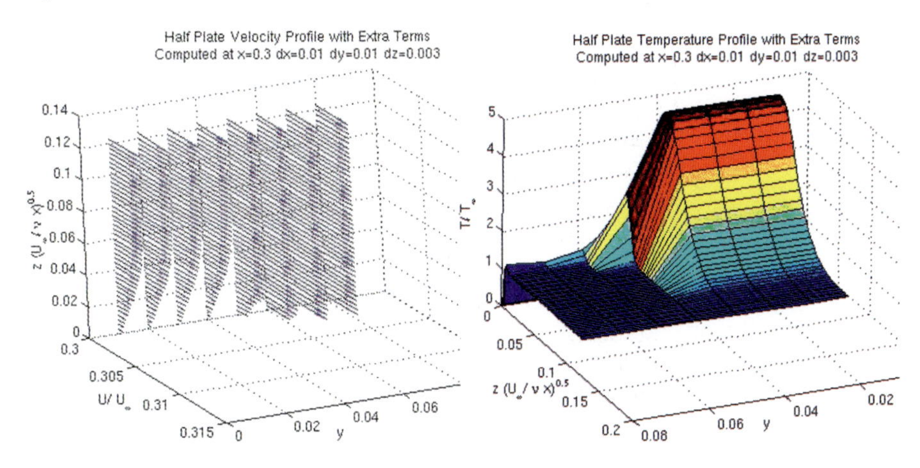


Figure 7. Velocity profiles with extra terms

Figure 8. Temperature profile

The accelerative effect of the Aitken's procedure is shown in Figures 9 and 10. These two figures indicate that with Aitken's acceleration 50% speed up is possible either with direct solver or with an iterative solver BGMRES, [6].

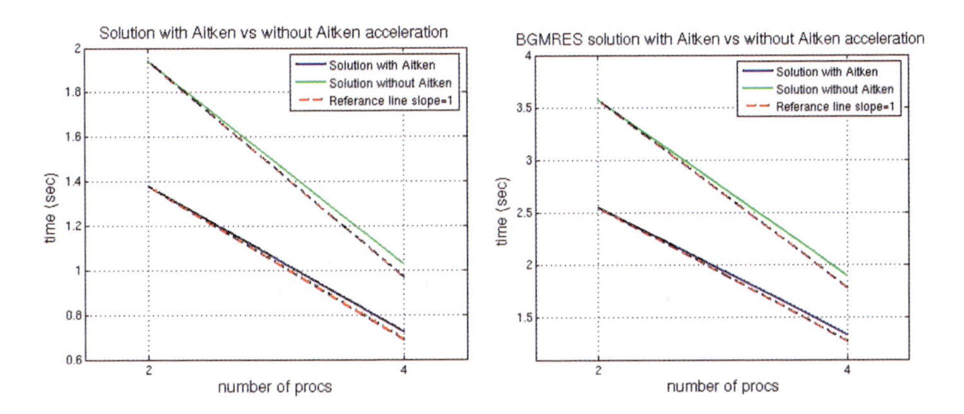


Figure 9. Aitken's for penta-diag.

Figure 10. Aitken's for BGMRES.

Finally, the speed up and the parallel efficiency curves for the procedure developed are provided in Figure 11 and 12. Figure 11 indicates that up to the 16 processors the code gains a super linear speed up. With 64 processors one can achieve 88% of parallel efficiency.

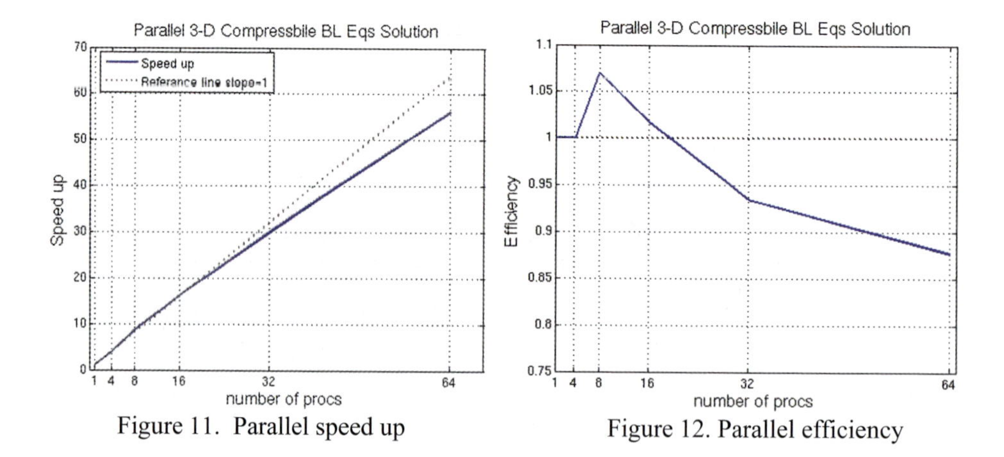


Figure 11. Parallel speed up

Figure 12. Parallel efficiency

## 6. CONCLUSION

The solutions with the extra terms included in the governing equations show that high temperature and enthalpy gradients near solid surfaces causes the boundary layer to grow twice as much thick as that of the solutions obtained without extra terms.
The Aitken's acceleration improves the solution speed about 50%.
The parallel efficiency of the present computations provides super linear speed up, up to 16 processors. The efficiency drops down to 88% for 64 processors.

## REFERENENCES


1. Durst, F., Transport Equations for Compressible Fluid Flows with Heat and Mass Transfer, Private communication, 2006.
2. Brenner, H., Navier-Stokes Revisited, Physica A, 349, 60, 2005.
3. Wu, J.C., The Solution of Laminar Boundary Layer Equations by the Finite Difference Method, Douglas Report SM-37484, 1960.
4. White, F.M.,Viscous Fluid Flow, McGraw-Hill, 2005.
5. Conte, S.D and C. de Boor, Elementary Numerical Analysis, McGraw Hill, 1972.
6. Saad, Y., Iterative Methods for Sparse Linear Systems, PWS Publishing Company, 1996.

# Numerical Simulation of Transonic Flows by a Flexible and Parallel Evolutionary Computation

G. Winter[a], B. González[a], B. Galván[a] and H. Carmona[a]

[a]Institute of Intelligent Systems and Numerical Applications in Engineering (IUSIANI), Evolutionary Computation and Applications Division (CEANI), University of Las Palmas de Gran Canaria. Edif. Central del Parque Científico y Tecnológico, $2^a$ Planta, Drcha. 35017 Las Palmas de Gran Canaria, Spain

In this work we propose a Flexible and Parallel Evolutionary Computation to solve a problem of Aeronautic interest, which is the non-linear full potential flow problem inside a nozzle. As the target to verify a numerical scheme for nodal points is equivalent to minimize a corresponding objective function, our methodology consists in using Evolutionary Algorithms as a numerical simulation method. Moreover, as Parallel Evolutionary Algorithms are demanded for an effective reduction of computation time, an Island or Coarse-Grained Parallel model is considered. The methodology proposed opens the possibility to solve the problem from a more local point of view as domain decomposition and permits to incorporate easily theoretical mathematical foundations of the problem. These are powerful characteristics of the method.

In its practical application, our aim is to obtain a good approximate solution of the exact solution, and from the good and useful information thereby made available, we can use a traditional simulation method to refine results, but starting the application of the method in a more appropriated way, for example, starting with an appropriate mesh. Here we continue our open research line [1,2] and get numerical results when a divergent nozzle has supersonic upstream flow at the entrance and subsonic flow at the exit. The steady state solution contains a transonic shock. The results agree with the exact position and conditions of the shock line.

## 1. THE PROBLEM TO SOLVE

We highlighted in previous works [1,2] the capability and applicability of the Evolutionary Computation in solving an non-linear Partial Differential Equation (PDE) boundary problem, as was the case when obtaining a solution for the stationary full potential flow problem. This involved the calculation of the speeds for transonic flow regime in the compressible and isentropic flow within a Laval nozzle, with maximum Mach number inside the nozzle approaching Mach 1 (and thus without the presence of a shock line). The problem was solved in 2D, and because the large search space required to find the optimal solution that minimized the objective function, we proposed consi-deration of our efficient optimiser as being appropriate for this case of large search space: namely, an Evolutionary Intelligent Agent-based software named Flexible Evolution Agent (FEA) [3]. A procedure associated with parallelization was made too in [2]. In all these previous works, the numerical scheme was a simple central difference approximation of first-order derivatives of the following non-linear differential equation for the nondimensional

I.H. Tuncer et al., *Parallel Computational Fluid Dynamics 2007*,
DOI: 10.1007/978-3-540-92744-0_54, 

velocity components $(u', v') = (u/c_0, v/c_0)$:

$$\left[\frac{\gamma+1}{2}u'^2 + \frac{\gamma-1}{2}v'^2 - 1\right]\frac{\partial u'}{\partial x} + u'v'\frac{\partial u'}{\partial y} + u'v'\frac{\partial v'}{\partial x} + \left[\frac{\gamma+1}{2}v'^2 + \frac{\gamma-1}{2}u'^2 - 1\right]\frac{\partial v'}{\partial y} = 0 \quad (1)$$

Our procedure was to generate candidate solutions for the velocity field and getting in many points to find the numerical values of the velocities that correspond to the global minimum of the fitness function (sum of all square numerical schemes associated with the points) for the whole domain.

We point out in passing an important technical difference between the situation considered in this paper and the one in our previous work [1,2,4]. Now we seek the capacity of our methodology to localize the position of the shock line - transonic shock and steady flow - when the nozzle has supersonic upstream flow at the entrance.

In the applications shown in this paper, the geometry of the nozzle is the same provided from H.M. Glaz and T.P. Liu [5] (see Figure 1), which is given by:

$$A(x) = 1.398 + 0.347\tanh(0.8x - 4),\ 0 \le x \le 10 \quad (2)$$

We consider in the computational implementation two subdomains:

- The left subdomain ($\Omega^-$) has supersonic flow at the entrance and $u \ge u_{entrance} > c_*$ on $\Omega^-$, where $c_*$ is the critical velocity.

- The right subdomain ($\Omega^+$) has subsonic flow at the exit with $u_{exit} \le u < c_*$ on $\Omega^+$, where $c_*$ is the critical velocity.

The size of both subdomains was chosen so as to be large enough for both to include the shock zone. For each subdomain, we solve the transonic flow with consideration of isentropic flow and using Evolutionary Computation. The target is to obtain the velocity and density fields on the whole nozzle and to determinate the localization of the shock line.

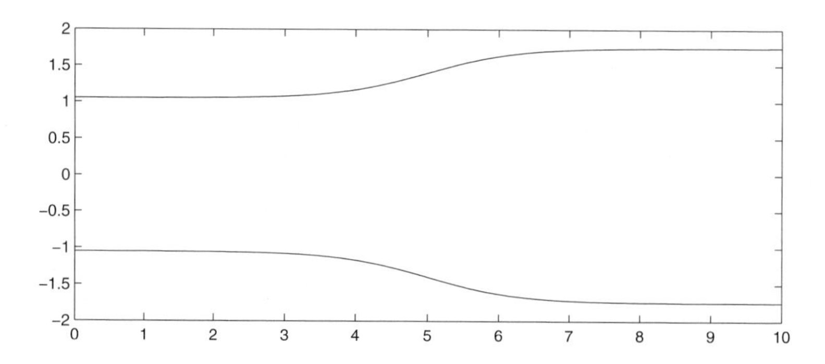


Figure 1. Geometry of the nozzle.

On the position of the shock line, it is well known that if $u$ has one discontinuity on the shock point $(x_s)$ then:

$$u^-(x_s)\; u^+(x_s) = c_*^2 \tag{3}$$

$$\rho(u^-(x_s))\; u^-(x_s) = \rho(u^+(x_s))\; u^+(x_s) \tag{4}$$

Eq. 3 is the Prandtl-Meyer relation.

Physical entropy condition [6]: The density increases across the shock in the flow direction: $\rho(u^-(x_s)) < \rho(u^+(x_s))$

The flow cases considered in this paper are close to being quasi-one-dimensional. Thus for the divergent nozzle with supersonic entrance considered, we can ensure that the mass flow: $\rho(u(x))u(x)A(x)$, is conserved at each point $x$.

## 2. FLEXIBLE EVOLUTION AGENT

Evolutionary algorithms (EAs) are efficient heuristic search methods with powerful characteristics of robustness and flexibility to solve complex optimization problems [7], even as numerical simulation method [1]. They have the ability to escape from local minima where deterministic optimization methods may fail or are not applicable. For computational cost, the merit of EAs is that they grow only linearly with problem size if they are well designed, while other methods usually grow exponentially.

However, the computational performance and efficiency of the EAs are closely related with the introduction of parameters and/or operators adaptation, which are specific for the problem to be solved. They are advanced methods, but normally they only are valid to solve efficiently the problem for which they were designed. Recently, a new, efficient and robust procedure for the optimization of complex problems, named by us as Flexible Evolution Agent (FEA), has been developed in the CEANI Division of IUSIANI [3] and applied to solve the application treated here, with serial computation.

The target of FEA is to get the biggest benefit from the explotation of the stored information, and in addition, to incorporate new procedures to facilitate the internal decision-making to be automatically made in the own optimization process. This Agent has a Dynamic Structure of Operators (DSO), Enlargement of the Genetic Code (EGC) of each candidate solution and uses a Central Control Mechanism (CCM). The DSO enables the use of any of the sampling operators at each step throughout every optimization run, depending on operator's previous contribution to the common task (to get the optimum). EGC includes useful information for the process control included in the CCM. After several implementations, the identification of the sampling operator used to obtain each variable of the candidate solutions has proved to be the most useful information, but only when a simple Probabilistic Control Mechanism (PCM) based on rules IF-THEN-ELSE is used as a CCM. The joint use of the DSO, the EGC and the CCM has permitted the elimination of the crossover and mutation probabilities. The PCM is responsible for reaching a trade-off between the exploration and the exploitation of the search space, which is equivalent to achieving a competitive and cooperative balance among the sampling operators.

## 3. NUMERICAL IMPLEMENTATION

We consider the formulation given by the equations:

$$div(\rho u A) = 0 \tag{5}$$

$$\rho(u) = \rho_0 \left[ 1 - \frac{\gamma - 1}{2c_0^2} u^2 \right]^{\frac{1}{\gamma - 1}} \tag{6}$$

where $c_0$ is the stagnation speed of sound and $\gamma = 1.4$ for the air.

### 3.1. Evolutionary Computation

To solve this problem with evolutionary algorithms we have designed the following algorithm [4]:

*Step 1*: Read the input data (boundary conditions and GA parameters).

*Step 2*: Set $cont = 0$;

*Step 3*: Set $cont = cont + 1$;

*Step 4*: Run the GA on $\Omega^-$;

*Step 5*: Run the GA on $\Omega^+$;

*Step 6*: Find the shock point $x_s \in \Omega^- \cap \Omega^+$ as the point where the following conditions are minima:

$$|u^-(x_s)u^+(x_s) - c_*^2| \tag{7}$$

$$|\rho(u^-(x_s))u^-(x_s) - \rho(u^+(x_s))u^+(x_s)| \tag{8}$$

*Step 7*: Set $\Omega_s^- = [0, x_s) \subset \Omega^-$ and $\Omega_s^+ = (x_s, 10] \subset \Omega^+$.

*Step 8*: Evaluate the fitness function $(ff)$ over all the points of the domain $\Omega_s = \Omega_s^- \cup \Omega_s^+$.

*Step 9*: If $cont < max\_cont$ or $ff \geq error$ go to Step 3. Otherwise go to Step 10.

*Step 10*: Run the FEA on the whole domain $\Omega = \Omega \cup \{x_s\}$.

*Step 11* Save results and stop.

The pseudocode of the GA is the following:

*Step* $1_{GA}$: The limit values of the variables are established (they are different depending on whether we are at Step 4 or Step 5 of the preceding algorithm).

*Step* $2_{GA}$: If $cont = 1$, the initial population is randomly generated. Otherwise, the initial population is generated from the best solution obtained at the corresponding Step 4 or Step 5 of the preceding algorithm when $cont = cont + 1$.

*Step* $3_{GA}$: Set $cont_{GA} = 0$;

*Step* $4_{GA}$: Set $cont_{GA} = cont_{GA} + 1$;

*Step* $5_{GA}$: The candidate solutions are repaired in such a way that they verify the hypotheses of the problem:

- If the algorithm is at Step 4: the velocities are ordered from lowest to highest and the densities from highest to lowest.

– If the algorithm is at Step 5: the velocities are ordered from highest to lowest and the densities from lowest to highest.

*Step* $6_{GA}$: Evaluate the fitness function.

*Step* $7_{GA}$: Order the population from lowest to highest value of the fitness function.

*Step* $8_{GA}$: Selection: Tournament selection operator (2:1).

*Step* $9_{GA}$: Crossover: Antithetic crossover operator.

*Step* $10_{GA}$: Mutation: Smooth mutation over one variable randomly chosen.

*Step* $11_{GA}$: If $cont_{GA} < max\_cont_{GA}$ go to Step $4_{GA}$. Otherwise leave.

The candidate solutions considered are:

- $(u_1, u_2, ..., u_{nleft-1}, \rho_1, \rho_2, ..., \rho_{nleft-1})$ when $\Omega^-$ is split in *nleft* points $x_i$ with $x_i = ih$, $0 \leq i < nleft$ and $h = |\Omega^-|/(nleft-1)$. Here $u(x_0) = u_{entrance}$ and $\rho(u(x_0)) = \rho_{entrance}$.
- $(u_1, u_2, ..., u_{nright-1}, \rho_1, \rho_2, ..., \rho_{nright-1})$ when $\Omega^+$ is split in *nright* points $x_i$ with $x_i = x^- + ih$, $0 < i \leq nright$, $x^- = 10 - |\Omega^+| - h$ and $h = |\Omega^+|/(nright-1)$. Here $u(x_{nright}) = u_{exit}$ and $\rho(u(x_{nright})) = \rho_{exit}$.

The fitness function considered is the following one:

$$\sum_{i=1}^{nvar} f(x_i) \tag{9}$$

where $nvar = \begin{cases} nleft - 1 \text{ if we are running the GA on } \Omega^- \\ nright - 1 \text{ if we are running the GA on } \Omega^+ \end{cases}$, and

$$f(x_i) = \omega_1 \left\{ \rho(u(x_i)) - \rho_0 \left[1 - \frac{\gamma - 1}{2c_o^2} u(x_i)^2\right]^{\frac{1}{\gamma-1}} \right\}^2 + \omega_2 \left[\rho(u(x_i))u(x_i)A(x_i) - \rho(u(0))u(0)A(0)\right]^2 \tag{10}$$

### 3.2. Parallel Evolutionary Computation

For an effective reduction of computation time are demanded Parallel Evolutionary Algorithms (PEAs), especially on asynchronous grid computing environments. In this work we present two level of parallelisation. The first one consits in running the GA on $\Omega^-$ and on $\Omega^+$ at the same time. The second one consits in an Island or Coarse-Grained Parallel model with Flexible Evolution Agent where the FEA of the master works with integer variables and the FEA of the islands works with real variables. The main difference from between both of them is the sampling engine. In the case of the master, the sampling methods are for integer variables, and in the case of the islands, the sampling methods are for real variables.

A master's candidate solution is:

$$(Migration\_period, Npop_1, ..., Npop_{nisland}, \%mig_1, ..., \%mig_{nisland})$$

where *Migration_period* is the period of migration for all islands, *nisland* is the number of islands considered and $Npop_i$ and $\%mig_i$ are the population size of the island $i$ and the percentage of candidate solutions that will migrate from the island $i$ $(i \in 1, 2, , nisland)$, respectively.

A island's candidate solution is:

$$(u_1, u_2, ..., u_{nvar}, \rho_1, \rho_2, ..., \rho_{nvar})$$

where $nvar$ is the number of points considered inside the whole domain.

The FEA algorithm considered is the following one:

Let $\mathbf{W} = (\mathbf{x}, \mathbf{y})$ be a candidate solution, where $\mathbf{x} = (x_1, x_2, \ldots, x_i, \ldots, x_{nvar})$ consists of $nvar$ variables, and $\mathbf{y} = (y_1, y_2, \ldots, y_i \ldots, y_{nvar})$ consists of the sampling methods used to sample each variable. Let $P^t(\mathbf{W})$ be the population and $P^t_\mu(\mathbf{W})$ the $\mu$ best candidate solutions of $P^t(\mathbf{W})$. Then the FEA algorithm is:

Establish population size & maximum number of generations;

$t = 0$;

Generate $P^t(\mathbf{W})$;

**While** not stop criterion **do**

- Evaluate $P^t(\mathbf{W})$;
- Extract $P^t_\mu(\mathbf{W}) \leftarrow [P^t(\mathbf{W})]$
- Set $P'^t(\mathbf{W}) \leftarrow$ selection$[P^t_\mu(\mathbf{W})]$
- Subject to the PCM from the Decision Engine: Generate $P^t_{new}(\mathbf{W}) \leftarrow s$ $(\forall s : s =$ sample $(x_i, y_i)$ o $s =$ sample $\mathbf{W})$;
- Set $P^{t+1}(\mathbf{W}) \leftarrow \{P^t_\mu(\mathbf{W}) \cup P^t_{new}(\mathbf{W})\}$;
- Set $P^t(\mathbf{W}) \leftarrow P^{t+1}(\mathbf{W})$

**od**

Print solutions & stored information;

The main difference between the island's FEA algorithm and the master's FEA one is the evaluation of the fitness function. In the latter a simple evaluation of the fitness function is not carried out (see Figure 2).

## 4. RESULTS

In order to validate our methodology, we consider the same data at the entrance and the exit of the divergent nozzle given in one example from Keppens [8]. Thus we have:

- Boundary conditions at the entrance: $M_{entrance} = 1.28$ (Mach number); $u_{entrance} = 2$ and $\rho_0 = 1.0156$.
- Boundary conditions at the exit: $\rho_{exit} = 0.8$.

From these boundary conditions, $c_0$, $c_*$ and $\rho_{entrance}$ are determined, and also the flow conditions at the exit of the nozzle under isentropic consideration.

In this test case we defined: $\Omega^- = [0, 6]$ and $\Omega^+ = [4, 10]$ and considered only 101 equidistant points on the whole domain, $nleft = nright = 61$, $max\_cont_{GA} = 100$, $max\_cont = 100000$ and $error = 10^{-9}$.

Figure 3 shows the solution obtained. The position of the shock point was well captured: $x_s = 5.1$, close to the exact position, and the results agree with the Prandtl-Meyer relation. Moreover, the range of values of the Mach number is compatible with the isentropic flow consideration. We have a shock in the numerical applications that is not strong.

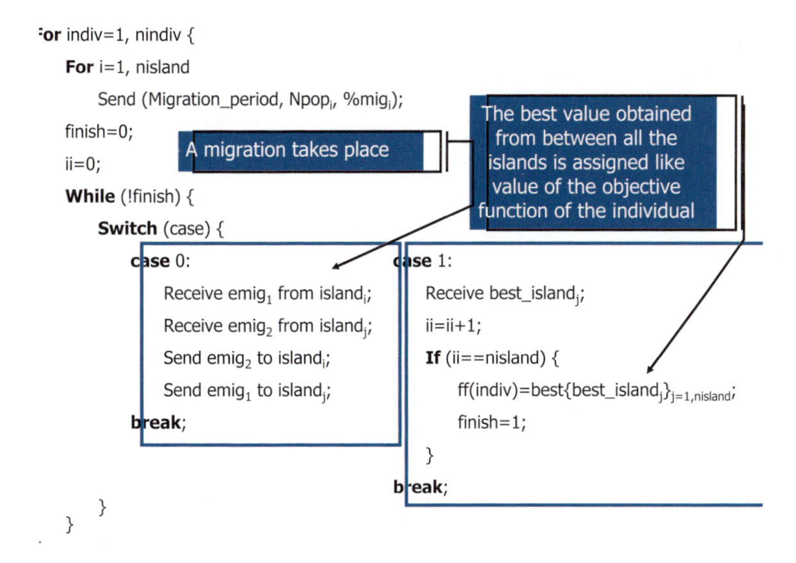


Figure 2. Evaluation of the fitness function for all candidate solutions of a population of the master's FEA.

## 5. CONCLUSIONS

We proposed a simple and efficient method for calculating transonic flows that captures the steady state solution and the position of the line shock with a good accuracy. The results of the numerical example shown in this paper provide strong evidence on the effectiveness of this new methodology as an alternative to use a traditional method. In this paper an effective reduction of computation time is achieved by the use of an Island or Coarse-Grained Parallel model with FEA.

## REFERENCES

1. G. Winter, J.C. Abderramán, J.A. Jiménez, B. González, P.D. Cuesta. Meshless Numerical Simulation of (full) potential flows in a nozzle by GAs. International Journal for Numerical Methods in Fluids, 43: 10-11, pp. 1167-1176. John Wiley & Sons (2003).
2. G. Winter, B. González, B. Galván, E. Benítez. Numerical Simulation of Transonic Flows by a Double Loop Flexible Evolution. Parallel Computational Fluid Dynamics. Theory and Applications, pp. 285-292. Elsevier B.V. (2006).
3. G. Winter, B. Galván, S. Alonso, B. González, D. Greiner, J.I. Jiménez. "Flexible Evolutionary Algorithms: cooperation and competition among real-coded evolutionary operators". Soft Computing - A Fusion of Foundations, Methodologies and Applications. 9 (4), pp. 299-323. Springer-Verlag (2005).

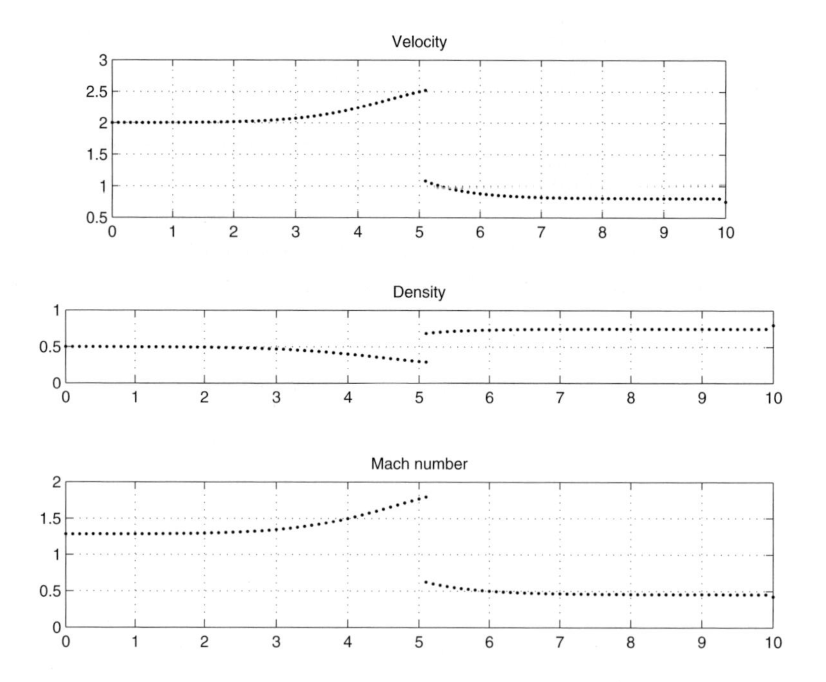


Figure 3. Velocity, density and Mach number in the nozzle.


4. B. González, B. Galván and G. Winter. "Numerical Simulation of Transonic Flows by a Flexible Optimiser Evolution Agent". CD-Rom of Proceedings of the ECCOMAS CFD 2006, (2006).
5. H.M. Glaz and T.P. Liu, "The asymptotic analysis of wave interactions and numerical calculations of transonic nozzle flow", Adv. in Appl. Math. 5, 111-146 (1984).
6. R. Courant and K.O. Friedrichs, Supersonic Flow and Shock Waves, Springer-Verlag: New York (1948).
7. T. Bäck and G. Winter. Mathematical Prospects on Evolutionary Algorithms, INGEnet Case Studies Open Day, Von Karmann Institute (summary in PowerPoint at www.ingenet.ulpgc.es), (2001).
8. Keppens (2003): http://www.rijnh.nl/n3/n2/f1234.htm

# Prediction of Ballistic Separation Effect by Direct Calculation of Incremental Coefficients

**Eugene Kim* and Jang Hyuk Kwon***

**Department of Aerospace Engineering, Korea Advanced Institute of Science and Technology, 373-1 Guseong-dong, Yuseong-gu, Daejeon, 305-701 Republic of Korea*

**Key Words**: BSE(Ballistic Separation Effect), Incremental coefficient, Unsteady calculation

## 1. Introduction

When a new weapon is introduced, target point estimation is one of important objectives in the flight test as well as safe separation. An aircraft OFP(Operational Flight Program) predicts the trajectory of unguided bomb and the ground impact point. A pipper positioned on the cockpit head-up display indicates to the pilot the predicted target point in real time, so the simulation algorithm must be as simple as possible. Actually, the fire control computer runs such a simulation during every computational cycle(0.02-0.25sec.)[1]. Therefore the OFP ballistic bombing algorithm is basically a three-degree-of-freedom point mass(drag and gravity) particle trajectory generation program.
If there are no influences, a bomb would fall along the OFP-predicted trajectory and impact at the target point. The actual dynamics of a released bomb, however, is much more complicated than that modeled in the fire control computer due to the fact that forces and moments act along all three axes of the bomb. The trajectory is perturbed by aircraft separation effects such as aircraft flow-field effects and ejector-induced forces and moments. The bombing algorithm must account for aircraft separation effects by means of initial position and velocity increment[2].
The aircraft flow field effect called BSE(Ballistic Separation Effect) is one of the largest miss-distance factors. Especially, in the case of an unguided bomb without control surfaces, the target point is controlled only by the input of pilot. Therefore, an accurate prediction model applied to OFP in the BSE region is important.
The best approach to generate a trajectory model is the flight test. It is a realistic method but difficult to gain dynamic data because of its excessive expense. Other approaches such as wind-tunnel testing and prediction methods can generate the trajectory model or help to design the flight test schedule. But the incremental aerodynamic coefficients in the aircraft flow field so-called BSE are difficult to predict. Generally, semi-empirical

I.H. Tuncer et al., *Parallel Computational Fluid Dynamics 2007*,
DOI: 10.1007/978-3-540-92744-0_55, © Springer-Verlag Berlin Heidelberg 2009

methods such as the grid methods, IFM(Influence Function Method) and Flow TGP(Trajectory Generation Program) using database are used for estimation of BSE. But these methods are quasi-steady methods using static aerodynamic loads[3]. Nowadays the time-accurate CFD method is often used to predict the store separation event[4]. The time accurate computation of aerodynamic loads is fully coupled with dynamics of the released store. This technique is commonly applied to predict short range after separation because of its computing cost. Recently the static flow field database generated by the CFD method substitutes for wind tunnel data in TGP[5].
Fig. 1 shows a general trajectory generating process using superposition of free stream aerodynamic loads and delta coefficients[6]. The interference database can be generated by wind tunnel tests, however, these are static data and must be interpolated.

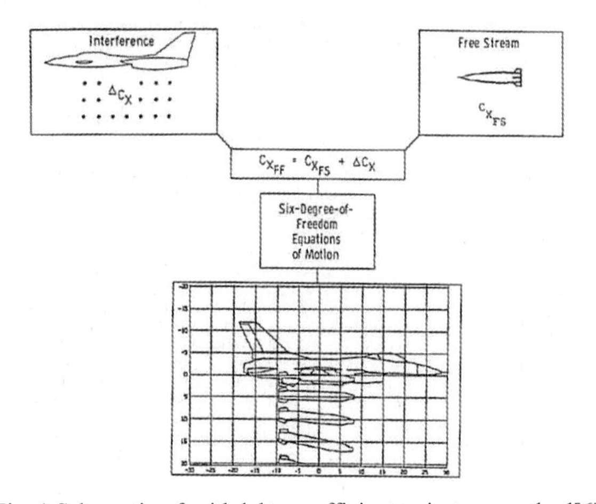


Fig. 1 Schematic of grid delta-coefficient trajectory method[6]

In this paper, a two step simulation method has been introduced. The first step is unsteady CFD calculation in BSE region and second step is off-line simulation using free stream bomb database. The focus of this work is how to switch these two methods. The delta coefficients can be used for the indicator. In the current process, the incremental aerodynamic coefficients in BSE regime are calculated directly and the elimination of those is checked simultaneously.

## 2. Direct calculation of aircraft flow-field effects

Fig. 2 shows the flow chart of developed algorithm for the direct calculation of incremental coefficients and BSE check. While the unsteady calculation is achieved to generate trajectory of released store, a sub-program is executed with another grid system in the free-stream condition in parallel. The time accurate grid velocities are applied to the sub-calculation. Only the rotation is updated for the attitude of store in free-stream condition because the altitude effect is neglected in this simulation. If the incremental coefficients converged to zero, we expect that the flow field effects fade out.

Therefore, we can determine whether the store gets out of BSE region or not by investigation of these delta coefficients.
A validation of this developed algorithm in two dimensional problems and application to a bomb released from aircraft will be shown in next sections. The stage when the store gets out of the BSE field can be used for the initial condition of Flow TGP with free-stream database. The off-line simulation in free flight condition is much simpler and cheaper than unsteady CFD calculation in the aircraft flow fields. Furthermore, it requires database of the bomb only in the free-stream condition and this doesn't require much time.

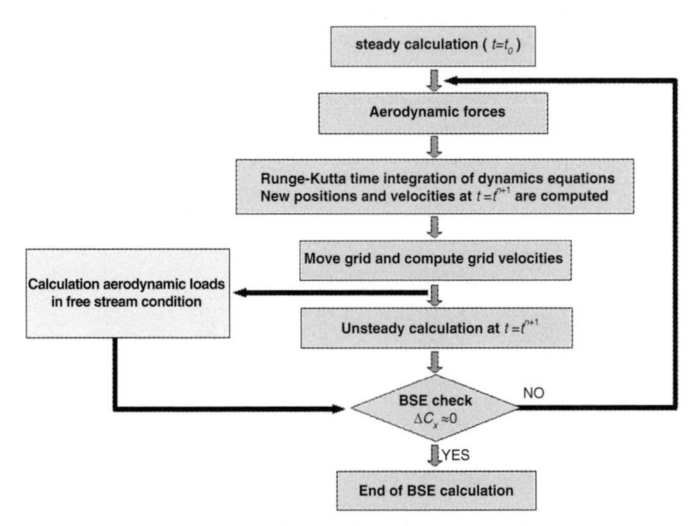

Fig. 2 Direct calculation of incremental coefficients

## 3. Validate the direct calculation of incremental coefficients

To validate the developed idea, two dimensional supersonic and subsonic store separation problems have been simulated and incremental coefficients are calculated. A parallel inviscid flow solver including a parallelized grid assembly is used for the unsteady calculation[7]. The numerical fluxes at cell interface are constructed using the Roe's flux difference splitting method and MUSCL(Monotone Upwind Scheme for Conservation Law) scheme. For the time integration, the diagonalized ADI method is used with the $2^{nd}$-order dual time stepping algorithm.
Fig. 3 shows aerodynamic force coefficients and incremental coefficients in the case of a submunition ejected from a mother missile with the free stream Mach number of 1.8. Two ejecting forces act on the moving body in short range. The positions are forward and aft of its center of gravity, and these make nose down moment. Fig. 4 shows temporal Mach number contours. The normal shock wave on the submunition is interacted with the shock wave from mother missile as in Fig. 4(a). This shock moves to the aft of body suddenly to become an oblique shock, so the incremental coefficients fade out suddenly as can be seen in Fig. 4(b).

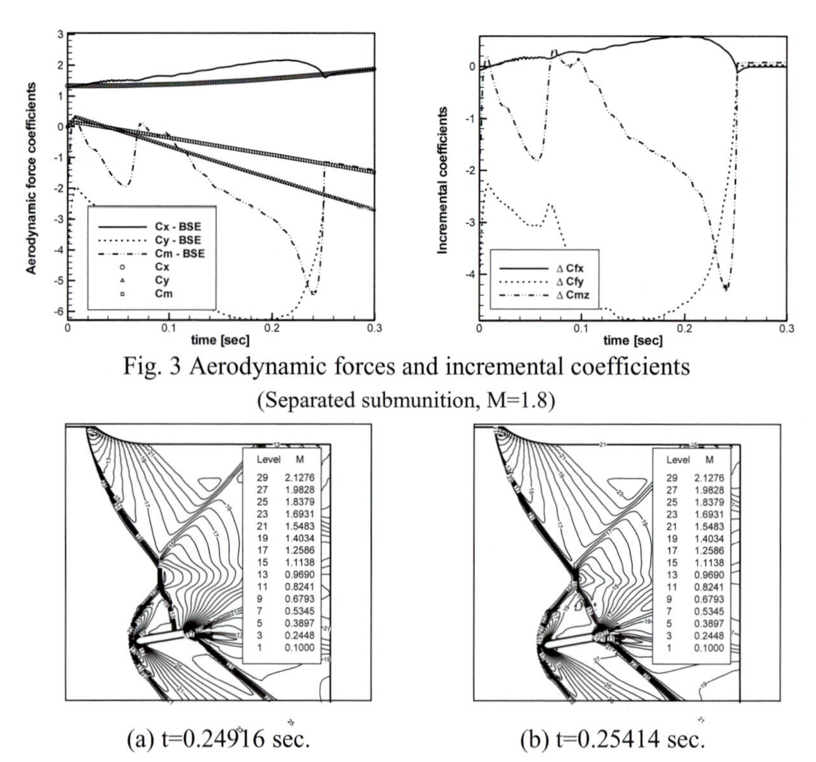


Fig. 3 Aerodynamic forces and incremental coefficients
(Separated submunition, M=1.8)

(a) t=0.24916 sec. (b) t=0.25414 sec.

Fig. 4 Temporal Mach number contours (Separated submunition, M=1.8)

Fig. 5 shows grid systems for the case of released store from an airfoil in subsonic flow regime. For the calculation in free-stream condition, the far-field boundary is extended as in Fig. 5(b). Fig. 6 shows aerodynamic forces and delta coefficients. In this free-drop case, a small but not negligible lifting force remains during the time longer than in the case of supersonic flow field. However, the time when the store gets out of BSE region can be estimated well from these data.

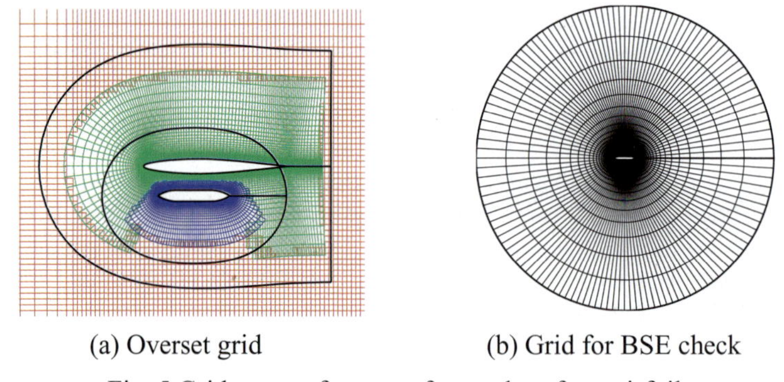

(a) Overset grid (b) Grid for BSE check

Fig. 5 Grid system for case of store drop from airfoil

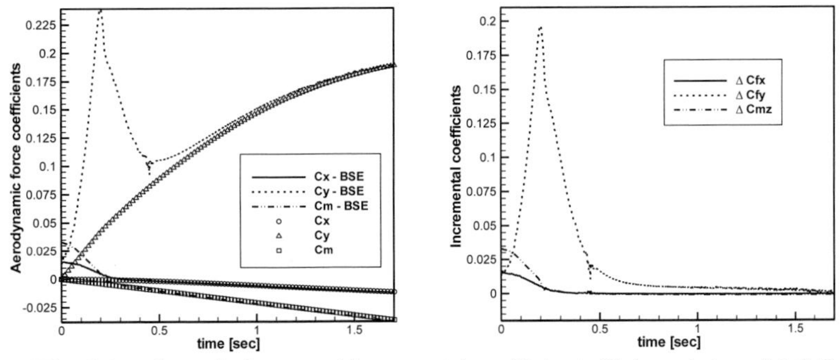


Fig. 6 Aerodynamic forces and incremental coefficients (Released store, M=0.6)

## 4. Application (Bomb separation from aircraft)

A generic store separation from the KT-1 aircraft(Korea Aerospace Industries, Ltd) has been simulated and the BSE field was calculated by the developed method. Fig. 7 is the surface mesh and the initial position of store. The test cases of store separation from inboard and outboard pylons are simulated each. The total grid system consists of 33 blocks and the number of node points including H-type block to improve the domain connection between sub grids is about 2.3 millions. The store is released at 5,000 ft height with no ejecting force. The free-stream Mach number is 0.4 and the aircraft angle of attack is zero. For the calculation, a linux PC cluster with P4-2.6GHz CPU in each node and linked by 100Mbps Ethernet network is used. The total computation time from the beginning to 1.256 seconds is about 8 hours with the unsteady time step of 0.0048 second when 31 processors are used. When t = 1.256 seconds, the store is departed from the aircraft about 1.52 times of half span length and delta coefficients are reduced to 2% of the initial(maximum) magnitudes. Figs. 8 and 9 show the computed aerodynamic force, moment coefficients and delta coefficients. The elimination of flow field effects is observed by measurement of two dominant aerodynamic force coefficients, lift and pitching moment only, since different grid systems may cause a little numerical difference in the force coefficients

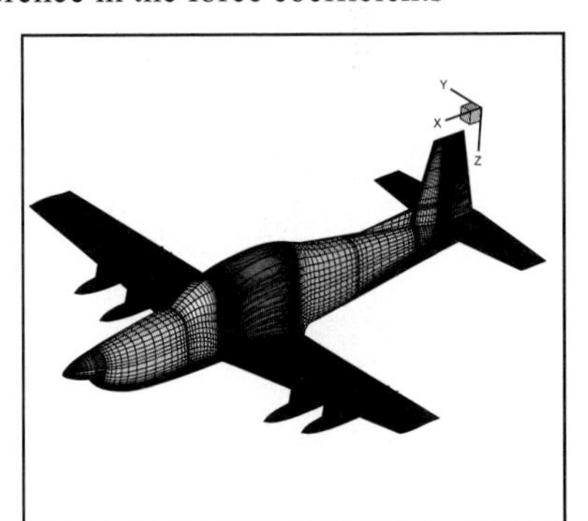

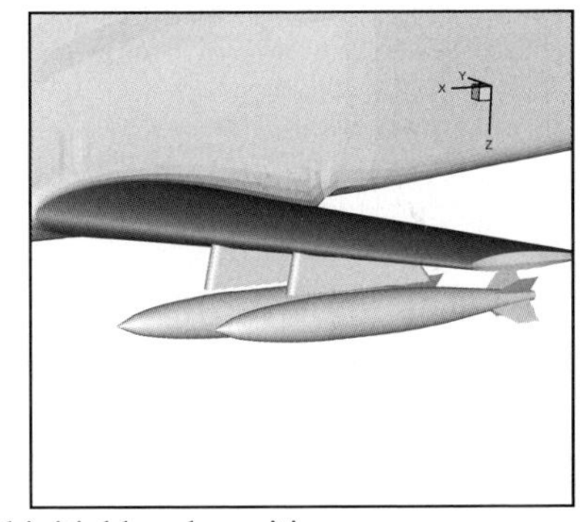

Fig. 7 Surface mesh and initial bomb positions

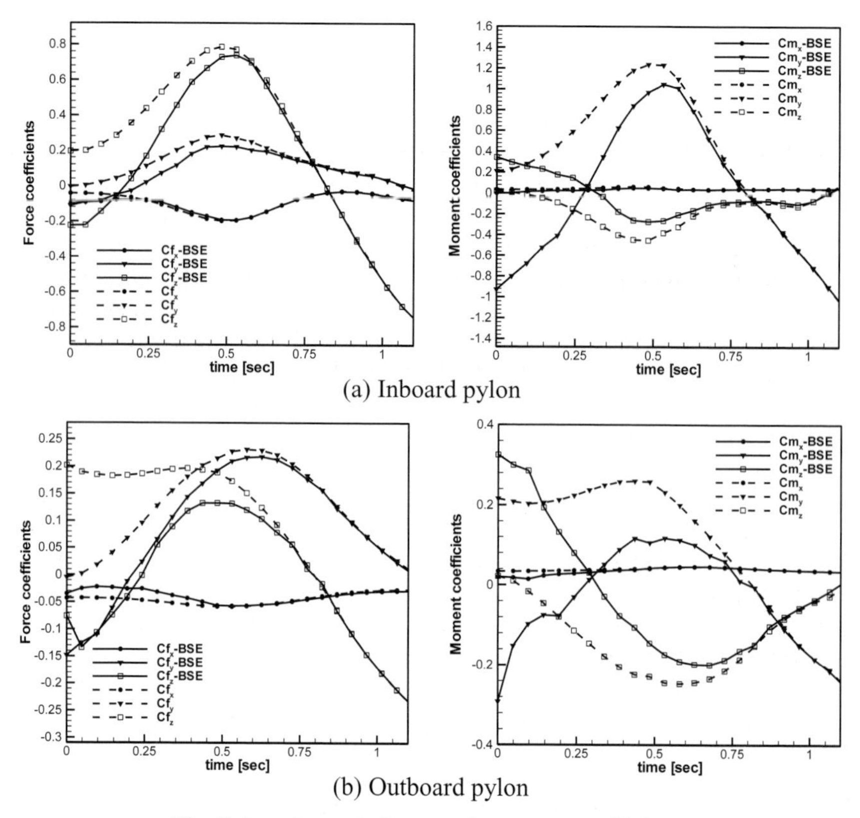


(a) Inboard pylon

(b) Outboard pylon

Fig. 8 Aerodynamic force and moment coefficients

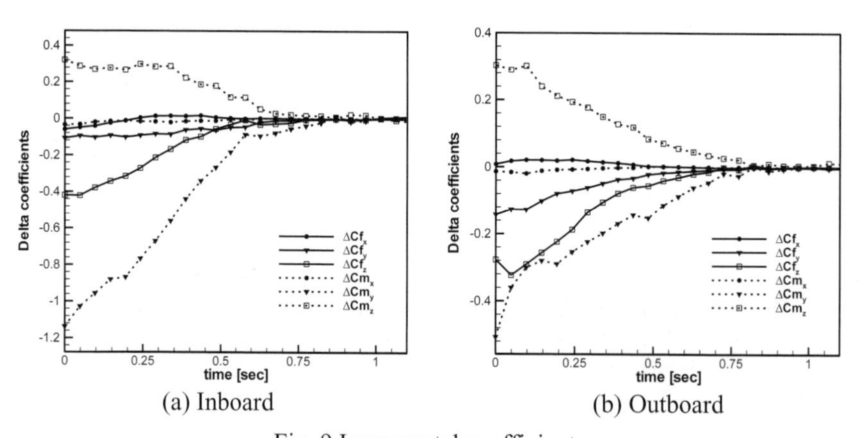


(a) Inboard (b) Outboard

Fig. 9 Incremental coefficients

## 5. Conclusion

The two step simulation can be used for the target point estimation and generation of bomb trajectory exactly. In the first step, the released bomb trajectory in aircraft flow-field has been computed and BSE has been check simultaneously using the developed algorithm to direct calculation aircraft flow-field effects. It can be used as the indicator to switch from an unsteady CFD calculation in aircraft flow field to the off-line simulation in free flight condition in the second step.

Two dimensional bomb separation problems in subsonic and supersonic flows have been performed. The trajectories of released store and ejected submunition have been computed including unsteady dynamics and ejector induced forces and moments in these cases. Furthermore, the BSE vanishing can be estimated by means of direct calculation of incremental coefficients. So the aircraft flow-field effects can be counted by this developed method. It has been applied to the generic bomb released from KT-1 aircraft and influences of aircraft flow-field effects are computed. When 1.256 seconds, the bomb is departed from aircraft about 1.52 times of half span length and incremental coefficients are almost decayed(below 2% of the maximum magnitudes). The unsteady trajectory calculations up to this time do not require too much resources and computing time. Furthermore, the off-line simulation in the free flight condition is very simple. Therefore, this two step simulation including BSE check is applicable generally.

The final trajectory and force coefficients in this paper can be used for the off-line simulation of TGP in the next step. The full trajectory to target points by this method can be compared with flight test data will be performed in the future.

**REFERENCES**

[1] Vlad Arie Gaton and Mark Harmatz, "Computational Estimate of the Separation Effect," *Journal of Aircraft*, Vol. 32, No. 6, Nov.-Dec. 1995.

[2] H. Paschall Massengill, Jr., "A Technique for Predicting Aircraft Flow-Field Effects Upon an Unguided Bomb Ballistic Trajectory and Comparison with Flight Test Results," AIAA paper 93-0856. Reno, NV, Jan. 11-14, 1993.

[3] S. Davids and A. Cenko, "Grid Based Approach to Store Separation," AIAA paper 2001-2418, Anaheim, CA, June 11-14, 2001.

[4] L.E. Lijewski and N.E. Suhs, "Time-Accurate Computational Fluid Dynamics Approach to Transonic Store Separation Trajectory Prediction," *Journal of Aircraft*, Vol.31, No.4, July-Aug, 1994.

[5] K.S. Dunworth, D.J. Atkins and J.M. Lee, "Incorporation of CFD Generated Aerodynamic Data in Store Separation Predictions," AIAA paper 2005-846, Reno, NV, Jan. 10-13, 2005.

[6] M.F.E. Dillenius, S.C. Perkins Jr. and D. Nixon, "Tactical missile aerodynamics," Chapter 13, pp. 575-666, Corporate Author American Institute of Aeronautics and Astronautics 2nd ed., 1992.

[7] Eugene Kim, Soo Hyung Park and Jang Hyuk Kwon, "Parallel Performance Assessment of Moving Body Overset Grid Application on PC cluster," Parallel CFD conference, Busan, South Korea, May 15-18, 2006.

# Parallel Implementation of a Gas-Kinetic BGK Method on Unstructured Grids for 3-D Inviscid Missile Flows

**Murat Ilgaz[a] and Ismail H. Tuncer[b]**

[a]Aerodynamics Division, Defense Industries Research and Development Institute, 06261, Ankara, Turkey

[b]Department of Aerospace Engineering, Middle East Technical University, 06531, Ankara, Turkey

A 3-D gas-kinetic BGK method and its parallel solution algorithm are developed for the computation of inviscid missile flows on unstructured grids. Flow solutions over a supersonic missile are presented to validate the accuracy and robustness of the method. It is shown that the computation time, which is an important deficiency of gas-kinetic BGK methods, may significantly be reduced by performing computations in parallel.

## 1. INTRODUCTION

The use of gas-kinetic methods for the simulation of compressible flows has become wide-spread in the two last decades. Among all the gas-kinetic methods, the most promising ones are the Equilibrium Flux Method (EFM)[1], the Kinetic Flux Vector Splitting (KFVS)[2,3] and the Gas-Kinetic BGK method [4,5]. Due to the inclusion of intermolecular collisions with BGK simplification, the gas-kinetic BGK method provides a more complete and a realistic description of flows and have been well studied [6–13] Although the gas-kinetic BGK method was first applied on structured grids, later, the method has been adapted on unstructured grids for flow predictions around complex bodies [14–16]. Recently, Ilgaz and Tuncer successfully applied the kinetic flux vector splitting and gas-kinetic BGK methods for 2-D flows on unstructured grids.[17] It has been shown that the gas-kinetic methods do not experience numerical instabilities as their classical counterparts.[17] In this work, the computation time, the basic deficiency of the gas-kinetic methods, has been improved by performing computations in parallel. Various flow problems were studied and it is shown that the computational efficiency of the gas-kinetic methods can be improved significantly.

While inviscid flow solutions can be obtained successfully on unstructured triangular grids, viscous flow solutions, especially the resolution of boundary layers is still not an easy task. Highly skew cells in the boundary layer regions similarly cause convergence difficulties in the numerical implementation of the gas-kinetic method similar to the those of classical methods. In addition, it is observed in the literature that the gas-kinetic BGK method is mostly applied to 1-D and 2-D flows, its applications to 3-D flow problems

I.H. Tuncer et al., *Parallel Computational Fluid Dynamics 2007*,
DOI: 10.1007/978-3-540-92744-0_56, © Springer-Verlag Berlin Heidelberg 2009

are quite limited. Recently, Ilgaz and Tuncer obtained 2-D viscous flow solutions on unstructured hybrid grids, and presented the high-order finite volume formulation of the gas-kinetic BGK method.[18]

In this study, the developmnent of a gas-kinetic BGK method for 3-D inviscid flows on unstructured grids is presented. The first order accurate finite volume formulations are given and the solutions are obtained in parallel. An inviscid flow over a missile geometry is presented to show the the numerical efficiency of the parallel computations and the the accuracy of the method developed.

## 2. GAS-KINETIC THEORY

In the gas-kinetic theory, gases are comprised of small particles and each particle has a mass and velocity. At standard conditions, the motion of large number of particles in a small volume is defined by the particle distribution function, which describes the probability of particles to be located in a certain velocity interval

$$f(x_i, t, u_i). \tag{1}$$

Here $x_i = (x, y, z)$ is the position, $t$ is the time and $u_i = (u, v, w)$ are the particle velocities. The macroscopic properties of the gas can be obtained as the moments of the distribution function. For example, the gas density can be written as

$$\rho = \sum_i m \cdot n_i \tag{2}$$

where $m$ is the particle mass, $n_i$ is the number density. Since, by definition, distribution function is the particle density in phase space, it is concluded that

$$m \cdot n_i = f(x_i, t, u_i), \tag{3}$$

$$\rho = \int\int\int f \cdot du \cdot dv \cdot dw. \tag{4}$$

The time evolution of the distribution function is governed by the Boltzmann equation

$$f_t + u_i \cdot f_{x_i} + a_i \cdot f_{u_i} = Q(f, f). \tag{5}$$

Here $a_i$ shows the external force on the particle in the $i$th direction and $Q(f, f)$ is the collision operator. When the collision operator is equal to zero, collisionless Boltzmann equation is obtained and the solution of this equation is given in terms of the Maxwellian (equilibrium) distribution function

$$g = \rho \cdot (\frac{\lambda}{\pi})^{\frac{N+3}{2}} \cdot exp\,\{-\lambda \cdot [(u_i - U_i)^2 + {\xi_i}^2]\} \tag{6}$$

where $\xi_i = (\xi_1, \xi_2, ...., \xi_N)$ are the particle internal velocities, $N$ is the internal degrees of freedom, $U_i = (U, V, W)$ are the macroscopic velocities of the gas and $\lambda$ is a function of temperature given by

$$\lambda = \frac{1}{2 \cdot R \cdot T} \tag{7}$$

$R$ being the gas constant.

### 2.1. Gas-Kinetic BGK Method

The gas-kinetic BGK method is based on the Boltzmann BGK equation where the collision operator (see Eq. (5)) is replaced by the Bhatnagar-Gross-Krook model [19]. The Boltzmann BGK equation in 3-D can be written as (ignoring external forces)

$$f_t + u \cdot f_x + v \cdot f_y + w \cdot f_z = \frac{g - f}{\tau} \tag{8}$$

where $f$ is the gas distribution function, $g$ is the equilibrium state approached by $f$ over particle collision time $\tau$; $u$, $v$ and $w$ are the particle velocities in x-, y- and z-directions, respectively. The equilibrium state is usually assumed to be a Maxwellian

$$g = \rho \cdot (\frac{\lambda}{\pi})^{\frac{K+3}{2}} \, exp\, \{-\lambda \cdot [(u-U)^2 + (v-V)^2 + (w-W)^2 + \xi^2]\} \tag{9}$$

where $\rho$ is the density, $U$, $V$ and $W$ are the macroscopic velocities in x-, y- and z-directions and $K = N$ is the dimension of the internal velocities. The general solution of the gas distribution function $f$ at the cell interface $ci$ and time $t$ is

$$\begin{aligned} f(s_{ci}, t, u, v, w, \xi) = \; & \tfrac{1}{\tau} \textstyle\int_0^t g(s', t', u, v, w, \xi) \cdot exp\, [-(t - t')/\tau] \cdot dt' \\ & + exp\, (-t/\tau) \cdot f_0(s_{ci} - u \cdot t - v \cdot t - w \cdot t). \end{aligned} \tag{10}$$

Here $s' = s_{ci} - u \cdot (t - t') - v \cdot (t - t') - w \cdot (t - t')$ is the particle trajectory of a particle and $f_0$ is the initial gas distribution function at the beginning of each time step. Since mass, momentum and energy are conserved during particle collisions, $f$ and $g$ must satisfy the conservation constraint of

$$\int (f - g) \cdot \psi \cdot d\Xi = 0, \tag{11}$$

at all $(x, y, z)$ and $t$, where

$$\psi = \begin{bmatrix} 1 & u & v & w & \frac{1}{2}(u^2 + v^2 + w^2 + \xi^2) \end{bmatrix}^T. \tag{12}$$

## 3. NUMERICAL METHOD

The numerical solution method is based on the cell-centered finite volume formulation. The numerical fluxes at the cell faces are evaluated using a first order gas-kinetic BGK method. Figure 1 shows sample tetrahedral control volumes, cell interfaces and the coordinate systems where *dots in black* represent cell centers, the *dot in red* shows the interface centroid, $L$ and $R$ the left and right states, $x$, $y$ and $z$ the local coordinate system normal and tangent to the cell interface $ci$, respectively, $X$, $Y$ and $Z$ the global coordinate system. $U_x$, $U_y$ and $U_z$ are the macroscopic velocity components in the local coordinate system.

In the present work, the initial gas distribution function $f_0$ and the equilibrium state $g$ are assumed to be

$$f_0 = \begin{cases} g^L, & \text{if } x \leq x_{ci} \\ g^R, & \text{if } x < x_{ci} \end{cases}, \qquad g = g^0 \tag{13}$$

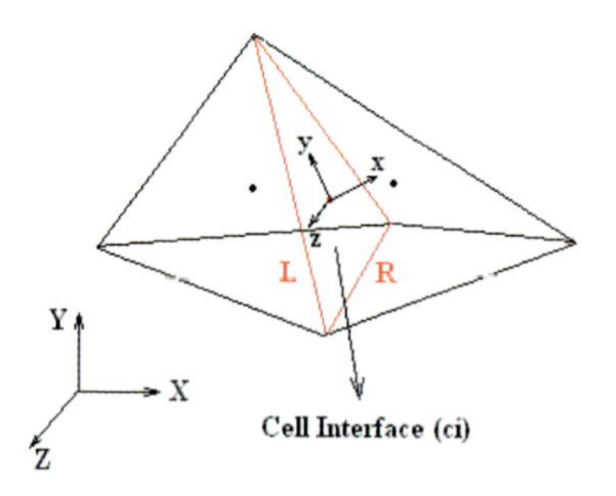


Figure 1. Sample control volumes, cell interface and coordinate systems.

where $g^L$, $g^R$ and $g^0$ are the equilibrium distribution functions to the left, right and at the centroid of the cell interface *ci*, respectively, of the form

$$g^* = \rho^* \cdot (\frac{\lambda^*}{\pi})^{\frac{K+3}{2}} \cdot exp\,\{-\lambda^* \cdot [(u_x - U_x^*)^2 + (u_y - U_y^*)^2 + (u_z - U_z^*)^2 + \xi^2]\}. \tag{14}$$

Here $u_x$, $u_y$ and $u_z$ are the local particle velocities normal and tangent to the cell interface, respectively. The equilibrium distribution functions to the left and right of the cell interface are obtained as

$$\int g^{L,R} \cdot \psi_{ci} \cdot d\Xi = \left[\begin{array}{ccccc} \rho^{L,R} & \rho U_x^{L,R} & \rho U_y^{L,R} & \rho U_z^{L,R} & \rho E^{L,R} \end{array}\right]^T . \tag{15}$$

The variables $\rho^*$, $U_x^*$, $U_y^*$, $U_z^*$ and $\lambda^*$ in $g^L$ and $g^R$ (Eq. (14)) can be uniquely determined from Eq. (15). The equilibrium state at the cell interface, $g^0$, can be evaluated using the compatibility condition

$$\int g^0 \cdot \psi_{ci} \cdot d\Xi = \int_{u \geq 0} g^L \cdot \psi_{ci} \cdot d\Xi + \int_{u<0} g^R \cdot \psi_{ci} \cdot d\Xi \tag{16}$$

from which the values of $\rho^0$, $U_x^0$, $U_y^0$, $U_z^0$ and $\lambda^0$ in $g^0$ are determined.

Substituting Eq. (13) into Eq. (10), the final gas distribution function at the cell interface *ci* is expressed as

$$\begin{aligned} f(x_{ci}, t, u, v, w, \xi) = \ & [1 - exp\,(-t/\tau)] \cdot g_0 \\ & + exp\,(-t/\tau) \cdot [H(u) \cdot g^L + (1 - H(u)) \cdot g^R] \end{aligned} \tag{17}$$

where $H(u)$ is the Heaviside function

$$H(u) = \begin{cases} 0, & \text{if } u < 0 \\ 1, & \text{if } u \geq 0 \end{cases} . \tag{18}$$

The local numerical fluxes for the mass, momentum and total energy across the cell interface *ci* can then be computed as

$$\begin{pmatrix} F_\rho \\ F_{\rho U_x} \\ F_{\rho U_y} \\ F_{\rho U_z} \\ F_{\rho E} \end{pmatrix} = \frac{1}{\Delta t} \cdot \int_0^{\Delta t} \int u_x \cdot f|_{x_{ci}} \cdot \psi_{ci} \cdot d\Xi \cdot dt. \tag{19}$$

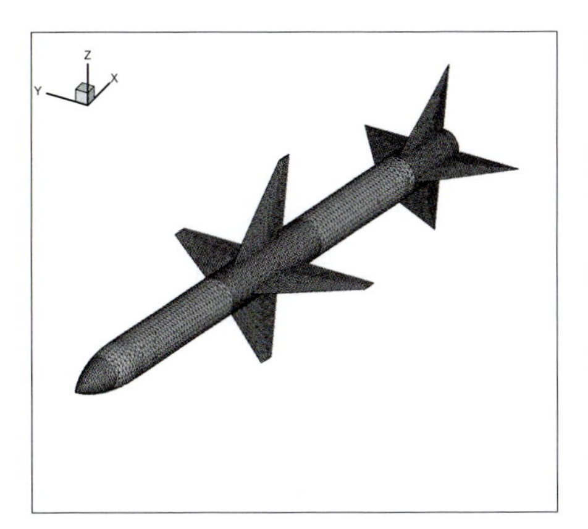


Figure 2. Surface grid.

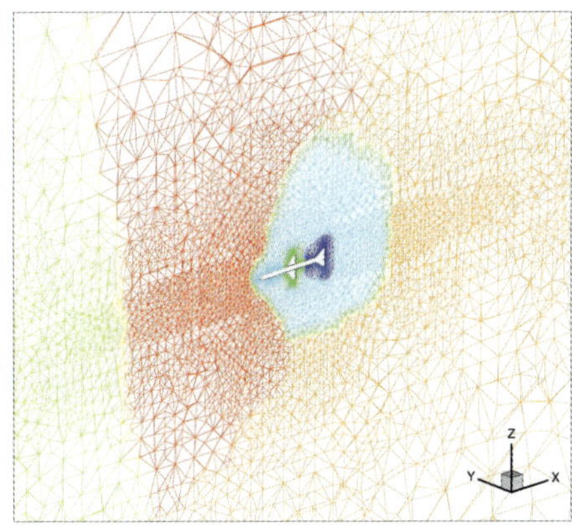


Figure 3. Partitions.

The collision time in Eq. (17) is taken as [13]

$$\tau = C_1 \cdot \Delta t + \frac{|p^L - p^R|}{p^L + p^R} \cdot \Delta t \tag{20}$$

where $C_1$ is a constant. The non-dimensional pressure jump in the second term corresponds to a numerical collision time. The second term is necessary especially in the under-resolved discontinuous regions and diminishes in smooth regions.

The fluxes at the cell faces are computed and the total flux for each cell is obtained by adding the flux contributions. The conservative variables at the cell centers are then updated using a third-order explicit Runge-Kutta time-stepping scheme.

### 3.1. Parallel Processing

Parallel processing is based on domain decomposition. Unstructured grids are partitioned using METIS. The inter-process communication between the partitions is performed using *Parallel Virtual Machine* (PVM) message-passing library routines.

Parallel computations are employed in a *master-worker* algorithm. The *master* process performs all the input-output, starts up PVM, spawns worker processes and sends the initial data to the workers. The *worker* processes first receive the initial data, apply the partition and flow boundary conditions, and solve the flow field within the partition. The flow variables at the partition boundaries are exchanged among the neighboring partitions at each time step.

## 4. RESULTS AND DISCUSSION

The validation of the gas-kinetic BGK method developed is shown for an inviscid supersonic flow over a Sparrow-type missile. The flow conditions are shown in Table 1 and the surface grid and the partitions are given in Figures 2 and 3, respectively. The computational grid consists of 214630 nodes and 1238305 cells. The flow solutions are obtained

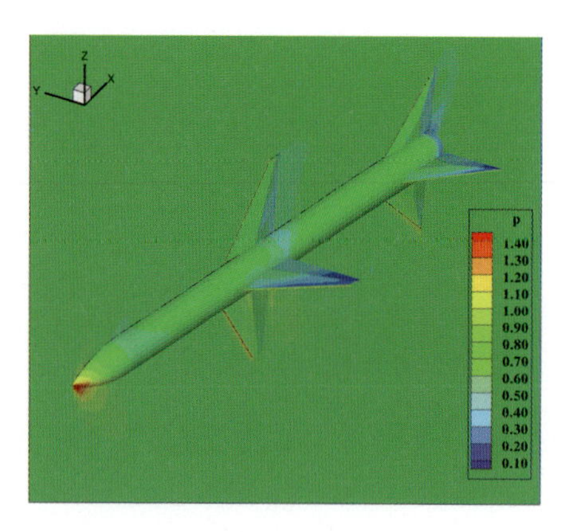


Figure 4. Pressure distribution at 10° angle of attack.

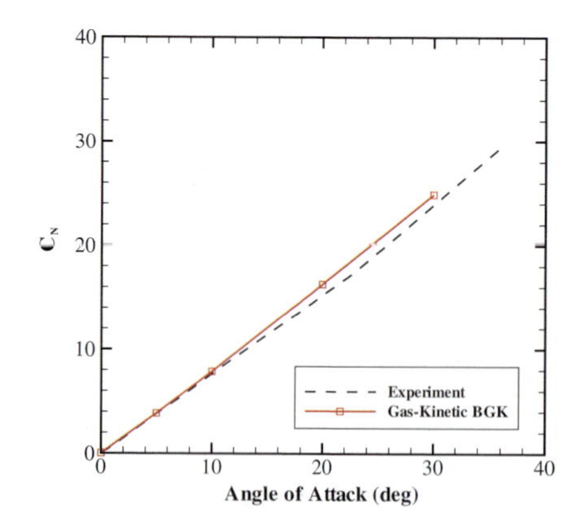


Figure 5. Normal force coefficient.

for various angles of attack and the results are compared to the available experimental data [20].

Table 1
Freestream conditions.

| Mach Number | Pressure, $Pa$ | Temperature, $K$ |
|---|---|---|
| 1.5 | 101325 | 288 |

Figure 4 shows the pressure distribution on the vertical symmetry plane of the missile at an angle of attack of 10°. The base pressure in the solution is corrected by setting the base pressure to the freestream pressure as specified in Reference [20]. The longitudinal aerodynamic load coefficients predicted by the gas-kinetic BGK method are presented in Figures 7, 5 and 6 along with the experimental data. The computed normal force and pitching moment coefficients agree well with the experimental values up to an angle of attack of approximately 12°. As expected, at higher angles of attack, the inviscid, attached flow assumption breaks down, and the predictions deviate from the experimental data. Similarly, the axial force, which strongly depends on the viscous effects and the base pressure, is underpredicted, and do not agree well with the experimental data.

The parallel computations are performed on the *AMD Opteron* and *Intel Itanium2* clusters operating on Linux kernel version 2.6 The dual AMD Opteron processors, each of which has a dual core, operate at 2.4Ghz with 1MB L2 cache and 2GB of memory per processor. The single core dual Intel Itanium2 processors operate at 1.3Ghz with 3MB L2 cache and 2GB of memory per processor. The parallel efficiency of the computations is shown in Table 2 and Figure 8 and the computational speed-up is given in Figure 9. It is observed that the computations are about twice as much faster on the AMD Opteron

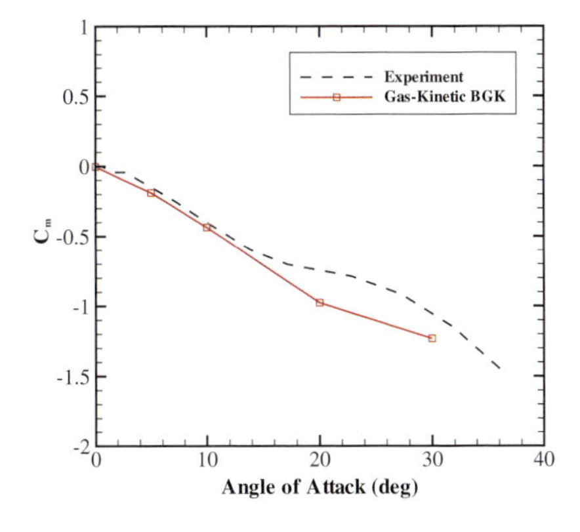


Figure 6. Pitching moment coefficient.

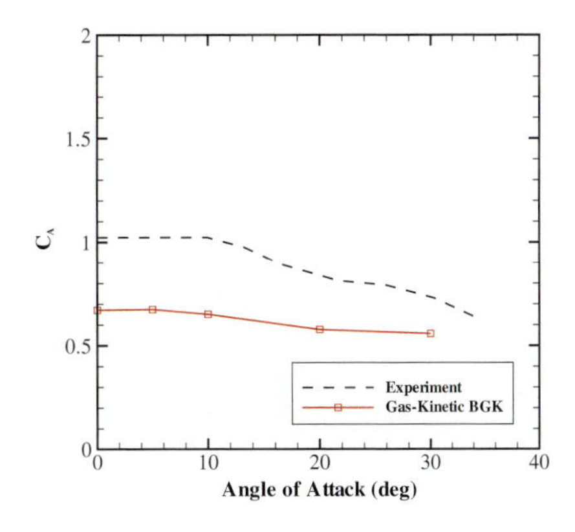


Figure 7. Axial force coefficient.

Table 2
Parallel Efficiency.

| Number of | AMD Opteron Processor | | Intel Itanium2 Processor | |
|---|---|---|---|---|
| Processors | *sec/iter* | % | *sec/iter* | % |
| 1 | 7.409 | 100 | 13.738 | 100 |
| 2 | 3.755 | 98.66 | 6.990 | 98.27 |
| 4 | 1.951 | 94.94 | 3.644 | 94.25 |
| 8 | 1.036 | 89.39 | 1.957 | 87.75 |
| 16 | 0.569 | 81.38 | 1.085 | 79.14 |

cluster than the Intel Itanium2 cluster. The parallel efficiency of computations stays high as the number of processors increases, which is attributed to the high computing to communication ratio in gas-kinetic BGK methods. It is also observed that the Opteron cluster has a slightly higher efficiency than the Itanium2 cluster.

## 5. CONCLUDING REMARKS

In this paper, the development of a parallel, 3-D flow solver based on a gas-kinetic BGK method on unstructured grids is presented. The method is validated for inviscid, supersonic missile flows successfully. The aerodynamic load predictions for mostly attached flows at angles of attack below 12° agree well with the experimental data except for the axial force prediction. The parallel computations are performed in two computer clusters based on AMD opteron and Intel Itanium2 processors. It is shown that parallel computations improve the computational efficiency of the gas-kinetic BGK method significantly. Such a high parallel efficiency and the improvement in the turn-around time of a flow

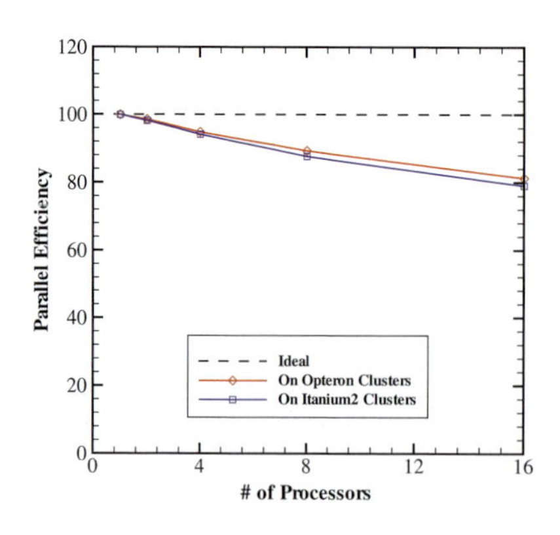


Figure 8. Parallel efficiency.

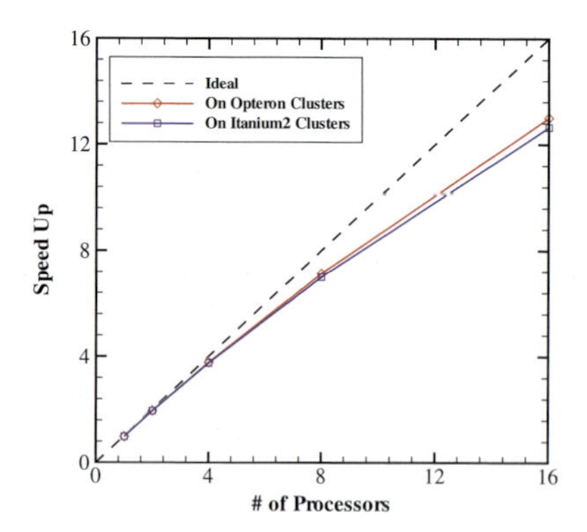


Figure 9. Computational speed up.

solution are expected to promote the use of gas-kinetic BGK methods for the solution of practical flow problems.

## REFERENCES

1. D. I. Pullin, J. Comp. Phys., 34, (1980), 231.
2. J. C. Mandal and S. M. Deshpande, Comp. Fluids, 23-2, (1994), 447.
3. S. Y. Chou and D. Baganoff, J. Comp. Phys., 130, (1997), 217.
4. K. H. Prendergast and K. Xu, J. Comp. Phys., 109, (1993), 53.
5. K. H. Prendergast and K. Xu, J. Comp. Phys., 114, (1994), 9.
6. K. Xu and A. Jameson, AIAA-95-1736, (1995).
7. K. Xu, L. Martinelli and A. Jameson, J. Comp. Phys., 120, (1995), 48.
8. K. Xu, J. Comp. Phys., 134, (1997), 122.
9. K. Xu, SIAM J. Sci. Comp., 20-4, (1997), 1317.
10. K. Xu and J. Hu, J. Comp. Phys., 142, (1998), 412.
11. D. Chae, C. Kim and O. Rho, J. Comp. Phys., 158, (2000), 1.
12. Y. S. Lian and K. Xu, J. Comp. Phys., 163, (2000), 349.
13. K. Xu, J. Comp. Phys., 171, (2001), 289.
14. C. Kim and A. Jameson, J. Comp. Phys., 143, (1998), 598.
15. G. May, B. Srinivasan and A. Jameson, AIAA-2005-1397, (2005).
16. G. May and A. Jameson, AIAA-2005-5106, (2005).
17. M. Ilgaz and I. H. Tuncer, AIAC-2005-018, (2005).
18. M. Ilgaz and I. H. Tuncer, AIAA-2006-3919, (2006).
19. P. L. Bhatnagar, E. P. Gross and M. Krook, Phys. Rev., 94, (1954), 511.
20. W. J. Monta, NASA TP 1078, (1977).

# 3-D Time-Accurate Inviscid and Viscous CFD Simulations of Wind Turbine Rotor Flow Fields

Nilay Sezer-Uzol[a*], Ankur Gupta[b†], and Lyle N. Long[a‡]

[a]Dept. of Aerospace Engineering, The Pennsylvania State University, University Park, PA, 16802, USA

[b]Dept. of Mechanical Engineering, The Pennsylvania State University, University Park, PA, 16802, USA

This paper presents the results of three-dimensional and time-accurate, parallel Computational Fluid Dynamics (CFD) simulations of the flow field around a Horizontal Axis Wind Turbine (HAWT) rotor. Large Eddy Simulation (LES) methodology with an instantaneous log-law wall model is used for the viscous computations. The 3-D, unsteady, parallel, finite volume flow solver, PUMA2, is used for the simulations. The solutions are obtained using unstructured moving grids rotating with the turbine blades. Three different flow cases with different wind speeds and wind yaw angles are investigated. Results for these three cases and comparisons with the experimental data are presented.

## 1. Introduction

Wind turbines, which offer the promise of less expensive and clean energy from a renewable energy source: the wind, have unique aerodynamic and aeroacoustic characteristics that make their prediction more challenging in many ways than already complicated problems of helicopter rotors or propellers. In particular, wind turbine blades can experience large changes in angle of attack associated with sudden large gusts, changes in wind direction (wind yaw), atmospheric boundary layer effects, atmospheric turbulence, or interaction with the unsteady wake shed from the tower support on downwind HAWTs. These three dimensional and unsteady blade/inflow/tower wake interactions can result in impulsive loading changes and dynamic stall over portions of the rotating blades [1]. Furthermore, rotor broadband noise sources (turbulence ingestion noise, airfoil self noise, tip-vortex noise, and other broadband noise sources) are considered as the primary noise source for wind turbines. In addition, the acceptance of wind turbines by the public depends strongly on achieving low noise levels in application. Hence, the accurate predictions of aerodynamic loads and noise are very important and challenging in the design process of wind turbines. These predictions could also lead to lighter and more flexible structures, which are necessary to reduce the cost of wind energy.

*Instructor, Dept. of Mechanical Engineering, TOBB University of Economics and Technology, Ankara, 06560, Turkey. E-mail: nsezeruzol@etu.edu.tr

†Graduate student. Dept. of Computer Science and Engineering, The Pennsylvania State University, University Park, PA, 16802, USA. E-mail: azg139@psu.edu

‡Professor, Dept. of Aerospace Engineering, The Pennsylvania State University, University Park, PA, 16802, USA. E-mail: lnl@psu.edu

I.H. Tuncer et al., *Parallel Computational Fluid Dynamics 2007*,
DOI: 10.1007/978-3-540-92744-0_57, © Springer-Verlag Berlin Heidelberg 2009

Three-dimensional flow properties of rotating blades are an essential feature of any wind turbine aerodynamic or aeroacoustic simulation. While important information can be learned from 2-D and non-rotating simulations, some aspects of the physics of wind turbine aerodynamics and noise must be obtained from rotating blade simulations. Three-dimensional flow over rotating blades can be significantly different than the flow over a wing, and there can also be dramatic differences between 2-D and 3-D simulations [2,3]. Rotating blades can have significant spanwise (or radial) flow. This radial flow can result in coriolis forces which may induce favorable pressure gradients and delay boundary layer separation [4,5]. Also, the blade speed varies linearly from root to tip. In addition, the three-dimensional wake of a rotating blade can remain in close proximity to the blade for a long period of time (compared to the wake of a wing). Using the time-dependent governing equations allows the simulation of a number of important phenomena: broadband noise, incoming atmospheric turbulence and gusts, wind shear (or atmospheric boundary layer). These are important for leading edge noise and tip noise prediction. By incorporating time dependent boundary conditions, either a gust or turbulent incoming flow can be introduced. Also, atmospheric gusts or wind turbine tower shadow effects can be simulated using source terms in the time-accurate, three-dimensional, governing momentum and energy equations. For these reasons, time-accurate, three-dimensional and compressible rotating blade simulations are essential.

Previous 3-D simulations include hybrid potential-RANS simulations [6] and incompressible RANS computations using a $k-\omega$ SST model [7] for the NREL blind code comparison [8] for an upwind Phase VI rotor for zero yaw for different wind speeds. Results of a DES computation of a non-rotating (parked) NREL Phase VI wind turbine blade undergoing pitching motion in the light stall region [9] have shown better accuracy for highly separated flows owing to the better resolution of three-dimensional flow structures, compared to the two-equation RANS calculations. For flows with unsteady separation and vortex shedding, Large Eddy Simulation, LES, generally gives better results compared to RANS predictions [10]. LES allows the representation of a wide range of scales of noise-generating eddies. Hence, LES is a valuable tool in calculating the noise sources at Reynolds numbers of engineering interest [11].

This paper presents the results of three-dimensional and time-accurate CFD simulations of the flow field around 2-bladed NREL Phase VI wind turbine rotor ([12–14]). The solutions are obtained using the parallel finite volume flow solver PUMA2 [13,15,16] with unstructured moving grids rotating with the turbine blades. Three different flow cases with different wind speeds and wind yaw angles are investigated: 7 m/s with 0° yaw (pre-stall case I), 7 m/s with 30° yaw (pre-stall, yawed case II), and 15 m/s with 0° yaw (post-stall case III). Results of the viscous simulations and comparisons with the inviscid results [12,13] and the experimental data are presented.

## 2. Computational Cases and Parallel Computations

The NREL Phase VI rotor model and the unstructured grids generated using Gridgen are described in detail in [12,13]. The Grid A with 3.6 million tetrahedral cells was used for all the inviscid computations whereas it was used only for cases I and II for the viscous computations. The Grid B with 4.2 million cells with more clustering around the blades is used for LES case III so that the average $y^+$ value of the first grid points away from the wall is small. The blades rotate about the $Y$-axis, in the negative direction at 72 rpm (7.54 rad/sec). Wind speed in the positive $Y$-direction with different yaw angles has been considered.

All the simulations were performed in parallel, mostly on the Linux Beowulf clusters available at Penn-State (LION-XL and MUFASA), and also on some of the supercomputers such as

Table 1
Characteristics of parallel computers and computational performance of simulations.

| | # of compute nodes | processors | memory | network |
|---|---|---|---|---|
| **PSU Lion-xl** | 128 | Dual 2.4 GHz Intel P4 | 4 GB of ECC RAM | Quadrics high-speed interconnect |
| **PSU Mufasa** | 81 | Dual 2.8 GHz AMD Athlon MP2200+ | 2 GB Memory (10 nodes) & 1 GB Memory (71 nodes). | Dolphin and fast ethernet networks |
| **NREL Lester** | STD = 56 HSN = 40 | Dual 1.8 GHz. AMD Opteron | 4 GB memory | STD = Gigabit Ethernet (50 MB/s) HSN = Dual Port Myrinet (800 MB/s) |
| **NCSA Tungsten** | 1280 | Dual 3.2 GHz Intel Xeon | 3 GB ECC DDR SDRAM memory | Gigabit Ethernet and Myrinet 2000 |
| **NAS Columbia** | 20 SGI Altix 3700 superclusters, each with 512 processors. | Dual 1.5 GHz Intel Itanium 2 | Global shared memory across 512 processors, 1 terabyte of memory per 512 processors | **Interconnect** SGI® NUMAlink™ InfiniBand network 10 gigabit Ethernet 1 gigabit Ethernet |

(a) Parallel computers

| # of processors | memory/ node [MB] | wall time required per revolution [days] | | | | |
|---|---|---|---|---|---|---|
| | | **Lion-xl** | **Mufasa** | **NREL cluster Std** | **NCSA cluster** | **NASA cluster** |
| 16 | 297 | 14 | 42 | 15 | 10 | 11 |
| 32 | 192 | 6 | 12 | 10 | 5 | 5.4 |
| 64 | 140 | - | 8 | - | 2.4 | 3.5 |
| 128 | 113 | - | 5 | - | 1.7 | 1.8 |

(b) Computational performance

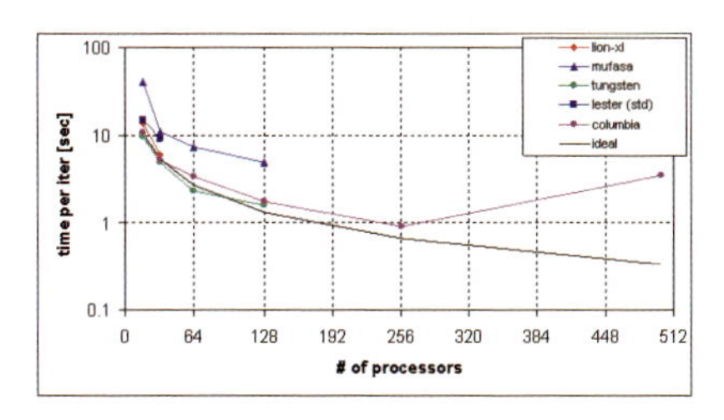


Figure 1. Computational performance of inviscid simulations on different clusters.

NREL, NCSA and NASA clusters (Table 1(a)). Table 1(b) and Figure 1 show the computational performance of PUMA2 inviscid computations on different clusters. The inviscid computation on LION-XL using 32 processors takes 6 days of wall-time for one full revolution (i.e. 0.8333 seconds) which requires 90,000 iterations. The LES computational cases were run on the NCSA Tungsten supercomputer using 64 processors. The time required for one revolution for the LES computations for cases I and II was 35 wall-time hours, whereas case III required 190 wall-time hours for a 90 degree rotation of the blades. This was due to the much smaller time step size and more cells in Grid B compared to Grid A. As one full revolution requires 0.833 sec, 104,167 iterations were required per revolution for cases I and II, and 1,800,000 iterations for case III.

For inviscid simulations, a 4-stage Runge-Kutta numerical time integration method and Roe's numerical flux scheme was used. Time-accurate computations were started from the freestream conditions (for a given wind speed). Because the time step for time-accurate computations using an explicit scheme is determined by the smallest cell in the volume grid, the minimum cell size and maximum number of cells were selected during the grid generation process by also considering the total computational time needed for the time-accurate simulations. The simulation time step size for the inviscid cases was selected as $9.26 \mu s$ so that the calculated a Courant-Friedrichs-Lewy (CFL) number will be less than 1.0 for the smallest cell size (0.0033 m).

For Large Eddy Simulations, Roe's scheme was used to compute the inviscid fluxes. Time integration was done using a two stage Runge-Kutta method. Riemann Inflow-Outflow boundary conditions were used at the outer surface and no-slip velocity boundary conditions are applied on the surface of the blades. Classical Smagorinsky model was used to model the sub grid-scale stresses and the value of $C_S$ was taken to be 0.10. An instantaneous log-law wall model was used to calculate the shear stress at the first grid point away from the blades. This local wall

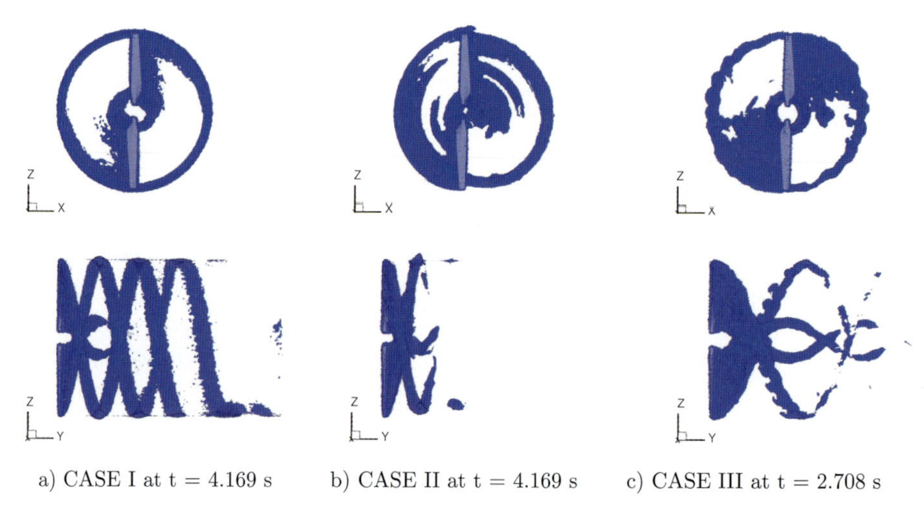


a) CASE I at t = 4.169 s b) CASE II at t = 4.169 s c) CASE III at t = 2.708 s

Figure 2. LES Results: Vorticity iso-surface for a) CASE I, b) CASE II, c) CASE III.

stress was then fed back to the outer LES in the form of proper momentum flux at the wall. For case I the time-accurate LES computations were started from the inviscid solution at the end of the 2nd revolution. For the other two cases, the computations were started from the inviscid solution at the end of the 3rd revolution. The maximum $y^+$ value at the first grid point away from the blades was found to be 985 for case I and 1043 for case II for Grid A. This corresponds to the upper edge of the logarithmic layer. The simulation time step was selected as $8\mu s$ for cases I and II so that the calculated CFL number and the Von Neumann number (for stability of viscous terms) for the smallest cell size will be less than one, and 0.5 respectively. The maximum $y^+$ value for case III was 576 for Grid B with the smallest cell size of 0.00017 m, and time step of $0.46\mu s$.

## 3. Results

From the results for instantaneous vorticity iso-surfaces presented in [12,13], it is observed that the flow is attached for inviscid case I and well defined vortical structures are shed from the blade tips. For case II, the flow is again attached, however the wake is asymmetrical because of the 30° yaw angle. For the higher wind speed case, i.e. case III, the flow is massively separated over the entire blade span. Although the tip vortices can still be depicted, the flow is highly unsteady because of the separation. This is much better observed when time-accurate results are visualized as an animation. In addition, the wake structures get convected at a higher speed compared to cases I and II. The vortical wake of case I was plotted at the end of 2nd revolution, and it is expected that it will continue to develop as the computation is continued for more revolutions. However, it is observed that the vortices are diffused for cases II and III. This may be due to the grid quality and numerical dissipation which need to be studied further. Also, the grid used for all three cases was created considering the 0° yaw cases, so it may not be the best choice for the yawed case II which has an unsymmetric skewed wake structure.

The plots of instantaneous vorticity iso-surfaces in Figure 2 for LES cases I and II are after the 5th revolution, whereas the plot for case III is at a 90 degree rotation after the 3rd revolution. In comparison to the inviscid results, we see that the vortices diffuse more in the stream-wise

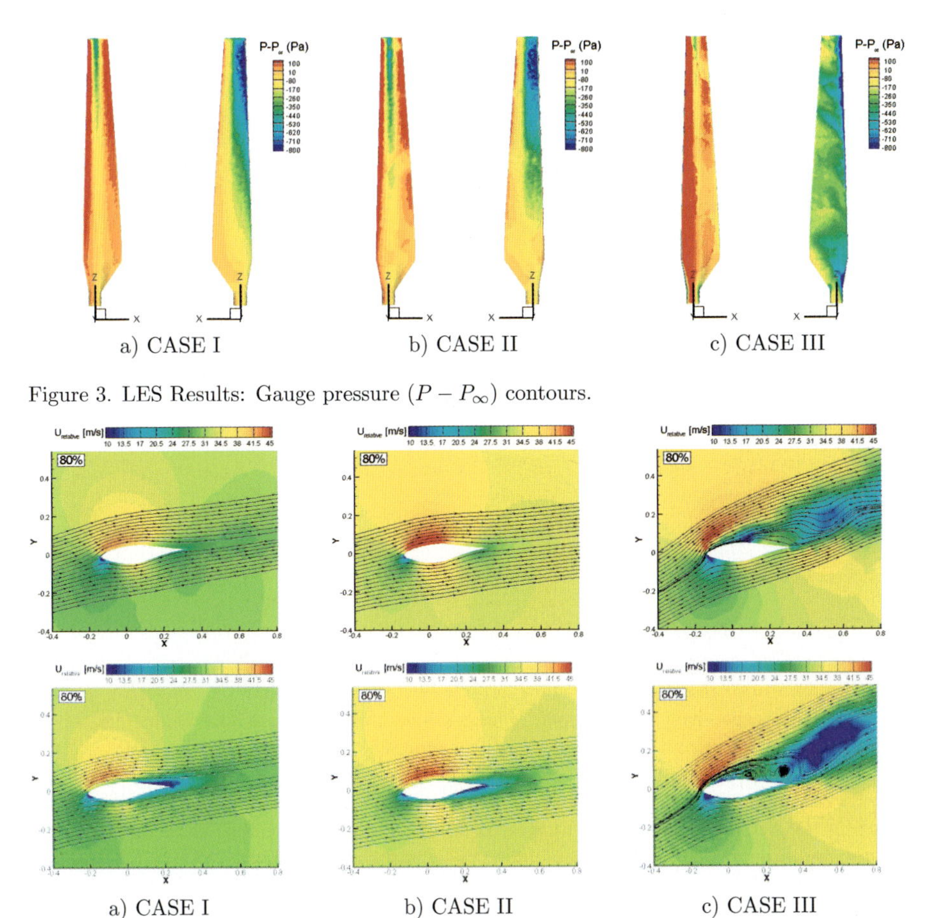


Figure 3. LES Results: Gauge pressure $(P - P_\infty)$ contours.



Figure 4. Relative velocity magnitude contours with streamlines around the airfoil sections at 80% span for inviscid results above and LES results below.

direction. In the inviscid cases, the factors that caused diffusion were the numerical scheme and the unstructured grid which becomes coarser in the stream-wise direction. For the viscous cases, in addition to the above factors we have viscous diffusion resulting in more diffusion of the vortices. It is observed that the vortices are more dissipated for case II as compared to the other cases. This could be attributed to the fact that the grid was refined so as to capture the wake shed in the positive Y-direction whereas in this case the wake shed would be inclined at an angle (the yaw angle). Further, as compared to the inviscid result for this case, a distinct vortical structure shed near the middle of the blade span is observed. For case III, we observe that the flow is massively separated over the span of the blades. The flow is highly unsteady and turbulent in this case. We can see blobs of vortical structures shed from the span of the blades. Also, as wind speed is higher in this case, the wake structure repeats itself over a greater length in the wind direction.

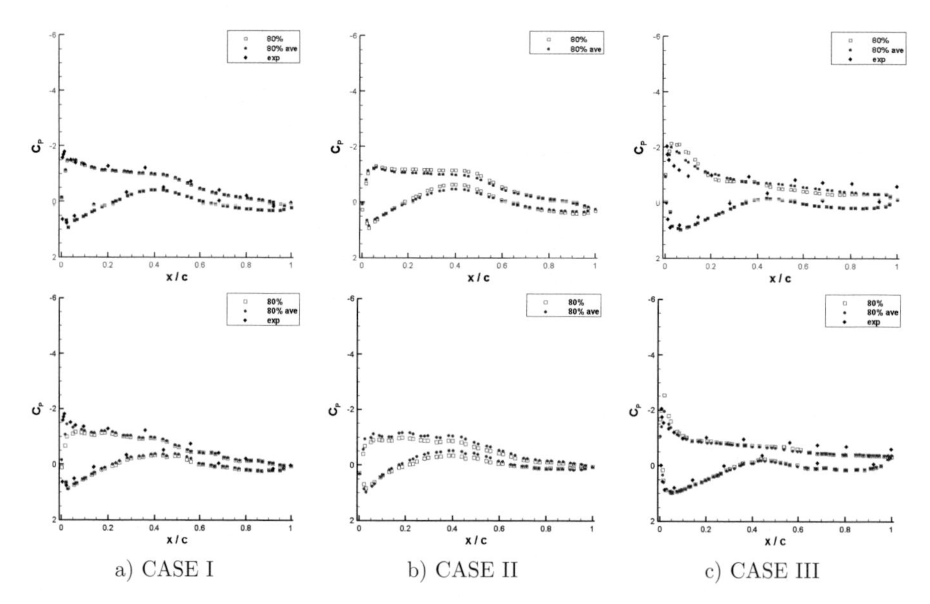


Figure 5. Chordwise pressure coefficients at 80% span (Inviscid: above & LES: below).

The instantaneous pressure contours on the rotor blade upper and lower surfaces (on the left and on the right, respectively, for each pair of blades) are shown in Figure 3 for blade 2 (at the 12 o'clock position where azimuth angle = 0°). For both inviscid and LES results, the contours show considerable spanwise pressure variations in addition to the chordwise variations. As compared to the inviscid results [12,13], we can see that the pressure is more irregular and fluctuating throughout the span of the blades for LES cases. For LES case II, the sweep effect of the vortex structure shed around the middle of the blade produces the distinct pressure variation. Similar vorticity "sweep effects" can be seen throughout the span of the blades for LES case III.

The instantaneous contours of relative velocity magnitude together with streamlines around the airfoil sections at 80% for blade 2 can be seen in Figure 4 for all LES cases. The flow is attached for both cases I and II, whereas it is massively separated for case III. This separation occurs due to the high angle of attack created by the higher wind speed than the other two cases while the rotational speed of the turbine is kept the same for all cases. As expected, in comparison to the inviscid results, the relative velocities are lower specifically near the leading edge and the trailing edge of the airfoil due to viscous effects.

Figure 5 shows the instantaneous and averaged chordwise pressure coefficient distributions at 80% spanwise station for blade 2, and other locations are presented in [12,13]. There is good agreement between the experimental data (from [17]) and the computations for inviscid case I. There is more deviation between the experiments and computations for inviscid case III, especially on the blade upper surface. However, the comparisons are still reasonable keeping in mind that the computations are inviscid on a relatively coarse grid. The highest differences between the instantaneous and averaged $C_P$ distributions are observed for the 30° yaw case (case II). This occurs due to the cyclic unsteady variation of blade surface pressure as the blade rotates. This is evident from the instantaneous surface pressure distributions on blade 1 and blade 2 [13], in which the blades are at different azimuthal positions. The unsteady pressure

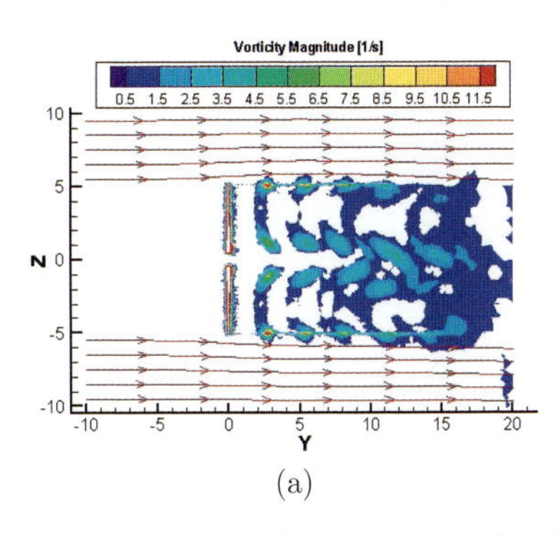


(a)

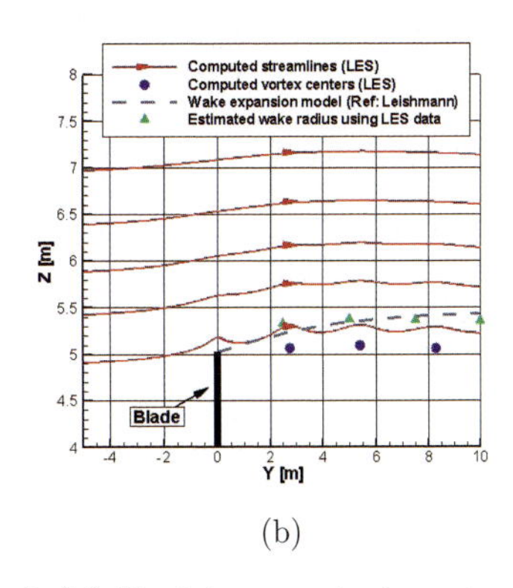


(b)

Figure 6. Rotor wake characteristics of LES Case I: (a) Vorticity magnitude contours and absolute streamlines. (b) Comparison of wake expansion characteristics.

variations also occurs for case III, however, there is no obvious cyclic variation as in case II. Consequently, the difference between the instantaneous and averaged pressure distributions is much less compared to case II, but still higher than case I which shows minimal level of cyclic variation as expected. For LES case III, we see that there is better agreement with experiments as compared to the inviscid results for 80% and also for 30% and 45% locations [13]. For the other spanwise locations, the agreement is not that good on parts of the upper blade surface. Given the fact that a simple Smagorinsky model with a log-law wall model was used, the computational results compare well with the experiments.

When the wake expansion characteristics of the LES results (for CASE I at t = 4.169 s), i.e. both the absolute streamlines and the estimated wake radii [13], are compared with the wake expansion model [18] as shown in Figure 6, the results are consistent with each other. The computed absolute streamlines clearly show the wake is expanding through the turbine rotor. Near the tip of the rotor blade, the streamlines become more wiggly due to the effect of the tip vortices. The cores of these vortices are also marked in this figure as blue circular symbols. The expansion in the case of LES seems to be less than the one predicted by the model when we get further away from the blade within the wake. This might be due to the strong root vortices shed from the rotor. Also, the numerical dissipation due to the computational method as well as the coarser grid after $Y = 15$ m might have an effect on the wake expansion. Also, further time-accurate computations for more revolutions might be necessary to see the effect of convergence. These points need further detailed analysis and investigation.

## 4. Conclusions

Time-accurate inviscid and LES simulations were performed for the 2-bladed NREL Phase VI wind turbine rotor using the 3-D, unsteady, parallel, finite volume flow solver, PUMA2, with rotating unstructured tetrahedral grids [12–14,19]. The LES simulations were performed for the quiescent air cases by using classical Smagorinsky model to model the sub grid-scale stresses. The instantaneous log-law wall model is implemented successfully to approximate the boundary layer effects near wall regions and applied in these rotating blade simulations. This presented an opportunity to simulate practical flows with high Reynolds number on complex bodies with reasonable accuracy.

Results from the inviscid and LES simulations for the three cases and comparisons with the experimental data were presented. The results show that the flow is attached for the pre-stall cases (I and II) with 0° and 30° yaw angles, with the latter having an asymmetrical wake structure, whereas there is massive separation over the entire blade span in the post-stall case (III), which has a higher wind speed of 15 m/s. Considerable spanwise pressure variations, in addition to the chordwise variations, are also observed in all three cases. The difference between the inviscid and the viscous computational results has been discussed in detail. Comparisons of sectional pressure coefficient distributions with experimental data show good agreement.

## REFERENCES

1. Robinson, M. C., Hand, M. M., Simms, D. A., and Schreck, S. J. NREL/CP-500-26337, 1999.
2. Corfeld, K., Strawn, R., and Long, L. N. Journal of the American Helicopter Society, Vol. 49, No. 3, 2004, pp. 350-356.
3. Modi, A., Sezer-Uzol, N., Long, L. N., and Plassmann, P. E. AIAA Paper 2002-2750, 2002.
4. Spera, D., Wind Turbine Technology: Fundamental Concepts of Wind Turbine Engineering, ASME Press, 1994.
5. Schreck, S. and Robinson, M. Wind Energy, Vol. 5, 2002, pp. 133-150.
6. Xu, G. and Sankar, L. N. Wind Energy, Vol. 5, 2002, pp. 171-183.
7. Sorensen, N. N., Michelsen, J. A., and Schreck, S. Wind Energy, Vol. 5, 2002, pp. 151-169.
8. Schreck, S. Wind Energy, Vol. 5, 2002, pp. 77-84.
9. Johansen, J., Sorensen, N. N., Michelsen, J. A., and Schreck, S. Wind Energy, Vol. 5, 2002, pp. 185-197.
10. Breuer, M., Jovicic, N., and Mazaev, K. International Journal for Numercial Methods in Fluids, Vol. 41, 2003, pp. 357-387.
11. Wang, M. and Moin, P. AIAA Journal, Vol. 38, No. 12, 2000, pp. 2201-2209.
12. Sezer-Uzol, N. and Long, L. N. AIAA Paper 2006-394, 2006.
13. Sezer-Uzol, N. PhD Thesis, Dept. of Aerospace Engineering, The Pennsylvania State University, Dec 2006.
14. Gupta, A. M.S. Thesis, Dept. of Mechanical and Nuclear Engineering, The Pennsylvania State University, Aug 2006.
15. Souliez, F. J. PhD Thesis, Dept. of Aerospace Engineering, The Pennsylvania State University, Aug 2002.
16. Jindal, S., Long, L. N., Plassmann, P. E., and Sezer-Uzol, N. AIAA Paper 2004-2228, 2004.
17. Duque, E. P. N., Burklund, M. D., and Johnson, W. AIAA Paper 2003-355, 2003.
18. Wilson, R. E. Collected Papers on Wind Turbine Technology, NASA CR-195432, May 1995.
19. Morris, P. J., Long, L. N., and Brentner, K. S. AIAA Paper 2004-1184, 2004.

# Modeling a Web Service-Based Decentralized Parallel Programming Environment

**Nihat Adar,** [a] **Selçuk Canbek,**[a] **Erol Seke,**[a] **Muammer Akçay,**[a]

[a] *Eskişehir Osmangazi University, Computer Engineering, Eskişehir 26480, Türkiye*

Keywords: Web service; grid; load distribution; cluster computers

## 1. Abstract

Heterogeneous computer clusters which deploy COTS PCs and/or high performance server computers are economic parallel systems. However, development and execution software should be installed on each cluster host along with the operating system. Cluster members can no longer be used as standalone PCs without interrupting cluster work. On the other hand, usability of hosts in both individual and cluster-related tasks at the same time is strongly desired. In this paper, we present a web service-based parallel heterogeneous model in which participation is conditional and dynamic. Grid functionality is provided by using service oriented software architecture. Since service architecture is widely supported by operating systems, the hosts can easily be adapted to the parallel environment. Completion of all jobs is guaranteed theoretically and shown practically on the laboratory implementation as long as at least one host is operational in the system.

## 2. Introduction

Clustered individual computers can make a virtual super computer. Parallel programming environments, PVM and MPI, are developed for such systems. Extensive use of the Internet allows similar strategies to be employed in larger scales motivated by extensive idle time of computers connected to the Internet [1]. There has been considerable research on grid computing in the last decade [2, 3]. In grid computing heterogeneity, resource management, fault management, reliability, scheduling and security problems must be addressed [4]. Much effort has been spent on named issues.
Generally in grid systems, special software tools such as Globus toolkit [5], need to be installed on participating hosts. Foster et. al. in their recent study, consider web services

I.H. Tuncer et al., *Parallel Computational Fluid Dynamics 2007*,
DOI: 10.1007/978-3-540-92744-0_58, 

and associated mechanisms for creating sophisticated distributed systems [6]. Some large private companies and interest groups announced strong support for Open Grid Software Architecture (OGSA) [6, 7]. Once mature grid protocols are incorporated into operating systems, grid studies are expected to step up.

Harnessing underutilized CPU resource of desktop PCs connected over the Internet is called enterprise/desktop grid computing [8]. Desktop grid computing may well be supported by service oriented software architecture. The web services approach has been widely accepted as a new way for easy and efficient resource sharing, and integration between systems and applications. However, a complete web service-based desktop grid environment has yet to be implemented. Recently, Huang [9] introduced the MPI message passing model using web services. Alchemi Project [10] utilized web services to interface its model to Globus grid system. Alchemi is a .NET-based framework that follows the master-worker parallel programming paradigm in which a central component dispatches independent units of parallel execution to host and manage them. Management module keeps system status, user requests, and utilization details of each participating host. Heavy network traffic caused by the centralized management is a weakness of such a system. A possible fault on the manager computer results in a system-wide failure causing the loss of all submitted jobs. In addition, dynamic participation of hosts is not possible in their model.

In this paper, a desktop grid environment on the Internet based clusters of heterogeneous desktop computers is presented. Integration of participating hosts is coordinated using web services. Any host, supporting Web services, can be a part of the system, conforming OGSAs proposal of conditional and dynamic resource sharing. The model handles changes in capabilities of and restrictions on the participating hosts anytime.

The model has distributed control mechanism where defined jobs can be submitted from any computer that shares its CPU cycles. The model defines how hosts interact to achieve grid functionality. It consists of a Web service protocol and two service applications. Users, who have these services running, can be part of the system.

One of the most desired attributes in grid systems is the tolerance to host failures [11]. Participating hosts may fail or unexpectedly leave the system. In our model, the jobs that have been already assigned to failing hosts are reassigned. Thus, completion of all jobs is guaranteed provided that at least one host is operational. This feature is proven with a theorem in section 4.3.

In section 3, the proposed architectural model, called Web Service-based Decentralized Parallel Programming Environment, WSDP$^2$E, is described. WSDP$^2$E software model is defined in section 4. Tests for evaluation of the operational parameters and their results are presented in section 5 followed by conclusions and future work.

## 3. Architecture

WSDP$^2$E's architecture constitutes Processors, the Depository, and Jobs. Processors (COTS PCs) and the Depository are hardware components of the WSDP$^2$E connected through a network. Cross Platform Connection unit is a Web service that translates

externally submitted jobs to WSDP$^2$E and vice versa. It may be running on any Processor in the system or on a separate PC.
Processors are networked computers with Web services. Processors are used as part of the system during their idle times beside their ordinary use. A Processor fault is defined as its hardware/software or network connection failure that prevents it to respond requests from the system. A participating Processor can also be prevented by an operator. This case is also considered as a Processor fault. Processors with different speed and features are assumed.
The Depository is a data storage unit that keeps copies of all jobs. Depository assures that once a job is submitted to WSDP$^2$E, it will not be lost even if any fault occurs. The Depository is not a management module. It is a safe storage unit that can be implemented by database servers. Having only database functionality, the Depository is less susceptible to failures than a centralized management computer which has a lot more tasks to handle. Utilizing RAID systems and redundant hardware components, almost fail safe storage servers can be easily constructed or readily bought from a market.
The Job is a work unit described as a command and its input/output data. A job can be submitted from any Processor. It is assumed that there is at least one Processor capable of executing any submitted job. In Depository, a job assumes one of the four possible states. These are *New*, *Assigned*, *Completed* and *Uncompleted.* A job with no Processor assignment is named *New* job. The Job that is transferred to a requesting Processor's input job queue is called *Assigned.* An executed job is marked as *Completed.* An assigned job's state must be updated to *Completed* earlier than its estimated completion time is over; otherwise its status is set to *Uncompleted.*

## 4. Implementation and Operation

WSDP$^2$E implementation consists of a web service-based protocol that specifies how distributed Processors interact with another in order to execute jobs. In addition to the web service protocol, two services are defined: access to computation power of participating Processors and access to data. Facilitating web services, WSDP$^2$E's protocol and services can be operated across different platforms, languages, and programming environments. Figure 1 provides an overview of the design and implementation.

### 4.1. Implementation

The model is implemented by two service applications, named as *Job_search* and *Executor*, and one web service protocol that constitutes three web services, named as *Job_submit*, *Job_request* and *Directory_update*. *Job_submit* web service and two applications (*Executor*, *Job_search*) are installed to all Processors. In Depository, only web service protocol is installed. Additionally, a cross platform exchange web service is implemented so that it can be installed to any Processor that functions as a cross platform exchange unit.

In the system, three parameters: *Current_load, Min_load* and *Max_load*, are employed on each participating Processor. *Current_load* parameter shows the instantaneous workload of a Processor. *Min_load* and *Max_load* parameters enable *Executer* to search or to reject jobs from the environment. Functions of three web services and two service applications are explained below.

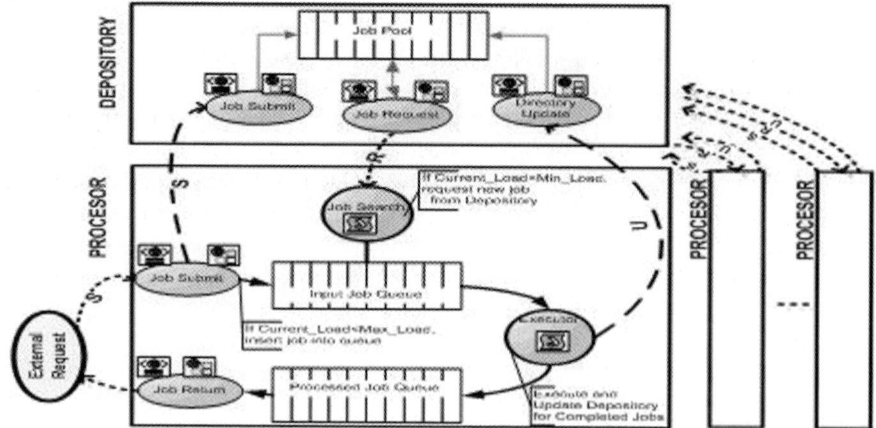


**Fig. 1.** WSDP²E architecture and interaction between its components

***Job_submit*** web service is used to submit a job to a Processor. A newly submitted job is inserted into Input job queue of the Processor if its *Current_load* value is less than its *Max_load* value. Consequently, *Job_submit* web service transfers a copy of the submitted job to Depository (path S in fig.1). The Processor, $P_i$, transferring a copy of the job, $T_n$, to the Depository, is called its owner (i.e. $T_n$ is assigned to $P_i$). Along with a copy of the job, its estimated execution time and its estimated completion time are also given to the Depository. Receiving its relative completion time, the Depository can calculate absolute completion time of the job. Therefore, a global time synchronization mechanism is not necessary. If Processor's *Current_load* value is greater than its *Max_load* value, newly submitted job is not included in Processor's input job queue and it is transferred to the Depository without an owner (i.e. this job's status is set to *New*).

***Job_request*** web service is used to ask for a job from the Depository (path R in fig.1). *Job_request* web service delivers *New* or *Uncompleted* jobs. If there is more than one Processor requesting a job, their requests are handled in order.

***Directory_update*** web service updates status of the jobs in the Depository. Once a Processor completes one of its jobs, it calls *Directory_update* web service from the Depository and changes the job's status to *Completed* (path U in fig.1).

***Job_search*** service is a timer-based application. It checks Processor's *Current_load* value at every tick of the timer. It requests a job from the Depository if *Current_load* value is less than *Min_load* value; otherwise it sleeps until next time (path R in fig.1).

***Executor*** service acquires a job from Processor's input job queue and issues to the Processor for execution. Once a job is executed, *Executor* transfers its results to processed job queue and changes its status to *Completed* via placing a *Directory_update* web service call to Depository (path U in fig.1).

### 4.2. Operation

A job can be given to the system through any Processor or Depository via *Job_submit.* A copy of the job is always transferred to the Depository. *Job_submit* checks *Current_load* value of its host, $P_i$. If it is less than *Max_load* value, the newly given job is inserted to the input job queue of $P_i$ and the job's status is set to *Assigned* for $P_i$; otherwise, the job is not inserted to the input job queue and job's status is set to *New* (i.e. a job without an owner). *Job_search* service continuously monitors *Current_load* value of its host. If a Processor's *Current_load* value is less than its *Min_load* value, then *Job_search* requests a job from the Depository. Depository delivers one of its *New* or *Uncompleted* jobs to the requesting Processor.

A Processor fault might occur any time. A faulty Processor can not change the status of its jobs in the Depository. In the case of fault, jobs' completion time, which is *Assigned* to faulty Processor, will elapse in the Depository and their status will be set to *Uncompleted.* Lightly loaded Processors (i.e. *Current_load* value smaller than *Min_load* value) will request and execute those *Uncompleted* jobs from the Depository. As a result, faulty Processor's jobs will also be completed by the other participating Processors.

### 4.3. Fault Tolerance

A **fault** is defined as a Processor's hardware/software or its network connection failure that prevents it to respond requests from the system. In the following section fault tolerance of the model is proven.

**Conditions:** Necessary conditions are;

1. There is at least one Processor capable of executing any selected job
2. A Processor can be in one of the two states; a) Normally executing its assigned jobs b)Being in fault state (i.e. does not respond)
3. The Depository does not fail, keeps copies of all jobs and has healthy network connection.

**Theorem:** When the conditions are satisfied, all jobs submitted to the system will be completed. If there are *Uncompleted* jobs (not finished) in the Depository, there must be no Processor capable of executing them.

**Proof:** Let $T$ be the list of the jobs kept in the Depository, and $P$ be the list of Processors. The healthy Processor $P_i$ in $P$ executes its assigned jobs from $T$. $P_i$ returns the outputs to the Depository after completing the job $T_k$. The Depository marks job $T_k$ as *Completed.* After completing all assigned jobs, $P_i$ requests a job from the Depository. In return, the Depository delivers requested amount of eligible jobs to $P_i$. In the Depository, newly delivered job $T_n$ with estimated completion time $\Delta t_n$ is assigned to $P_i$. $P_i$ completes its assigned jobs provided that no Processor fault occurs. As a result, the number of the jobs waiting to be completed in the list $T$ is decreased. As long as there are working Processors, number of *Uncompleted* and *New* jobs decreases monotonically and reaches to zero in finite time interval.

If a Processor fault occurs, $P_i$ can not return any result to the Depository. Therefore the Depository marks jobs already assigned to $P_i$ as *Uncompleted* after their estimated completion time elapses. *Uncompleted* jobs of $P_i$ will be requested and executed by another capable Processor. According to condition 1, there must be at least one.
Thus all submitted jobs will be completed that there is at least one Processor is operational and enough time is given. Condition 3 assures that no jobs are lost from the system because of Processor faults.

## 5. Tests and Evaluation

The test bed is a cluster of 7 PCs (Intel Pentium IV 3.2 GHz, 1024 MB DDR Ram) connected by a gigabit switch. One PC is used as the Depository and 6 PCs as the Processors. All computers have running web services and Executors. 1000 dummy jobs are created which do nothing but keep the Executor busy. Each job's execution time is set to a random value from 5 to 15 time units. Average execution time, avex, is therefore approximately 10 time units. 5 different distributions (named c1, c2, c3, c4, c5) of 1000 jobs are tested.
Each distribution is tested with 7 different pairs of *Min_load*, *Max_load*; (*avex**1, *avex**20)-*Min_Load*=1, (*avex**3, *avex**20)-*Min_Load*=3, (*avex**10, *avex**20)-*Min_Load*=10, (*avex**15, *avex**20)-*Min_Load*=15, (*avex**3, *avex**5)-*Max_Load*=5, (*avex**3, *avex**30) -*Max_Load*=30 and (*avex**3, *avex**50)-*Max_Load*=50. The results are collected and the following values are calculated for each initial distribution case;
*Theoretical Overall Execution Time*: Sum of job execution times divided by the number of Processors (6), *Real Overall Execution Time*: Total operation time of the Processor that executes the last job, *Total Busy and Idle Times*: total execution and idle time of each processor. Test results for 5 initial distributions and 7 different pairs of *Min_load*, *Max_load* are given in figures 2.a and 2.b. In figure 2, each sample point is *Real Overall Execution Time* for given initial distribution. "Teo" stands for *Theoretical Overall Execution Time* in both figures.

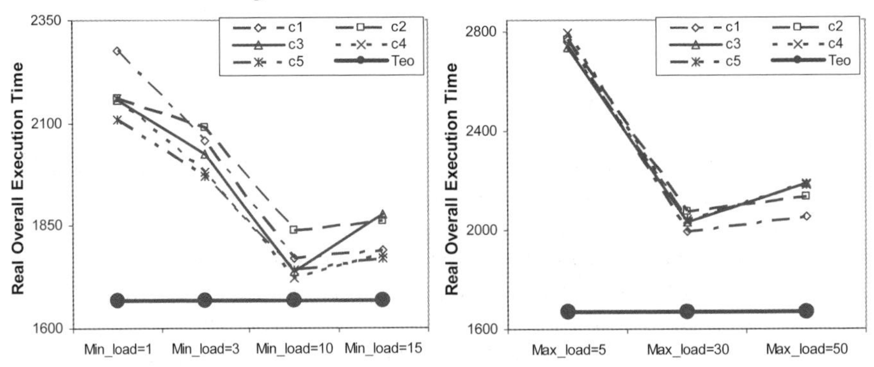


**Fig. 2.** a) *Real Overall Execution Times* for 5 initial distributions and 4 *Min_load* values for each b) *Real Overall Execution Times* for 5 initial distributions and 3 *Max_load* values for each.

Processors with very small *Min_load* values request new jobs as they become idle. In that case *Real Overall Execution Time* increases. Also, tests show that high *Min_load,* high *Max_load*, and low *Max_load* values, causing load imbalance among Processors, degrading overall performance. *Min_load*=10 allows Processors to request new jobs before finishing all its current jobs, and *Max_load*=30 prevents disproportionate loading of individual processors. *Min_load*=10 and *Max_load*=30 give best *Real Overall Execution Time*s for all initial job distributions. In figure 3, Processor execution times for two tests are shown.

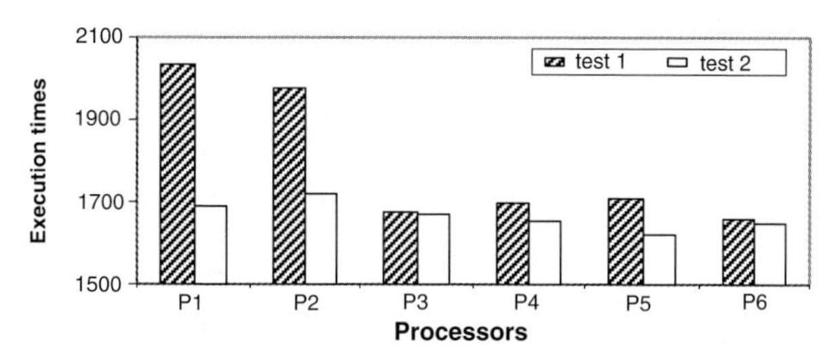


**Fig. 3.** Processor execution times for two tests of Min_load, Max_load pairs

Test-1 shows that high *Min_load* and *Max_load* values cause erratically varying Processor execution times causing degraded performance. Test-2 demonstrates balanced job allocation with correct *Min_load* and *Max_load* values.
Additionally, the following tests are performed to confirm that $WSDP^2E$ completes all submitted jobs.

1. All initial distribution cases are applied and the system is tested with no fault. It is observed that all jobs are completed in each test.
2. Initial distribution case 1 is applied and the system is tested with one random Processor fault generation during operation. It is observed that all jobs, including faulty Processor's, are completed.
3. Initial distribution case 3 is applied and Processors 1 through 5 are stopped at randomly selected points in time during operation. It is observed that all jobs are completed by the remaining single Processor (Processor 6).

## 4. Conclusions and Future Work

Web Service-based Decentralized Parallel Programming Environment, $WSDP^2E$, provides an integrated approach to the coordinated use of hosts (Processors) at multiple sites for computation. In the model, a web service-based protocol is defined that specifies how distributed Processors interact with each other in order to execute jobs. $WSDP^2E$ utilizes standard computers as a part of parallel system. The model handles changes in capabilities of and restrictions on participating hosts anytime. Deploying or removing a host does not disrupt the system's operation. Therefore, Processors can be utilized as an individual computer or as a part of parallel system. A theorem and its proof stating that "completion of all submitted jobs is guaranteed as long as there is at least one Processor to execute them" are included.

After implementing WSDP$^2$E, extensive tests are performed to select correct system parameters (*Min_load* and *Max_load*) and to observe system behaviors. *Min_load* and *Max_load* values should be selected proportional to the average execution time, avex. *Min_load*=10 and *Max_load*=30 provide best *Real Overall Execution Time*s for different initial job distributions. WSDP$^2$E's job completion assurance is also tested practically for different fault simulation cases.
We are currently studying to embed diffusion load balancing method and to replace atomic operation *Job_request* web service with auction based *Job_request* web service so that systems' performance can further be increased. In a parallel path, we are studying to add our web service protocol and service applications into .NET framework. Thus, any PC, running MS Windows .NET OS including our protocol and services, can be a part of WSDP$^2$E. This integration will allow a networked host to participate and share its computing resources (or to access another participating host's computing resource) without any pre-registration.

**References**

1. M. Mutka and M. Livny, The Available Capacity of a Privately Owned Workstation Environment, Journal of Performance Evaluation, Volume 12, Issue 4, , 269-284pp, Elsevier Science, The Netherlands, July 1991.
2. Ian Foster and Carl Kesselman (editors), The Grid: Blueprint for a Future Computing Infrastructure, Morgan Kaufmann Publishers, USA, 1999.
3. Ian Foster, Carl Kesselman, and S. Tuecke, The Anatomy of the Grid: Enabling Scalable Virtual Organizations, International Journal of Supercomputer applications, 15(3), Sage Publications, 2001, USA.
4. Rajkumar Buyya, Economic-based Distributed Resource Management and Scheduling for Grid Computing, Ph.D. Thesis, Monash University Australia, April 2002.
5. Ian Foster and Carl Kesselman, Globus: A Metacomputing Infrastructure Toolkit, International Journal of Supercomputer Applications, 11(2): 115-128, 1997.
6. Ian Foster, Carl Kesselma, Jeffrey M. Nick, Steven Tuecke, The Physiology of the Grid:An Open Grid Services Architecture for Distributed Systems Integration, Open Grid Service Infrastructure WG, Global Grid Forum, June 22, 2002.
7. Open Grid Service Architecture (OGSA) Working Group, http://forge.gridforum.org/projects/ogsa-wg
8. Andrew Chien, Brad Calder, Stephen Elbert, and Karan Bhatia, Entropia: architecture and performance of an enterprise desktop grid system, Journal of Parallel and Distributed Computing, Volume 63, Issue 5, Academic Press, USA, May 2003.
9. Huang, Y., Huang, Q., WS-Based Workflow Description Language for Message Passing, IEEE International Symposium on Cluster Computing and the Grid,2005.
10. Akshay Luther, Rajkumar Buyya, Rajiv Ranjan, and Srikumar Venugopal, Alchemi: A .NET-based Enterprise Grid Computing System, IC2005-2
11. Limaye, K., Leangsuksun, B., Greenwood, Z., Scott, S. L., Engelmann, C., Libby, R., and Chanchio, K., Job-Site Level Fault Tolerance for Cluster and Grid environments, IEEE International Conference on Cluster Computing (Cluster 2005), Boston, Massachusetts, USA, September 27 - 30, 2005.

# Computation of unsteady hovering flapping motion in parallel environment


**Ebru Sarıgöl[a] and Nafiz Alemdaroğlu[b]**

[a] *Dr.,METU Aerospace Engineering Department, 06531 Ankara, Türkiye*

[b] *Prof. Dr., METU Aerospace Engineering Department, 06531 Ankara, Türkiye*



Keywords: unsteady flow, flapping motion, parallel solution, DNS, low Reynolds number


## 1. INTRODUCTION

Flight Vehicles using flapping wing motion hold several distinct advantages over conventional platforms in low Reynolds number regimes. These advantages are mainly due to flapping wing's ability to achieve very high unsteady lifts as well as the ability to hover. The ability of small birds and insects to quickly change direction and to achieve extremely high accelerations and decelerations has been observed in nature [1].

The theory of force production in insect flight is thwarted partly due to insufficient knowledge on the unsteady forces generated by the wings. Small insects fly in the intermediate Reynolds number regime (10-1000) where the data are extremely sparse. Dickinson et al. aimed to expand the knowledge of unsteady mechanisms that might be employed by insects during flight by studying the impulsively moved wings [2].

Studies based on the use of robotic models which mimic the flapping motion of real fliers, investigated the aerodynamic mechanisms during flapping flight and obtained the aerodynamic forces generated [3-7]. Visualizations of flow during flapping motion are also performed in order to gain insight about the leading edge vortices which are responsible for the generation of aerodynamic forces [8-10]. Various computational studies are performed to investigate a wide range of unsteady aerodynamics of model fruit flies, three-dimensional modeling of Drosophila wings and the oscillating wings [11-15].

This paper investigates the normal hovering flapping motion in laminar, incompressible flow for a Reynolds number of 1000. The pressure and the velocity fields are obtained by solving the Navier-Stokes equations numerically and the resulting aerodynamic forces are calculated. The vortex regions are analyzed and compared to experimental



DOI: 10.1007/978-3-540-92744-0_59,

results obtained by the PIV technique for 2D calculations. The three-dimensional flow field for hovering flight is analyzed to understand the vortical flow structures around a three-dimensional wing during the translational and rotational phases of the flapping motion and the aerodynamic forces calculated numerically are compared with the two-dimensional results. The parallel computational efficiency during the two- and three-dimensional computations is also compared.

## 2. NUMERICAL METHODS

### 2.1. Definition of Motion

The flapping motion is divided into four kinematical phases: two translational phases (up-stroke and down-stroke) where the wing sweeps through air with a high angle of attack and two rotational phases (pronation and supination) where the wing rapidly rotates and reverses its direction of motion. Each region is composed of a translational and a rotational phase (Figure 1). The rotation is such that the leading edge stays as leading edge during all phases of the motion [14-15]. Hovering is an extreme mode of flight where the forward velocity is zero. The characteristic speed and length are taken as the maximum translational speed and the chord of the airfoil respectively for the calculation of flow Reynolds number in both 2D and 3D solutions [15, 22].

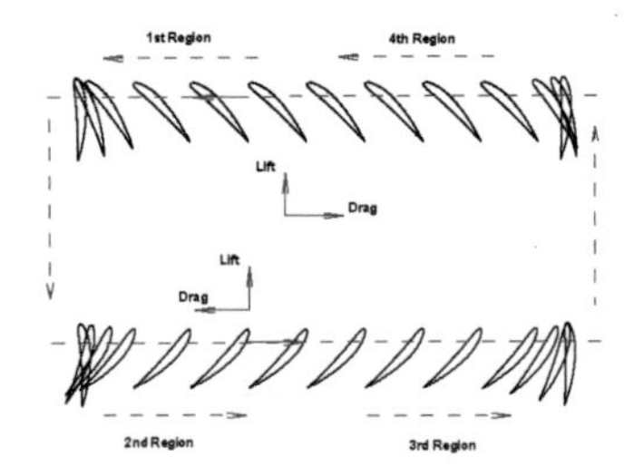


Figure 1 Definition of normal hovering flapping motion

### 2.2. Flow Solver and Governing Equations

The unsteady laminar flow around a moving wing with a prescribed motion is solved by the commercial CFD solver, Fluent v6.22. 2D time-dependent Navier-Stokes equations are solved using the finite-volume method for an incompressible flow. The governing equations are described by:

$$\nabla.\vec{V} = 0 \qquad (1)$$

$$\frac{\partial}{\partial t}\left(\vec{V}\right)+\left(\vec{V}.\nabla\right)\vec{V} = -\left(\frac{1}{\rho}\right)\nabla p + \nu\nabla^2\vec{V} \qquad (2)$$

where $\vec{V}$ is the flow velocity vector, $\rho$ is constant density, $\nu$ is the kinematic viscosity and p is the static pressure [16].

At Re=1000, the flow is assumed to be laminar and incompressible (Mach=$O(10^{-3})$). Therefore no turbulence model is used so the simulation is time-accurate or Direct

Numerical Simulation (DNS). The space discretization scheme used is the first order upwind as well as the time discretization which is the only way to use the dynamic mesh module implemented in Fluent v6.22. The pressure-velocity coupling in incompressible flow simulations was obtained using the iterative PISO scheme with under-relaxation coefficients for pressure, momentum and body forces equal to 0.3, 0.7 and 1.0 respectively. The accuracy was set to double-precision. The convergence criterion for the iterative method was satisfied with mass and momentum residues dropping four orders of magnitude, i.e. $O(10^{-4})$.

*2.2.1. Parallel Computation*

Parallel algorithm used in the solutions splits up the grid and data into multiple partitions, then assigns each grid partition to a different compute process (or node). The compute-node processes can be executed on a massively-parallel computer, a multiple-CPU workstation, or a network of workstations using the same or different operating systems. FLUENT uses a host process that does not contain any grid data. Instead, the host process only interprets commands from FLUENT's graphics-related interface, cortex. The host distributes those commands to the other compute nodes via a socket communicator to a single designated compute node called compute-node-0. This specialized compute node distributes the host commands to the other compute nodes. Each compute node simultaneously executes the same program on its own data set. Communication from the compute nodes to the host is possible only through compute-node-0 and only when all compute nodes have synchronized with each other. A FLUENT communicator is a message-passing library which could be a vendor implementation of the Message Passing Interface (MPI) standard [16]. Computations are performed on a 64-bit HP workstation having two dual core processors, total four cores, on Linux operating system. Each core has a CPU speed of 3.0 GHz.

**2.3. Mesh Generation and Boundary Conditions**

O-type, hyperbolically generated single block computational grid is used for 2D computations. The grid consists of 14388 cells and the outer boundary is located at approximately 15c. The velocity of the motion is defined with respect to the center of rotation, which is located at the quarter chord location. On the airfoil surface no-slip boundary condition is applied. The farfield boundary condition is implied by setting the pressure outlet for incompressible flow with constant density and viscosity. The direction of the flow is computed from the neighboring cells at the pressure boundary.

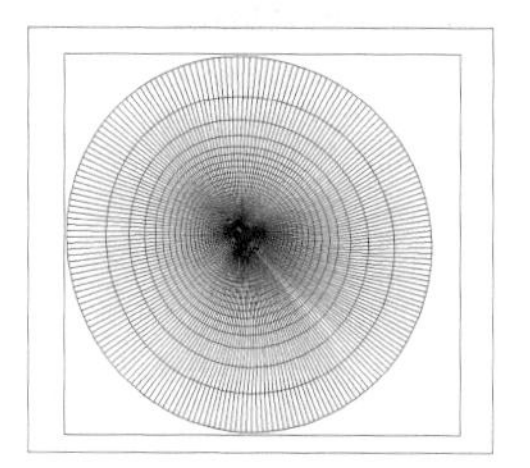
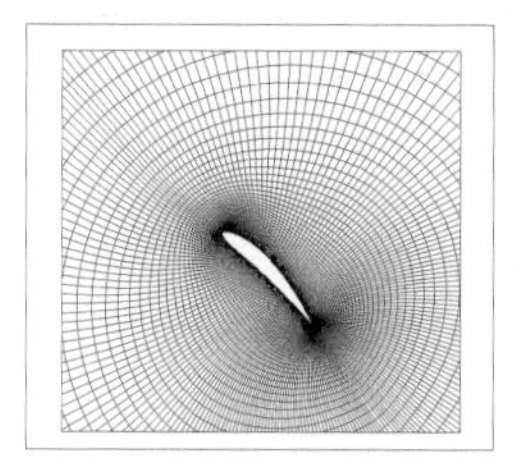

Figure 2 Two-dimensional computational mesh.

The airfoil profile is a cambered one, NACA 6412, of chord length 1cm for both 2D and 3D numerical solutions. Normally, the aspect ratio for most insects which are capable of hovering is 10-11 for full wing configuration. In our 3D calculations, the aspect ratio is chosen as 5 (i.e. half wing configuration with the use of symmetry condition at the root section). The finite wing has no taper.

### 2.4. Vortex Identification

Although vortical flows can be seen in almost every branch of fluid mechanics, there is no agreed definition of this basic vortical flow structure. However as suggested by Jeong and Hussain [17]; the geometrical characteristics of the identified vortex core must be Galilean invariant; i.e . independent of the coordinate system chosen.
The definitions of a vortex based on the analysis of the velocity gradient tensor $\nabla u$ [17-19,20,21] are Galilean invariant, and are less intuitive physically. The issue of vortex identification is by no means trivial, and in fact several of the existing techniques can give good results in many situations, but all of them can be shown to fail (or, at least, to produce ambiguous answers) in particular conditions [20]. The most popular local criteria are the Q criterion proposed by Hunt et al. [19], the $\lambda_2$ criterion proposed by Jeong et al. [19] and the $\Delta$ criterion as proposed by Chong et al. [20]. Scalar methods such as the vorticity magnitude and the Q criterion are used during the analyses.

## 3. RESULTS

Efficiency of parallel computations is measured by the wall clock time needed to complete the computations. *Speed-up* is the criterion which shows how fast the parallel code is and is defined as the ratio to serial counterpart. The higher this ratio is, the better is the parallel algorithm. Figure 3 (a) shows the speedup and Figure 3 (b) shows the efficiency obtained by 2D parallel computations. 3D speedup and efficiency plots which are obtained for one period of iterations are shown in Figure 4. Note that the speedup ratios for both 2D and 3D solutions are less than 3 using 4 CPUs which is normally around 4. The reason that the ratio is smaller than the usual value may be the partitioning technique of the computational grid.

The results are analyzed at specified non dimensional time steps. Figure 5 shows the lift and drag coefficient variations with respect to time . There are two means of generating the lift during the downstroke cycle. The first one is the 'delayed stall' mechanism during the translational motion which in two-dimensional case, produces high lift by causing a leading-edge vortex which reduces the pressure on the wings until the inevitable flow separation. In three-dimensional case this stall is prevented by the spanwise flow. The second mechanism is the so called "Rotational lift" which is created when the angle of attack of the wing changes resulting in a repelling force. During upstroke, wake capture is yet another rotational mechanism to generate lift that depends on pronation (0.25T) and supination (0.75T) during stroke reversal [12]. This results in an extra lift gained by recapturing the energy lost in the wake. As the wing moves through the air, it leaves whirlpools, or vortices, of air behind it. If the wing is rotated, the wing can intersect its own wake and capture its energy in the form of lift.

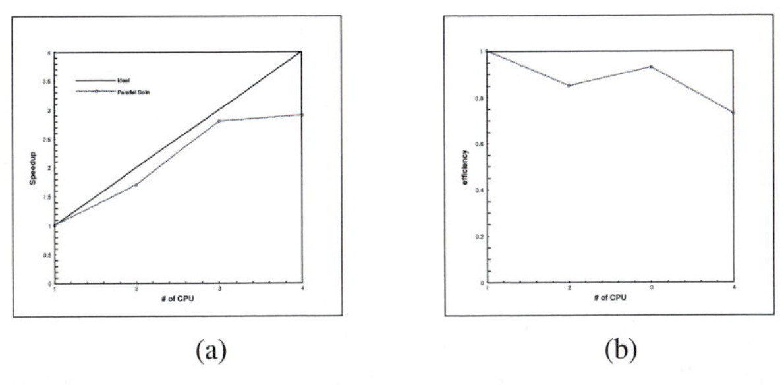


(a) (b)

Figure 3 (a) Speed up and (b) efficiency of two-dimensional numerical solutions.

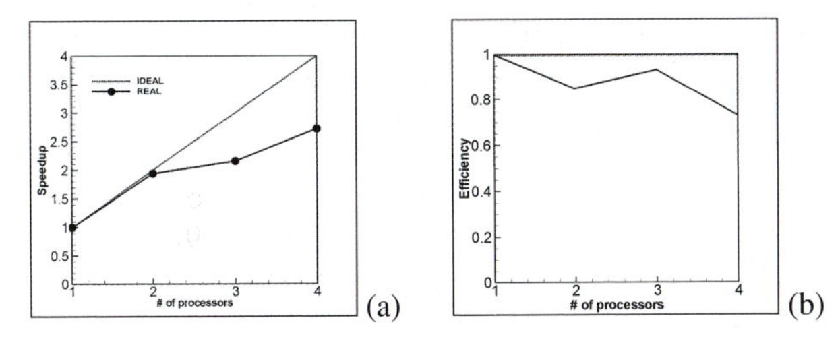


(a) (b)

Figure 4 (a) Speed up and (b) efficiency of three-dimensional numerical solutions.

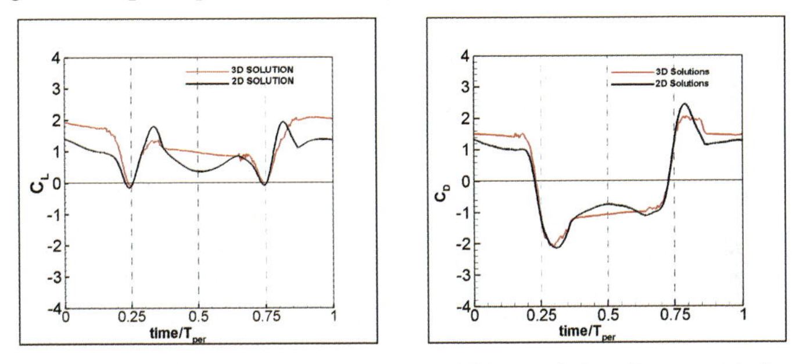


Figure 5 2D and 3D numerical lift and drag coefficients variations in one period at $\alpha$=45°.

Figure 6 shows the vortex regions responsible for aerodynamic force generation which are identified by using the non-dimensional vorticity contours for 2D case. The generated leading and trailing edge vortices and the shedding of these vortices as time evolves are clearly seen. The sense of rotations of the vortices changes when the direction of motion of the airfoil is changed.

Comparison of 2D flow field to that of 3D flow field at the midspan location during hovering flapping motion by using the Q criterion is shown in Figure 7. At each non-dimensional time step, the upper and lower figures show the 2D and the 3D flow fields respectively at midspan location. 2D leading and trailing edge vortices merge much earlier than the 3D flow case. In 3D flow, the trailing edge vortex diffuses more and the shear regions, identified by cold colors, are larger compared to 2D flow solutions. On the other hand, the locations and the shapes of vortex regions are very similar in both cases as expected.

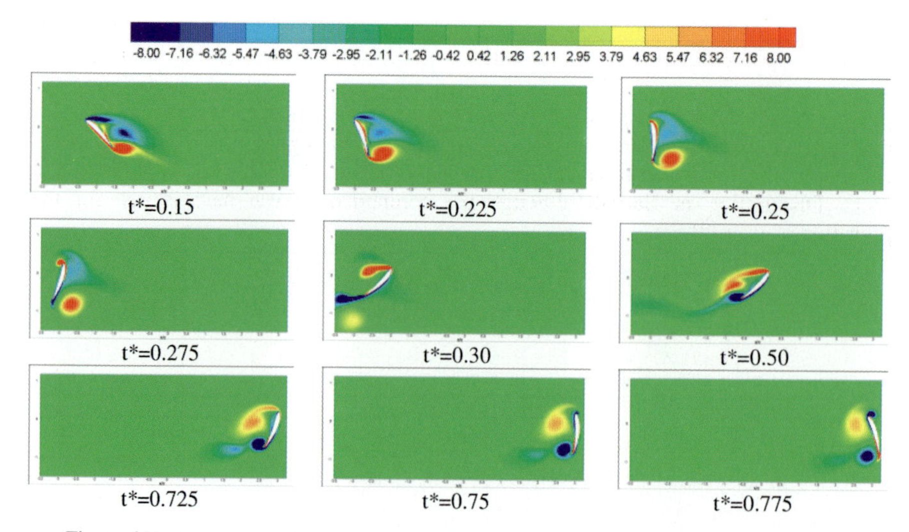


Figure 6 Vortex regions identified by non-dimensional vorticty contours of two-dimensional NACA 6412 airfoil in one period at Re=1000, α=45°.

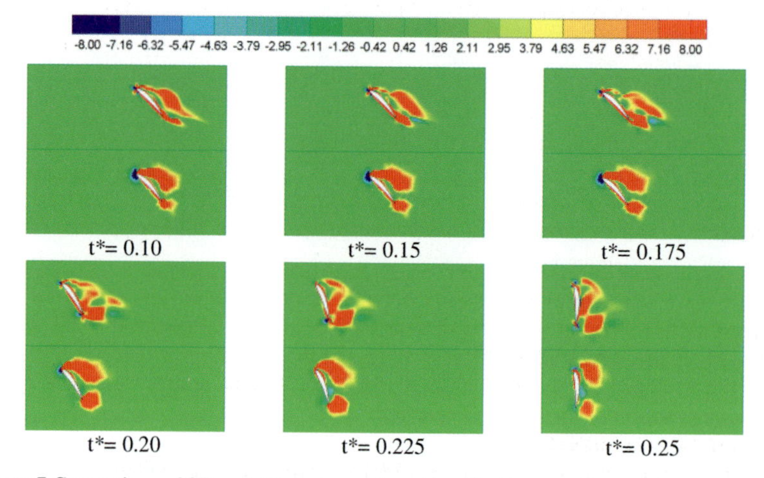


Figure 7 Comparison of 2D and 3D vortex regions identified by non-dimensional contours of second invariant of velocity gradient, Q criterion during the first quarter period at α=45°.

2D computations are compared to experimental results qualitatively in Figure 8 using the Q criterion. In these figures warm colors (red) indicate the vortex dominated regions whereas the cold colors indicate the shear dominated regions. As it can be seen, numerical computations give reliable results in capturing the correct locations of vortices. The leading edge vortex (black circle) and the trailing edge vortex (white circle) regions as well as the vortex regions from previous time steps (blue circle) are shown for comparison. It is clearly seen that all vortex regions in the experimental results are a little bit apart from the profile when compared to vortex regions indicated by the numerical results. The unrealistic vortex regions seen at the middle of the pictures are due to camera positioning during the experiments which are not evitable for the time being. The explanation for these unreal vortex regions and a comparison of experimental and numerical results for a complete period of motion can be found in the Ph.D. Dissertation of Sarıgöl although the comparison given here is just for the pronation phase [22].

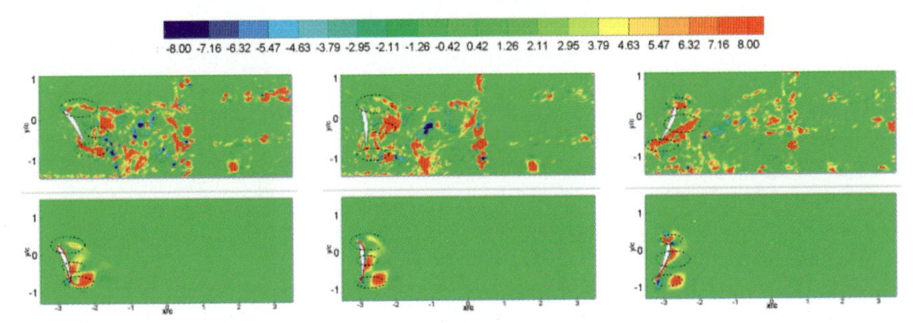


Figure 8 Comparison of 2D experimental (top) and numerical results during the pronation using non-dimensional contours of second invariant of velocity gradient, Q criterion at $\alpha=45°$.

## 4. CONCLUSIONS

This paper investigated the parallel solutions of hovering flapping motion for two- and three-dimensional flowfields. The parallel efficiency during the computations is investigated and it is seen that the parallel computations improve the computational time significantly. There are small discrepancies at the peak force coefficients' locations. Although the agreement between the 2D and 3D force coefficients is good in general, at these peak forces coefficient locations 2D solutions overestimate those of 3D. Moreover, comparison of the 2D and 3D vortex regions using a Galilean invariant vortex identification method shows that the 2D vortex regions exhibit the main characteristics of those of 3D vortex regions at the midspan location.

**ACKNOWLEDGEMENTS**

This research was supported by TUBİTAK under project number 104M417.

**REFERENCES**

1. Ames, R.G., "On the Flowfield and Forces Generated by a Rectangular Wing Undergoing Moderate Reduced Frequency Flapping at Low Reynolds Number", PhD Dissertation, Georgia Institute of Technology Department of Aerospace Engineering, April 2001.

2. Dickinson, M.H. and Götz, K., "Unsteady Aerodynamic Performance of Model Wings at Low Reynolds Numbers", J. Exp. Biol., 174, pp.45-64, 1993.
3. Wang, Z.J., Birch, J.M. and Dickinson, M.H., "Unsteady forces and flows in low Reynolds number hovering flight: two-dimensional somputations vs robotic wing experiments", J. Exp. Biol., 207, pp.449-460, 2004.
4. Dickinson, M.H., Lehmann, F.O. and Sane S.P., "Wing rotation and the aerodynamic basis of insect flight", Science,284, pp.1954-1960,1999.
5. Ellington, C.P., vanderberg C., Willmott, A.P. and Thomas, A.L.R., "Leading-edge vortices in insect flight", Nature 384, pp.626-630, 1996.
6. Dickson, W.B. and Dickinson, M.H., "The effect of advance ratio on the aerodynamics of revolving wings", J. Exp. Biol. 207, pp.4250-4281, 2004.
7. Sane, S.P. and Dickinson, M.H., "The control of flight force by a flapping wing: lift and drag production", J. Exp. Biol. 204, pp.2607-2626, 2001.
8. Birch, J.M., Dickson, W.B. and Dickinson, M.H., "Force production and flow structure of the leading-edge vortex on flapping wings at high and low Reynolds numbers", J. Exp. Biol. 207, pp.1063-1072, 2004.
9. Thomas, A.L.R., Taylor, G.K., Srygley, R.B., Nudds, R.L. and Bomphrey, R.J., "Dragonfly fligh: free-flight and tehered flow visualizations reveal a diverse array of unsteady lift-generating mechansims, controlled primarily via angle of attack", J. Exp. Biol. 207, pp.4299-4323, 2004.
10. Sun, M. and Tang, J., "Unsteady aerodynamic force generation by a model fruit fly wing in flapping motion", J. Exp. Biol. 205, pp.55-70, 2002.
11. Sun, M. and Tang, J., "Aerodyanmic force generation and power requirements in forward flight in a fruit fly with modeled wing motion", J. Exp. Biol. 206, pp.3065-3083, 2003.
12. Ramamurti, Ravi and Sandberg, William C., "A Three-dimensional computational study of the aerodynamic mechanisms of insect flight", J. Exp. Biol. 205, 1507-1518, 2002.
13. Miller, Laura A. and Peskin, Charles S., "A computational fluid dynamics of 'clap and fling' in the smallest insects", J. Exp. Biol. 208, 195-212, 2005.
14. Sarıgöl, E. and Alemdaroğlu,N., "A parametric study on two dimensional flapping motion", Proc. of 2nd European Micro-Air Vehicles Conference EMAV 2006, Braunschweig, Germany, 2006.
15. Kurtuluş, D.F., "Experimental and Numerical Analysis of lift with flapping wing. Application to micro air vehicles",PhD. Dissertation, METU Aerospace Engineering Department, 2005.
16. Fluent 6.3 User's Guide, 2006.
17. Jeong, J. And Hussain F., "On the identification of a vortex", J. Fluid Mech. Vol.285, pp.69-94, 1995.
18. Chong M.S., Perry A.E. and Cantwell B.J., "A general classification of three-dimensional flows",Phys. Fluids A 2, 765, 1990.
19. Hunt J.C.R., Wray A.A. and Moin P., "Eddies, stream and convergence zones in turbulent flows", Center for Turbulence Research Report CTR-S88, 193, 1998.
20. Cucitore R., Quadrio M. And Baron A., "On the effectiveness and limitations of local criteria for the identification of a vortex", Eur. J. Mech. B/Fluids, 18 n°2, pp.261-282, 1999
21. Chakraborty P., Balachandar S. And Adrian R.J., "On the relationships between local vortex identification schemes", J. Fluid Mech., Vol. 535, pp.189-214, 2005.
22. Sarıgöl, E., "Numerical and Experimental Analysis of Flapping Wing Motion", Ph.D. Dissertation, METU Aerospace Engineering Department, July 2007.

# *Editorial Policy*

1. Volumes in the following three categories will be published in LNCSE:

i) Research monographs
ii) Lecture and seminar notes
iii) Conference proceedings

Those considering a book which might be suitable for the series are strongly advised to contact the publisher or the series editors at an early stage.

2. Categories i) and ii). These categories will be emphasized by Lecture Notes in Computational Science and Engineering. **Submissions by interdisciplinary teams of authors are encouraged.** The goal is to report new developments – quickly, informally, and in a way that will make them accessible to non-specialists. In the evaluation of submissions timeliness of the work is an important criterion. Texts should be well-rounded, well-written and reasonably self-contained. In most cases the work will contain results of others as well as those of the author(s). In each case the author(s) should provide sufficient motivation, examples, and applications. In this respect, Ph.D. theses will usually be deemed unsuitable for the Lecture Notes series. Proposals for volumes in these categories should be submitted either to one of the series editors or to Springer-Verlag, Heidelberg, and will be refereed. A provisional judgment on the acceptability of a project can be based on partial information about the work: a detailed outline describing the contents of each chapter, the estimated length, a bibliography, and one or two sample chapters – or a first draft. A final decision whether to accept will rest on an evaluation of the completed work which should include

- at least 100 pages of text;
- a table of contents;
- an informative introduction perhaps with some historical remarks which should be accessible to readers unfamiliar with the topic treated;
- a subject index.

3. Category iii). Conference proceedings will be considered for publication provided that they are both of exceptional interest and devoted to a single topic. One (or more) expert participants will act as the scientific editor(s) of the volume. They select the papers which are suitable for inclusion and have them individually refereed as for a journal. Papers not closely related to the central topic are to be excluded. Organizers should contact Lecture Notes in Computational Science and Engineering at the planning stage.

In exceptional cases some other multi-author-volumes may be considered in this category.

4. Format. Only works in English are considered. They should be submitted in camera-ready form according to Springer-Verlag's specifications.
Electronic material can be included if appropriate. Please contact the publisher.
Technical instructions and/or LaTeX macros are available via http://www.springer.com/authors/book+authors?SGWID=0-154102-12-417900-0. The macros can also be sent on request.

## *General Remarks*

Lecture Notes are printed by photo-offset from the master-copy delivered in camera-ready form by the authors. For this purpose Springer-Verlag provides technical instructions for the preparation of manuscripts. See also *Editorial Policy.*

Careful preparation of manuscripts will help keep production time short and ensure a satisfactory appearance of the finished book.

The following terms and conditions hold:

Categories i), ii), and iii):
Authors receive 50 free copies of their book. No royalty is paid. Commitment to publish is made by letter of intent rather than by signing a formal contract. Springer- Verlag secures the copyright for each volume.

For conference proceedings, editors receive a total of 50 free copies of their volume for distribution to the contributing authors.

All categories:
Authors are entitled to purchase further copies of their book and other Springer mathematics books for their personal use, at a discount of 33.3% directly from Springer-Verlag.

Addresses:

Timothy J. Barth
NASA Ames Research Center
NAS Division
Moffett Field, CA 94035, USA
e-mail: barth@nas.nasa.gov

Michael Griebel
Institut für Numerische Simulation
der Universität Bonn
Wegelerstr. 6
53115 Bonn, Germany
e-mail: griebel@ins.uni-bonn.de

David E. Keyes
Department of Applied Physics
and Applied Mathematics
Columbia University
200 S. W. Mudd Building
500 W. 120th Street
New York, NY 10027, USA
e-mail: david.keyes@columbia.edu

Risto M. Nieminen
Laboratory of Physics
Helsinki University of Technology
02150 Espoo, Finland
e-mail: rni@fyslab.hut.fi

Dirk Roose
Department of Computer Science
Katholieke Universiteit Leuven
Celestijnenlaan 200A
3001 Leuven-Heverlee, Belgium
e-mail: dirk.roose@cs.kuleuven.ac.be

Tamar Schlick
Department of Chemistry
Courant Institute of Mathematical
Sciences
New York University
and Howard Hughes Medical Institute
251 Mercer Street
New York, NY 10012, USA
e-mail: schlick@nyu.edu

Mathematics Editor at Springer:
Martin Peters
Springer-Verlag
Mathematics Editorial IV
Tiergartenstrasse 17
D-69121 Heidelberg, Germany
Tel.: *49 (6221) 487-8185
Fax: *49 (6221) 487-8355
e-mail: martin.peters@springer.com

# Lecture Notes in Computational Science and Engineering

1. D. Funaro, *Spectral Elements for Transport-Dominated Equations.*
2. H. P. Langtangen, *Computational Partial Differential Equations.* Numerical Methods and Diffpack Programming.
3. W. Hackbusch, G. Wittum (eds.), *Multigrid Methods V.*
4. P. Deuflhard, J. Hermans, B. Leimkuhler, A. E. Mark, S. Reich, R. D. Skeel (eds.), *Computational Molecular Dynamics: Challenges, Methods, Ideas.*
5. D. Kröner, M. Ohlberger, C. Rohde (eds.), *An Introduction to Recent Developments in Theory and Numerics for Conservation Laws.*
6. S. Turek, *Efficient Solvers for Incompressible Flow Problems.* An Algorithmic and Computational Approach.
7. R. von Schwerin, ***M**ulti **B**ody **S**ystem **SIM**ulation.* Numerical Methods, Algorithms, and Software.
8. H.-J. Bungartz, F. Durst, C. Zenger (eds.), *High Performance Scientific and Engineering Computing.*
9. T. J. Barth, H. Deconinck (eds.), *High-Order Methods for Computational Physics.*
10. H. P. Langtangen, A. M. Bruaset, E. Quak (eds.), *Advances in Software Tools for Scientific Computing.*
11. B. Cockburn, G. E. Karniadakis, C.-W. Shu (eds.), *Discontinuous Galerkin Methods.* Theory, Computation and Applications.
12. U. van Rienen, *Numerical Methods in Computational Electrodynamics.* Linear Systems in Practical Applications.
13. B. Engquist, L. Johnsson, M. Hammill, F. Short (eds.), *Simulation and Visualization on the Grid.*
14. E. Dick, K. Riemslagh, J. Vierendeels (eds.), *Multigrid Methods VI.*
15. A. Frommer, T. Lippert, B. Medeke, K. Schilling (eds.), *Numerical Challenges in Lattice Quantum Chromodynamics.*
16. J. Lang, *Adaptive Multilevel Solution of Nonlinear Parabolic PDE Systems.* Theory, Algorithm, and Applications.
17. B. I. Wohlmuth, *Discretization Methods and Iterative Solvers Based on Domain Decomposition.*
18. U. van Rienen, M. Günther, D. Hecht (eds.), *Scientific Computing in Electrical Engineering.*
19. I. Babuška, P. G. Ciarlet, T. Miyoshi (eds.), *Mathematical Modeling and Numerical Simulation in Continuum Mechanics.*
20. T. J. Barth, T. Chan, R. Haimes (eds.), *Multiscale and Multiresolution Methods.* Theory and Applications.
21. M. Breuer, F. Durst, C. Zenger (eds.), *High Performance Scientific and Engineering Computing.*
22. K. Urban, *Wavelets in Numerical Simulation.* Problem Adapted Construction and Applications.

23. L. F. Pavarino, A. Toselli (eds.), *Recent Developments in Domain Decomposition Methods.*

24. T. Schlick, H. H. Gan (eds.), *Computational Methods for Macromolecules: Challenges and Applications.*

25. T. J. Barth, H. Deconinck (eds.), *Error Estimation and Adaptive Discretization Methods in Computational Fluid Dynamics.*

26. M. Griebel, M. A. Schweitzer (eds.), *Meshfree Methods for Partial Differential Equations.*

27. S. Müller, *Adaptive Multiscale Schemes for Conservation Laws.*

28. C. Carstensen, S. Funken, W. Hackbusch, R. H. W. Hoppe, P. Monk (eds.), *Computational Electromagnetics.*

29. M. A. Schweitzer, *A Parallel Multilevel Partition of Unity Method for Elliptic Partial Differential Equations.*

30. T. Biegler, O. Ghattas, M. Heinkenschloss, B. van Bloemen Waanders (eds.), *Large-Scale PDE-Constrained Optimization.*

31. M. Ainsworth, P. Davies, D. Duncan, P. Martin, B. Rynne (eds.), *Topics in Computational Wave Propagation.* Direct and Inverse Problems.

32. H. Emmerich, B. Nestler, M. Schreckenberg (eds.), *Interface and Transport Dynamics.* Computational Modelling.

33. H. P. Langtangen, A. Tveito (eds.), *Advanced Topics in Computational Partial Differential Equations.* Numerical Methods and Diffpack Programming.

34. V. John, *Large Eddy Simulation of Turbulent Incompressible Flows.* Analytical and Numerical Results for a Class of LES Models.

35. E. Bänsch (ed.), *Challenges in Scientific Computing - CISC 2002.*

36. B. N. Khoromskij, G. Wittum, *Numerical Solution of Elliptic Differential Equations by Reduction to the Interface.*

37. A. Iske, *Multiresolution Methods in Scattered Data Modelling.*

38. S.-I. Niculescu, K. Gu (eds.), *Advances in Time-Delay Systems.*

39. S. Attinger, P. Koumoutsakos (eds.), *Multiscale Modelling and Simulation.*

40. R. Kornhuber, R. Hoppe, J. Périaux, O. Pironneau, O. Wildlund, J. Xu (eds.), *Domain Decomposition Methods in Science and Engineering.*

41. T. Plewa, T. Linde, V.G. Weirs (eds.), *Adaptive Mesh Refinement – Theory and Applications.*

42. A. Schmidt, K.G. Siebert, *Design of Adaptive Finite Element Software.* The Finite Element Toolbox ALBERTA.

43. M. Griebel, M.A. Schweitzer (eds.), *Meshfree Methods for Partial Differential Equations II.*

44. B. Engquist, P. Lötstedt, O. Runborg (eds.), *Multiscale Methods in Science and Engineering.*

45. P. Benner, V. Mehrmann, D.C. Sorensen (eds.), *Dimension Reduction of Large-Scale Systems.*

46. D. Kressner, *Numerical Methods for General and Structured Eigenvalue Problems.*

47. A. Boriçi, A. Frommer, B. Joó, A. Kennedy, B. Pendleton (eds.), *QCD and Numerical Analysis III.*

48. F. Graziani (ed.), *Computational Methods in Transport.*

49. B. Leimkuhler, C. Chipot, R. Elber, A. Laaksonen, A. Mark, T. Schlick, C. Schütte, R. Skeel (eds.), *New Algorithms for Macromolecular Simulation.*

50. M. Bücker, G. Corliss, P. Hovland, U. Naumann, B. Norris (eds.), *Automatic Differentiation: Applications, Theory, and Implementations.*

51. A.M. Bruaset, A. Tveito (eds.), *Numerical Solution of Partial Differential Equations on Parallel Computers.*

52. K.H. Hoffmann, A. Meyer (eds.), *Parallel Algorithms and Cluster Computing.*

53. H.-J. Bungartz, M. Schäfer (eds.), *Fluid-Structure Interaction.*

54. J. Behrens, *Adaptive Atmospheric Modeling.*

55. O. Widlund, D. Keyes (eds.), *Domain Decomposition Methods in Science and Engineering XVI.*

56. S. Kassinos, C. Langer, G. Iaccarino, P. Moin (eds.), *Complex Effects in Large Eddy Simulations.*

57. M. Griebel, M.A Schweitzer (eds.), *Meshfree Methods for Partial Differential Equations III.*

58. A.N. Gorban, B. Kégl, D.C. Wunsch, A. Zinovyev (eds.), *Principal Manifolds for Data Visualization and Dimension Reduction.*

59. H. Ammari (ed.), *Modeling and Computations in Electromagnetics: A Volume Dedicated to Jean-Claude Nédélec.*

60. U. Langer, M. Discacciati, D. Keyes, O. Widlund, W. Zulehner (eds.), *Domain Decomposition Methods in Science and Engineering XVII.*

61. T. Mathew, *Domain Decomposition Methods for the Numerical Solution of Partial Differential Equations.*

62. F. Graziani (ed.), *Computational Methods in Transport: Verification and Validation.*

63. M. Bebendorf, *Hierarchical Matrices.* A Means to Efficiently Solve Elliptic Boundary Value Problems.

64. C.H. Bischof, H.M. Bücker, P. Hovland, U. Naumann, J. Utke (eds.), *Advances in Automatic Differentiation.*

65. M. Griebel, M.A. Schweitzer (eds.), *Meshfree Methods for Partial Differential Equations IV.*

66. B. Engquist, P. Lötstedt, O. Runborg (eds.), *Multiscale Modeling and Simulation in Science.*

67. I.H. Tuncer, Ü. Gülcat, D.R. Emerson, K. Matsuno (eds.), *Parallel Computational Fluid Dynamics.*

*For further information on these books please have a look at our mathematics catalogue at the following URL:* `www.springer.com/series/3527`

# Monographs in Computational Science and Engineering

1. J. Sundnes, G.T. Lines, X. Cai, B.F. Nielsen, K.-A. Mardal, A. Tveito, *Computing the Electrical Activity in the Heart.*

*For further information on this book, please have a look at our mathematics catalogue at the following URL:* `www.springer.com/series/7417`

# Texts in Computational Science and Engineering

1. H. P. Langtangen, *Computational Partial Differential Equations.* Numerical Methods and Diffpack Programming. 2nd Edition
2. A. Quarteroni, F. Saleri, *Scientific Computing with MATLAB and Octave.* 2nd Edition
3. H. P. Langtangen, *Python Scripting for Computational Science*. 3rd Edition
4. H. Gardner, G. Manduchi, *Design Patterns for e-Science.*
5. M. Griebel, S. Knapek, G. Zumbusch, *Numerical Simulation in Molecular Dynamics.*

*For further information on these books please have a look at our mathematics catalogue at the following URL:* `www.springer.com/series/5151`

Printing: Krips bv, Meppel, The Netherlands
Binding: Stürtz, Würzburg, Germany